Studies in Computational Intelligence

Volume 749

The series "Studies in Computational Intelligence" (SCI) publishes new developments and advances in the various areas of computational intelligence—quickly and with a high quality. The intent is to cover the theory, applications, and design methods of computational intelligence, as embedded in the fields of engineering, computer science, physics and life sciences, as well as the methodologies behind them. The series contains monographs, lecture notes and edited volumes in computational intelligence spanning the areas of neural networks, connectionist systems, genetic algorithms, evolutionary computation, artificial intelligence, cellular automata, self-organizing systems, soft computing, fuzzy systems, and hybrid intelligent systems. Of particular value to both the contributors and the readership are the short publication timeframe and the world-wide distribution, which enable both wide and rapid dissemination of research output.

More information about this series at http://www.springer.com/series/7092

Oscar Castillo · Patricia Melin
Janusz Kacprzyk
Editors

Fuzzy Logic Augmentation of Neural and Optimization Algorithms: Theoretical Aspects and Real Applications

Editors
Oscar Castillo
Division of Graduate Studies and Research
Tijuana Institute of Technology
Tijuana, Baja California
Mexico

Janusz Kacprzyk
Systems Research Institute
Polish Academy of Sciences
Warsaw
Poland

Patricia Melin
Division of Graduate Studies and Research
Tijuana Institute of Technology
Tijuana, Baja California
Mexico

ISSN 1860-949X ISSN 1860-9503 (electronic)
Studies in Computational Intelligence
ISBN 978-3-319-89028-9 ISBN 978-3-319-71008-2 (eBook)
https://doi.org/10.1007/978-3-319-71008-2

Printed on acid-free paper

This Springer imprint is published by Springer Nature
The registered company is Springer International Publishing AG
The registered company address is: Gewerbestrasse 11, 6330 Cham, Switzerland

Preface

We describe in this book recent advances in the fuzzy logic-based augmentation of neural networks and in optimization algorithms and their application in areas such as, to just mention a few, intelligent control and robotics, pattern recognition, medical diagnosis, time series prediction and optimization of complex problems and system. The book is organized in seven main parts which contain groups of papers focused on a similar subject.

The first part consists of papers in which the main theme is the use of type-2 fuzzy logic in meta-heuristics. It basically consists of papers that propose new concepts and algorithms based on the type-2 fuzzy sets for dynamic parameter adaptation in meta-heuristics. The second part contains papers which are mainly concerned with the theory and applications of neural networks. It includes basically papers dealing with new concepts and algorithms in neural networks. The second part also contains papers describing applications of neural networks in diverse areas, such as the time series prediction and pattern recognition. The third part contains papers that present the theory and practice of meta-heuristics in different areas of application. The fourth part presents diverse applications of fuzzy logic in the control area which can be considered to be intelligent controllers. The fifth part contains papers describing applications of fuzzy logic in diverse areas such as the time series prediction and pattern recognition. The sixth part includes papers describing new optimization and evolutionary algorithms and their applications in different areas. Finally, the seventh part contains papers that present the design and application of different hybrid intelligent systems.

In the first part of the book with five papers, we are concerned with theoretical aspects and applications of the type-2 fuzzy logic in meta-heuristics which basically consists of papers that propose new concepts and algorithms based on the type-2 fuzzy logic for dynamic parameter adaptation in meta-heuristics. The aim of using the type-2 fuzzy logic is to provide a better uncertainty management in the process of searching for an optimal solution to a problem considered.

In the second part dealing with the theory and applications of neural networks, there are four papers that describe different contributions which propose new models, concepts and algorithms centred on the neural networks. The aim of using the neural networks is to provide learning and adaptive capabilities to intelligent systems. There are also papers that describe different contributions on the application of these kinds of neural models to solve complex real-world problems, such as the time series prediction and pattern recognition.

In the third part, devoted to the theory and practice of meta-heuristics in different areas of application, there are five papers that describe different contributions that propose new models and concepts which are also applied in diverse areas of applications. The nature-inspired methods include variations of different methods as well as new nature-inspired paradigms.

In the fourth part, diverse applications of fuzzy logic in the control area, which can be considered intelligent controllers, are presented in five papers that describe different contributions on the application of these kinds of fuzzy systems to solve complex real-world control problems, such as in robotics.

In the fifth part, on applications of fuzzy logic, there are six papers that describe different contributions on the application of these kinds of fuzzy logic models to solve complex real-world problems exemplified by the time series prediction, medical diagnosis, fuzzy control and pattern recognition.

In the sixth part, focused on the optimization, there are five papers that describe different contributions which propose new models, concepts and algorithms for the optimization inspired by different paradigms. The aim of using these algorithms is to provide general optimization methods and solution to some real-world problem in areas such as scheduling, planning and project portfolios.

In the seventh part, there are eight papers that present the nature-inspired design and applications of different hybrid intelligent systems. There are also papers that describe different contributions to the application of these kinds of hybrid intelligent systems to solve complex real-world problems such as the time series prediction, medical diagnosis and pattern recognition.

In conclusion, the edited book comprises papers on diverse aspects of fuzzy logic, neural networks and nature-inspired optimization meta-heuristics and their application in various areas exemplified by intelligent control and robotics, pattern recognition, time series prediction and optimization of complex problems. There are theoretical aspects as well as application papers.

We wish to thank the authors for their excellent and inspiring contributions and anonymous peer reviewers whose insight and suggestions have helped a lot to improve the contributions. And last but not least, we wish to thank Dr. Tom Ditzinger, Dr. Leontina di Cecco and Mr. Holger Schaepe for their dedication and help to implement and finish this large and ambitious publication project.

Tijuana, Mexico Oscar Castillo
Tijuana, Mexico Patricia Melin
Warsaw, Poland Janusz Kacprzyk
August 2017

Contents

Part I
Type-2 Fuzzy Logic in Metaheuristics

A Comparative Study of Dynamic Adaptation of Parameters in the GWO Algorithm Using Type-1 and Interval Type-2 Fuzzy Logic

Luis Rodríguez, Oscar Castillo, Mario García and José Soria

Abstract The main goal of this paper is to present a comparative study of dynamic adjustment of parameters in the Grey Wolf Optimizer algorithm using type-1 and interval type-2 fuzzy logic respectively. We proposed the fuzzy inference system for both types of fuzzy logic and we present the performance of these proposed methods with a set of 13 benchmark functions that we are presenting in this paper.

Keywords Grey wolf optimizer · Fuzzy logic · Interval type-2
Benchmark functions · Dynamic · Optimization

1 Introduction

Everyday humans are optimizing in each action that are considered "normal" activities, in other words, in an indirect way we are always searching for the best route for driving, the best price of the products, or we want to obtain the minimum time to do some activity etc.

It is important to remember that one of the main objectives of computational intelligence is simulate or trying to do activities, actions and features that humans perform in a natural way.

Computer science has areas for optimizing some problems when the main goal is to maximize and minimize, problems that we can find in simulation, benchmark problems, plants and applications in a real world.

We can define a metaheuristic as an optimization technique, which main features are the following: use of a stochastic model or randomness in the algorithm, the majority are bioinspired and specially because of this they have the ability to avoid local optima and we can mention that they can be classified as evolutionary [1], Based on physics [2] and Swarm Intelligence [3].

L. Rodríguez · O. Castillo (✉) · M. García · J. Soria
Tijuana Institute of Technology, Tijuana, BC, Mexico
e-mail: ocastillo@tectijuana.mx

O. Castillo et al. (eds.), *Fuzzy Logic Augmentation of Neural and Optimization Algorithms: Theoretical Aspects and Real Applications*, Studies in Computational Intelligence 749, https://doi.org/10.1007/978-3-319-71008-2_1

In addition we can mention that recently fuzzy logic has shown that it has better performance in some applications areas that conventional mathematical models, especially when we need to adjust the parameters according to the performance of the critical values in the algorithm and this improvement is because fuzzy logic can have soft changes that help to improve the results comparing with the conventional mathematical models.

Also in this research we are presenting interval type-2 fuzzy logic that have the same abilities of type-1 fuzzy logic, but includes a higher degree of uncertainty in making decisions and this helps a lot when the problem is more complex.

This paper is organized as follows: Sect. 2 shows a brief explanation of the Grey Wolf Optimizer algorithm, Sect. 3 describes the proposed methods, in Sect. 4 we are presenting the simulation results and finally in Sect. 5 we present some conclusions.

2 Grey Wolf Optimizer Algorithm

The Grey Wolf Optimizer (GWO) [4] algorithm was proposed by Seyedali Mirjalili in 2014 based on the behavior of the Canis lupus or grey wolf and his justification was because in the literature and specifically in the Swarm Intelligence area there was not a technique that mimics the behavior of this species, and also based on the NFL Theorem [5].

In the GWO the original author designed the mathematical model based on two main features of the grey wolf that are the following: the hunting process and the hierarchy in the members of the pack.

According to Muro [6] the hunting mechanisms of the grey wolves has 3 main phases and are the following: follow and approach the prey, after, encircling, and harassing the prey until it stops moving and finally attack the prey.

In addition we can find the hierarchy pyramid of the pack as a leadership strategy, in other words the best wolf is called alpha (α) and is considered as the best solution in the algorithm, the second and third best solutions are called beta (β) and delta (δ) respectively and finally, the rest of the candidate solution are consider as omega (ω) wolves.

Finally in order to mathematically model the two main inspirations described above, we present the following equations:

$$D = \left\| C \cdot \mathbf{X_p}(t) - \mathbf{X}(t) \right\| \tag{1}$$

$$X(t + 1) = \mathbf{X_p}(t) - A\mathbf{D} \tag{2}$$

We can provide a brief explanation of the mathematical model. Equation 1 represents the distance between the best solution with a random motion and the current individual that we are analyzing. Equation 2 represents the next position of the current individual and is defined as the difference between the position of the best individual and the distance obtained in Eq. 1 multiplied by a weight that is defined as a coefficient.

$$A = 2a \cdot r_1 - a \quad (3)$$

$$C = 2 \cdot r_2 \quad (4)$$

The "A" coefficient in Eq. 3 represents the weight that directly affects the distance of Eq. 2, where "a" is linearly decreasing through of the iterations and r_1 is a random value between 0 and 1.

The "C" parameter represents the randomness in the algorithm, because is the random motion of the best solution in Eq. 1, where r_2 is a random value in the range [0, 1].

$$\mathbf{D}_\alpha = \|C_1 \cdot \mathbf{X}_\alpha - \mathbf{X}\|, \quad \mathbf{D}_\beta = C_2 \cdot \mathbf{X}_\beta - \mathbf{X}, \quad \mathbf{D}_\delta = C_3 \cdot \mathbf{X}_\delta - \mathbf{X} \quad (5)$$

$$\mathbf{X}_1 = \mathbf{X}_\alpha - A_1 \cdot (\mathbf{D}_\alpha), \quad \mathbf{X}_2 = \mathbf{X}_\beta - A_2 \cdot (\mathbf{D}_\beta), \quad \mathbf{X}_3 = \mathbf{X}_\delta - A_3 \cdot (\mathbf{D}_\delta) \quad (6)$$

$$\mathbf{X}(t+1) = \frac{\mathbf{X}_1 + \mathbf{X}_2 + \mathbf{X}_3}{3} \quad (7)$$

Equations 5 and 6 are the same equations as Eqs. 1 and 2 respectively, but in this case the results are based on the best three individuals in the algorithm that are called alpha, beta and delta respectively as we mention above.

Finally, the next position of the current individual that we are analyzing is represented in Eq. 7, which we can described as a weighted average based on the results obtained in Eq. 6 that represent the leaders of the pack.

3 Proposed Method

In this work we are presenting the dynamic adaptation of parameters in the algorithm using type-1 [7] and interval type-2 fuzzy logic [8] for the "a" and "C" parameters respectively.

Figure 1 shows the general flowchart of the GWO algorithm, and also we can find the blue block that is where we are introducing the proposed methods.

It is important to say that in this work the parameters that are dynamically adjusted are the "a" and "C" parameters with a range of [0.5, 2.5] and [0, 0.5] respectively, based on a previous study of parameters [9].

In Fig. 2 we are presenting the general structure of the FIS for type-1 fuzzy logic [10], where we have a simple fuzzy inference system with one input, that in this case is the current iteration in the algorithm and one output that in this case is the "a" parameter. It is important to mention that the structure for the "C" parameter is the same, but the output of the FIS in this case is the value for the "C" parameter.

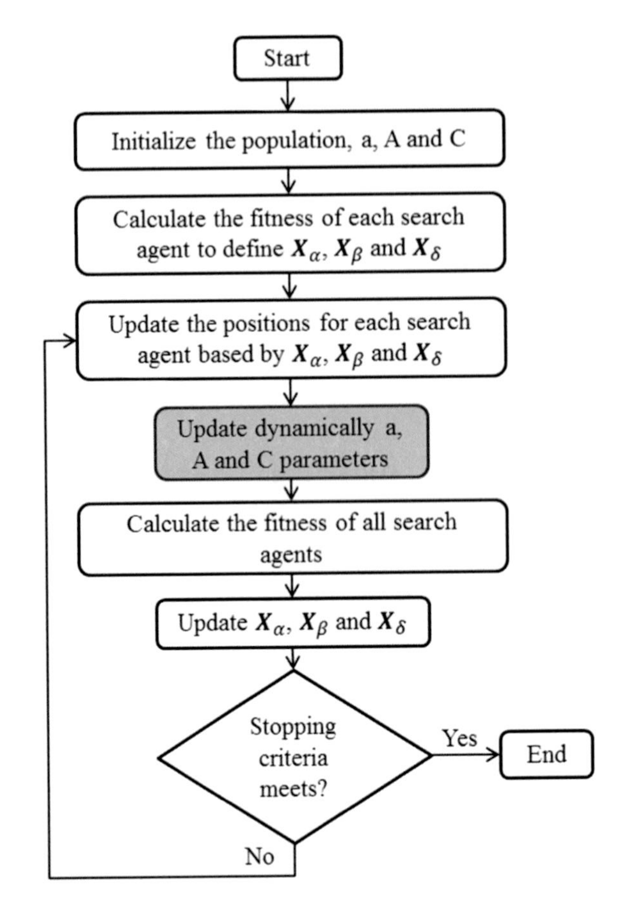

Fig. 1 General flowchart of the GWO

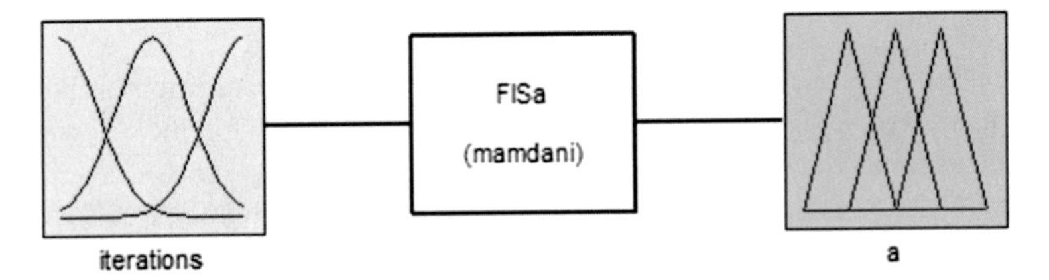

Fig. 2 General structure of the type-1 FIS

In addition we can mention that the fuzzy inference systems including type-1 and type-2 that we are presenting are of Mandani type and have centroid defuzzification [11].

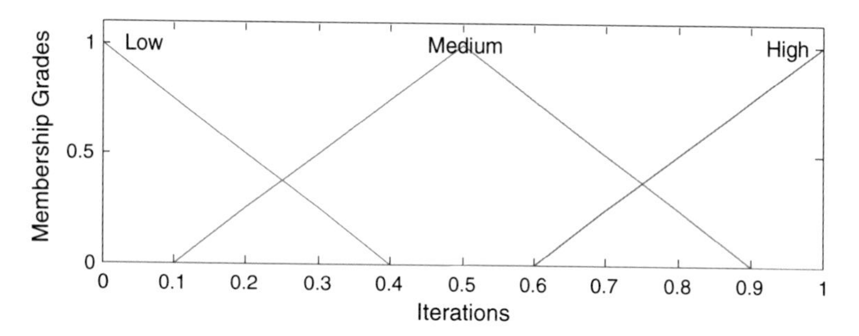

Fig. 3 Input of the type-1 FIS

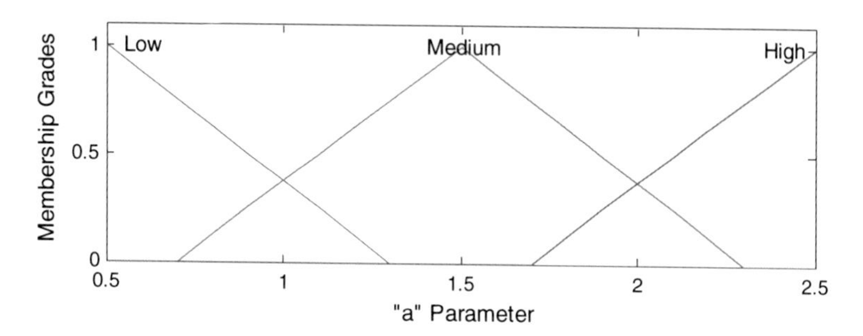

Fig. 4 Example of output in FIS of the "a" parameter

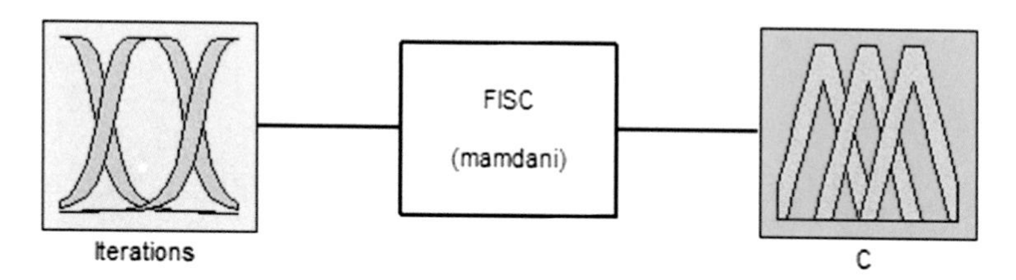

Fig. 5 General structure of the type-2 FIS

In Fig. 3 we can find the input for both fuzzy systems ("a" and "C" parameters) that is the iteration (normalized between 0 and 1) and has three triangular membership functions classified as: low, medium and high respectively.

Figure 4 shows an example of the output, in this case we are presenting for the "a" parameter and for the parameter "C" is similar, but the difference is the range that we mentioned above.

Also we are presenting a type-2 fuzzy system [12] and for this work basically is the same idea, just adding the footprint of uncertainty (FOU). Figure 5 shows the general structure for the type-2 FIS [13–16].

In Fig. 5 we can find the general structure for an interval type-2 fuzzy system for the "C" parameter (is the same for "a" parameter), also has one input and one output as the FIS of type-1 fuzzy logic.

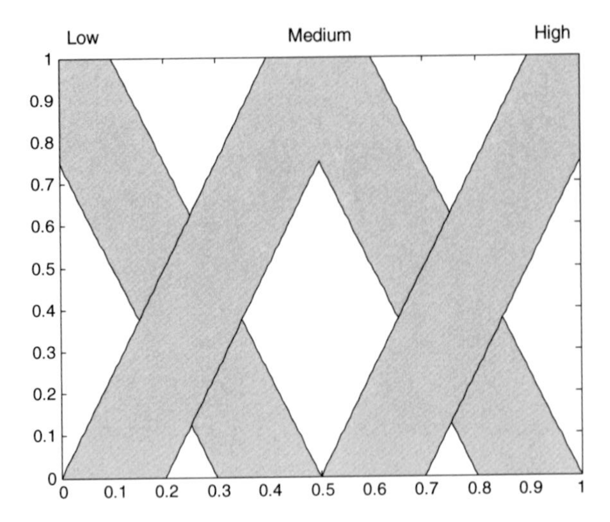

Fig. 6 Input of type-2 FIS

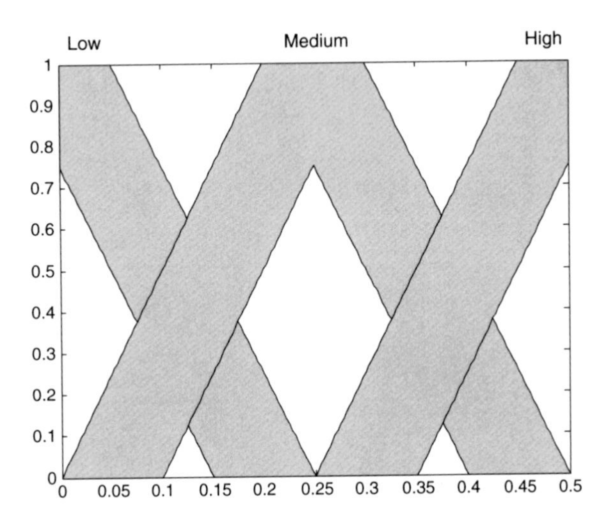

Fig. 7 Example of output in FIS of "C" parameter

Figures 6 and 7 show an example of the input and output for the type-2 FIS and the membership functions that are triangular and are granulated into low, medium and high respectively. In addition we can mention that the FOU of these membership functions that are proposed in this work are not optimized.

Finally, we present the rules for the fuzzy systems that are proposed in this paper.

In Table 1 we can find the rules for the "a" parameter for both proposed methods, type-1 and interval type-2 fuzzy logic, and we can note that the adjustment of the parameter is in a decrease fashion.

Table 2 shows the rules for the "C" parameter and in this case the rules are in an increase fashion, contrary to the "a" parameter.

Table 1 Rules for the "a" parameter in type-1 and type-2 fuzzy logic

Rules in decrease for the "a" parameter
1. if(iterations is low) then (a is high)
2. if(iterations is medium) then (a is medium)
3. if(iterations is high) then (a is low)

Table 2 Rules for the "C" parameter in type-1 and type-2 fuzzy logic

Rules in increase for the "C" parameter
1. if(iterations is low) then (C is low)
2. if(iterations is medium) then (C is medium)
3. if(iterations is high) then (C is high)

Table 3 Unimodal benchmark functions

Function	Range	f_{min}
$f_1(x) = \sum_{i=1}^{n} x_i^2$	[−100, 100]	0
$f_2 = \sum_{i=1}^{n} \lvert x_1 \rvert + \prod_{i=1}^{n} \lvert x_1 \rvert$	[−10, 10]	0
$f_3 = \sum_{i=1}^{n} \left(\sum_{j-1}^{i} x_j \right)^2$	[−100, 100]	0
$f_4 = \max_i \{ \lvert x_1 \rvert, 1 \le i \le n \}$	[−100, 100]	0
$f_5(x) = \sum_{i=1}^{n-1} \left[100 \left(x_{i+1} - x_i^2 \right)^2 + (x_1 - 1)^2 \right]$	[−30, 30]	0
$f_6(x) = \sum_{i=1}^{n} ([x_1 + 0.5])^2$	[−100, 100]	0
$f_7(x) = \sum_{i=1}^{n} i x_i^4 + random[0, 1]$	[−1.28, 1.28]	0

4 Simulations Results

In this paper we are presenting results for 13 benchmark functions [17–19] that are used for test the proposed methods.

In this paper we are presenting benchmark functions that are classified as unimodal and multimodal problems respectively. In addition we can mention that for these benchmark functions we are testing with 128, 256 and 512 dimensions respectively.

Table 3 shows the Equations for the first 7 functions that are the unimodal benchmark functions, this table also shows the search space for each equation and finally, the optimal value respectively.

Finally, in Fig. 8 we are presenting the 3-D surfaces of each unimodal benchmark function that we are using to test the proposed methods in this paper.

In Table 4 we can find the last 6 benchmark functions that were used to test the type-1 and interval type-2 fuzzy systems, these functions are called multimodal benchmark functions and we can also note the range and the optimal values for each function respectively.

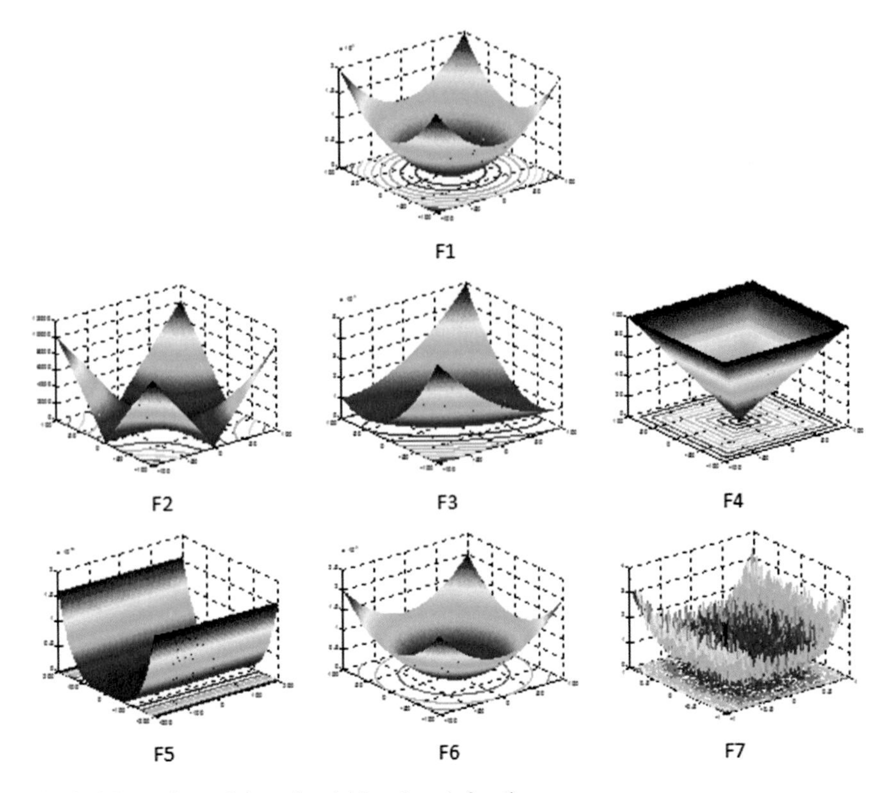

Fig. 8 3-D versions of the unimodal benchmark functions

According with the equations, Fig. 9 shows the surfaces for the multimodal benchmark functions in 3-D in order to show in a graphical way the problems that we are considering.

In addition we are presenting a hypothesis test [20], specifically the z-test in order to show statistically a comparison between the performance of the algorithm when we use type-1 and interval type-2 fuzzy logic.

The main goal of hypothesis testing is demonstrate statistically if using interval type-2 fuzzy logic in the algorithm has better performance than type-1 fuzzy logic and as a brief explanation, we can mention that if the z-value is less than -1.645 we can conclude that interval type-2 fuzzy logic has better performance than type-1 fuzzy logic.

For the simulation results, we are presenting the averages and standard deviations for the following tables, as the result of 50 independent executions for each benchmark function. In addition it is important to mention that for these experiments the algorithm was executed with 500 iterations and 30 individuals having a total of 15,000 function evaluations.

Table 5 shows the result of hypothesis testing among the proposed methods for dynamic adjustment the "a" parameter with 128 dimensions and we can conclude

Table 4 Multimodal benchmark functions

Function	Range	f_{min}
$f_8(x) = \sum_{i=1}^{n} -x_i \sin\left(\sqrt{\lvert x_i \rvert}\right)$	[−500, 500]	−2094.9
$f_9(x) = \sum_{i=1}^{n} \left[x_i^2 - 10\cos(2\pi x_i) + 10\right]$	[−5.12, 5.12]	0
$f_{10}(x) = 20\exp\left(-0.2\sqrt{\frac{1}{n}\sum_{i=1}^{n} x_i^2}\right) - \exp\left(\frac{1}{n}\sum_{i=1}^{n}\cos(2\pi x_i)\right) + 20 + e$	[−32, 32]	0
$f_{11}(x) = \frac{1}{4000}\sum_{i=1}^{n} x_1^2 - \prod_{i=1}^{n}\cos\left(\frac{x_i}{\sqrt{i}}\right) + 1$	[−600, 600]	0
$f_{12}(x) = \frac{\pi}{n}\{10\sin(\pi y_1) + \sum_{i=1}^{n-1}(y_i - 1)^2[1 + 10\sin^2(\pi y_{y+1})] + (y_n - 1)^2\} + \sum_{i=1}^{n} u(x_i, 10, 100, 4)$ $y_1 = 1 + \frac{x_i + 1}{4}$ $u(x_1, a, k, m) = \begin{cases} k(x_i - a)^m, & x_i a \\ 0, & -a x_i a \\ k(-x_i - a)^m, & x_i - a \end{cases}$	[−50, 50]	0
$f_{13}(x) = 0.1\{\sin^2(3\pi x_i) + \sum_{i=1}^{n}(x_1 - 1)^2[1 + \sin^2(3\pi x_i + 1)] + (x_n - 1)^2[1 + \sin^2(2\pi x_n)]\} + \sum_{i=1}^{n} u(x_i, 5, 100, 4)$	[−50, 50]	0

that the interval type-2 fuzzy logic is better in 5 of the 13 benchmark functions that were analyzed in this paper.

In Table 6 we can find the results of the hypothesis testing between type-1 and type-2 fuzzy logic respectively, when we dynamically perform adaptation of the "C" parameter also with 128 dimensions and we can conclude that in this case the type-2 fuzzy logic has poor performance because does not better performance than type-1 fuzzy logic in the 13 benchmark functions that were analyzed.

Table 7 shows the hypothesis testing among the two proposed methods in this paper, so according to table we can conclude that for 256 dimensions, when we dynamically adjustment the "a" parameter the interval type-2 fuzzy logic has better performance that type-1 fuzzy logic in 4 of the 13 benchmark functions that were analyzed in this work.

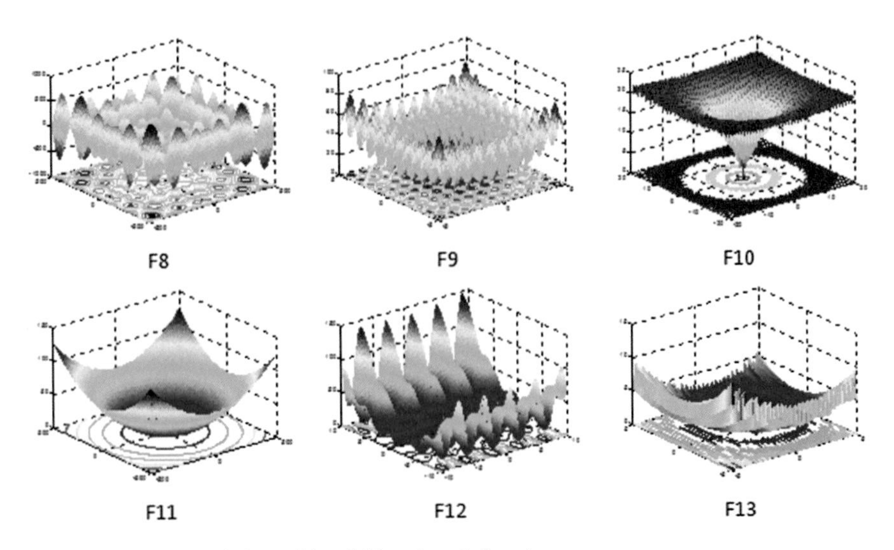

Fig. 9 3-D versions of the multimodal benchmark functions

Table 5 Hypothesis testing between type-1 and interval type-2 fuzzy logic with the "a" parameter in 128 dimensions

128 dimensions					
Function	FL T1	STD	FL T2	STD	Z-value
F1	1.79E−17	1.45E−17	**8.10E−19**	**8.73E−19**	−8.3255
F2	2.60E−11	1.14E−11	**4.41E−12**	**2.07E−12**	−13.1782
F3	**5872.53**	**6689.02**	6564.15	8269.90	0.4598
F4	1.1046	1.6699	**0.6467**	**0.7299**	−1.7766
F5	**125.9661**	**0.5333**	125.8880	0.6268	−0.6708
F6	**15.7500**	**0.9952**	15.9032	1.0510	0.7485
F7	**0.0063**	**0.0031**	0.0062	0.0027	−0.1547
F8	−8788.95	1366.50	**−7856.3784**	**889.3207**	−4.0446
F9	**7.5547**	**8.6716**	6.0275	11.2503	−0.7603
F10	3.48E−10	1.42E−10	**9.716E−11**	**4.576E−11**	−11.9177
F11	**0.0040**	**0.0096**	0.0047	0.0116	0.3323
F12	**0.4256**	**0.0615**	0.4250	0.0639	−0.0433
F13	**10.2908**	**0.4235**	10.3134	0.4029	0.2739

Bold indicates best values

On the other hand in Table 8 we can find the results of the hypothesis testing between both proposed methods in this paper, with the same number of dimensions for each benchmark functions (256), but in this case with the dynamically adaptation of the "C" parameter and we can conclude that for this experiment the type-2 is better only in one function that were analyzed in this work, comparing with type-1 fuzzy logic.

Table 6 Hypothesis testing between type-1 and interval type-2 fuzzy logic with the "C" parameter in 128 dimensions

128 dimensions					
Function	FL T1	STD	FL T2	STD	Z-value
F1	**2.99E−15**	**4.04E−15**	2.09E−15	2.00E−15	−1.4125
F2	**1.20E−09**	**5.68E−10**	1.31E−09	6.38E−10	0.8656
F3	**589.7345**	**893.3437**	612.7560	677.0520	0.1452
F4	**0.4762**	**1.9129**	0.2944	0.8453	−0.6147
F5	**126.0945**	**0.5462**	126.1093	0.5091	0.1407
F6	**18.1829**	**1.1047**	18.0757	1.0942	−0.4873
F7	**0.0038**	**0.0017**	0.0042	0.0017	1.1995
F8	**−15,078.9856**	**4757.0132**	−13,812.44	5352.02	−1.2507
F9	**1.0104**	**2.9878**	1.3828	3.9834	0.5288
F10	**4.14E−09**	**1.80E−09**	4.31E−09	1.95E−09	0.4337
F11	**8.96E−04**	**0.0045**	0.0028	0.0108	1.1390
F12	**0.4825**	**0.0621**	0.4842	0.0673	0.1369
F13	**9.9477**	**0.4474**	9.9570	0.4487	0.1033

Bold indicates best values

Table 7 Hypothesis testing between type-1 and interval type-2 fuzzy logic with the "a" parameter in 256 dimensions

256 dimensions					
Function	FL T1	STD	FL T2	STD	Z-value
F1	1.22E−11	7.76E−12	**1.49E−12**	**1.10E−12**	−9.6153
F2	8.53E−08	2.71E−08	**2.68E−08**	**9.03E−09**	−14.4686
F3	**102,235.09**	**50,582.66**	105,307.06	58,999.50	0.2795
F4	**68.1519**	**9.9724**	75.3357	9.8049	3.6322
F5	**253.9446**	**0.4636**	254.0442	0.2909	1.2864
F6	**42.8669**	**1.5121**	43.2155	1.2363	1.2622
F7	**0.0132**	**0.0048**	0.0121	0.0041	−1.2829
F8	−12,891.59	1934.58	**−11,206.62**	**1272.93**	−5.1449
F9	**12.7538**	**14.8321**	14.6011	16.9402	0.5802
F10	1.97E−07	5.24E−08	**8.17E−08**	**2.91E−08**	−13.6094
F11	**0.0087**	**0.0184**	0.0046	0.0146	−1.2103
F12	**0.6282**	**0.0401**	0.6527	0.0417	3.0044
F13	**23.1226**	**0.4127**	23.1845	0.3214	0.8368

Bold indicates best values

Table 9 shows the hypothesis testing when the problems have 512 dimensions respectively and when we are dynamically adjusting the "a" parameter for type-1 and type-2 fuzzy logic respectively and with helps of the table we can mention that in this case the interval type-2 fuzzy logic has better performance in 5 of the 13 benchmark functions that were presenting in this paper.

Table 8 Hypothesis testing between type-1 and interval type-2 fuzzy logic with the "C" parameter in 256 dimensions

256 dimensions					
Function	FL T1	STD	FL T2	STD	Z-value
F1	**2.34E−10**	**2.27E−10**	2.45E−10	2.28E−10	0.2493
F2	**1.00E−06**	**3.26E−07**	1.01E−06	2.37E−07	0.1215
F3	**21,372.60**	**13,544.66**	22,110.02	12,412.46	0.2838
F4	**26.4670**	**11.1172**	28.3710	10.4282	0.8832
F5	**253.9959**	**0.4049**	253.9431	0.3823	−0.6698
F6	**44.9579**	**1.3003**	44.6937	1.5777	−0.9135
F7	**0.0081**	**0.0025**	0.0080	0.0039	−0.1658
F8	**−25,339.60**	**9360.52**	−23,829.50	10,906.03	−0.7430
F9	**3.0436**	**7.9925**	3.0232	6.5101	−0.0140
F10	**9.16E−07**	**3.15E−07**	9.88E−07	3.43E−07	1.0973
F11	**0.0041**	**0.0121**	0.0024	0.0099	−0.7959
F12	**0.6615**	**0.0525**	0.6664	0.0467	0.4887
F13	22.5809	0.4468	**22.4255**	**0.4279**	−1.7755

Bold indicates best values

Table 9 Hypothesis testing between type-1 and interval type-2 fuzzy logic with the "a" parameter in 512 dimensions

512 dimensions					
Function	FL T1	STD	FL T2	STD	Z-value
F1	1.09E−07	5.46E−08	**2.06E−08**	**1.15E−08**	−11.1677
F2	2.59E−05	5.47E−06	**9.31E−06**	**1.97E−06**	−20.1416
F3	**591,612.62**	**194,708.87**	619,435.78	164,818.33	0.7712
F4	**90.4063**	**3.6096**	92.6810	3.9953	2.9872
F5	**509.9102**	**0.1494**	509.9175	0.1517	0.2425
F6	**100.6532**	**1.8356**	101.6776	1.3740	3.1591
F7	0.0256	0.0086	**0.0223**	**0.0078**	−2.0164
F8	−17,689.52	2801.84	**−15,843.39**	**1754.69**	−3.9487
F9	**19.2024**	**18.4240**	27.0614	25.5493	1.7642
F10	1.50E−05	4.04E−06	**6.57E−06**	**1.58E−06**	−13.7849
F11	**0.0104**	**0.0249**	0.0078	0.0217	−0.5587
F12	**0.8130**	**0.0284**	0.8235	0.0290	1.8273
F13	**49.1345**	**0.4406**	49.3115	0.5076	1.8618

Bold indicates best values

Finally, in Table 10 we can find the last simulation results that were executed in this work and are the problems with 512 dimensions and when we use dynamically adaptation for the "C" parameter. In addition we can conclude that in this case the performance of interval type-2 fuzzy logic is poor as we presented in the experiment for 128 dimensions.

Table 10 Hypothesis testing between type-1 and interval type-2 fuzzy logic with the "C" parameter in 512 dimensions

512 dimensions					
Function	FL T1	STD	FL T2	STD	Z-value
F1	**7.87E−07**	**3.83E−07**	9.26E−07	5.18E−07	1.5316
F2	**1.37E−04**	**3.10E−05**	1.4E−04	3.2E−05	0.9815
F3	**161,824.99**	**67,228.76**	163,037.88	60,394.62	0.0949
F4	**64.5026**	**6.5495**	62.6761	7.7842	−1.2695
F5	**509.7852**	**0.1756**	509.7893	0.1548	0.1238
F6	**101.8930**	**1.5077**	101.4486	1.8288	−1.3256
F7	**0.0183**	**0.0066**	0.0182	0.0054	−0.0622
F8	**−43,582.17**	**16,881.39**	−46.028.82	13,519.47	0.7999
F9	**9.7099**	**20.0149**	7.9232	9.3187	−0.5723
F10	**4.13E−05**	**9.38E−06**	4.31E−05	1.33E−05	0.7669
F11	**0.0097**	**0.0240**	0.0045	0.0148	−1.3060
F12	**0.8060**	**0.0299**	0.8083	0.0285	0.3969
F13	**48.0424**	**0.5669**	48.1184	0.5782	0.6638

Bold indicates best values

5 Conclusions

In this work we presented a comparative study between type-1 and interval type-2 for dynamically adaptation of the "a" and "C" parameters in the GWO algorithm with 13 benchmark functions that were tested with 128, 256 and 512 dimensions respectively.

In addition we presented a hypothesis test in order to demonstrate statistically which proposed method has better performance with the problems that we presented in this research, and for these problems we can conclude that in general the type-1 fuzzy logic has better performance that interval type-2 fuzzy logic.

It is important to mention that when we dynamically adjustment the "a" parameter the interval type-2 fuzzy logic has better performance in some benchmark functions than when we dynamically adaptation the "C" parameter.

In addition we can mention that the FOUs for the fuzzy inference system type-2 are not optimized, because as a study is important that the fuzzy systems were the most possible similar, so this fuzzy inference system can be improve with a different types of membership functions or different number of partitions for example. In addition we can mention that although the fuzzy system type-2 is not optimized, with the "a" parameter show a good performance in some functions.

Finally, with these results we can conclude that interval type-2 fuzzy logic has better performance in other types of problems, especially in more complex problems, as a future work we proposed to test these methods with benchmark functions that include other type of complexity for example: white Gauss noise and show which method is better with these problems if type-1 or interval type-2 fuzzy logic.

References

1. H.R. Maier, Z. Kapelan, Evolutionary algorithms and other metaheuritics in water resources: current status, research challenges and future directions. Environ. Model Softw. **62**, 271–299 (2014)
2. U. Can, B. Alatas, Physics based metaheuristic algorithms for global optimization. Am. J. Inf. Sci. Comput. Eng. **1**, 94–106 (2015)
3. X. Yang, M. Karamanoglu, in *Swarm Intelligence and Bio-Inspired Computation: an Overview*. Swarm intelligence and bio-inspired computation (2013), pp. 3–23
4. S. Mirjalili, M. Mirjalili, Lewis A: Grey Wolf optimizer. Adv. Eng. Softw. **69**, 46–61 (2014)
5. D.H. Wolpert, W.G. Macready, No free lunch theorems for optimization. Evolut. Comput. IEEE Trans. **1**, 67–82 (1997)
6. C. Muro, R. Escobedo, L. Spector, R. Coppinger, Wolf-pack (*Canis lupus*) hunting strategies emerge from simple rules in computational simulations. Behav. Process. **88**, 192–197 (2011)
7. L. Zadeh, Fuzzy sets. Inf. Control **8**, 338–353 (1965)
8. J. Mendel, G.J. Mouzouris, Type-2 fuzzy logic systems. IEEE Trans. Fuzzy Syst. nº 7, 643–658 (1999)
9. L. Rodríguez, O. Castillo, J. Soria, in *A Study of Parameters of the Grey Wolf Optimizer Algorithm for Dynamic Adaptation with Fuzzy Logic*. Nature-inspired design of hybrid intelligent systems (2017), pp. 371–390
10. L. Rodriguez, O. Castillo, J. Soria, Grey Wolf Optimizer (GWO) with dynamic adaptation of parameters using fuzzy logic. IEEE CEC **3116**, 3123 (2016)
11. J. Barraza, P. Melin, F. Valdez, C. Gonzalez, Fuzzy FWA with dynamic adaptation of parameters. IEEE CEC 4053–4060 (2016)
12. E. Rubio, O. Castillo, F. Valdez, P. Melin, I. Gonzalez, G. Martinez, An extension of the fuzzy possibilistic clustering algorithm using type-2 fuzzy logic techniques. Adv. Fuzzy Syst. 7094046:1–7094046:23 (2017)
13. F. Olivas, F. Valdez, O. Castillo, C. González, G. Martinez, P. Melin, Ant colony optimization with dynamic parameter adaptation based on interval type-2 fuzzy logic systems. Appl. Soft Comput. **53**, 74–87 (2017)
14. J. Pérez, F. Valdez, O. Castillo, P. Melin, C. González, G. Martinez, Interval type-2 fuzzy logic for dynamic parameter adaptation in the bat algorithm. Soft. Comput. **21**(3), 667–685 (2017)
15. B. González, F. Valdez, P. Melin, in *A Gravitational Search Algorithm Using Type-2 Fuzzy Logic for Parameter Adaptation*. Nature-inspired design of hybrid intelligent systems (2017), pp. 127–138
16. O.D. De la, O. Castillo, J. Soria, in *Nature-Inspired Design of Hybrid Intelligent Systems*. Optimization of reactive control for mobile robots based on the CRA using type-2 fuzzy logic (2017), pp. 505–515
17. J. Digalakis, K. Margaritis, On benchmarking functions for genetic algorithms. Int. J. Comput. Math. **77**, 481–506 (2001)
18. M. Molga, C. Smutnicki, Test functions for optimization needs (2005)
19. X.-S. Yang, Test problems in optimization, arXiv, preprint arXiv: 1008.0549; 2010
20. R. Larson, B. Farber, *Elementary Statistics Picturing the World* (Pearson Education Inc. 2003), pp. 428–433

Ensemble Neural Network Optimization Using a Gravitational Search Algorithm with Interval Type-1 and Type-2 Fuzzy Parameter Adaptation in Pattern Recognition Applications

Beatriz González, Patricia Melin, Fevrier Valdez and German Prado-Arechiga

Abstract In this paper we consider the problem of optimizing ensemble neural networks for pattern recognition with Type-1 and Type-2 fuzzy logic for parameter adaptation in the gravitational search algorithm. The database to be used is of echocardiography images, since these images are very important in clinical echocardiography, and these images help the doctors to diagnose cardiac diseases, as well as to prevent this type of diseases in patient treatment.

Keywords Gravitational search algorithm · Type-2 fuzzy logic · Pattern recognition

1 Introduction

The Gravitational Search Algorithm (GSA) proposed in 2009 by Esmat Rashedi. This algorithm is based on populations and is inspired in physical rules: law of gravitation and motion [1].

Some examples of applications are: optimization of modular neural networks [2–4], ensembles [5–8], medical images [9–12], fuzzy logic [13–15, 20], Type-2 fuzzy logic [16–18], etc.

B. González · P. Melin (✉) · F. Valdez
Tijuana Institute of Technology, Calzada Tecnologico s/n, Tijuana, Mexico
e-mail: pmelin@tectijuana.mx

B. González
e-mail: betygm8@hotmail.com

F. Valdez
e-mail: fevrier@tectijuana.mx

G. Prado-Arechiga
Cardio-Diagnostico, Paseo de los Heroes No. 2507, Zona Rio, Tijuana, Mexico

O. Castillo et al. (eds.), *Fuzzy Logic Augmentation of Neural and Optimization Algorithms: Theoretical Aspects and Real Applications*, Studies in Computational Intelligence 749, https://doi.org/10.1007/978-3-319-71008-2_2

We are using the GSA algorithm for optimization of ensemble neural networks with Type-1 and Type-2 fuzzy logic for parameter adaptation.

The echocardiogram is a type of ultrasound test and is used to help diagnose problems with the heart.

This paper is organized in this way: in Sect. 2 shows the basic concepts, in Sect. 3 the method process is described, in Sect. 4 the architecture of the ensemble neural network is presented, in Sect. 5 the results are shown, and in Sect. 6 the conclusions are offered.

2 Basic Concepts

2.1 The Laws of Newton

The laws of Newton are the following:

Law of gravity "The gravitational force between two particles is directly proportional to the product of their masses and inversely proportional to the square of the distance between them".

Law of motion The acceleration of an object is directly proportional to the net force acting on it and inversely proportional to its mass [19].

2.2 The Process of the Gravitational Search Algorithm Is the Following

2.2.1 Randomized Initialization

The initial population and its position is represented by Eq. 1, where N is the number of agents:

$$X_i = \left(X_i^1, \ldots, X_i^d, \ldots X_i^n\right) \text{ for } i = 1, 2, \ldots, N, \quad (1)$$

where X_i^d:

i = represents the position from agent,
d = represents dimension.

2.2.2 Fitness Evaluation of Agents

The agents are evaluated and in Eq. 2 the best is calculated and with Eq. 3 the worst is calculated and this is presented for a minimization problem.

$$\text{best}(t) = \min_{j \in \{1,\ldots,N\}} \text{fit}_j(t) \tag{2}$$

$$\text{worst}(t) = \max_{j \in \{1,\ldots,N\}} \text{fit}_j(t) \tag{3}$$

where $\text{fit}_j(t)$ represents the fitness value of the *j*th agent at iteration t.

2.2.3 Update G (t) and M

The G value is calculated in Eq 4.

$$G(t) = G_0 e^{-\alpha t/T} \tag{4}$$

The inertial mass is calculated as follows in Eqs. 5 and 6:

$$\text{m}_i(t) = \frac{\text{fit}_i(t) - \text{worst}(t)}{\text{best}(t) - \text{worst}(t)} \tag{5}$$

$$M_i(t) = \frac{m_i(t)}{\sum_{j=1}^{N} m_j(t)} \tag{6}$$

2.2.4 Calculation of the Total Force in Different Directions

Where F_i^d is the total force acting on the ith agent as follows:

$$F_i^d(t) = \sum_{j \in K\,best, j \neq 1} rand_j F_{i,j}^d(t) \tag{7}$$

where K best is the set of first K agents with the best fitness value and biggest mass.

When agent j acts on agent i with a particular force, at a specific time (t) the force is calculated as follows:

$$F_{i,j}^d(t) = G(t) \frac{M_{pi}(t) x M_{aj}(t)}{R_{ij}(t) + s} \left(x_j^d(t) - x_i^d(t) \right) \tag{8}$$

where:

M_{aj} = the active gravitational mass of object j,
M_{pi} = the passive gravitational mass of object i,
G(t) = gravitational constant at time t.

2.2.5 Calculation of Acceleration and Velocity

The equations for calculating the acceleration and velocity are:

$$a_i^d(t) = \frac{F_i^d(t)}{M\ddot{u}(t)} \tag{9}$$

$$V_i^d(t+1) = rand_i x V_i^d(t) + a_i^d(t) \tag{10}$$

2.2.6 Updating Agents' Position

The equation for updating the position is:

$$X_i^d(t+1) = X_i^d(t) + V_i^d(t+1) \tag{11}$$

2.2.7 Repeat Steps 2.2.2 to G Until the Stop Criteria Is Reached

2.2.8 End

3 Proposed Method

In the initial iterations we have a small alpha and a high G value in order to apply a force and acceleration in large quantities, helping the exploration, and in later iterations we have to provide a high alpha value and small G value to help the algorithm with the exploitation task [19].

In this case if we change the alpha value, the algorithm applies a different G value, and to change its acceleration, for the agents to explore other good solutions in the search space and improve the final result.

3.1 Proposed Method with Type-1 Fuzzy Logic

Figure 1 shows the proposed method of a Mamdani system with interval type-1 fuzzy logic for the value of alpha.

Figure 2 shows the input variable as the number of iterations.

Figure 3 shows the output variable as the alpha parameter.

Figure 4 shows the fuzzy rules.

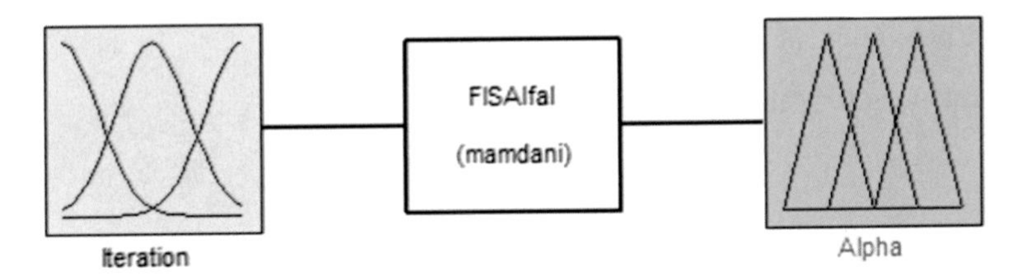

Fig. 1 Proprosed method of a Mamdani fuzzy system value of alpha with type-1 fuzzy logic

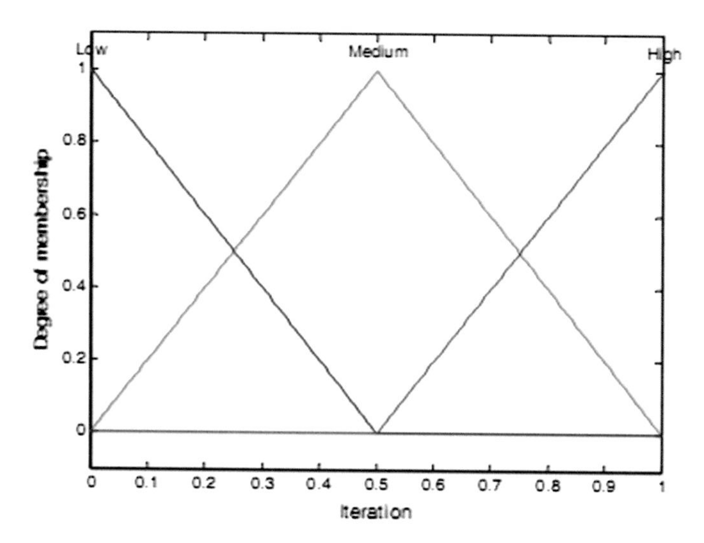

Fig. 2 The input variable

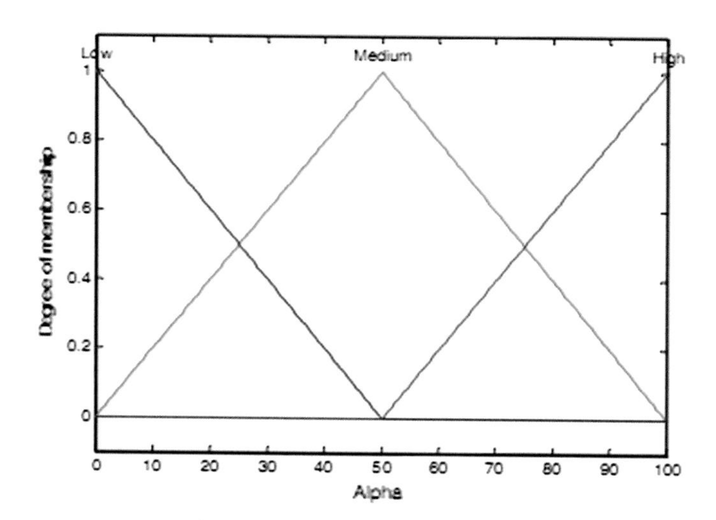

Fig. 3 The output variable

Fig. 4 Rules

1. If (Iteration is Low) then (Alpha is Low) (1)
2. If (Iteration is Medium) then (Alpha is Medium) (1)
3. If (Iteration is High) then (Alpha is High) (1)

3.2 Proposed Method with Interval Type-2 Fuzzy Logic

In Fig. 5 we show the proposed method of a Mamdani system with interval type-2 fuzzy logic to adapt the value of alpha.

In Fig. 6 the input variable of the number of iterations is represented.

In Fig. 7 we show the output variable and is the alpha value.

In Fig. 8 the fuzzy rules are presented.

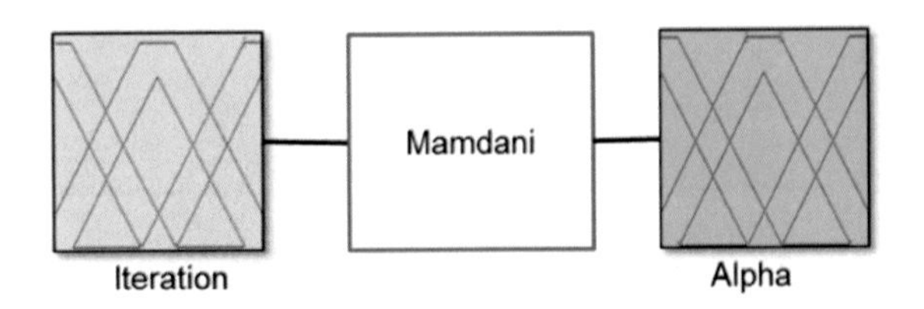

Fig. 5 Proposed method of a Mamdani fuzzy system for the value of alpha with type-2 fuzzy logic

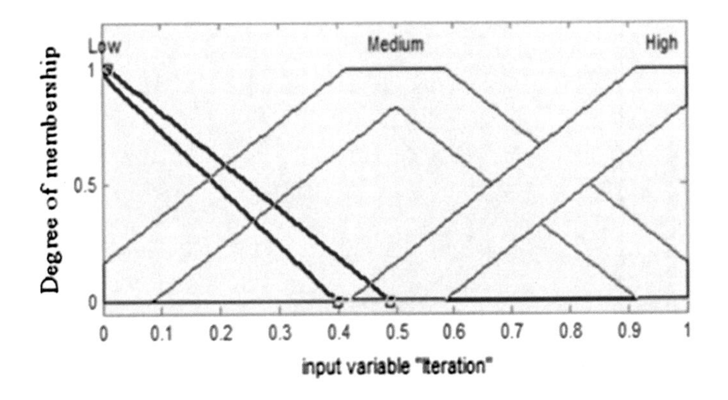

Fig. 6 Input variable as the iterations

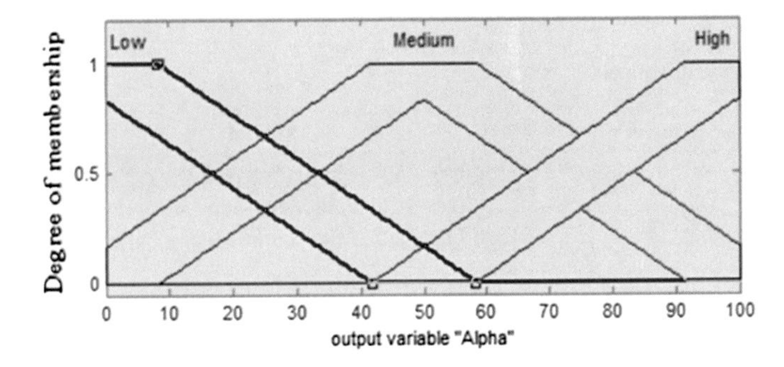

Fig. 7 Output variable the alpha

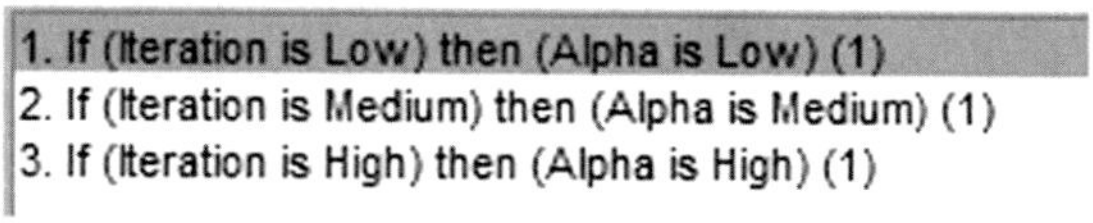

1. If (Iteration is Low) then (Alpha is Low) (1)
2. If (Iteration is Medium) then (Alpha is Medium) (1)
3. If (Iteration is High) then (Alpha is High) (1)

Fig. 8 Fuzzy rules

Database
Module 1
Pre-
preprocessing
Integrator
Disease
patient or not
Module 2

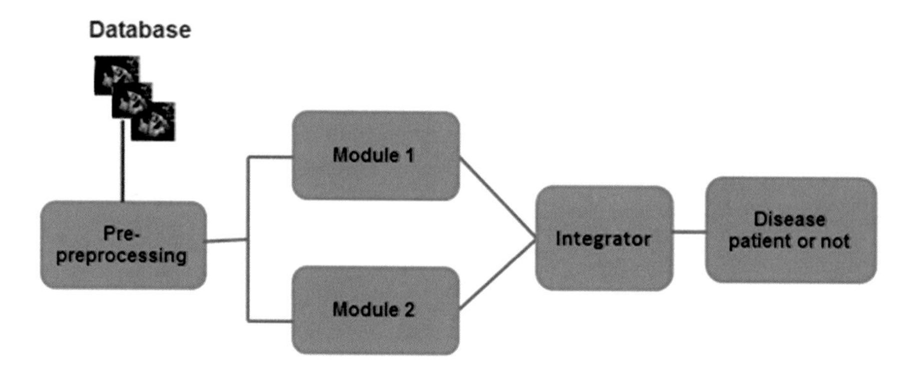

Fig. 9 The architecture of the ensemble neural networks

4 Architecture of the Ensemble Neural Networks

Figure 9 shows the architecture of an ensemble neural network.

4.1 Database of Echocardiograms

The database includes echocardiograms from 18 patients, and there are 10 images per patient, in bmp.

5 Experimental Results

We describe below the simulation results of the proposed approach for echocardiogram recognition with ensemble neural networks to find out the optimal number of layers and nodes of the neural network.

In Table 1 we show results from the "trainscg" method with 2 layers and another one with 3 layers.

In Table 2 we show results of the "traingda" method with 2 layers and another one with 3 layers.

Table 1 Results of the FGSA with type-1 fuzzy logic and the trainscg method

Training	FGSA with 2 layers		FGSA with 3 layers	
	Neurons	% Rec.	Neurons	% Rec.
1	18, 23	100	15, 13, 11	97.2
2	23, 15	98.6	18, 16, 7	87.5
3	15, 25	100	17, 16, 10	98.6
4	22, 15	98.6	17, 7, 13	83.3
5	28, 12	98.6	18, 14, 11	91.6
6	9, 23	97.2	12, 8, 22	100
7	23, 19	100	10, 13, 20	94.4
8	21, 18	100	7, 8, 18	79
9	24, 22	98.6	18, 10, 13	100
10	24, 10	100	13, 14, 18	91.6
11	23, 25	100	18, 8, 17	97.2
12	15, 17	98.6	12, 7, 19	93
13	23, 17	98.6	18, 11, 18	77.7
14	24, 13	98.6	17, 4, 9	94.4
15	26, 28	100	19, 7, 15	97.2

Table 2 Results of the FGSA with type-1 fuzzy logic and the traingda method

Training	FGSA with 2 layers		FGSA with 3 layers	
	Neurons	% Rec.	Neurons	% Rec.
1	23, 23	100	19, 12, 16	51.3
2	23, 26	100	12, 17, 18	76.3
3	16, 23	95.8	21, 14, 9	87.5
4	21, 26	98.6	12, 13, 22	80.5
5	22, 17	97.2	22, 14, 20	94.4
6	25, 28	97.2	23, 19, 11	93
7	25, 19	94.4	21, 9, 20	100
8	22, 9	91.6	23, 18, 15	81.94
9	26, 24	97.2	08:21:22	100
10	25, 19	94.4	23, 13, 20	97.2
11	22, 11	91.6	19, 8, 12	80.5
12	25, 19	94.4	21, 4, 15	87.5
13	18, 10	80.5	28, 21, 23	100
14	23, 23	93.05	22, 13, 27	97.2
15	11, 20	93	27, 7, 14	98.6

In Table 3 we show results for the "trainscg" method with 2 layers and another one with 3 layers.

In Table 4 we show results for the "traingda" method with 2 layers and another one with 3 layers.

Table 3 Results of the FGSA with type-2 fuzzy logic and the trainscg method

Training	FGSA with 2 layers		FGSA with 3 layers	
	Neurons	% Rec.	Neurons	% Rec.
1	12, 15	100	11, 13, 12	94.4
2	10, 10	86.11	14, 5, 10	100
3	14, 17	100	14, 10, 10	98.6
4	10, 9	86.11	14, 10, 13	98.6
5	11, 8	97.22	16, 17, 14	100
6	8, 13	98.61	15, 7, 18	100
7	12, 13	100	12, 14, 12	98.6
8	5, 14	98.61	16, 13, 18	100
9	13, 14	93.05	14, 13, 16	100
10	11, 14	98.61	16, 17, 14	100
11	9, 9	81.94	15, 7, 18	100
12	13, 13	100	12, 14, 12	98.1
13	14, 14	93.05	16, 13, 18	100
14	13, 9	95.83	14, 13, 16	100
15	9, 9	94.44	10, 4, 17	100

Table 4 Results of the FGSA with type-2 fuzzy logic and the traingda method

Training	FGSA with 2 layers		FGSA with 3 layers	
	Neurons	% Rec.	Neurons	% Rec.
1	16, 17	88.88	18, 16, 15	94.4
2	16, 13	80.55	15, 14, 20	97.2
3	14, 16	88.88	14, 10, 19	94.4
4	12, 12	70.83	11, 9, 19	80.5
5	12, 19	91.66	19, 7, 16	91.6
6	19, 16	98.61	19, 18, 18	97.2
7	19, 14	90.27	16, 19, 17	91.6
8	16, 17	95.83	17, 9, 15	84.7
9	19, 19	97.22	18, 11, 16	95.8
10	19, 18	97.22	17, 10, 14	93
11	17, 18	97.22	19, 5, 15	100
12	12, 16	86.11	14, 20, 17	94.4
13	15, 14	84.72	16, 17, 18	90.2
14	19, 12	81.94	19, 2, 15	94.4
15	14, 17	84.72	17, 16, 18	94.4

6 Conclusions

The results obtained with the proposed FGSA algorithm with type-1 fuzzy logic and type-2 fuzzy logic are good; the best learning algorithm in this case was the scaled conjugate gradient with 99.4%, 3 layers and the FGSA algorithm with type-2 fuzzy logic. Type-2 fuzzy logic was better when comparing with method FGSA algorithm with type-1 fuzzy logic. Type-2 fuzzy logic has achieved many good applications, like in [21, 22, 26]. As future works we would like to consider applying the proposed approach in problems, like the ones in [20, 23–25, 27].

References

1. E. Rashedi, H. Nezamabadi-pour, S. Saryazdi, GSA: a gravitational search algorithm. Inf. Sci. **179**, 2232–2248 (2009)
2. D. Sánchez, P. Melin, O. Castillo, Optimization of modular granular neural networks using a hierarchical genetic algorithm based on the database complexity applied to human recognition. Inf. Sci. **309**(10), 73–101 (2015)
3. D. Sánchez, P. Melin, O. Castillo, Optimization of modular granular neural networks using hierarchical genetic algorithms for human recognition using the ear biometric measure. Eng. Appl. Artif. Intell. **27**, 41–56 (2014)
4. F. Valdez, P. Melin, O. Castillo, Modular Neural Networks architecture optimization with a new nature inspired method using a fuzzy combination of particle swarm optimization and genetic algorithms. Inf. Sci. **270**(20), 143–153 (2014)
5. A. Zameer, J. Arshad, A. Khan, M. Asif, Intelligent and robust prediction of short term wind power using genetic programming based ensemble of neural networks. Energy Convers. Manag. **134**(15), 361–372 (2017)
6. H. Li, X. Wang, S. Ding, Research of multi-sided multi-granular neural network ensemble optimization method. Neurocomputing **197**(12), 78–85 (2016)
7. M. Pulido, P. Melin, O. Castillo, Particle swarm optimization of ensemble neural networks with fuzzy aggregation for time series prediction of the Mexican Stock Exchange. Inf. Sci. **280**(1), 188–204 (2014)
8. Z. Zhao, X. Feng, Y. Lin, F. Wei, S. Wang, T. Xiao, M. Cao, Z. Hou, Evolved neural network ensemble by multiple heterogeneous swarm intelligence, Neurocomputing **149**(Part A, 3), 29–38 (2015)
9. G. Gao, X. Wan, S. Yao, Z. Cui, C. Zhou, X. Sun, Reversible data hiding with contrast enhancement and tamper localization for medical images. Inf. Sci. **385–386**, 250–265 (2017)
10. M. Arsalan, A. Qureshi, A. Khan, M. Rajarajan, Protection of medical images and patient related information in healthcare: using an intelligent and reversible watermarking technique. Appl. Soft Comput. **51**, 168–179 (2017)
11. Y. Yang, W. Zhang, D. Liang, N. Yu, Reversible data hiding in medical images with enhanced contrast in texture area. Digit. Signal Proc. **52**, 13–24 (2016)
12. J. Oliva, H. Lee, N. Spolaôr, C. Coy, F. Wu, Prototype system for feature extraction, classification and study of medical images. Expert Syst. Appl. **63**(30), 267–283 (2016)
13. F. Valdez, J.C. Vazquez, P. Melin, O. Castillo, Comparative study of the use of fuzzy logic in improving particle swarm optimization variants for mathematical functions using co-evolution. Appl. Soft Comput. **52**, 1070–1083 (2017)

14. A. Bakdi, A. Hentout, H. Boutami, A. Maoudj, O. Hachour, B. Bouzouia, Optimal path planning and execution for mobile robots using genetic algorithm and adaptive fuzzy-logic control. Robot. Auton. Sys. **89**, 95–109 (2017)
15. S. Rajak, P. Parthiban, R. Dhanalakshmi, Sustainable transportation systems performance evaluation using fuzzy logic. Ecol. Ind. **71**, 503–513 (2016)
16. L. Cervantes, O. Castillo, Type-2 fuzzy logic aggregation of multiple fuzzy controllers for airplane flight control. Inf. Sci. **324**(10), 247–256 (2015)
17. C. Ulu, Exact analytical inversion of interval type-2 TSK fuzzy logic systems with closed form inference methods. Appl. Soft Comput. **37**, 60–67 (2015)
18. F. Olivas, F. Valdez, O. Castillo, C.I. Gonzalez, G. Martinez, P. Melin, Ant colony optimization with dynamic parameter adaptation based on interval type-2 fuzzy logic systems. Appl. Soft Comput. **53**, 74–87 (2017)
19. A. Sombra, F. Valdez, P. Melin, O. Castillo, A new gravitational search algorithm using fuzzy logic to parameter adaptation, in *IEEE Congress on Evolutionary Computation* (Cancun, México, 2013), pp. 1068–1074
20. F. Valdez, P. Melin, O. Castillo, Evolutionary method combining particle swarm optimization and genetic algorithms using fuzzy logic for decision making, in *IEEE International Conference on Fuzzy Systems* (2009), pp. 2114–2119
21. G.M. Mendez, O. Castillo, Interval type-2 TSK fuzzy logic systems using hybrid learning algorithm, in *Fuzzy Systems, FUZZ'05. The 14th IEEE International Conference on* (2005), pp. 230–235
22. O. Castillo, P. Melin, Design of intelligent systems with interval type-2 fuzzy logic, in *Type-2 Fuzzy Logic: Theory and Applications* (2008), pp. 53–76
23. O. Castillo, P. Melin, E. Ramírez, J. Soria, Hybrid intelligent system for cardiac arrhythmia classification with Fuzzy K-Nearest Neighbors and neural networks combined with a fuzzy system, Expert Sys. Appl. **39**(3), 2947–2955 (2012)
24. L. Aguilar, P. Melin, O. Castillo, Intelligent control of a stepping motor drive using a hybrid neuro-fuzzy ANFIS approach, Appl. Soft Comput. **3**(3), 209–219 (2003)
25. P. Melin, O. Castillo, Modelling, simulation and control of non-linear dynamical systems: an intelligent approach using soft computing and fractal theory, CRC Press, (2001)
26. P. Melin, CI Gonzalez, JR Castro, O. Mendoza, O. Castillo, Edge-detection method for image processing based on generalized type-2 fuzzy logic, IEEE Trans. Fuzzy Sys. **22**(6), 1515–1525 (2014)
27. P. Melin, O. Castillo, Intelligent control of complex electrochemical systems with a neuro-fuzzy-genetic approach, IEEE Trans. Industr. Electron. **48**(5), 951–955 (2001)

Improved Method Based on Type-2 Fuzzy Logic for the Adaptive Harmony Search Algorithm

Cinthia Peraza, Fevrier Valdez and Oscar Castillo

Abstract This paper proposes a novel method based on interval type-2 fuzzy logic for the adaptive harmony search algorithm. Based on a study carried out previously it is decided to use a second input that we will be called diversity, and this in order to obtain that so close or far they are the harmonies of the solution. The method is applied to 11 mathematical benchmark functions using 2 and 10 variables to test the proposed method and present a comparison with the original method and with harmony search using type-1 fuzzy logic. In previous works we used the type-1 and type-2 fuzzy logic to dynamically adjust the parameters of the algorithm, such as the number of improvisations or the iterations, but adjusting each parameter separately. In this case we use as a second input the diversity and as output the harmony memory accepting parameter to achieve a control of the exploration and exploitation of the search space. We can say that this is the difference between the previous works and this proposed method.

Keywords Harmony search · Type-1 fuzzy logic · Type-2 fuzzy logic
Dynamic parameter adaptation · Diversity

1 Introduction

Metaheuristic algorithms are at the forefront to solve optimization problems, in engineering, cost reduction among others. These algorithms use exploration techniques and exploit the search space in order to obtain better solutions within a given range or problem. In this case, a metaheuristic algorithm based on the improvisation of music, specifically jazz, is referred to as the harmony search algorithm [5].

The method has been studied in previous works [12–14] as well as its variants. This algorithm has been very useful to solve a great variety of problems [1, 4, 8, 15, 18] and there are hybrid methods that perform optimization such as [19–21]. It is decided to

C. Peraza · F. Valdez (✉) · O. Castillo
Tijuana Institute of Technology, Tijuana, BC, Mexico
e-mail: fevrier@tectijuana.mx

O. Castillo et al. (eds.), *Fuzzy Logic Augmentation of Neural and Optimization Algorithms: Theoretical Aspects and Real Applications*, Studies in Computational Intelligence 749, https://doi.org/10.1007/978-3-319-71008-2_3

consider in this work a new variable, which is in charge of measuring the distance between the harmonies and this variable will be called diversity, which has been used in several works such as in [3].

Fuzzy logic is currently related and based on the theory of fuzzy sets [9, 10]. According to this theory, the degree of membership of an element to a set will be determined by a membership function, which can have real values included in the Interval [0, 1].

Unlike the existing variants of this algorithm, in this paper fuzzy logic is used to dynamically change the parameters as the improvisations in the algorithm advances, thus avoiding trial and error experiments, and it is possible to maintain control of the exploration at the beginning to end with the exploitation within the search space [2, 6, 7, 11, 16, 17].

The proposed method is tested with 11 benchmark mathematical functions with 2 and 10 variables or dimensions, each one tested with the original method and the proposed methods with type-1 and type-2 fuzzy logic.

This work is organized in the following form: Sect. 2 shows the completed methodology, Sect. 3 presented the results of the experiments, and Sect. 4 presented the conclusions.

2 Methodology

This part shows the proposed method known as fuzzy harmony search algorithm using type-1 and type-2 fuzzy logic. This method is founded on the original harmony search algorithm, which seeks to find the perfect harmony measured by the esthetic standards. In this case fuzzy logic is responsible for the change of the parameters within the algorithm, as the improvisations advance.

This proposed method, unlike the existing ones, uses the improvisations and a new measure as an input called diversity, which is responsible for calculating the distance between the harmonies and based on if/then rules the adjustment of these parameters is performed to achieve the control of the exploration and exploitation of the search space. The part focused on this proposal is highlighted with green in the flow diagram as shown in Fig. 1.

Figure 1 shows the proposal, in the improvisation step, the *HMR*(Harmony memory accepting) represents the exploitation and *PArate* (Pitch Adjustment) represents the exploration of this algorithm and these are dynamically changed by means of a Type-1 fuzzy system and intervals type-2.

This method uses the improvisations or the iterations defined in Eq. 1 and the diversity defined in Eq. 2 are the two measures that are taken into account for the inputs of the fuzzy system to obtain the overall optimum.

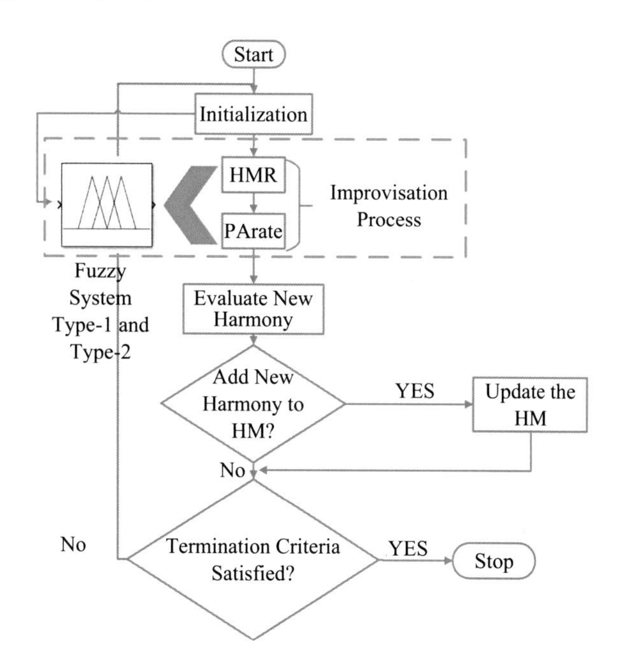

Fig. 1 scheme of the proposed method

$$\text{Improvisation} = \frac{\textit{Current Improvisations}}{\textit{Maximun of Improvisations}} \tag{1}$$

$$\text{Diversity}(\text{S}(\text{t})) = \frac{1}{n_S}\sum_{i=1}^{n_S}\sqrt{\sum_{j=1}^{n_x}\left(x_{ij}(t) - \bar{x}_j(t)\right)^2} \tag{2}$$

Where Eq. 1, is the current improvisation is defined by the number of improvisations elapsed and maximum of improvisations is defined by the number of improvisations established for HS to find the best solution. In Eq. 2, S is the harmonies or the population of HS; t is the current improvisation or time, n_s is the size of the harmonies, i is the number of the harmony, n_x is the total number of dimensions, j is the number of the dimension, x_{ij} is the j dimension of the harmony i, $\bar{x}_j$ is the j dimension of the current best harmony of the harmonies.

The fuzzy systems used in this case are shown in Fig. 2 (Type-1 fuzzy system) and Fig. 3 (Type-2 fuzzy system). In the two proposed fuzzy systems we use as inputs the improvisations and the diversity and as output the *HMR* parameter. In this case triangular membership functions were used in all fuzzy systems and all are granulated into three membership functions.

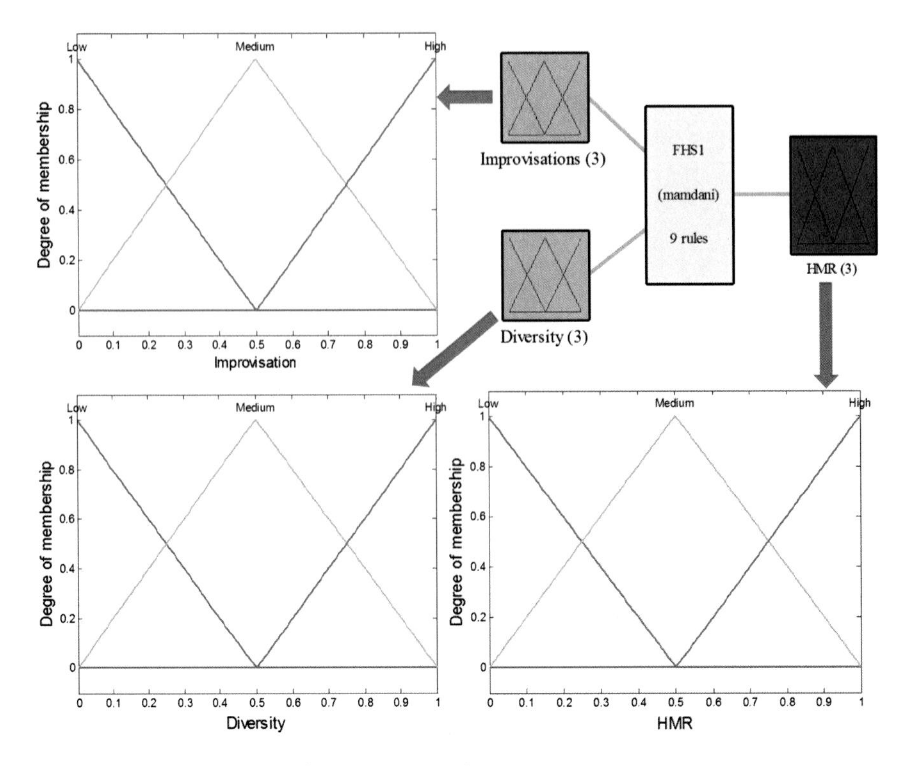

Fig. 2 Diagram of the type-1 fuzzy system (FHS1)

It is observed in Fig. 2 that the membership functions for the input are the "improvisation" and "diversity" and the membership functions for the output of the fuzzy system is the "HMR" parameter, it is important to note that this first proposal uses type-1 fuzzy logic.

It is observed in Fig. 3 that the membership functions for the inputs are the "improvisation" and "diversity" and the membership functions for the output of the fuzzy system are for the "HMR" parameter, it is important to note that this second proposal uses type-2 fuzzy logic.

The rules that are used in the FHS1 and FHS2 method are presented in Fig. 4. These rules were constructed based on previous experiments in other works and based on the operation of the original method to control the exploration and exploitation of the search space, to obtain the global optimum.

In Fig. 4 show the rules used of the first and second fuzzy system, which are increasing as the improvisations and diversity of the algorithm advance and with this achieve an equilibrium. Figure 5 shows the surface plot of the behavior of the fuzzy rules.

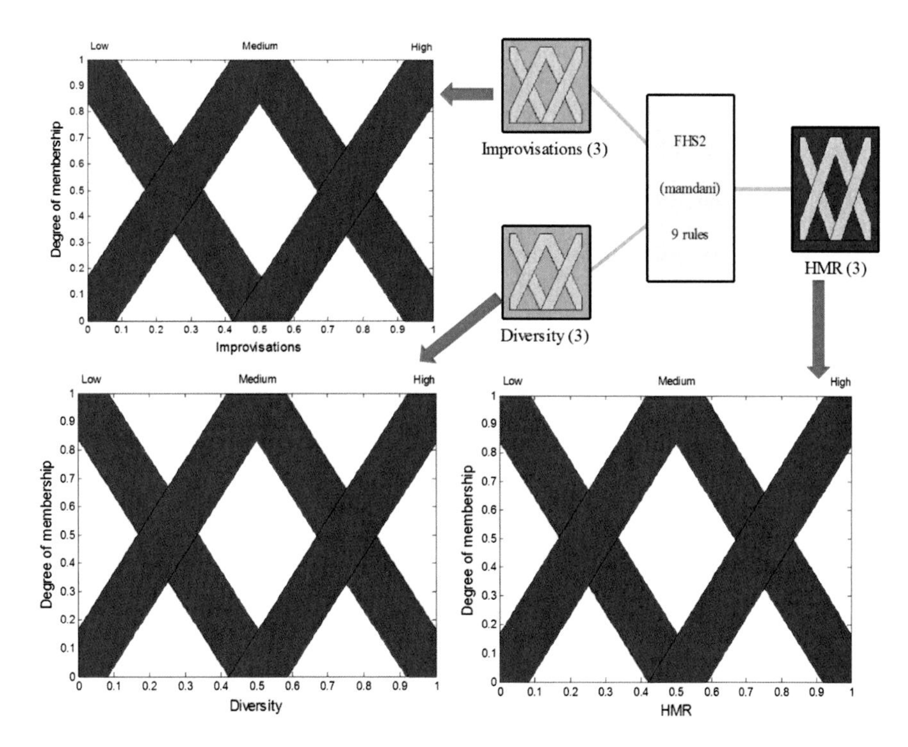

Fig. 3 Diagram of the type-2 fuzzy system (FHS2)

```
1. If (Improvisation is Low) and (Diversity is Low) then (HMR is Low) (1)
2. If (Improvisation is Medium) and (Diversity is Medium) then (HMR is Medium) (1)
3. If (Improvisation is High) and (Diversity is High) then (HMR is High) (1)
4. If (Improvisation is Low) and (Diversity is Low) then (HMR is High) (1)
5. If (Improvisation is Medium) and (Diversity is Medium) then (HMR is Low) (1)
6. If (Improvisation is High) and (Diversity is High) then (HMR is High) (1)
7. If (Improvisation is Low) and (Diversity is Medium) then (HMR is High) (1)
8. If (Improvisation is Medium) and (Diversity is High) then (HMR is Medium) (1)
9. If (Improvisation is High) and (Diversity is Low) then (HMR is High) (1)
```

Fig. 4 Rules for FHS1 and FHS2

In Fig. 5 we can observe the behavior of the proposed rules for these two fuzzy systems, and these rules achieve a control of the exploration and exploitation of the search space as the improvisations progress and with the diversity measure the distance between the harmonies. These techniques are combined to achieve the goal of finding the global minimum.

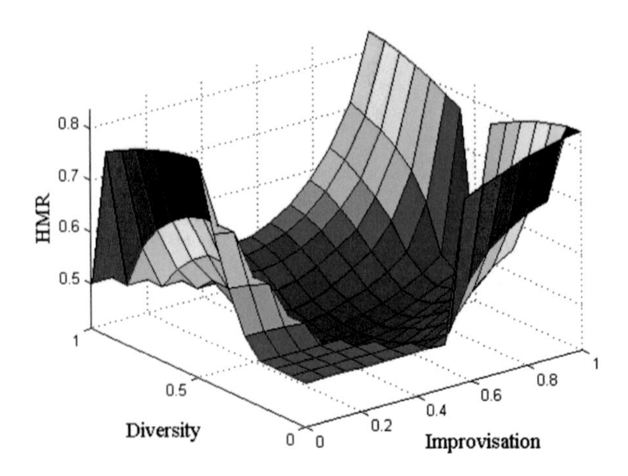

Fig. 5 Plot of the model of the increasing rules

3 Results of the Experiments

This section demonstrates the experiments performed and the results obtained. In this paper, several proposals were presented using fuzzy logic, these proposed methods were applied to 11 mathematical functions with a different number of dimensions, as shown in Table 1.

In Table 1, the benchmark functions with which the proposed method was tested, also the used range, dimensions, and the global minimum for each function are shown.

50 runs were performed for each mathematical function and an average was obtained for each function as presented in Table 2.

In Table 2 we are presenting the average for each mathematical function, the HS column is the original method, FHS1 is composed of the two inputs and one output,

Table 1 Functions Benchmark and parameters

Function	Dimension	Search domain	Global minimum
Rosenbrock	10	[−5,10]	0
Sphere	10	[−5.12,5.12]	0
Hump	10	[−5,5]	0
Rastrigin	10	[−5.12,5.12]	0
Schwefel	10	[−500,500]	0
Shubert	2	[−10,10]	−186.7309
Sum square	10	[−10,10]	0
Zakharov	10	[−5,10]	0
Griewank	10	[−600,600]	0
Powel	10	[−4,5]	0
Trid	6	[−36,36]	−50
Trid	10	[−100,100]	−200

Table 2 Results for each function in each method

Function	HS	FHS 1	FHS 2
Rosenbrock	2.57E−02	3.81E−02	5.63E−08
Sphere	1.00E+01	0.00E+00	0.00E+00
Hump	−1.02E+00	−1.02E+00	−1.02E+00
Rastrigin	1.07E+00	6.57E−03	8.72E−07
Schwefel	1.87E+01	5.07E−02	8.46E−05
Shubert	−1.85E+02	−1.86E+02	−1.86E+02
Sum square	1.47E−01	4.37E−05	1.60E−10
Zakharov	1.65E−01	2.93E−05	7.20E−09
Griewank	3.90E−01	1.62E−02	9.41E−12
Powel	2.66E+00	0.00E+00	0.00E+00
Trid	−1.51E+01	−1.96E+00	−3.34E+01
Trid	6.83E−01	1.89E+03	1.21E−01

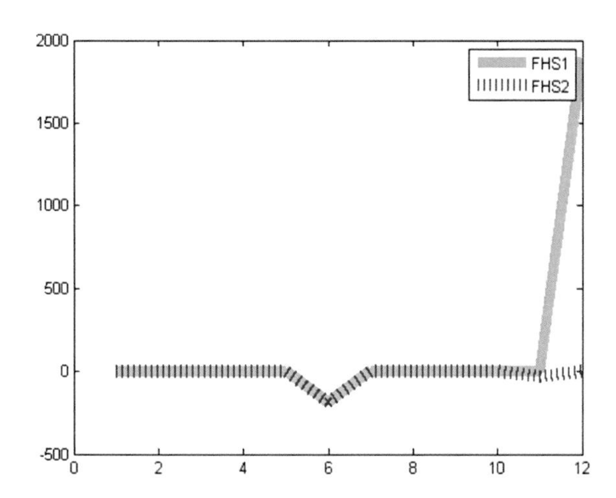

Fig. 6 Comparison between type-1 and type-2 methods

used the type-1 fuzzy logic, FHS2 is composed of the two inputs and one output, used the type-2 fuzzy logic. It should be mentioned that it is an average of 50 experiments. Figure 6 shows the behavior of the methods, the pink line indicates the average obtained with FHS1 method and the blue dotted line shows the data obtained with the FHS2 method.

4 Conclusions

This paper used the original HS algorithm and combined it with type-1 and interval type-2 fuzzy logic to realize the optimization of the benchmark mathematical functions, and find the global optimum, it also considers a second measure as input, which "diversity" that produced best results. The obtained results confirm the efficacy of the algorithm in most of the reference functions.

We can note that by using the type-2 fuzzy logic we can get even better results in most mathematical functions applied and achieves greater stability for the set of reference functions. In future work the method will be applied to other cases, such as control or more complex benchmark functions, additional outputs will be appended to the fuzzy system.

References

1. A. Assad, K. Deep, Applications of harmony search algorithm in data mining: a survey, in *Proceedings of Fifth International Conference on Soft Computing for Problem Solving* (Springer, Singapore, 2016)
2. E. Bernal, O. Castillo, J. Soria, F. Valdez, Imperialist competitive algorithm with dynamic parameter adaptation using fuzzy logic applied to the optimization of mathematical functions. Algorithms **10**(1), 18 (2017)
3. C. Caraveo, F. Valdez, O. Castillo, Optimization mathematical functions for multiple variables using the algorithm of self-defense of the plants, in *Nature-Inspired Design of Hybrid Intelligent Systems* (Springer International Publishing, 2017), pp. 631–640
4. K.Z. Gao et al., Discrete harmony search algorithm for flexible job shop scheduling problem with multiple objectives. J. Intell. Manuf. **27**(2), 363–374 (2016)
5. Z. Geem, *Music Inspired Harmony Search Algorithm Theory and Applications, Studies in Computational Intelligence* (Springer, Heidelberg, Germany 2009), pp. 8–121
6. B. González, F. Valdez, P. Melin, A gravitational search algorithm using type-2 fuzzy logic for parameter adaptation, in *Nature-Inspired Design of Hybrid Intelligent Systems* (Springer International Publishing, 2017), pp. 127–138
7. J.C. Guzmán, P. Melin, G. Prado-Arechiga, Neuro-fuzzy hybrid model for the diagnosis of blood pressure, in *Nature-Inspired Design of Hybrid Intelligent Systems* (Springer International Publishing, 2017), pp. 573–582
8. P. Kar, S.C. Swain, A harmony search-firefly algorithm based controller for damping power oscillations, in *Computational Intelligence & Communication Technology (CICT), 2016 Second International Conference on* (IEEE, 2016)
9. Q. Liang, J.M. Mendel, Interval type-2 fuzzy logic systems: theory and design. IEEE Trans. Fuzzy Syst. **8**(5), 535–550 (2000)
10. J.M. Mendel, *Uncertain Rule-Based Fuzzy Logic Systems: Introduction and New Directions* (Prentice Hall PTR, Upper Saddle River, 2001)
11. P. Ochoa, O. Castillo, J. Soria, Differential evolution using fuzzy logic and a comparative study with other metaheuristics, in *Nature-Inspired Design of Hybrid Intelligent Systems* (Springer International Publishing, 2017), pp. 257–268
12. C. Peraza, F. Valdez, M. Garcia, P. Melin, O. Castillo, A new fuzzy harmony search algorithm using fuzzy logic for dynamic parameter adaptation. Algorithms **9**(4), 69 (2016)
13. C. Peraza, F. Valdez, O. Castillo, An adaptive fuzzy control based on harmony search and its application to optimization, in *Nature-Inspired Design of Hybrid Intelligent Systems* (Springer International Publishing, 2017), pp. 269–283
14. C. Peraza, F. Valdez, O. Castillo, Interval type-2 fuzzy logic for dynamic parameter adaptation in the harmony search algorithm, in *Intelligent Systems (IS), 2016 IEEE 8th International Conference on* (IEEE, 2016)
15. M.P. Saka, O. Hasançebi, Z.W. Geem, Metaheuristics in structural optimization and discussions on harmony search algorithm. Swarm Evol. Comput. **28**, 88–97 (2016)

16. A. Uriarte, P. Melin, F. Valdez, A new hybrid PSO method applied to benchmark functions, in *Nature-Inspired Design of Hybrid Intelligent Systems* (Springer International Publishing, 2017), pp. 423–430
17. F. Valdez, P. Melin, O. Castillo, Evolutionary method combining particle swarm optimization and genetic algorithms using fuzzy logic for decision making, in *IEEE International Conference on Fuzzy Systems* (2009), pp. 2114–2119
18. G.G. Wang, A.H. Gandomi, X. Zhao, H.C.E. Chu, Hybridizing harmony search algorithm with cuckoo search for global numerical optimization. Soft. Comput. **20**(1), 273–285 (2016)
19. G. Wang, L. Guo, A novel hybrid bat algorithm with harmony search for global numerical optimization. J. Appl. Math. (2013)
20. G.G. Wang, A. Hossein Gandomi, A. Hossein Alavi, A chaotic particle-swarm krill herd algorithm for global numerical optimization. Kybernetes **42**(6), 962–978 (2013)
21. G.G. Wang, A.H. Gandomi, A.H. Alavi, Stud krill herd algorithm. Neurocomputing **128**, 363–370 (2014)

Comparison of Bio-Inspired Methods with Parameter Adaptation Through Interval Type-2 Fuzzy Logic

Frumen Olivas, Fevrier Valdez and Oscar Castillo

Abstract In the development of this paper we perform a comparison with two bio-inspired methods, Ant Colony Optimization (ACO) and Gravitational Search Algorithm (GSA). Each one of these methods use our methodology for parameter adaptation using interval type-2 fuzzy logic, where based on some metrics about the algorithm, like the percentage of iterations elapsed or the diversity of the population, we try to control their behavior and therefore control their abilities to perform a global or a local search. To test these methods two problems were used in which a fuzzy controller is optimized to minimize the error in the simulation with nonlinear complex plants.

Keywords Interval type-2 fuzzy logic · Ant Colony Optimization
Gravitational Search Algorithm · Dynamic parameter adaptation

1 Introduction

Bio-inspired optimization algorithms can be applied to most combinatorial problems but for different problems needs different parameters, in order to obtain better results. There are in literature several methods to model a better behavior of these algorithms by adapting some parameters [18], introducing different parameters in the equations of the algorithms [4], performing a hybridization with other algorithm [17], and using fuzzy logic [5–9, 14, 16].

F. Olivas · F. Valdez · O. Castillo (✉)
Tijuana Institute of Technology, Tijuana, Mexico
e-mail: ocastillo@tectijuana.mx

F. Olivas
e-mail: frumen@msn.com

F. Valdez
e-mail: fevrier@tectijuana.mx

O. Castillo et al. (eds.), *Fuzzy Logic Augmentation of Neural and Optimization Algorithms: Theoretical Aspects and Real Applications*, Studies in Computational Intelligence 749, https://doi.org/10.1007/978-3-319-71008-2_4

In this paper is presented a methodology for parameter adaptation using an interval type-2 fuzzy system, where on each method it tries to model a better behavior in order to obtain better quality results.

The proposed methodology was successfully applied to different bioinspired optimization methods like BCO (Bee Colony Optimization) in [1], CSA (Cuckoo Search Algorithm) in [3], PSO (Particle Swarm Optimization) in [5, 7], ACO (Ant Colony Optimization) in [6, 8], GSA (Gravitational Search Algorithm) in [9, 16], DE (Differential Evolution) in [10], HSA (Harmony Search Algorithm) in [11], BA (Bat Algorithm) in [12] and in FA (Firefly Algorithm) in [15].

The algorithms used in this research are ACO (Ant Colony Optimization) from [8] and GSA (Gravitational Search Algorithm) from [9], each one with dynamic parameter adaptation using an interval type-2 fuzzy system.

Fuzzy logic proposed by Zadeh in [19–21] help us to model a complex problem, with the use of membership functions and fuzzy rules, with the knowledge of a problem from an expert, fuzzy logic can bring tools to create a model and attack a complex problem.

The contribution of this paper is the comparison between the bioinspired methods which use an interval type-2 fuzzy system for dynamic parameter adaptation, in the optimization of fuzzy controllers for nonlinear complex plants. The adaptation of parameters with fuzzy logic helps to perform a better design of the fuzzy controllers, based on the results which are better than the original algorithms.

2 Bioinspired Optimization Methods

ACO is a bioinspired algorithm based on swarm intelligence of the ants, proposed by Dorigo in [2], where each individual helps each other to find the best route from their nest to a food source. Artificial ants represent the solutions to a problem, where each ant is a tour and each node is a dimension or a component of the problem. Biological ants use pheromone trails to communicate to other ants which path is the best and the artificial ants tries to mimic that behavior.

Artificial ants use probability to select the next node using Eq. 1, where with this equation calculate the probability of an ant k to select the node j from node i.

$$P_{ij}^{k} = \frac{[\tau_{ij}]^{\alpha}[\eta_{ij}]^{\beta}}{\sum_{l} \in N_{i}^{k}[\tau_{il}]^{\alpha}[\eta_{il}]^{\beta}}, \quad if\ j \in N_{i}^{k} \tag{1}$$

The components of Eq. 1 are: P^k is the probability of an ant k to select the node j from node i, τ_{ij} represent the pheromone in the arc that joins the nodes i and j and η_{ij} represent the visibility from node i to node j, with the condition that node j must be in the neighborhood of node i.

Also like in nature the pheromone trail evaporate over time, and ACO use Eq. 2 to simulate the evaporation of pheromone in the trails.

$$\tau_{ij} \leftarrow (1-\rho)\,\tau_{ij}, \quad \forall (i,j) \in L \tag{2}$$

The components of Eq. 2 are: τ_{ij} represent the pheromone trail in the arc that joins the nodes i and j, ρ represent the percentage of evaporation of pheromone, this equation is applied to all arcs in the graph L.

There are more equations of ACO, but these two equations are the most important in the dynamics of the algorithm, also these equations contains the parameters used to model a better behavior of the algorithm using an interval type-2 fuzzy system.

GSA proposed by Rashedi in [13], is a population based algorithm that use laws of physics to update its individuals, more particularly use the Newtonian law of gravity and the second motion law, in this algorithm each individual is considered as agent where each one represent a solution to a problem, each agent has its own mass and can move to another agent. The mass of an agent is given by the fitness function, agents with bigger mass are better.

Each agent apply some gravitational force to all other agents, and is calculated using Eq. 3.

$$F_{ij}^{d}(t) = G(t)\frac{M_{pi}(t) \times M_{aj}(t)}{R_{ij}(t) + \varepsilon}\left(x_{j}^{d}(t) - x_{i}^{d}(t)\right) \tag{3}$$

The components of Eq. 3 are: F_{ij}^{d} is the gravity force between agents i and j, G is the gravitational constant, M_{pi} is the mass of agent i or passive mass, and M_{aj} is the mass of agent j or active mass, R_{ij} is the distance between agents i and j, ε is an small number used to avoid division by zero, x_{j}^{d} is the position of agent j and x_{i}^{d} is the position of agent j.

The gravitational force is used to calculate the acceleration of the agent using Eq. 4.

$$a_{i}^{d}(t) = \frac{F_{i}^{d}(t)}{M_{ii}(t)} \tag{4}$$

The components of Eq. 4 are: a_{i}^{d} is the acceleration force of agent i, F_{i}^{d} is the gravitational force of agent i, and M_{ii} is the inertial mass of agent i.

In GSA the gravitational constant G from Eq. 3 unlike real life here can be variable and is given by Eq. 5.

$$G(t) = G_{0}^{-\alpha\, t/T} \tag{5}$$

The components of Eq. 5 are: G is the gravitational constant, G_0 is the initial gravitational constant, α is a parameter defined by the user of GSA and is used to control the change in the gravitational constant, t is the actual iteration and T is the total iterations.

To control the elitism GSA uses Eq. 6 to allow only the best agents to apply their force to other agents, in initial iterations all the agents apply their force but Kbest will decrease over time until to only a few agents are allowed to apply their force.

$$F_i^d(t) = \sum_{j \in Kbest, j \neq 1} rand_i F_{ij}^d(t) \tag{6}$$

The components of Eq. 6 are: F_i^d is the new gravity force of agent i, *Kbest* is the number of agents allowed to apply their force, sorted by their fitness the best *Kbest* agent can apply their force to all other agents, in this equation j is the number of dimension of agent i.

3 Methodology for Parameter Adaptation

The optimization methods involved in this comparison have dynamic parameter adaptation through an interval type-2 fuzzy system, and each of this adaptation are described in details for ACO in [8] and for GSA in [9].

The way in which this adaptation of parameter was done is, first a metric about the performance of the algorithms need to be created, in this case the metrics are a percentage of iteration elapsed described by Eq. 7 and the diversity of individuals described by Eq. 8, after the metrics are defined we need to select the best parameters to be dynamically adjusted, and this was done based on experimentation with different levels of all the parameters of each optimization method.

$$Iteration = \frac{Current\ Iteration}{Maximum\ of\ Iterations} \tag{7}$$

The components of Eq. 7 are: *Iteration* is a percentage of the elapsed iterations, *current iteration* is the number of elapsed iterations, and *maximum of iterations* is the total number iterations set for the optimization algorithm to find the best possible solution.

$$Diversity\ (S(t)) = \frac{1}{n_s} \sum_{i=1}^{n_s} \sqrt{\sum_{j=1}^{n_x} \left(x_{ij}(t) - \bar{x}_j(t)\right)^2} \tag{8}$$

The components of Eq. 8 are: *Diversity(S)* is a degree of dispersion of the population S, n_s is the number of individual in the population S, n_x is the number of dimensions in each individual from the population, x_{ij} is the j dimension of the individual i, *tested* x_j is the j dimension of the best individual in the population.

After the metrics are defined and the parameters selected, a fuzzy system is created to adjust one parameter, and with this obtain a fuzzy rule set to control this parameter, and for all the parameters we need to do the same, and at the end only one fuzzy system will be created to control all the parameters at the same time combining all the created fuzzy systems.

The proposed methodology for parameter adaptation is illustrated in Fig. 1, where it has the optimization method, which has an interval type-2 fuzzy system for parameter adaptation.

Figure 1 illustrate the general scheme for parameter adaptation, in which the bio-inspired optimization algorithm is evaluated by the metrics and used these as inputs for the interval type-2 fuzzy system which will adapt some parameters of the optimization algorithm based on the metrics and the fuzzy rules, then this method with parameter adaptation will provide the parameters or solutions for a problem, in this case the parameters for the fuzzy system used for control.

The final interval type-2 fuzzy systems for each optimization method are illustrated in Figs. 2 and 3, for ACO and GSA correspondingly. Each one of these fuzzy systems has iteration and diversity as inputs, with a range from 0 to 1 using Eqs. 7 and 8 correspondingly to each input, and two outputs but these differs from each optimization method because each one has its own parameters to be adjusted dynamically.

The interval type-2 fuzzy system from Fig. 2 has two inputs and two outputs, the inputs are granulated into three type-2 triangular membership functions and the outputs into five type-2 triangular membership functions, and nine rules, in this case the parameters to be adjusted dynamically over the iterations are α *(alpha)* and ρ *(rho)* from Eqs. 1 and 2 respectively, both with a range from 0 to 1.

The interval type-2 fuzzy system from Fig. 3 has *iteration* and *diversity* as inputs with three type-2 triangular membership functions and two outputs which are the parameters to be adjusted in this case, α *(alpha)* with a range from 0 to 100 and

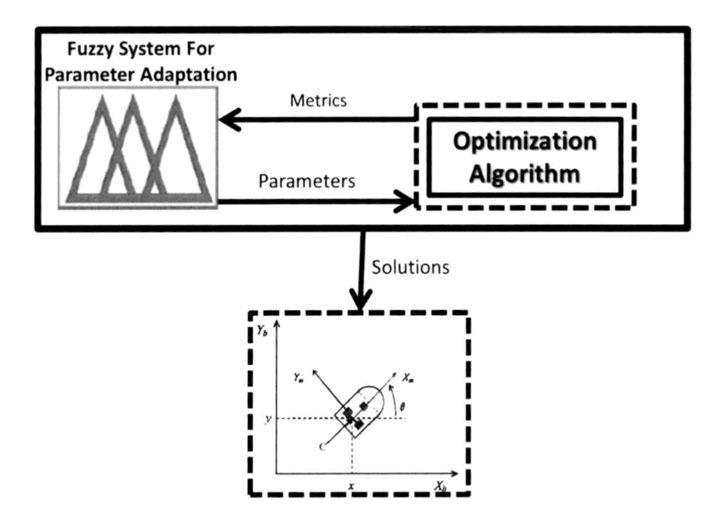

Fig. 1 General scheme of the proposal for parameter adaptation

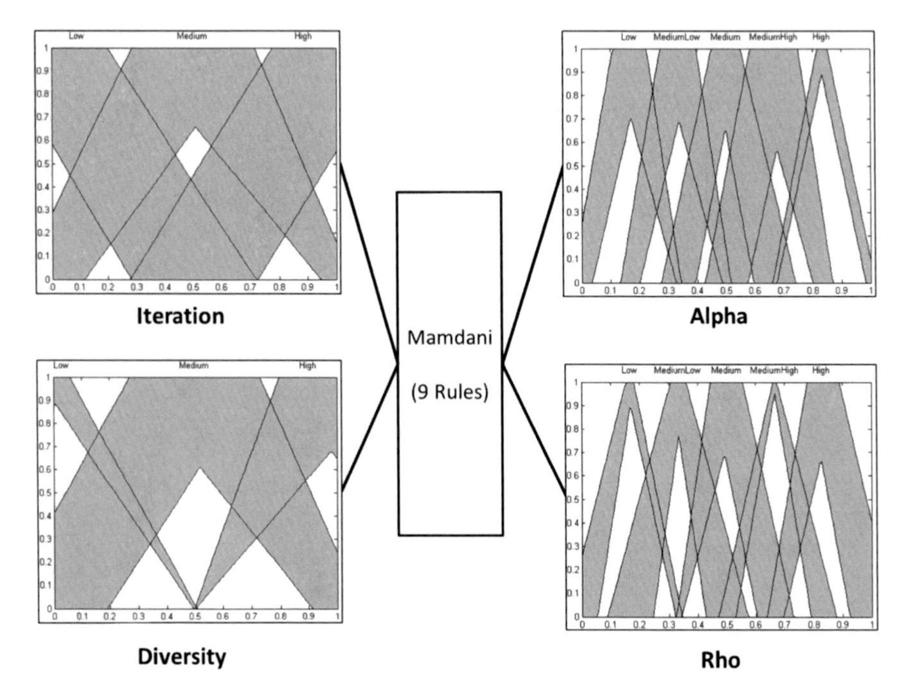

Fig. 2 Interval type-2 fuzzy system for parameter adaptation in ACO

Kbest from 0 to 1, each output is granulated into five type-2 triangular membership functions with a fuzzy rule set of nine rules. The parameters α *(alpha)* and *Kbest* are from Eqs. 5 and 6 respectively.

4 Problems Statement

The comparison of ACO and GSA is through the optimization of a fuzzy controller from two different non-linear complex plants, where this two problems use a fuzzy system for control.

The first problem is the optimization of the trajectory of an autonomous mobile robot and the objective is minimize the error in the trajectory, the robot has two wheeled motors and one stabilization wheel, it can move in any direction.

The desired trajectory is illustrated in Fig. 4, where first the robot must start from point (0, 0) and it needs to follow the reference using the fuzzy system from Fig. 5 as controller.

The reference illustrated in Fig. 4 helps in the design of a good controller because it use only nonlinear trajectories, to insurance that the robot can follow any trajectory.

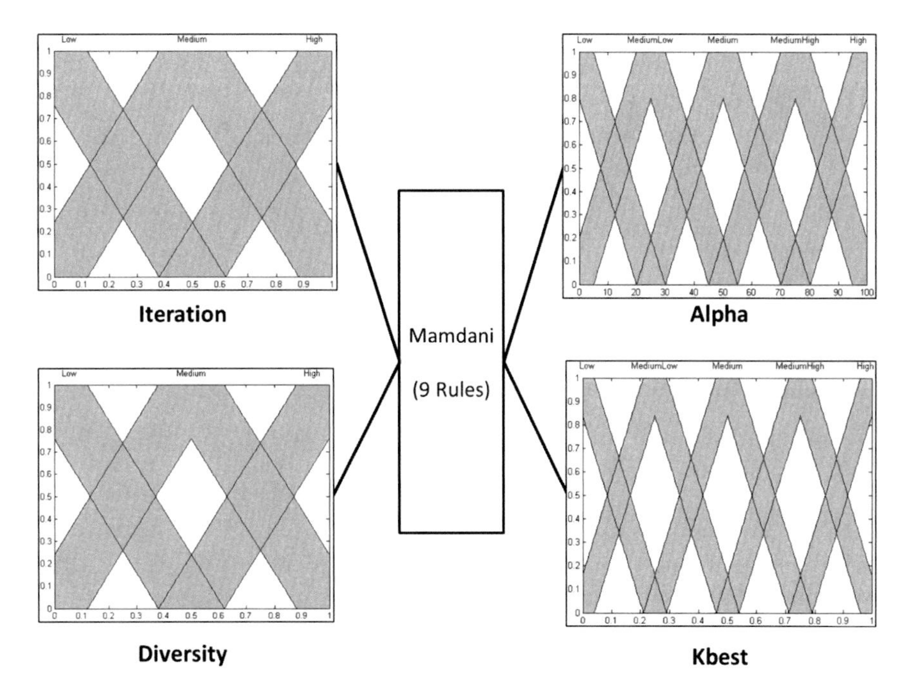

Fig. 3 Interval type-2 fuzzy system for parameter adaptation in GSA

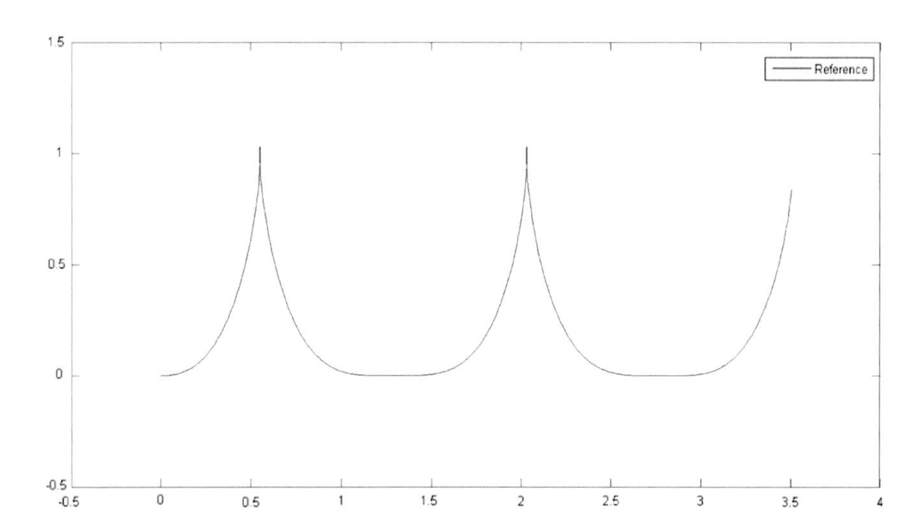

Fig. 4 Trajectory for the autonomous mobile robot

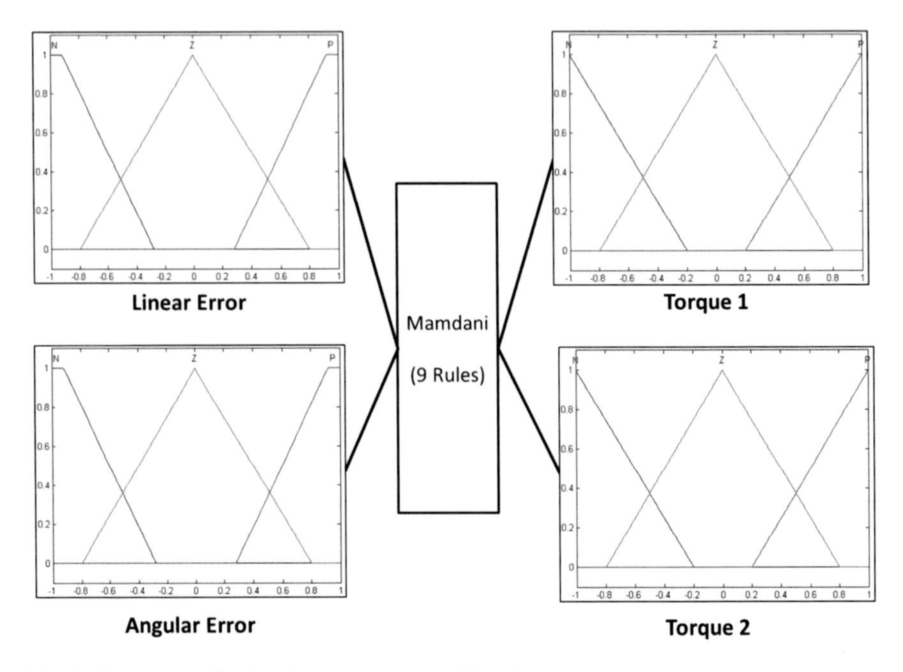

Fig. 5 Fuzzy controller for the autonomous mobile robot

The fuzzy system used for control illustrated in Fig. 5, use the linear and angular errors to control the motorized wheels of the robot.

In this problem the optimization methods will try to find better parameters for the membership functions, using the same fuzzy rule set.

The second problem is the automatic temperature control in a shower, and the optimization method will try to optimize the fuzzy controller illustrated in Fig. 8, which will try to follow the flow and temperature references illustrated in Figs. 6 and 7 respectively.

The flow reference illustrated in Fig. 6 has a variation from 0.5 to 0.9 and is changing over time to simulate the open-close mechanism in a shower.

The temperature reference illustrated in Fig. 7 is changing over time from 19 to 27 degrees in temperature, to simulate the change in the desired temperature.

The fuzzy system used as control illustrated in Fig. 8 has two input variables, temperature and flow, a fuzzy rule set of nine rules and two outputs cold and hot. The fuzzy system use the inputs and with the fuzzy rules try to control the open-close mechanism of the cold and hot water.

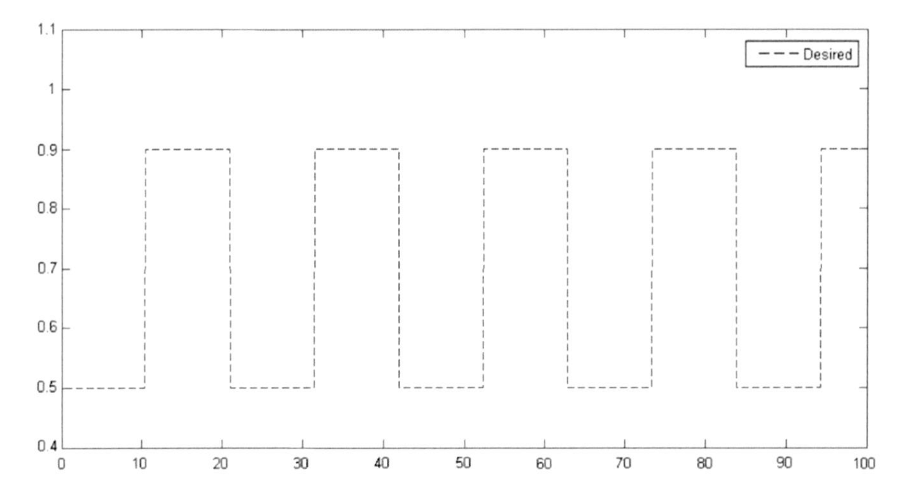

Fig. 6 Flow reference

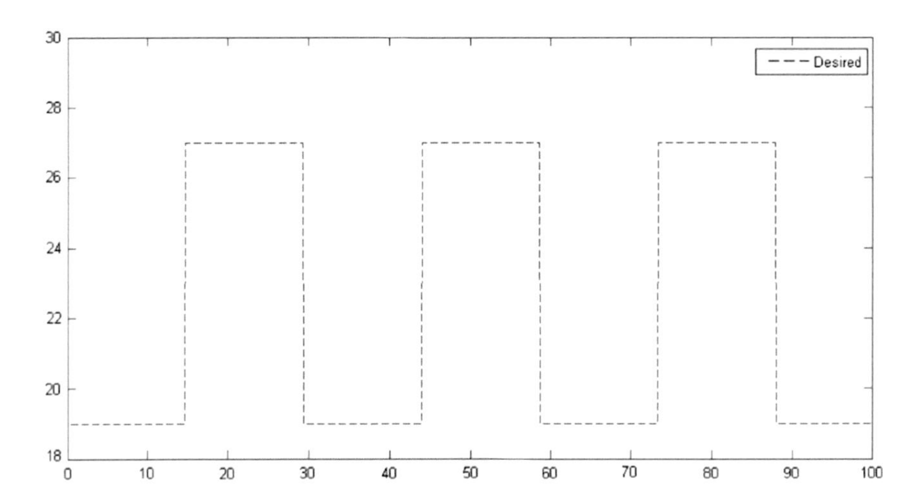

Fig. 7 Temperature reference

5 Simulations, Experiments and Results

The optimization methods were applied to the optimization of the membership functions of the fuzzy system used as controllers for the two problems described in Sect. 4.

Using the parameters from Table 1, each method was applied to both problems. In the case of the problem of the trajectory of an autonomous mobile robot there are 40 points to be search for all the membership functions, and in the problem of the automatic temperature control in a shower there are 52 points.

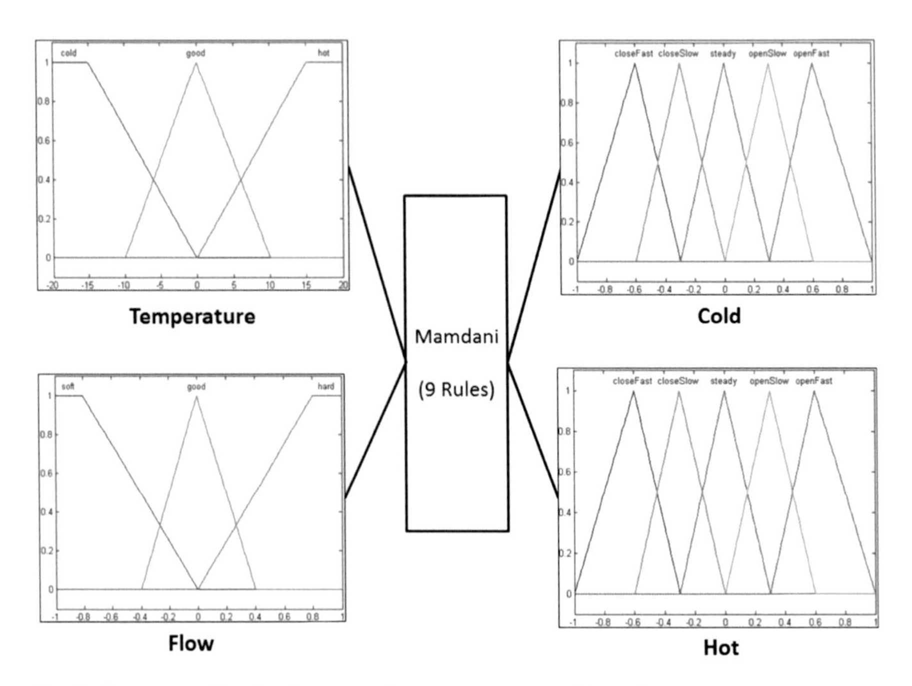

Fig. 8 Fuzzy controller for the automatic temperature control in a shower

Table 1 Parameters for each optimization method

Parameter	Original ACO	ACO with parameter adaptation	Original GSA	GSA with parameter adaptation
Population	30	30	30	30
Iterations	50	50	50	50
α (Alpha)	1	Dynamic	40	Dynamic
β (Beta)	2	2		
ρ (Rho)	0.1	Dynamic		
Kbest			Linear decreasing from 100 to 2%	Dynamic
G_0			100	100

The methods to be compared are: the original ACO method, ACO with parameter adaptation, original GSA method and GSA with parameter adaptation.

The parameters from Table 1 are a challenge for the optimization methods, because there are only 50 iterations to found the best possible fuzzy controller for each problem. This is a good manner to show the advantages of the proposed methodology for parameter adaptation using an interval type-2 fuzzy system.

Table 2 contains the results of applying all the optimization methods to the optimization of the fuzzy controller for an autonomous mobile robot, the average is

Table 2 Results of the simulations with the robot problem

MSE	Original ACO	ACO with parameter adaptation	Original GSA	GSA with parameter adaptation
Average	0.4641	**0.0418**	36.4831	15.4646
Best	0.1285	**0.0048**	10.4751	3.2375
Worst	0.9128	**0.1276**	76.0243	30.8511
Standard deviation	0.2110	**0.0314**	15.8073	8.6371

Bold indicates best values

Table 3 Results of the simulations with the shower problem

MSE	Original ACO	ACO with parameter adaptation	Original GSA	GSA with parameter adaptation
Average	0.6005	0.4894	3.8611	**0.1151**
Best	0.5407	0.3980	1.9227	**0.0106**
Worst	0.9036	0.5437	6.5659	**0.3960**
Standard deviation	0.0696	0.0378	1.0860	**0.0913**

Bold indicates best values

from 30 experiments [with the metric of Mean Square Error (MSE)] and results in bold are best, the 30 experiments means that each method was applied to the optimization of the fuzzy controller 30 times resulting in 30 different fuzzy controller for each method.

From Table 2 the optimization method that obtain better results is ACO with parameter adaptation using the proposed methodology using an interval type-2 fuzzy system, also it can be seen that the results of GSA with parameter adaptation are better that the original GSA but ACO is better.

The results contained in Table 3 are from applying all the methods to optimize the fuzzy controller for the automatic temperature control in a shower, the average is from 30 experiments [with the metric Mean Square Error (MSE)] and also the results in bold are best, same as the first problem the 30 experiments means that each method was applied to the optimization of the fuzzy controller and obtaining 30 different fuzzy controller for each method.

From the results in Table 3 in this case GSA with parameter adaptation using the proposed methodology using an interval type-2 fuzzy system can obtains better results than the other methods. Also it can be seen that ACO with parameter adaptation can obtain better results than the original ACO method and the original GSA method.

6 Statistical Comparison

The Z-test is a tool to prove that the methods with parameter adaptation can obtain on average better results than its counterparts the original methods, also to know what method is better on certain problem by comparing its results with all of the other methods.

The comparison between the methods is using the statistical test Z-test, using the parameters from Table 4 and the results of the comparisons are in Tables 5 and 6 for the robot and shower problems, respectively.

The results in Table 5 are using the parameters in Table 4 for the Z-test, where it claims that a method (μ_1) has on average better results (we are comparing errors, so minimum is better) than the other method (μ_2), in Tables 5 and 6 the first column correspond to the methods as μ_1 and the first row correspond to the methods as μ_2,

Table 4 Parameters for the statistical Z-test

Parameter	Value
Level of significance	95%
Alpha (α)	5%
Alternative hypothesis (H_a)	$\mu_1 < \mu_2$ (claim)
Null hypothesis (H_0)	$\mu_1 \geq \mu_2$
Critical value	−1.645

Table 5 Results of the Z-test for comparison in the robot problem

μ_2 / μ_1	Original ACO	ACO with parameter adaptation	Original GSA	GSA with parameter adaptation
Original ACO		10.8415	**−12.4795**	**−9.5098**
ACO with parameter adaptation	**−10.8415**		**−12.6269**	**−9.7803**
Original GSA	12.4795	12.6269		6.3911
GSA with parameter adaptation	9.5098	9.7803	**−6.3911**	

Bold indicates best values

Table 6 Results of the Z-test for comparison in the shower problem

μ_2 / μ_1	Original ACO	ACO with parameter adaptation	Original GSA	GSA with parameter adaptation
Original ACO		7.6813	**−16.4115**	23.1516
ACO with parameter adaptation	**−7.6813**		**−16.9950**	20.7332
Original GSA	16.4115	16.9950		18.8264
GSA with parameter adaptation	**−23.1516**	**−20.7332**	**−18.8264**	

Bold indicates best values

also we are not comparing the same method with itself, results in bold means that there are enough evidence to reject the null hypothesis.

From the results in Table 5, which correspond to the optimization of a fuzzy controller for the trajectory of an autonomous mobile robot, there is enough evidence that ACO method with parameter adaptation can obtain on average better results than all of the other methods. There is enough evidence that the original ACO method can obtain on average better results than the original GSA and GSA with parameter adaptation. There is also enough evidence that GSA with parameter adaptation can obtain on average better results than the original GSA method.

From the results in Table 6, which correspond to the optimization of a fuzzy controller for the automatic temperature control in a shower, there is enough evidence that GSA with parameter adaptation can obtain on average better results than all of the other methods. There is enough evidence that ACO with parameter adaptation can obtain on average better results than the original ACO method and the original GSA method. There is also enough evidence that the original ACO method can obtain on average better results than the original GSA method.

7 Conclusions

The optimization of a fuzzy controller is a complex task, because require the search of several parameters in infinite possibilities in the range of each input or output variables. The bioinspired optimization methods help in the search because is guided by some kind of intelligence, from swarm intelligence or from laws of physics and can make a better search of parameters.

With the inclusion of a fuzzy system in this case an interval type-2, the bioinspired methods can search even in a better way, because is guided by the knowledge of an expert system that model a proper behavior in determined states of the search, in the beginning improves the global search or exploration of the search space and in final improves the local search or the exploitation of the best area found so far of the entire search space.

From the results with the MSE there is clearly that ACO with parameter adaptation has the best results in the robot problem, and GSA with parameter adaptation has the best results in the shower problem, but with the statistical test it confirm these affirmations.

The statistical comparison shows that the methods with parameter adaptation are better than their counterparts the original methods. Also ACO is a better method with the robot problem, but GSA is better in the shower problem.

References

1. L. Amador-Angulo, O. Castillo, Statistical analysis of Type-1 and Interval Type-2 Fuzzy Logic in dynamic parameter adaptation of the BCO, in *2015 Conference of the International Fuzzy Systems Association and the European Society for Fuzzy Logic and Technology (IFSA-EUSFLAT-15)*, Atlantis Press, June 2015
2. M. Dorigo, Optimization, learning and natural algorithms, Ph.D. Thesis, Dipartimento di Elettronica, Politechico di Milano, Italy, 1992
3. M. Guerrero, O. Castillo, M. Garcia, Fuzzy dynamic parameters adaptation in the Cuckoo Search Algorithm using fuzzy logic, in *2015 IEEE Congress on Evolutionary Computation (CEC)* (IEEE, New York, May 2015), pp. 441–448
4. L. Hongbo, A. Ajith, A fuzzy adaptive turbulent particle swarm optimization. Int. J. Innov. Comput. Appl. **1**(1), 39–47 (2007)
5. P. Melin, F. Olivas, O. Castillo, F. Valdez, J. Soria, J. Garcia, Optimal design of fuzzy classification systems using PSO with dynamic parameter adaptation through fuzzy logic. Exp. Syst. Appl. **40**(8), 3196–3206 (2013)
6. H. Neyoy, O. Castillo, J. Soria, in *Dynamic Fuzzy Logic Parameter Tuning for ACO and Its Application in TSP Problems*. Studies in Computational Intelligence vol. **451** (Springer, Berlin, 2012), pp. 259–271
7. F. Olivas, F. Valdez, O. Castillo, P. Melin, Dynamic parameter adaptation in particle swarm optimization using interval type-2 fuzzy logic. Soft. Comput. **20**(3), 1057–1070 (2016)
8. F. Olivas, F. Valdez, O. Castillo, C. Gonzalez, G. Martinez, P. Melin, Ant colony optimization with dynamic parameter adaptation based on interval type-2 fuzzy logic systems. Appl. Soft Comput. (2016)
9. F. Olivas, F. Valdez, O. Castillo, P. Melin, Interval type-2 fuzzy logic for dynamic parameter adaptation in a modified gravitational search algorithm. Eng. Appl. Artif. Intell. (2017, Under review)
10. P. Ochoa, O. Castillo, J. Soria, Differential evolution with dynamic adaptation of parameters for the optimization of fuzzy controllers, in *Recent Advances on Hybrid Approaches for Designing Intelligent Systems* (Springer International Publishing, 2014), pp. 275–288
11. C. Peraza, F. Valdez, O. Castillo, An improved harmony search algorithm using fuzzy logic for the optimization of mathematical functions, in *Design of Intelligent Systems Based on Fuzzy Logic, Neural Networks and Nature-Inspired Optimization* (Springer International Publishing, 2015), pp. 605–615
12. J. Perez, F. Valdez, O. Castillo, P. Melin, C. Gonzalez, G. Martinez, Interval type-2 fuzzy logic for dynamic parameter adaptation in the bat algorithm. Soft Comput. 1–19 (2016)
13. E. Rashedi, H. Nezamabadi-pour, S. Saryazdi, GSA: a gravitational search algorithm. Inf. Sci. **179**(13), 2232–2248. (2009)
14. Y. Shi, R. Eberhart, Fuzzy adaptive particle swarm optimization, in *Proceeding of IEEE International Conference on Evolutionary Computation*, Seoul, Korea (IEEE Service Center, Piscataway, NJ, 2001), pp. 101–106
15. C. Solano-Aragon, O. Castillo, Optimization of benchmark mathematical functions using the firefly algorithm with dynamic parameters, in *Fuzzy Logic Augmentation of Nature-Inspired Optimization Metaheuristics* (Springer International Publishing, 2015), pp. 81–89
16. A. Sombra, F. Valdez, P. Melin, O. Castillo, A new gravitational search algorithm using fuzzy logic to parameter adaptation, in *2013 IEEE Congress on Evolutionary Computation (CEC)* (IEEE, New York, June 2013), (pp. 1068–1074)

17. N. Taher, A. Ehsan, J. Masoud, A new hybrid evolutionary algorithm based on new fuzzy adaptive PSO and NM algorithms for distribution feeder reconfiguration. Energy Convers. Manag. **54**, 7–16 (2012)
18. B. Wang, G. Liang, W. Chan Lin, D. Yunlong, A new kind of fuzzy particle swarm optimization fuzzy_PSO algorithm, in *1st International Symposium on Systems and Control in Aerospace and Astronautics, ISSCAA 2006*, pp 309–311 (2006)
19. L. Zadeh, Fuzzy sets. Inf. Control **8** (1965)
20. Zadeh, L. Fuzzy logic. IEEE Comput., 83–92 (1965)
21. L. Zadeh, The concept of a linguistic variable and its application to approximate reasoning—I. Inform. Sci. **8**, 199–249 (1975)

Differential Evolution Algorithm with Interval Type-2 Fuzzy Logic for the Optimization of the Mutation Parameter

Patricia Ochoa, Oscar Castillo and José Soria

Abstract In this paper we propose using interval type-2 fuzzy logic for the optimization of parameters the form dynamic using the Differential Evolution algorithm. For this particular work we use Benchmark mathematical functions for the experiments that were performed adhering to the rules of the competition for the IEEE Congress on Evolutionary Computation (CEC) benchmark set of 2015. We are presenting a comparison against the winning paper of the competition IEEE Congress on Evolutionary Computation (CEC) to verify how good the proposed method Fuzzy Differential Evolution algorithm really is.

Keywords Differential Evolution algorithm · Fuzzy Differential Evolution Type-2 fuzzy logic

1 Introduction

The use of algorithms based on nature has become very common in evolutionary computation and metaheuristics. In this paper we propose to use one of these algorithms, in particular the Differential Evolution integrating fuzzy logic for dynamically adapting its parameters.

Differential Evolution (DE) is one of the latest evolutionary algorithms that has been proposed in the literature. It was created in 1994 by Price and Storn in an attempt to solve the Chebychev polynomial problem. The following years after that these two authors also proposed the DE for optimization of nonlinear and non-differentiable functions on continuous spaces [20].

Fuzzy logic or multi-valued logic is based on the fuzzy set theory proposed by Zadeh in 1965, which can help us with modeling expert knowledge, through the use of if-then fuzzy rules. Fuzzy set theory provides a systematic calculus to deal with

P. Ochoa · O. Castillo (✉) · J. Soria
Tijuana Institute of Technology, Tijuana, BC, Mexico
e-mail: ocastillo@tectijuana.mx

O. Castillo et al. (eds.), *Fuzzy Logic Augmentation of Neural and Optimization Algorithms: Theoretical Aspects and Real Applications*, Studies in Computational Intelligence 749, https://doi.org/10.1007/978-3-319-71008-2_5

linguistic information, and improves the numerical computation by using linguistic labels stipulated by membership functions [16].

In addition, in reviewing the recent literature there are papers on Differential Evolution applications that use this algorithm to solve real world problems [23, 24]. In the last years the concept of fuzzy logic has been used to adapt certain parameters in metaheuristic algorithms [15, 17, 18, 26–28], which demonstrate the importance and the improvement of these algorithms.

Regarding related works that use fuzzy logic to optimize the performance of metaheuristic algorithms we can find some relevant ones in [4, 5, 11–13].

In addition, related work with regards to the Special Session & Competition on Real-Parameter Single Objective Optimization at CEC-2015 can also be mentioned, like the works that obtained the first place: A Self-Optimization Approach for L-SHADE Incorporated with Eigenvector-Based Crossover and Successful-Parent-Selecting Framework on CEC 2015 Benchmark Set [8] (rank 1).Finally, the other works presented in the Special Session & Competition on Real-Parameter Single Objective Optimization at CEC-2015 [1–3, 6, 8, 19, 21, 22, 29, 30].

The rest of the paper is organized in the following form: Sect. 2 describes the Differential Evolution algorithm. Section 3 describes the proposed methods using the fuzzy logic approach. Section 4 presents the experimentation with the Benchmark functions. Finally, Sect. 5 offers the Conclusions.

2 The Differential Evolution Algorithm

Differential Evolution algorithm (DE) is an optimization method belonging to the category of evolutionary computation that can be applied in solving complex optimization problems. The differential evolution consists mainly of 4 steps [20]:

- Initialization
- Mutation
- Crossing
- Selection

This is the mathematical form of the Differential Evolution algorithm [13, 14]:

Population structure

$$P_{x,g} = \left(\mathbf{x}_{i,g}\right), \quad i = 0, 1, \ldots, Np, \;\; g = 0, 1, \ldots, g_{max} \tag{1}$$

$$\mathbf{x}_{i,g} = \left(x_{j,i,g}\right), \quad j = 0, 1, \ldots, D - 1 \tag{2}$$

$$P_{v,g} = \left(\boldsymbol{v}_{i,g}\right), \quad i = 0, 1, \ldots, Np - 1, \;\; g = 0, 1, \ldots, g_{max} \tag{3}$$

$$\boldsymbol{v}_{\mathrm{i,g}} = \left(v_{\mathrm{j,I,g}}\right), \quad \mathrm{j} = 0, 1, \ldots, \mathrm{D} - 1 \tag{4}$$

$$\mathrm{P}_{\mathrm{v,g}} = \left(\boldsymbol{u}_{\mathrm{i,g}}\right), \quad \mathrm{i} = 0, 1, \ldots, \mathrm{Np} - 1, \;\; \mathrm{g} = 0, 1, \ldots, \mathrm{g}_{\max} \tag{5}$$

$$\boldsymbol{u}_{\mathrm{i,g}} = \left(u_{\mathrm{j,I,g}}\right), \quad \mathrm{j} = 0, 1, \ldots, \mathrm{D} - 1 \tag{6}$$

Initialization

$$\mathrm{x}_{\mathrm{j,i,0}} = \mathrm{rand}_{\mathrm{j}}(0, 1) \cdot \left(\mathrm{b}_{\mathrm{j,U}} - \mathrm{b}_{\mathrm{j,L}}\right) + \mathrm{b}_{\mathrm{j,L}} \tag{7}$$

Mutation

$$\boldsymbol{v}_{\mathrm{i,g}} = \mathbf{x}_{\mathrm{r0,g}} + \mathrm{F} \cdot \left(\mathbf{x}_{\mathrm{r1,g}} - \mathbf{x}_{\mathrm{r2,g}}\right) \tag{8}$$

Crossover

$$U_{i,g} = \left(u_{j,i,g}\right) = \left\{ \begin{array}{ll} v_{j,i,g} & \textit{if } \left(rand_j(0,1) \leq Cr \textit{ or } j = j_{rand}\right) \\ x_{j,i,g} & \textit{otherwise.} \end{array} \right\} \tag{9}$$

Selection

$$X_{i,g+1} = \left\{ \begin{array}{ll} U_{i,g} & \textit{if } f\left(U_{i,g}\right) \leq f\left(X_{i,g}\right) \\ X_{i,g} & \textit{otherwise.} \end{array} \right\} \tag{10}$$

3 Proposed Method

We propose using DE enhanced with fuzzy logic to dynamically modify the F parameter (mutation) during execution of the algorithm. We have previously work with the Differential Evolution algorithm using fuzzy logic and this method was called Fuzzy Differential Evolution, and we are now extending this previous work by using new and more complex functions to verify in more detail the efficiency of the proposed algorithm [14]. The way in which we integrate a fuzzy system algorithm can be found in Fig. 1, where we have the flowchart of the Differential Evolution algorithm and we have a component with Interval-type 2 fuzzy logic, which dynamically calculates the F parameter (mutation) to then make calculations in the algorithm. Figure 2 shows the structure for the Interval-type 2 fuzzy system, which contains one input and one output.

Figure 3 shows the input variable (Generation) that has 3 membership functions and is granulated in the Low, Medium and High values that were used by each membership function are symmetric the ranges are as follows:

- Low: $[-0.5859 \quad -0.08598 \quad 0.4141 \quad -0.4193 \quad 0.08068 \quad 0.5807]$

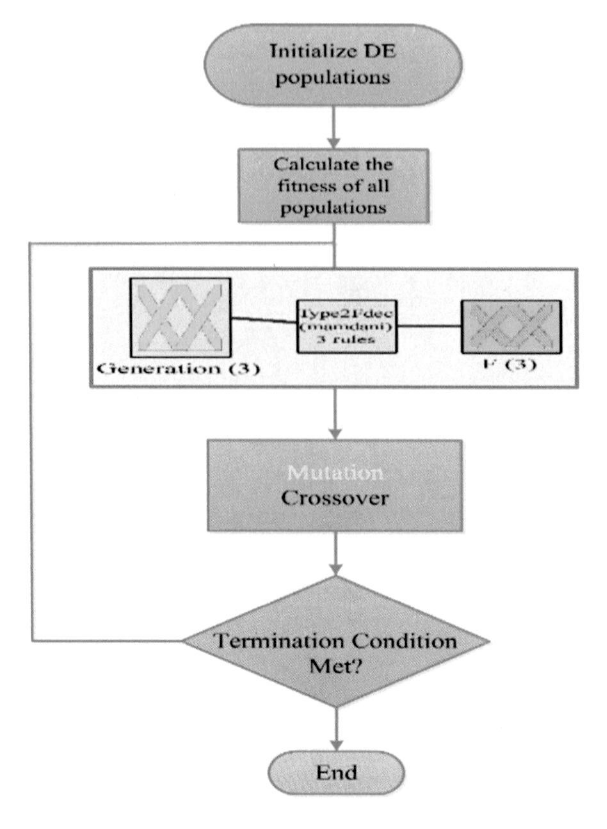

Fig. 1 Propose method with Interval-type 2 fuzzy logic

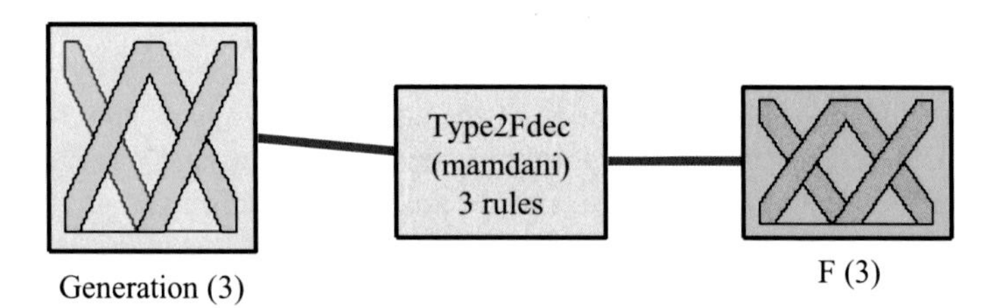

Fig. 2 Fuzzy system

- Medium: $[-0.08333 \quad 0.4167 \quad 0.9167 \quad 0.08333 \quad 0.5833 \quad 1.083]$
- High: $[0.4167 \quad 0.9167 \quad 1.417 \quad 0.5833 \quad 1.083 \quad 1.583]$

Figure 4 presents the output variable (F) has 3 membership functions and is granulated in Low, Medium and High values that were used by each membership function are symmetric the ranges are as follows:

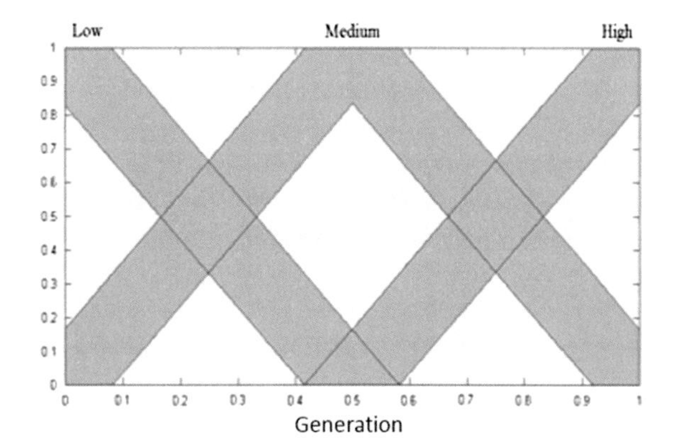

Fig. 3 Input generation

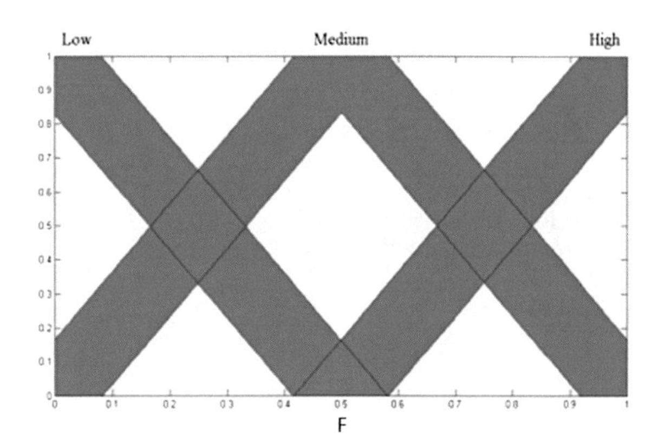

Fig. 4 Output F parameter

1. - If (Generations is Low) then (F is High) (1)

2. - If (Generations is Medium) then (F is Medium) (1)

3. - If (Generations is High) then (F is Low) (1)

Fig. 5 Rules for the fuzzy system

- Low: $[-0.5859 \quad -0.08598 \quad 0.4141 \quad -0.4193 \quad 0.08068 \quad 0.5807]$
- Medium: $[-0.0807 \quad 0.419 \quad 0.919 \quad 0.086 \quad 0.586 \quad 1.09]$
- High: $[0.4167 \quad 0.9167 \quad 1.417 \quad 0.5833 \quad 1.083 \quad 1.583]$

Figure 5 represents the rules of the interval-type 2 fuzzy logic and the Fig. 6 shows the surface of the interval-type 2 fuzzy system.

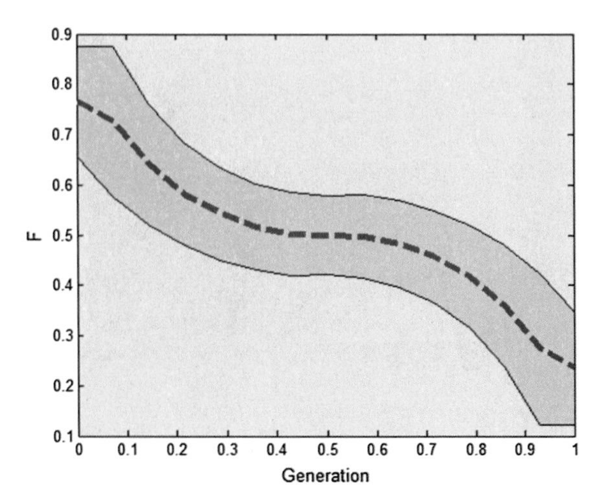

Fig. 6 Surface of the fuzzy system

4 Results of the Experiments

In this paper we consider the set of CEC 2015 Benchmark Functions to perform the experiments and thereby be to [7, 9, 10, 25].

Below a brief description of the set of CEC 2015 Benchmark Functions is presented.

Table 1 shows the set of 15 CEC 2015 Benchmark Functions used in this paper.

For the experiments we use the metrics given by the comments of CEC'15 Benchmark Functions, are comprised of 15 minimization problems and the experiments for all functions are performed with different dimensions and for this competition are of D = 10, 30, 50, 100.

Tables 2 and 3 show the comparison between the article that obtained the first place in the competition of the CEC′15 and our proposed method [8].

Figures 7, 8, 9 and 10 represent the comparison graph for each dimension D = 10, D = 30, D = 50 and D = 100.

Table 1 Summary of the CEC'15 learning-based benchmark suite

	No.	Functions	*Fi** = *Fi* (*x**)
Unimodal functions	1	Rotated high conditioned elliptic function	100
	2	Rotated cigar function	200
Simple multimodal functions	3	Shifted and rotated Ackley's function	300
	4	Shifted and rotated Rastrigin's function	400
	5	Shifted and rotated Schwefel's function	500
Hybrid functions	6	Hybrid function 1 ($N = 3$)	600
	7	Hybrid function 2 ($N = 4$)	700
	8	Hybrid function 3 ($N = 5$)	800
Composition functions	9	Composition function 1 ($N = 3$)	900
	10	Composition function 2 ($N = 3$)	1000
	11	Composition function 3 ($N = 5$)	1100
	12	Composition function 4 ($N = 5$)	1200
	13	Composition function 5 ($N = 5$)	1300
	14	Composition function 6 ($N = 7$)	1400
	15	Composition function 7 ($N = 10$)	1500
Search range: $[-100,100]^D$			

Table 2 Comparison between the references and the propose method with D = 10 and D = 30

	D = 10		D = 30	
	Best SPS-L-SHADE-EIG [8]	Best (FDE IT2)	Best SPS-L-SHADE-EIG [8]	Best (FDE IT2)
f1	0.00E+00	3.50E−03	0.00E+00	2.20E−02
f2	0.00E+00	1.00E+00	0.00E+00	1.11E+00
f3	0.00E+00	0.00E+00	2.00E+01	2.10E+01
f4	0.00E+00	0.00E+00	1.05E−02	5.12E+02
f5	3.12E−01	0.00E+00	6.58E+02	8.18E+03
f6	0.00E+00	0.00E+00	2.68E+01	5.18E−03
f7	0.00E+00	0.00E+00	6.23E−01	3.23E+02
f8	6.02E−08	0.00E+00	2.07E+00	2.40E−03
f9	1.00E+02	1.20E+02	1.02E+02	1.20E+02
f10	2.17E+02	2.04E+03	1.48E+02	3.25E−03
f11	2.61E–02	2.01E+02	3.00E+02	1.70E+03
f12	1.00E+02	1.34E+02	1.02E+02	2.14E+02
f13	3.03E−02	5.09E+01	2.56E−02	1.57E+02
f14	1.00E+02	1.15E+02	3.11E+04	8.35E+04
f15	1.00E+02	1.63E+01	1.00E+02	4.99E+04

Table 3 Comparison between the references and proposes method D = 50 and D = 100

	D = 50		D = 100	
	Best SPS-L-SHADE-EIG [8]	Best (FDE IT2)	Best SPS-L-SHADE-EIG [8]	Best (FDE IT2)
f1	0.00E+00	5.28E−02	0.00E+00	1.21E−01
f2	0.00E+00	2.01E+00	0.00E+00	3.81E+00
f3	2.00E+01	2.13E+01	2.00E+01	2.01E+01
f4	9.01E–05	1.06E+03	1.59E+01	2.29E+03
f5	1.38E+03	1.40E+04	3.98E+03	4.23E+04
f6	5.18E+01	3.11E−03	9.61E+02	9.82E+02
f7	6.49E+00	1.58E+03	9.06E+01	1.52E+04
f8	1.00E+01	1.35E−03	6.87E+02	7.12E+02
f9	1.03E+02	1.20E+02	1.05E+02	3.20E+03
f10	6.79E+02	2.10E−03	1.59E+03	1.78E+03
f11	3.00E+02	2.85E+03	3.01E+02	4.08E+03
f12	1.03E+02	2.97E+02	1.11E+02	3.28E+02
f13	6.99E–02	9.09E+02	5.96E–02	7.11E+02
f14	4.95E+04	2.82E+04	1.09E+05	1.15E+05
f15	1.00E+02	3.78E+05	1.00E+02	2.00E+04

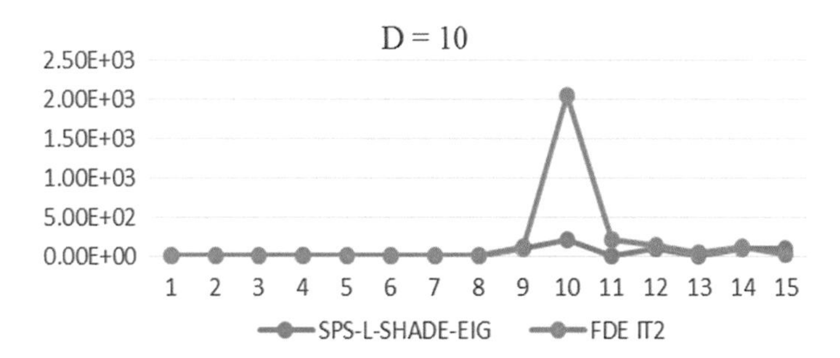

Fig. 7 Comparison with D = 10

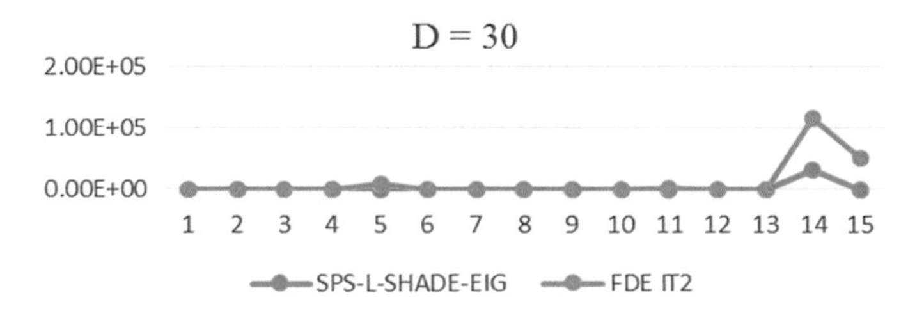

Fig. 8 Comparison with D = 30

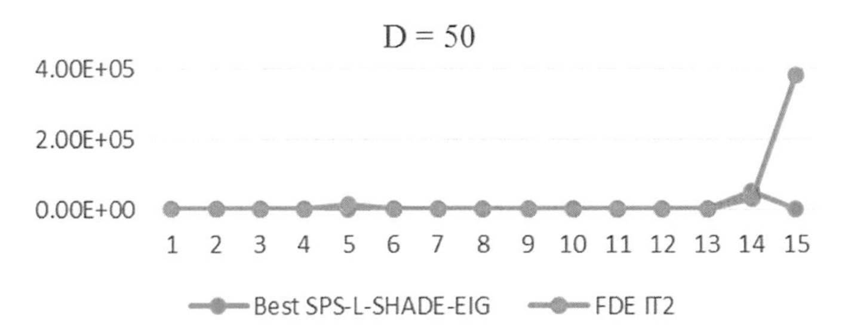

Fig. 9 Comparison with D = 50

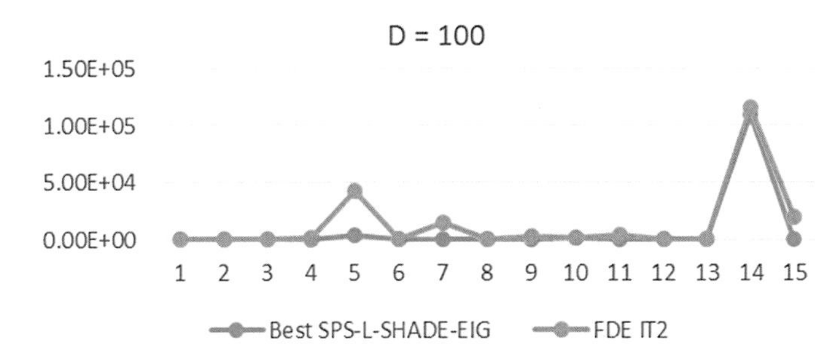

Fig. 10 Comparison with D = 100

5 Conclusions

In conclusion, we can notice that the results obtained using interval-type 2 fuzzy logic are not as good as we expected, most likely because the fuzzy system is a simple one input and one output and another part of the footprint uncertainty membership functions the input and the output is symmetric and not optimized.

We will try to improve our fuzzy system and/or we will make the footprint uncertainty trail be optimized by some algorithm and thus be able to make a comparison with statistical test to be able to affirm or reject if our proposed method is good with the functions Benchmark, But we can also conclude that our diffuse system for some functions if it is competitive and for those cases means that the uncertainty footprint is the one indicated for certain functions Benchmark CEC'15.

References

1. A. Al-Dujaili, K. Subramanian, S. Suresh, HumanCog: a cognitive architecture for solving optimization problems, in *2015 IEEE Congress on Evolutionary Computation (CEC)* (IEEE, New York, May 2015), pp. 3220–3227

2. N. Awad, M.Z. Ali, R.G. Reynolds, A differential evolution algorithm with success-based parameter adaptation for CEC2015 learning-based optimization, in *2015 IEEE Congress on Evolutionary Computation (CEC)* (IEEE, New York, May 2015), pp. 1098–1105
3. D. Aydın, T. Stutzle, A configurable generalized artificial bee colony algorithm with local search strategies, in *2015 IEEE Congress on Evolutionary Computation (CEC)* (IEEE, New York, May 2015), pp. 1067–1074
4. O. Castillo, H. Neyoy, J. Soria, P. Melin, F. Valdez, A new approach for dynamic fuzzy logic parameter tuning in Ant Colony Optimization and its application in fuzzy control of a mobile robot. Appl. Soft Comput. **28**, 150–159 (2015)
5. O. Castillo, P. Melin, Intelligent adaptive model-based control of robotic dynamic systems with a hybrid fuzzy-neural approach. Appl. Soft Comput. **3**(4), 363–378 (2003)
6. Q. Chen, B. Liu, Q. Zhang, J.J. Liang, P.N. Suganthan, B.Y. Qu, Problem definition and evaluation criteria for CEC 2015 special session and competition on bound constrained single-objective computationally expensive numerical optimization. Computational Intelligence Laboratory, Zhengzhou University, China and Nanyang Technological University, Singapore, Technical report (2014)
7. L. Chen, C. Peng, H.L. Liu, S. Xie, An improved covariance matrix leaning and searching preference algorithm for solving CEC 2015 benchmark problems, in *2015 IEEE Congress on Evolutionary Computation (CEC)* (IEEE, New York, May 2015), pp. 1041–1045
8. S.M. Guo, J.S.H. Tsai, C.C. Yang, P.H. Hsu, A self-optimization approach for L-SHADE incorporated with eigenvector-based crossover and successful-parent-selecting framework on CEC 2015 benchmark set. In *2015 IEEE Congress on Evolutionary Computation (CEC)* (IEEE, New York, May 2015), pp. 1003–1010
9. X. Li, Decomposition and cooperative coevolution techniques for large scale global optimization. In *Proceedings of the Companion Publication of the 2014 Annual Conference on Genetic and Evolutionary Computation* (ACM, New York, July 2014), pp. 819–838
10. J.J. Liang, L. Guo, R. Liu, B.Y. Qu, A self-adaptive dynamic particle swarm optimizer, in *2015 IEEE Congress on Evolutionary Computation (CEC)* (IEEE, New York, May 2015), pp. 3206–3213
11. R. Martínez-Soto, O. Castillo, L.T. Aguilar, Type-1 and Type-2 fuzzy logic controller design using a Hybrid PSO–GA optimization method. Inf. Sci. **285**, 35–49 (2014)
12. P. Melin, O. Castillo, A review on type-2 fuzzy logic applications in clustering, classification and pattern recognition. Appl. Soft Comput. **21**, 568–577 (2014)
13. P. Ochoa, O. Castillo, J. Soria, A fuzzy differential evolution method with dynamic adaptation of parameters for the optimization of fuzzy controllers, in *2014 IEEE Conference on Norbert Wiener in the 21st Century (21CW)* (IEEE, June 2014), pp. 1–6
14. P. Ochoa, O. Castillo, J. Soria, Differential evolution with dynamic adaptation of parameters for the optimization of fuzzy controllers, in *Recent Advances on Hybrid Approaches for designing intelligent systems* (Springer International Publishing, 2014), pp. 275–288
15. F. Olivas, F. Valdez, O. Castillo, P. Melin, Dynamic parameter adaptation in particle swarm optimization using interval type-2 fuzzy logic. Soft. Comput. **20**(3), 1057–1070 (2016)
16. F. Olivas, O. Castillo, Particle swarm optimization with dynamic parameter adaptation using fuzzy logic for benchmark mathematical functions, in *Recent Advances on Hybrid Intelligent Systems* (Springer, Berlin, 2013), pp. 247–258
17. C. Peraza, F. Valdez, O. Castillo, A Harmony search algorithm comparison with genetic algorithms, in *Fuzzy Logic Augmentation of Nature-Inspired Optimization Metaheuristics* (Springer International Publishing, 2015), pp. 105–123
18. C. Peraza, F. Valdez, O. Castillo, Fuzzy control of parameters to dynamically adapt the HS algorithm for optimization, in *Fuzzy Information Processing Society (NAFIPS) held jointly with 2015 5th World Conference on Soft Computing (WConSC), 2015 Annual Conference of the North American* (IEEE, New York, Aug 2015), pp. 1–6
19. R. Poláková, J. Tvrdík, P. Bujok, Cooperation of optimization algorithms: a simple hierarchical model, in *2015 IEEE Congress on Evolutionary Computation (CEC)* (IEEE, New York, May 2015), pp. 1046–1052

20. K. Price, R.M. Storn, J.A. Lampinen, in *Differential Evolution: A Practical Approach to Global Optimization* (Springer Science & Business Media, 2006)
21. J.L. Rueda, I. Erlich, Testing MVMO on learning-based real-parameter single objective benchmark optimization problems, in *2015 IEEE Congress on Evolutionary Computation (CEC)* (IEEE, May 2015), pp. 1025–1032
22. K.M. Sallam, R.A. Sarker, D.L. Essam, S.M. Elsayed, Neurodynamic differential evolution algorithm and solving CEC2015 competition problems, in *2015 IEEE Congress on Evolutionary Computation (CEC)* (IEEE, New York, May 2015), pp. 1033–1040
23. R. Storn, On the usage of differential evolution for function optimization, in *Fuzzy Information Processing Society, 1996. NAFIPS. 1996 Biennial Conference of the North American* (IEEE, New York, June 1996), pp. 519–523
24. R. Storn, K. Price, *Differential Evolution—A Simple and Efficient Adaptive Scheme for Global Optimization Over Continuous Spaces*, vol. 3 (ICSI, Berkeley, 1995)
25. R. Tanabe, A. Fukunaga, Success-history based parameter adaptation for differential evolution, in *2013 IEEE Congress on Evolutionary Computation* (IEEE, New York, June 2013), pp. 71–78
26. F. Valdez, P. Melin, O. Castillo, Evolutionary method combining particle swarm optimization and genetic algorithms using fuzzy logic for decision making, in *IEEE International Conference on Fuzzy Systems* (2009), pp. 2114–2119
27. F. Valdez, P. Melin, O. Castillo, Evolutionary method combining particle swarm optimisation and genetic algorithms using fuzzy logic for parameter adaptation and aggregation: the case neural network optimisation for face recognition. Int. J. Artif. Intell. Soft Comput. **2**(1–2), 77–102 (2010)
28. F. Valdez, P. Melin, O. Castillo, An improved evolutionary method with fuzzy logic for combining particle swarm optimization and genetic algorithms. Appl. Soft Comput. **11**(2), 2625–2632 (2011)
29. C. Yu, L.C. Kelley, Y. Tan, Dynamic search fireworks algorithm with covariance mutation for solving the CEC 2015 learning based competition problems, in *2015 IEEE Congress on Evolutionary Computation (CEC)* (IEEE, New York, May 2015), pp. 1106–1112
30. Y.J. Zheng, X.B. Wu, Tuning maturity model of ecogeography-based optimization on CEC 2015 single-objective optimization test problems, in *2015 IEEE Congress on Evolutionary Computation (CEC)* (IEEE, New York, May 2015), pp. 1018–1024

Part II
Neural Networks Theory and Applications

Person Recognition with Modular Deep Neural Network Using the Iris Biometric Measure

Fernando Gaxiola, Patricia Melin, Fevrier Valdez and Juan Ramón Castro

Abstract In this paper a modular deep neural network architecture are applied for recognize persons based on the iris biometric measurement of humans. The modular neural network consists of three modules, each module work with a deep neural network. This paper works with the human iris database improved with image preprocessing methods, these methods make a cut of the area of interest allowing remove the noise around the human iris. The input to the modular deep neural network is the preprocessed iris images and the output is the person identified. The "Gating Network" integrator is used for the integration of the modules for obtain the final results.

Keywords Deep neural networks · Face recognition · Biometric

1 Introduction

The recognition using biometrics measures are an area of investigation that has been very exploited in many years, the investigation was performed utilizing the many biometrics measures, like the fingerprint, the face of the humans, the characteristics of the ears, the hand palm, etc.; taking advantage of the unique characteristics of these measures it is possible to ensure that the recognition of the persons are the correct with an highly probability of success.

In this paper, we make the recognition of persons using the human iris biometric measure, the recognition is performed with the deep neural network model in an ensemble architecture. The deep neural network models are an methodology that

F. Gaxiola · P. Melin · F. Valdez (✉) · J. R. Castro
Tijuana Institute of Technology, Autonomous University
of Baja California, Tijuana, Mexico
e-mail: fevrier@tectijuana.edu.mx

F. Gaxiola
e-mail: fergaor_29@hotmail.com

P. Melin
e-mail: pmelin@tectijuana.mx

O. Castillo et al. (eds.), *Fuzzy Logic Augmentation of Neural and Optimization Algorithms: Theoretical Aspects and Real Applications*, Studies in Computational Intelligence 749, https://doi.org/10.1007/978-3-319-71008-2_6

has a great impact in works of distinct areas of the intelligent systems, like classification, detection, patrons recognition, and others [7, 14, 18].

We are presenting an ensemble with three deep neural networks for the experiments. The final result for the ensemble was obtained with type-1 and type-2 fuzzy integration and the integrator "gating network" and "Winner takes all".

This research uses the method of deep learning taking in consideration the robustness and effectivity with which this method is performed in many works. We work in developed an architecture of modular deep neural network that allowed to obtain good results for the recognition of persons using the human iris biometrics.

The next section presents basic concepts of genetic algorithms, particle swarm optimization, type-2 fuzzy logic and neural networks, and explains background of research about modifications of the backpropagation algorithm, different management strategies of weights in neural networks and optimization with genetics algorithm and particle swarm optimization. Section 3 explains the proposed method and the problem description. Section 4 presents the scheme of optimization of type-1 and type-2 fuzzy integrator for the ensemble neural network with genetic algorithm (GA) and particle swarm optimization (PSO). Section 5 describes the simulation results for the type-1 and type-2 fuzzy integrator optimized with GA and PSO proposed in this paper. Finally, in Sect. 6, some conclusions are presented.

2 Background and Basic Concepts

2.1 *Basic Concepts*

A. **Human Iris Biometrics**

The use of the human iris like measure for identification of persons is one of the most robustness and effective in comparison with others biometric measure. In literature, the human have the lowest probability of two human iris produced the same code, which is 1 in 10^{78} [15].

The properties of the human iris describes it like an internal organ of the eye, located behind the cornea and the aqueous humor, which consists of characteristics distinctive that allow the uniqueness between the persons, like screening connective tissue, fibers, rings and colors (see Fig. 1) [12].

B. **Modular Deep Neural Network**

The theory of modular deep neural network consists of use the structure of the module neural network, which consists of two o more neural networks with different inputs in each module and obtaining the final output using a response integrator for the final output. In this case, we use a deep neural network for the modules, the deep neural network consists in train a neural network for phases [5].

First train the inputs with the neurons for the hidden layer, adapting the weights for these connections; in the second phase, the outputs of the first phase are used

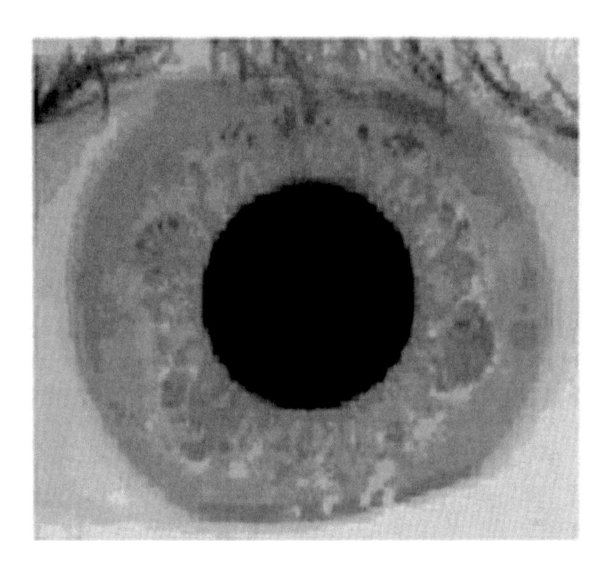

Fig. 1 Example of a human iris

like inputs for the neurons of the output layer, adapting the weights for these connections; and finally, the weights acquired in the first and second phase for the connections between the input and hidden layers, and the hidden and output layer, are used to train a final neural network. The process of the deep neural network allows in a certain way to optimize the weights for the hidden and output layer [4, 9, 10].

2.2 *Overview of Related Works*

The French ophthalmologist Bertillon (1880) [17] used the iris for classified criminals in base at the color of their eyes. Flom and Safir [8], presented the idea of utilized the iris as a tool to identify persons. Daugman [3], developed an algorithm that performed the identification of persons using the biometric measurement of human iris.

Birajadar et al. [1], performed the recognition using the monogenic wavelets and the Gabor wavelets. Cruz et al. [2], implemented the algorithm developed by Daugman on Raspberry pi. Risk et al. [13], used the particle swarm optimization (PSO) and gravitational search algorithm (GSA) to optimize the weights and biases of a forward neural network.

In addition, recent works deep neural network have been developed, like Rhee et al. [14], that presented a deep convolutional neural network for face recognition using synthesized three-dimensional (3-D) shape information together with two-dimensional (2-D) color; Hinton et al. [6], works with deep neural network for speech recognition; Simonyan and Zisserman [16], used deep convolutional neural network in the recognition for large-scale images.

3 Proposed Method and Problem Description

The recognition of persons is the principal objective of this work. In this problem innumerable researches have been developed, considering the use of different biometric measure to achieve the recognition, like the fingerprint, voice, face, ear, signature, etc. and applying various methods with the goal of high percent of identification, of which this work is focused in neural network applications.

The specific problem considered in this work is: "obtain a high percent of person's recognition applying modular deep neural network with the implementation of the human iris biometric measure".

Aiming to achieve the high rate of identification of "n" images of "m" persons, we propose to work with modular deep neural networks, with 3 modules. By simulating the trained deep neural network for "pi" images of "m" persons to identify, we obtain a vector of results "R" of "a" output neurons, then one can identify:

$$if\ Pa \approx 1\ in\ R\ then\ P = Pa$$

where

Pa output neuron person.
R Vector of results output neurons a.
P person identified by the neural network.

To determine that the network identification is correct for "pi" images the following condition must be met:

$$\mathrm{P} = \mathrm{m}$$

Then the rate of identification of "m" persons will come from the sum of the allocations 1–0 (correct–incorrect identification for p images of m people to identify) between the total number of images to identify.

In base to test the accuracy of the proposed method, we chose to use the database of human Iris from the Institute of Automation Chinese Academy of Sciences (CASIA) (see Fig. 2). The database consist of a total of 1386 images, which are obtained of 99 individuals and each individual have 14 images, 7 of the right eye and 7 of the left eye. We use 4 images of each eye for the inputs to utilize in the training the deep neural network, and 3 images of each eye for test the proposed method. The images have dimensions of 320 pixels per 280 pixels, and in JPEG format.

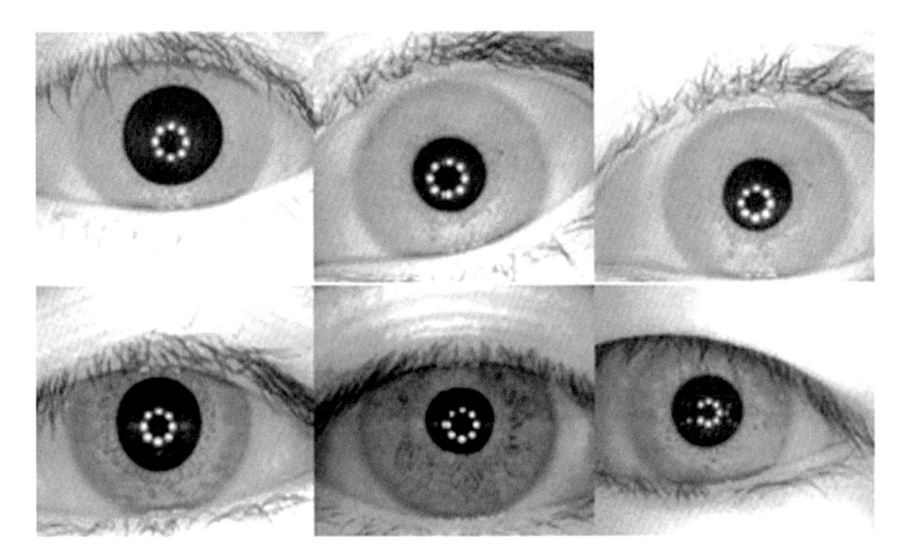

Fig. 2 Illustrations of images by The CASIA database of human iris

4 Image Pre-processing

Before the images are used as inputs for the deep neural networks a pre-processing step is necessary. The pre-processing is as follows:

(1) The method developed by Masek and Kovesi is applied to obtain the coordinates and radius of the iris and pupil [11]. This method consists in applied a series of filters and mathematical calculations to achieve acquire the coordinates and radius of the iris and pupil; First it is applied the Canny method for edge detection (a), the next step consist in used a gamma adjustment of the images (b), after, the process continue applying to the image a no maxima suppression (c), and subsequently, we performed to the image a threshold method (d). This process is shown in Fig. 3.
    Finally, we used the Hough transform to localize the maximum in the Hough space and, therefore, the row and column at the center of the iris and the radius. In base at the parameters of the iris, the same process is applied to obtain the co-ordinates of the center and radius of the pupil, but delimiting the process at the end of the center coordinates and radius of the iris to identify the pupil parameters.
(2) Make a cut around the human iris, to make this cut we uses the coordinates of the Iris for obtain the upper right and lower left points to perform a square cut around the iris, that allow eliminate some noise like part of eyelash, eyelid and cornea. We show an example of the cut of human iris image in Fig. 4.
(3) We perform a resize of the cut of the iris to 21 pixels per 21 pixels, it is implemented with the idea of work with the same size for all the images.

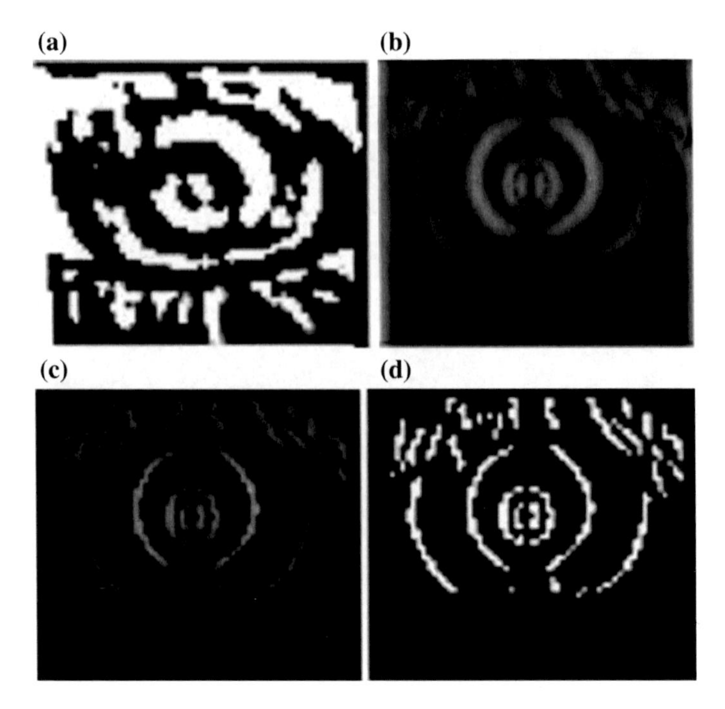

Fig. 3 Process in sequence of the method of Masek and Kovesi

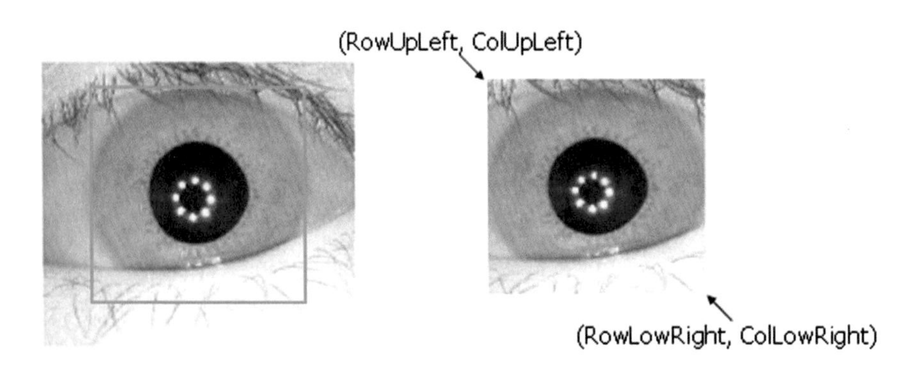

Fig. 4 Example of the cut around the human iris image

(4) To manage inputs more pertinent for the deep neural network, we convert the images from matrix to vector.
(5) Finally, process matrices by mapping each row's means to 0 and standard deviations to 1 are applied to normalize the images.

5 Modular Deep Neural Network Architecture

The work was concentrated on the person's recognition using a modular deep neural network. We implemented the architecture with 3 modules, the inputs for each module consists of 33 individuals, considering 264 images for training and 198 images for testing the deep neural network.

In the integration, we used the method called "Gating Network" for obtain the person result to identify.

The deep neural network architecture is described in Fig. 5.

In each module work with one deep neural network and for training we used the scaled conjugate gradient algorithm.

In the three deep neural network we used the sparse auto encoder in the first layer with the parameters: L2 Weight Regularization = 0.004, Sparsity Regularization = 4, and Sparsity Proportion = 0.15.

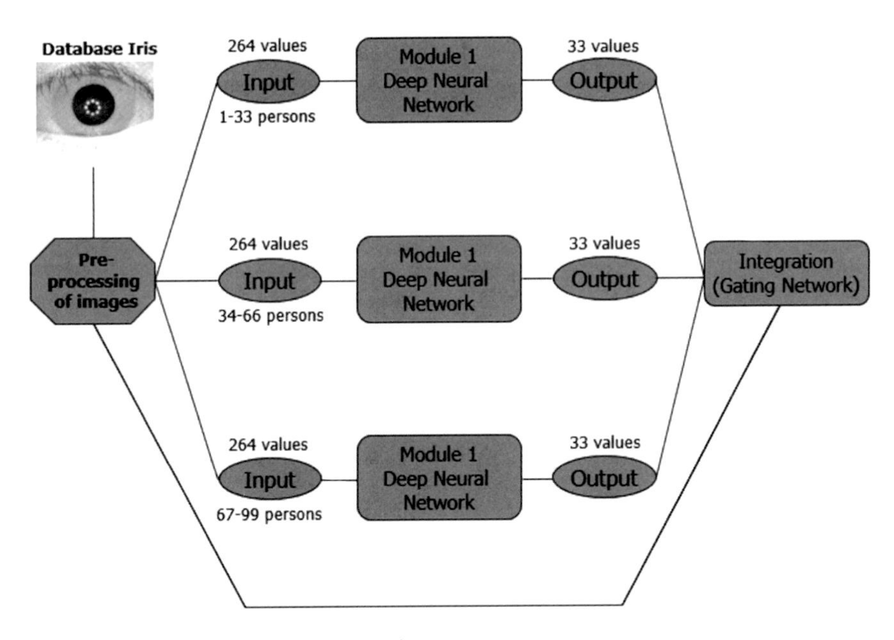

Fig. 5 Modular deep neural network architecture

6 Simulation Results

Experiments were performed with the modular deep neural network architecture described in the previous section. We work with a deep neural network with two hidden layers and we performed experiments for optimize manually the number of neurons in the hidden layers.

The following results were achieved by each of the 3 modules in terms of percentage of identification.

Module 1

The results of this module found that the best result was achieved with 350 neurons in the first hidden layer and 175 neurons in the second hidden layer with a 92.93% of identification rate (184/198) (see Table 1).

We presented 10 experiments, but we performed the optimization manually starting from 100 until 2000 neurons in the first hidden layer and the neurons of the second hidden layer was obtained by divided the neurons for the first hidden layer.

We used 400 epochs in the encoder for the inputs to the first hidden layer, 100 epochs in the encoder for the first hidden layer to the second hidden layer, and 400 epochs for the hidden layer to the output layer.

We presented only the best 10 experiments for each module. We take the architecture with the lowest number of neurons in the hidden layers.

The deep neural network has the characteristic that almost all the experiments enforced obtain the same results of person recognition.

Module 2

The results of this module found that the best result was achieved with 615 neurons in the first hidden layer and 308 neurons in the second hidden layer with a 95.45% of identification rate (189/198) (see Table 2).

Table 1 Results for the deep neural network in the module 1

No.	Neurons in hidden layer 1	Neurons in hidden layer 2	Recognition	Identification	Percent of identification
E1	350	175	264/264	184/198	92.93
E2	595	298	264/264	184/198	92.93
E3	655	328	264/264	184/198	92.93
E4	740	370	264/264	184/198	92.93
E5	1465	733	264/264	184/198	92.93
E6	690	345	264/264	183/198	92.42
E7	1215	608	264/264	182/198	91.93
E8	1335	668	264/264	182/198	91.93
E9	1345	673	264/264	182/198	91.93
E10	1915	958	264/264	182/198	91.93

Table 2 Results for the deep neural network in the module 2

No.	Neurons in hidden layer 1	Neurons in hidden layer 2	Recognition	Identification	Percent of identification
E1	615	308	264/264	189/198	95.45
E2	640	320	264/264	189/198	95.45
E3	665	333	264/264	189/198	95.45
E4	825	413	264/264	188/198	94.95
E5	920	460	264/264	188/198	94.95
E6	1215	608	264/264	188/198	94.95
E7	1330	665	264/264	188/198	94.95
E8	850	425	264/264	187/198	94.45
E9	765	383	264/264	186/198	93.93
E10	1080	540	264/264	186/198	93.93

Table 3 Results for the deep neural network in the module 3

No.	Neurons in hidden layer 1	Neurons in hidden layer 2	Recognition	Identification	Percent of identification
E1	940	470	264/264	181/198	91.41
E2	1190	595	264/264	181/198	91.41
E3	1280	640	264/264	181/198	91.41
E4	1305	653	264/264	181/198	91.41
E5	1485	743	264/264	181/198	91.41
E6	1510	755	264/264	181/198	91.41
E7	1630	815	264/264	181/198	91.41
E8	1910	955	264/264	181/198	91.41
E9	475	238	264/264	180/198	90.91
E10	365	183	264/264	179/198	90.40

Module 3

The results of this module found that the best result was achieved with 940 neurons in the first hidden layer and 470 neurons in the second hidden layer with a 91.41% of identification rate (181/198) (see Table 3).

Integration

Analyzing the results obtained in the three modules, we see that in the first module the deep neural network with 350 neurons in the first hidden layer and 175 in the second hidden layer showed better results and the lowest neurons in comparison with all experiments for this module.

In the second module, the deep neural network with 615 neurons in the first hidden layer and 308 in the second hidden layer showed better results and the lowest neurons in comparison with all experiments for this module.

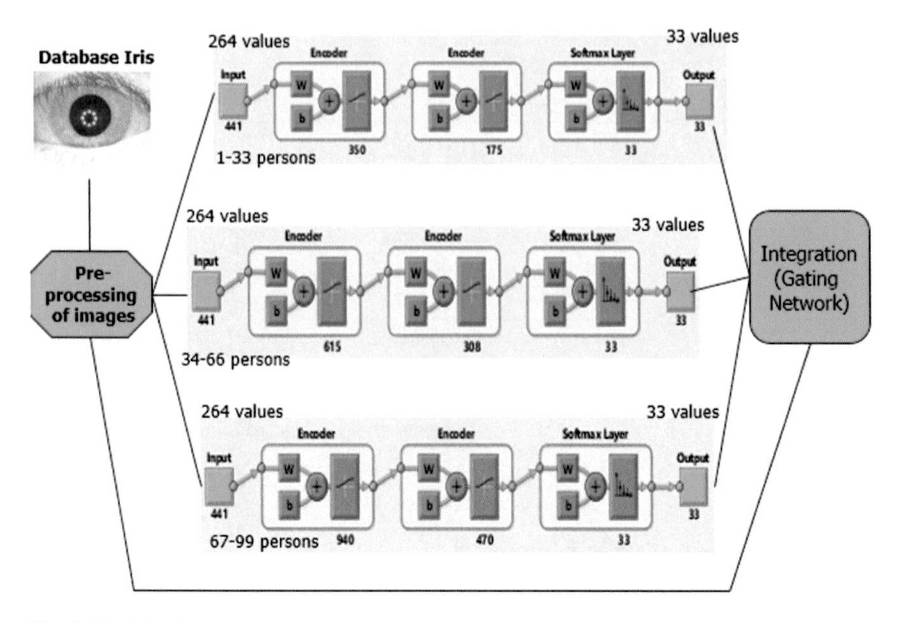

Fig. 6 Modular deep neural network architecture optimized manually

Table 4 Results for the integration with "Gating Network" integrator for the modular deep neural network architecture

Integrator	Module 1	Module 2	Module 3	Recognition	Identification	Percent of Identification
Gating Network	184/198 92.93%	189/198 95.45%	181/198 91.41%	792/792	554/594	93.27

In the third module, the deep neural network with 940 neurons in the first hidden layer and 470 in the second hidden layer showed better results and the lowest neurons in comparison with all experiments for this module.

The modular deep neural network architecture with the optimization manually of the neurons in the first hidden layer and second hidden layer is show in Fig. 6.

The integration response using "gating network" integrator is done with architecture of the modular deep neural network mentioned above. The final result is 93.27% of identification rate (575/594) (see Table 4).

7 Conclusions

In this paper we presented a modular deep neural network architecture with three modules, which has as input the database of human iris images. The images are used for training the deep neural network for each module. In this work, several methods were used to make a square cut around the iris utilizing the coordinates of the center and radius to allow eliminate noise of the images outside of the iris.

The results achieving of 93.27% identification rate (554 images of 594 test images) with the "Gating Network" integrator are good, due the parameters of the three deep neural networks are not optimized.

These results evidence that the use of human iris biometric measure implemented with modular deep neural networks obtained favorable results of person recognition.

The robustness to obtain almost in all experiments the same results of person recognition for the deep neural network is a good performance for this method.

Future work consists in use a method for optimization for the modular deep neural network, with which you can get beyond the rate of recognition obtained in the results shown in this paper and considered other images pre-processing methods different or that complements the current one.

References

1. P. Birajadar, P. Shirvalkar, S. Gupta, V. Patidar, U. Sharma, A. Naik, V. Gadre, A novel iris recognition technique using monogenic wavelet phase encoding, in *2016 International Conference on Signal and Information Processing (IConSIP)*, pp. 1–6 (2016)
2. F.R.G. Cruz, C.C.Hortinela, B.E. Redosendo, B.K. Asuncion, C.J. Leoncio, N.B. Linsangan, W. Chung, Iris recognition using Daugman algorithm on Raspberry Pi, in *2016 IEEE Region 10 Conference (TENCON)*, pp. 2126–2129 (2016)
3. J. Daugman, Statistical richness of visual phase information: update on recognizing persons by iris patterns. Int. J. Comput. Vis. **45**(1), 25–38 (2001)
4. D. Erhan, P.A. Manzagol, Y. Bengio, S. Bengio, P. Vincent, The difficulty of training deep architectures and the effect of unsupervised pre-training, in *AISTATS'2009*, pp. 153–160 (2009)
5. F. Gaxiola, P. Melin, M. Lopez, Modular neural networks for person recognition using the contour segmentation of the human iris biometric measurement. Stud. Comput. Intell. **312**, 137–153 (2010)
6. G. Hinton, L. Deng, D. Yu, G. Dahl, A. Mohamed, N. Jaitly, A. Senior, V. Vanhoucke, P. Nguyen, T. Sainath, B. Kingsbury, Deep neural networks for acoustic modeling in speech recognition: the shared views of four research groups. IEEE Signal Process. Mag. **29**(6), 82–97 (2012)
7. Q. Jiang, L. Cao, M. Cheng, C. Wang, J. Li, Deep neural networks-based vehicle detection in satellite images, in 2015 International Symposium on Bioelectronics and Bioinformatics (ISBB), pp. 184–187 (2015)
8. L. Flom, A. Safir, Iris recognition system. U.S. Patent 4,641,349 (1987)

9. H. Larochelle, Y. Bengio, J. Louradour, P. Lamblin, Exploring strategies for training deep neural networks. J. Mach. Learn. Res. **10**, 1–40 (2009)
10. D. Li, G. Hinton, B. Kingsbury, New types of deep neural network learning for speech recognition and related applications: an overview, in *2013 IEEE International Conference on Acoustics, Speech and Signal Processing (ICASSP)*, pp. 8599–8603 (2013)
11. L. Masek, P. Kovesi, MATLAB source code for a biometric identification system based on iris patterns. The School of Computer Science and Software Engineering the University of Western Australia (2003)
12. A. Muroó, J. Pospisil, The human iris structure and its usages. Physica **39**, 89–95 (2000)
13. M. Risk, H. Farag, L. Said, Neural network classification for iris recognition using both particle swarm optimization and gravitational search algorithm, in *2016 World Symposium on Computer Applications & Research (WSCAR)*, pp. 12–17 (2016)
14. S.M. Rhee, B. Yoo, J.J. Han, W. Hwang, Deep neural network using color and synthesized three-dimensional shape for face recognition. J. Electron. Imaging, **26**(2) (2017)
15. O. Sánchez, J. González, Access control based on iris recognition, Technological University Corporation of Bolívar, Faculty of Electrical Engineering, Electronics and Mechatronics, Cartagena de Indias, Colombia, pp. 1–137 (2003)
16. K. Simonyan, A. Zisserman, Very deep convolutional networks for large-scale image recognition, in *Conference on ICLR 2015*, pp. 1–13 (2015)
17. C. Tisse, L. Martin, L. Torres, M. Robert, Person identification technique using human iris recognition, in *Canadian Image Processing and Pattern Recognition Society (CIPPRS) 15th International Conference on Vision Interface*, pp. 294–299 (2002)
18. Z. Zhang, C. Xu, W. Feng, Road vehicle detection and classification based on deep neural network, in *2016 7th IEEE International Conference on Software Engineering and Service Science (ICSESS)* (2017)

Neuro-evolutionary Neural Network for the Estimation of Melting Point of Ionic Liquids

Jorge A. Cerecedo-Cordoba, Juan Javier González Barbosa, J. David Terán-Villanueva and Juan Frausto-Solís

Abstract Ionic Liquids (ILs) are salts known for their low melting point, wide liquid phase, and their low toxicity. Also, ILs have an extensive range of applications. Choosing the "best" IL for an application requires the prior knowledge of the physicochemical properties of all the existing ILs which is currently inadequate, furthermore, the synthesis of ILs is generally expensive and time-consuming; thus, a large-scale study is infeasible. Therefore, an estimation system of the melting points could solve partially this problem, the estimation is complex since the ILs exhibit unconventional behavior and the information available may be inaccurate. This paper presents a neuro-evolution neural network for the estimation of the melting point of ILs.

Keywords Ionic liquids · Neuro-evolution · QSPR · Melting point

1 Introduction

The Ionic liquids (ILs) are compounds formed exclusively by ions and have a lower melting point than regular salts, usually lower than room temperature. Some benefits offered by the ILs are: They have excellent performance as solvents for a wide range of materials, have great polarity, low viscosity, high thermal stability and are immiscible with large amounts of organic compounds [1, 2].

J. A. Cerecedo-Cordoba (✉) · J. J. González Barbosa · J. D. Terán-Villanueva
J. Frausto-Solís
TecNM/Instituto Tecnológico de Ciudad Madero, Ciudad Madero, México
e-mail: joalceco@gmail.com

J. J. González Barbosa
e-mail: jjgonzalezbarbosa@itcm.edu.mx

J. D. Terán-Villanueva
e-mail: david_teran00@yahoo.com.mx

J. Frausto-Solís
e-mail: juan.frausto@itcm.edu.mx

O. Castillo et al. (eds.), *Fuzzy Logic Augmentation of Neural and Optimization Algorithms: Theoretical Aspects and Real Applications*, Studies in Computational Intelligence 749, https://doi.org/10.1007/978-3-319-71008-2_7

One of the advantages of ILs versus traditional solvents is their capability to be designed for specific applications. Some examples of applications are as solvents on lithium batteries, additives in paint, and solvents in reaction systems.

Choosing the "best" ILs for a specific application is difficult due to:

- The number of binary combinations that could form an ILs is a range from 10^{12} to 10^{18} [1, 3].
- Many of their physical properties are unknown or do not have the same values among different sources [1, 4].
- All the fragments of the molecules contribute to the observed physical properties [1].
- Minor differences in the molecular structure could change greatly their properties [5].
- The process for analyzing all the ILs is costly and slowly.

Fortunately, a prediction model of the physical properties of ILs could reduce the analysis cost to only the most promising ones.

Generally, the prediction models use numeric vector data, an encoding process is needed for ILs since the input data is a molecule. There are mainly three methods to achieve the transformation:

- Quantitative structure–activity relationship (QSAR): In this method, the molecular descriptors are measured and are data taken from the molecule and its different properties, including physical and chemical.
- Group Contribution: This method operates on the principle that all the fragments of the molecule contribute to the final melting point temperature. The encoding is the counting of different fragments contained in the molecule.
- Graph Encoding: This technique refers to the transformation of the molecule into a graph representation to be feed directly to the model [6].

Typically, computational methods are used in the construction of predictive models; these can learn from complex data and discover their implicit patterns. The most popular algorithms for data regression are multiple linear regression [1, 7, 8], support vector machines [9], regression trees [5], and neural networks [5, 9].

2 ANNs and Neuroevolution

Artificial Neural networks (ANNs) are simplified representations of the brain and its interconnections. The basic component of a neural network is the artificial neuron, a set of incoming connections called dendrites stimulate the neuron, then the neuron accumulates all the stimulus using an accumulation function and then it fires an activation function. The most popular activation functions are the sigmoid and hyperbolic tangent since they are easily differentiable which is helpful for the training of neural networks.

The most common training method of neural networks is the backpropagation technique. The algorithm consists of two phases. In the First phase, an input activates the network and the information pass from the input layer, then to the inner layers to the output layer. The second phase is responsible for the propagation of the calculated error in the inverse order of the first phase; i.e., the error is propagated from the output layer towards the previous layers.

ANNs can learn from nonlinear data [10] and are resistant to missing or erroneous data [11]. They are highly parallelizable [12] and can learn highly complex patterns. A disadvantage of ANNs is that relevant inputs must be selected [10]. Also, the topology must be defined previously and are likely to fall into local optima caused by the training [10]. An extensive trial and error test is needed in order to define the "best" topology or architecture for a particular problem.

Neuroevolution is the training of neural networks through evolutionary methods. Depending on the heuristic used, this provides an escape mechanism of local optima. Also, neuroevolution refers to the training of topologies of ANNs, this is particularly helpful since the methodology does not need a predefined topology.

3 Methodology

The data set used for the training was obtained from the literature [13]. The data is composed of 43 ILs. Then were transformed to a digital format using the software JChem v16.8.8.0 [14]. The data was then encoded to numerical values through the calculation of molecular descriptors. The software used was PaDel [15]. For the final model, we selected four descriptors; the ones with the highest correlation to the melting point. The general structure of the experiments is presented in Fig. (1).

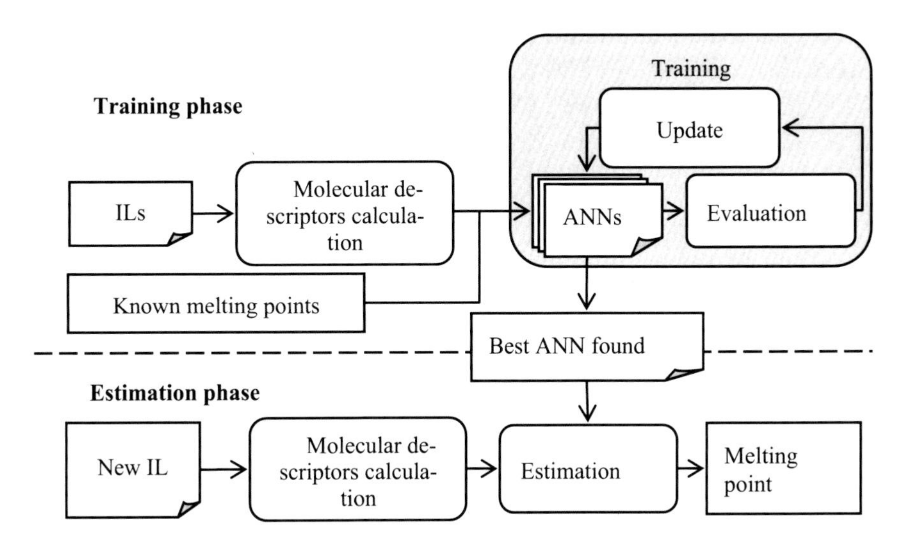

Fig. 1 The general structure for melting point prediction

Three training schemes that were selected for the experimentation was:

- BackPropagation (BP) is by far the most popular training method for neural networks, the version implemented uses an adaptive learning rate and adaptive momentum.
- Resilient backpropagation (Rprop) is a heuristic that uses only the sign of the gradient and ignores the magnitude, the algorithm is responsible for updating each weight multiplying it by a positive or negative factor depending on the sign of the previous iteration [16].
- NeuroEvolution of Augmenting Topologies (NEAT) is a genetic algorithm that trains the weights alongside the network architecture [17]. Therefore, the definition of a previous architecture is discarded.

The three implementations were made with the framework for machine learning Encog [18].

4 Experimental Results

A topology search for BP and Rprop was performed. The analyzed topologies have from 1 to 30 neurons per layer. In these experiments, only the first two layers were explored. Each of the ANN configurations was executed 30 times to ensure reliable results. The median values were used to obtain a representative result of the experiments. The molecular descriptors and the melting point were normalized for training purposes. The Percentage Split method was used, the prediction model was built using the 70% of data and we validate with the remaining 30%. The Mean Squared Error (MSE) was used for measuring the quality of the prediction on training phase and is defined in Eq. (1). The errors were calculated with Root Mean Squared Error (RMSE) and Mean Absolute Error (MAE) shown in Eqs. (2) and (3). Also, the errors were calculated on the training set and validation set.

$$MSE = \frac{1}{n}\sum_{i=1}^{n}\left(\widehat{Y}_i - Y_i\right)^2 \tag{1}$$

$$RMSE = \sqrt{\frac{1}{n}\sum_{i=1}^{n}\left(\widehat{Y}_i - Y_i\right)^2} \tag{2}$$

$$MAE = \frac{1}{n}\sum_{i=1}^{n}\left|\widehat{Y}_i - Y_i\right| \tag{3}$$

Two experiments were performed with different stop criteria. The first one was made with a maximum number of iterations set to 1,000 and the second experiment with 10,000 iterations. This was done to compare the different training schemes over time. The results of the training with 1,000 epoch are presented in Table 1.

Table 1 Experimental results for ANNs with 1,000 iterations

Training method	Size layer 1	Size layer 2	RMSE (Training)	MAE (Training)	RMSE (Validation)	MAE (Validation)
BP	15	0	31.30366	24.41300	42.62099	30.34775
NEAT	–	–	37.0524	30.5629	43.0115	35.5530
BP	25	0	30.35049	23.90755	44.11757	32.81350
Rprop	20	0	30.5397	23.9939	44.1433	32.0341
BP	10	0	31.37968	24.39260	44.24867	32.02195
Rprop	10	0	31.7857	24.7292	44.3233	32.4717
Rprop	30	0	30.2415	23.7057	44.4239	33.8029
BP	1	20	37.81944	31.63675	44.49533	37.06780
BP	1	25	38.01665	31.58825	44.49535	37.06780

The experimental results were sort from lowest to highest RMSE. The data obtained indicate that the BP and NEAT algorithms obtain an error smaller than Rprop in validation. Note that in the case of BP networks the topologies of a single hidden layer yield better results than the networks of two hidden layers, this can be due to a series of causes:

- It is possible that the descriptors used are not suitable because they do not correctly reflect the characteristics necessary to make a prediction.
- Very likely a data cleansing is needed, this is difficult since the real values are only obtainable making a synthesis of the ILs, Related to the issue, Valderrama proposed an error detection with neural networks [19]; however, more work on this subject is needed.
- The data may be too few to represent the generality of liquids, hence learns of details very fast, in other words, the model overfits to fast.

An extended experimentation was made with 10,000 iterations to validate the obtained results. The results of the training are presented in Table 2.

Table 2 Experimental results for ANNs with 10,000 iterations sort by RMSE on validation

Training method	Size layer 1	Size layer 2	RMSE (Training)	MAE (Training)	RMSE (Validation)	MAE (Validation)
NEAT	–	–	35.5909	29.8966	38.5873	31.0942
BP	1	20	40.1705	33.8434	44.4955	37.0680
BP	1	10	40.1705	33.4929	44.4955	37.0679
BP	1	5	40.1705	33.8435	44.4956	37.0680
BP	1	15	39.5494	32.5842	44.4957	37.0680
BP	1	25	40.1705	33.8435	44.4957	37.0681
BP	10	5	40.4485	33.6978	47.9882	39.6965
Rprop	45	0	24.1203	15.4509	49.7605	35.0104
Rprop	45	0	24.2251	15.7359	49.8734	36.7426

Table 3 Experimental results for ANNs with 10,000 iterations sort by RMSE on training

Training method	Size layer 1	Size layer 2	RMSE (Training)	MAE (Training)	RMSE (Validation)	MAE (Validation)
Rprop	55	30	22.3660	10.7261	58.6325	43.3040
Rprop	45	60	22.3664	10.7439	58.6025	41.5811
Rprop	85	30	22.3667	10.7905	58.3745	42.3569
Rprop	95	50	22.3675	10.7831	55.9334	38.7640
Rprop	65	30	22.3677	10.7858	55.5953	38.6843
Rprop	85	70	22.3680	10.7890	56.2693	39.5761
Rprop	45	40	22.3684	10.8176	62.5797	46.9853
Rprop	65	40	22.3686	10.8526	60.2870	44.9735
Rprop	75	20	22.3687	10.7916	68.6918	49.0397

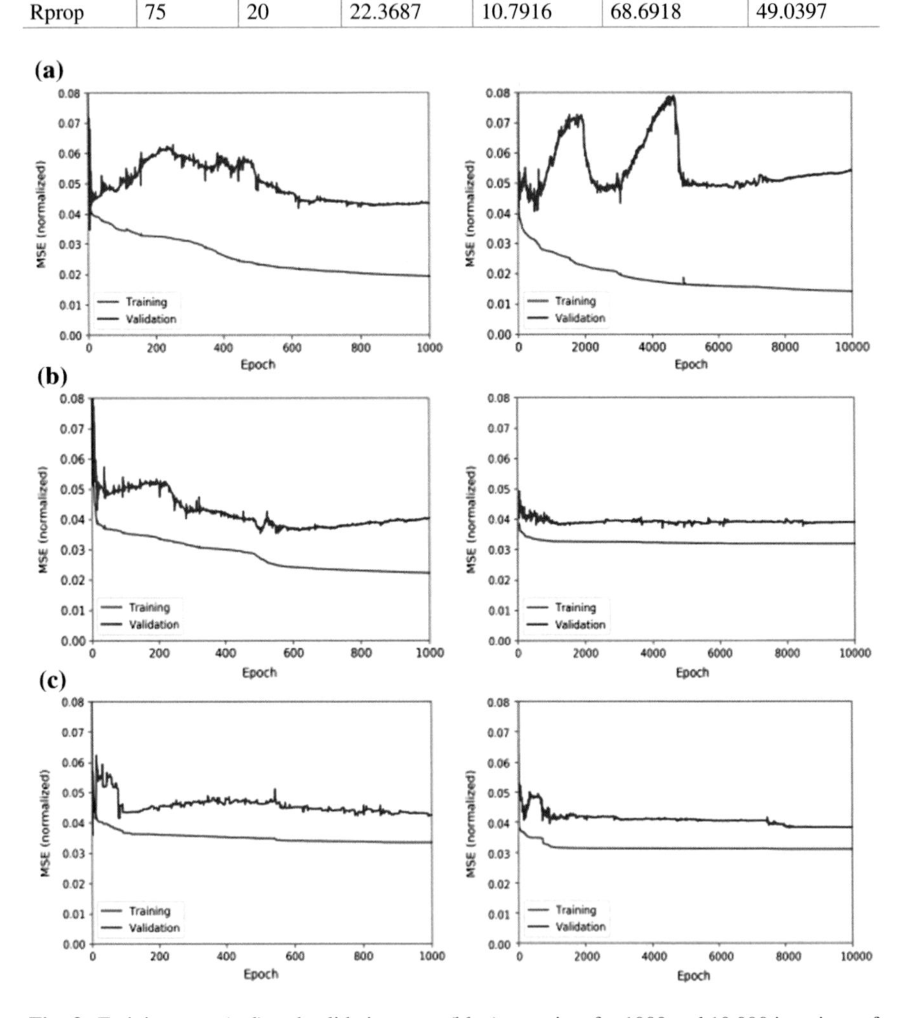

Fig. 2 Training error (red) and validation error (blue) over time for 1000 and 10,000 iterations of: **a** Rprop, **b** BP, **c** NEAT

The experimental data confirms that NEAT and BP have a better generalization of the dataset archiving a better score in the validation set than Rprop. However, Rprop achieves a better score in training set with high numbers of neurons per layer. Table 3 shows results sorted by training RMSE. Here we can see apparent signs of overfitting, which lead to poor performance on the Validation sets.

Figure 2 shows the training error over time. Here we can see that NEAT shows a slower training than BP. However, it achieves a lower error than BP. Also, Rprop has a faster training that NEAT and BP achieving better scores in training but it falls quickly in overtraining.

5 Conclusions and Future Work

In this paper, we showed the differences in training of backpropagation neural networks and neuroevolutionary neural networks on the estimation of the melting point of ionic liquids. The obtained results indicate that the neuroevolution is a promising approach to melting point prediction since they achieved a more general model with the dataset used. We are currently working on expanding the dataset used to obtain a more representative behavior of ionic liquids. Another key future work is a detail selection of molecular descriptors alongside statistical methods to support the selection. We are currently working on improving the heuristic training in topology selection method.

Acknowledgements The authors would like to acknowledge with appreciation and gratitude to CONACYT, TECNM and PRODEP. Also, acknowledge to Laboratorio Nacional de Tecnologías de la Información in the Instituto Tecnológico de Ciudad Madero for the access to the cluster. This work has been partial supported by CONACYT Project 254498. Jorge A. Cerecedo-Cordoba and J. David Terán-Villanueva would like to thank the supports 434694 and 177007.

References

1. A.R. Katritzky, A. Lomaka, R. Petrukhin, R. Jain, M. Karelson, A.E. Visser, R.D. Rogers, QSPR correlation of the melting point for pyridinium bromides, potential ionic liquids. J. Chem. Inf. Comput. Sci. **42**(1), 71–74 (2002)
2. P. Wasserscheid, T. Welton, *Ionic Liquids in Synthesis*, vol. 1, no. 10 (Wiley, 2008)
3. M. Alvarez-Guerra, P. Luis, A. Irabien, Modelo de contribución de grupos para la estimación de la ecotoxicidad de líquidos iónicos. Afinidad **68**(551), 20–24 (2011)
4. J.O. Valderrama, R.A. Campusano, Melting properties of molten salts and ionic liquids. Chemical homology, correlation, and prediction. C. R. Chim. **19**(5), 654–664 (2016)
5. G. Carrera, J. Aires-de-Sousa, Estimation of melting points of pyridinium bromide ionic liquids with decision trees and neural networks. Green Chem. **7**(1), 20 (2004)
6. R. Bini, C. Chiappe, C. Duce, A. Micheli, A. Starita, R. Solaro, M.R. Tine, Ionic liquids: prediction of their melting points by a recursive neural network model. Green Chem. **10**, 306–309 (2008)

7. S. Trohalaki, R. Pachter, Prediction of melting points for ionic liquids. QSAR Comb. Sci. **24** (4), 485–490 (2005)
8. N. Sun, X. He, K. Dong, X. Zhang, X. Lu, H. He, S. Zhang, Prediction of the melting points for two kinds of room temperature ionic liquids. Fluid Phase Equilib. **246**(1–2), 137–142 (2006)
9. A. Varnek, N. Kireeva, I.V Tetko, I.I. Baskin, V.P. Solov'ev, Exhaustive QSPR studies of a large diverse set of ionic liquids: how accurately can we predict melting points? J. Chem. Inf. Mod. **47**(3), pp. 1111–1122 (2007)
10. G. Deyfus, *Neural Networks* (2004)
11. B. Kosko, *Neuronal Networks and Fuzzy Systems* (1992)
12. C. Fyfe, Artificial neural networks and information theory. 1–204 (2000)
13. S. Zhang, X. Lu, Q. Zhou, X. Li, X. Zhang, S. Li, *Ionic Liquids Physicochemical Properties* (2009)
14. ChemAxon, MarvinSketch (JChem Base) version 16.8.8, http://www.chemaxon.com/products/marvin/marvinsketch/ (2016)
15. C.W. Yap, PaDEL-descriptor: an open source software to calculate molecular descriptors and fingerprints. J. Comput. Chem. **32**(7), 1466–1474 (2011)
16. M. Riedmiller, Advanced supervised learning in multi-layer perceptrons—from backpropagation to adaptive learning algorithms. Computer Standards and Interfaces **16**(3), 265–278 (1994)
17. K.O. Stanley, R. Miikkulainen, Evolving neural networks through augmenting topologies. Evol. Comput. **10**(2), 99–127 (2002)
18. J. Heaton, Encog: library of interchangeable machine learning models for java and C#. J. Mach. Learn. Res. **16**, 1243–1247 (2015)
19. J.O. Valderrama, R.E. Rojas, Redes Neuronales Artificiales como Herramienta para detectardatos Erróneos de Temperatura de Fusión de Líquidos Iónicos, in *XXVI Congreso Interamericano de Ing. Química* (2012)

A Proposal to Classify Ways of Walking Patterns Using Spiking Neural Networks

Karen Fabiola Mares, Rosario Baltazar, Miguel Ángel Casillas, Víctor Zamudio and Lenin Lemus

Abstract In this work the Spiking Neural Networks (SNNs) for the classification of ways of walking patterns is presented. The Differential Evolution (DE) Algorithm as an optimization technique was used for weights and delays settings. Two accelerometers, each one with three axes, were used to obtain simultaneous information on both legs. The information formed by nine features has been stored in a database: the first three correspond to the accelerations of x, y and z axis, next three correspond to the velocities which are obtained by doing an integration of the acceleration data for each axis and finally the positions x, y and z are calculated by the integration of velocities respectively.

Keywords Spiking neural networks · Ambient assited living
Walking patterns

1 Introduction

In ambient assisted living there are several works that study the recognition of physical activity, with the purpose of caring for the elderly and to assist them [1]. The activity recognition that implements the data extraction of accelerometers can

K. F. Mares (✉) · R. Baltazar · M. Á. Casillas · V. Zamudio
TecNM-Leon Institute of Technology, Av. Tecnologico s/n, 37290 Leon, Guanajuato, Mexico
e-mail: karenmares90@gmail.com

R. Baltazar
e-mail: r.baltazar@ieee.org

M. Á. Casillas
e-mail: miguel.casillas@itleon.edu.mx

V. Zamudio
e-mail: vic.zamudio@ieee.org

L. Lemus
Polytechnic University of Valencia, Valencia, Spain
e-mail: lemus@upv.es

O. Castillo et al. (eds.), *Fuzzy Logic Augmentation of Neural and Optimization Algorithms: Theoretical Aspects and Real Applications*, Studies in Computational Intelligence 749, https://doi.org/10.1007/978-3-319-71008-2_8

be observed in [2]. In this work is shown a study of the recognition of physical activity implementing five biaxial accelerometers placed in: thigh, wrist, arm, hip and foot. Different algorithms were implemented, where the decision trees achieved better classification efficiencies. In [3] a system of monitoring and classification of activities is presented by developing a device called *eWatch* that works with one biaxial accelerometer and illumination sensor and presents an investigation of classification dependence, using the *eWatch* in different parts as belt, clothes pockets, backpacks and necklaces. The results show a better performance with the decision trees in terms of effectiveness and efficiency. In [4] the recognition is realized in a device that is integrated by a microprocessor and a triaxial accelerometer. For classification they implemented a decision tree classifier, integrated to an intelligent environment that allows to react depending on the activity and to help in the detection of falls emitting an alert to request urgent support.

The purpose of this work is to study the effectiveness of the SNNs for an activity classification, in a system of multiple sensors placed on both legs. The main objective is to classify the variation of walking on elderly people and detect alterations. This article is structured as follows. In Sect. 2 we provide theoretical information of SNNs and DE Algorithm as an optimization technique. The methodology of the proposed method for the classification of variation of walking is shown in Sect. 3. The results obtained from our experiments are explained in Sect. 4 and finally Sect. 5 presents our conclusions and future work.

2 Theoretical Framework

This section provide a brief description of each technique used in this work, elementary definitions of Artificial Neural Networks (ANNs), SNNs, neural models like Leaky Integrate and Fire (LIF) and Spike Response Model (SRM) and the DE Algorithm is shown.

2.1 Artificial Neural Networks

An ANN is a computational model of information processing that simulates the behavior of biological brain systems. It is composed of a set of elements called neurons that are connected to each other through connections with a specific weight, as shown in Fig. 1 [5]. An ANN works as follows: a set of neurons (Input Layer) receive a signal from the environment and propagate it, with a *propagation rule*, to the neurons with which they are directly connected (Hidden Layer), which they apply an *activation function* to their input signal to obtain an output, which propagates to the last connected neurons (Output Layer), which will apply the same activation function to obtain its output signal.

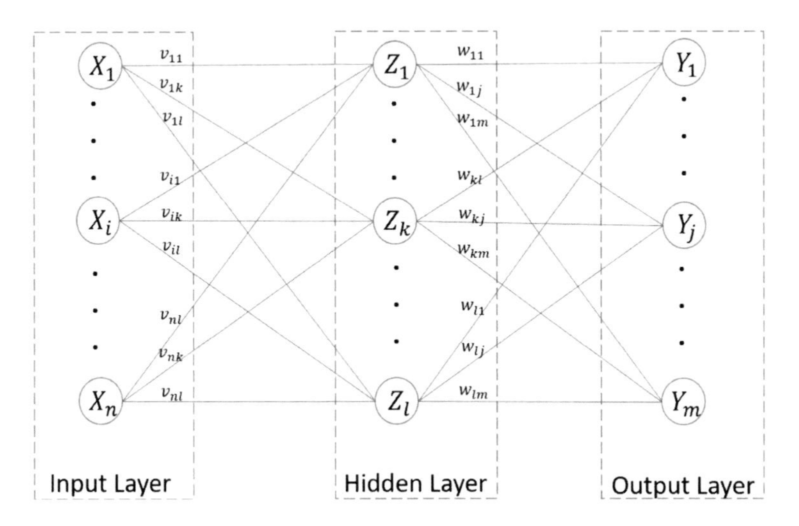

Fig. 1 General structure of an ANN

The propagation rule usually consists of the dot product between the feature vector **x** and the weights vector **w**, corresponding to the connections of the neurons, added to the weight vector of a neuron known as *bias*. The propagation rule is illustrated in Eq. (1).

$$Y_{input,j} = b_j + \sum_{i=1}^{N} x_i \cdot w_{ij}, \tag{1}$$

where N is the number of neurons on each layer. The activation function is usually a non-linear function and depends on the problem. Some commonly used activation functions are shown in Table 1 [5].

The ANNs have been widely used in the area of pattern recognition in their classification phase [6, 7], in which input neurons represent the feature vector and output neurons the number of classes that are attempted to separate. A different generation from the classical neural networks is explained in the next section.

Table 1 Usual activation functions

Identity function	$f(x) = x$	The input signal is transmitted directly to the output
Binary step function	$f(x) = \begin{cases} 1, x \geq \theta \\ 0, x < \theta \end{cases}$	The output is 1 if x reaches a threshold θ, or 0 otherwise
Bipolar step function	$f(x) = \begin{cases} 1, x \geq \theta \\ -1, x < \theta \end{cases}$	The output is 1 if x reaches a threshold θ, or -1 otherwise
Binary sigmoid	$f(x) = \frac{1}{1+e^{-\sigma x}},\ \sigma > 0$	Used when there is a training and are evaluated both $f(x)$ and $f'(x)$

2.2 Spiking Neural Networks

SNNs represent a closer analogy to biological neural networks, which process the information as electrical pulses, known as *action potential* or *spikes*, through the time. A post-synaptic neuron receives from a pre-synaptic neuron a series of electrical pulses, which are transmitted to the soma to process the received signal, generating an output signal if during the process if the input signal has reached a threshold; the output pulses are delivered through the axon (to play now the role of a pre-synaptic neuron) to the dendrites of the next directly connected neuron (post-synaptic neuron), which takes the signal as input and performs the same process [8].

Analogous to this process, the SNNs add the concept of *simulation time* in its operation, during which the calculation of the output signal of the neurons is performed. All information is encoded in the action potential of the neurons, which is received as a sequence of action potentials or *spike train*, by the neurons directly connected. Because of this, in the input layer of the neural network, a feature vector with real values must be encoded at a pulse rate. There are several models to perform the conversion of information to firing rates, one of the easiest to implement is the 1D Coding, which uses a linear function and a temporal coding interval, subtracted from the total simulation time. The formulation of the 1D coding is shown in Eq. (2) [9]:

$$X_i(f) = \frac{(b-a)}{range} \times f + \frac{(a \times max - b \times min)}{range}, \tag{2}$$

where f is the actual value of the feature, $[a, b]$ is the time coding interval and $range = max - min$, is the range of the feature being calculated. The input spike train is transmitted by the connections of the neural network, the following neurons receive this signal and process it during the simulation time, through a *spiking neuron model* they determine the firing rate that generated the input signal, and send it to the following neurons directly connected.

There are several spiking neuron models to propagate the signal received as firing rates, some as LIF and the SRM are explained in following section.

2.2.1 Leaky Integrate and Fire Model

LIF model can be seen in two phases, depending on the voltage changes that occur throughout the simulation time: it is initially considered that the neuron behaves passively, when there is no voltage injection, it decays to a certain value of rest, from this state, are injected voltages; when the action potential reaches a threshold, an activation of the action potential (neuron spike) is generated at the current time and immediately a reset is made to the state of rest of the neuron. The really important information of the process are the times in which occur an activation of the action potential. The LIF model is defined in the following equation:

$$v' = I(t) + a - bv, \quad (3)$$

where v' is the state of the neuron at time t which is changing by the injection of a voltage $I = \mathbf{x} \cdot \mathbf{w} \times \gamma$, where $\mathbf{x}$ is the input feature vector, $\mathbf{w}$ is a vector of synaptic weights of the connections and γ is a gain factor; so, if it happens that $v > v_{threshold}$ then the state of the neuron is reset to a state of rest c. In the classification process, it is expected that patterns belonging to the same class will obtain a similar firing rate, so the Average Firing Rate (AFR) and the Standard Deviation of Firing Rate (SDFR) are calculated for each class to be classified. To decide whether a feature vector $\mathbf{x}$ belongs to one class or another, Eq. (4) is used:

$$class = \arg min_{k=1}^{K}(|AFR_k - fr|), \quad (4)$$

where fr is the firing rate generated by the input feature vector $\mathbf{x}$, and K is the number of classes to be classified [10].

2.2.2 Spike Response Model

SRM is a generalization of the LIF model, since the state of the neuron is not dependent on the injected voltage but of the last activation of the action potential on the neuron. Initially neuron state v_i is set to 0; the entrance of spike trains from pre-synaptic neurons alters the state of the post-synaptic neuron, which is updated with Eq. (5):

$$v_i(t) = \sum w_{ij} y_i(t), \quad (5)$$

where $y_i(t) = \varepsilon(t - t_i - d_{ij})$ is a function that describes, through the time, the response of an incoming spike trains at the simulation time, t is the current time, t_i is the firing rate of the pre-synaptic neuron, which have a synaptic delay d_{ij}. If after the entrance of the spike trains of several pre-synaptic neurons the state of the neuron has reached the threshold, an activation of the action potential of the post-synaptic neuron is generated. The function that models this behavior is indicated in Eq. (6):

$$\varepsilon(t) = \frac{t}{\tau} e^{1-\left(\frac{t}{\tau}\right)} \quad (6)$$

which returns the corresponding value when $t > 0$, otherwise 0 is returned. In Eq. (6), τ is the time constant of the neuron potential. The synaptic weights in the LIF model, and the weights and delays in the SRM model affect the performance of the neural network, because of this, metaheuristic algorithms have been used to find weights and delays that allow to obtain optimal values of classification. One of these algorithms for adjusting weights of a neural network is explained below.

2.3 *Differential Evolution Algorithm*

DE is an evolutionary algorithm that is based on the evolution of a population of candidate solutions (individuals) spread in a search space. DE uses some control parameters, which must be initialized before beginning the process of evolution of individuals; the parameters are as follows: a mutation value *F*, a cross parameter *Cr*, as well as the number of *N* individuals and *GEN* generations. At first a population of individuals $\mathbf{x}_i$ is initialized with random values in a *D*-dimensional space, which are evaluated with an objective function *f(x)*; in each generation three random individuals of the population, different from the individual being evaluated, are chosen; which are used to create a mutated individual. This individual is evaluated with the objective function and replaces the current individual if he obtains a better fitness value. This process is performed for the *N* individuals, until reaching the *GEN* generations. The DE algorithm is summarized in Algorithm 1 [11].

Algorithm 1: DE

1. Define the objective function $f(\mathbf{x})$, $\mathbf{x} = (x_1, \ldots, x_D)$
2. Define the control parameters *F* and *Cr*
3. Initialize the population of individuals$\mathbf{x}_i (i = 1,2, \ldots, N)$
4. Calculate $f(\mathbf{x}_i)$ and assign the best individual to $\mathbf{x}_*$
5. **WHILE** (*g*<*GEN*)
6.   **For**$i = 1$ to N
7.   Choose randomly three individuals with indexes$r1, r2, r3 \in [1, \ldots, N], r1 \neq r2 \neq r3 \neq i$
8.   Generate a new individual $\mathbf{x}_{new}$ according to:
9. $\mathbf{x}_{new} = \begin{cases} \mathbf{x}_{r3} + F(\mathbf{x}_{r1} - \mathbf{x}_{r2}), & \text{If}(rand[0,1] < Cr)\ OR\ (Rnd[1, D] = i) \\ \mathbf{x}_i, & \text{otherwise} \end{cases}$
10.   **If**$(f(\mathbf{x}_{new}) < f(\mathbf{x}_i))$
11.   Replace $\mathbf{x}_i$ with $\mathbf{x}_{new}$
12.   **End If**
13.   **End For**
14. **End WHILE**

3 Experimental Methodology

This section describes the methodology used in the collection of information and classification process. Two MMA7361 three-axis modules (x, y and z) were implemented for data collection, which were placed on the legs, under the knee, of a person to collect samples for a given time and form the feature vectors.

The feature vector is formed by nine features (three of acceleration, three of velocity and three of position). To form the vectors the following procedure was performed: while the person walked following a specific pattern, seven samples were taken for each 0.7 s of walking. Then, the average was calculated using these seven samples, obtaining the first three features of the vector corresponding to the acceleration; Using a numerical integration technique of Simpson's Rule, a process of integration of the seven acceleration samples was carried out. Next, one velocity sample was obtained for each group of three accelerations (the grouping is shown in Fig. 2), the average of the three velocity samples was calculated to obtain the following three features. A final integration process was performed with the three velocity samples, obtaining the last three features of the vector, corresponding to the position. This process was carried out for a time of 5.8 min to obtain a total of 500 vectors per class. The detection of walking was studied with two variations, class one: walking in a circle and class two: walking in a straight line with recoil. The process to form the database is illustrated in Fig. 2.

For the classification process, cross validation was used with a $k = 10$; to convert the input vectors to pulses we used the 1D coding; the calculation of output pulses of neurons was performed using the models LIF and SRM; the DE algorithm was implemented for the adjustment of weights and delays. For each test the following process was followed: the dataset was divided using 10-cross validation to obtain the test and training sets, later a population of 40 individuals was created, corresponding to the weights and delays of the SNN connections (weights for the LIF model and weights and delays for the SRM), this population was sent to the DE

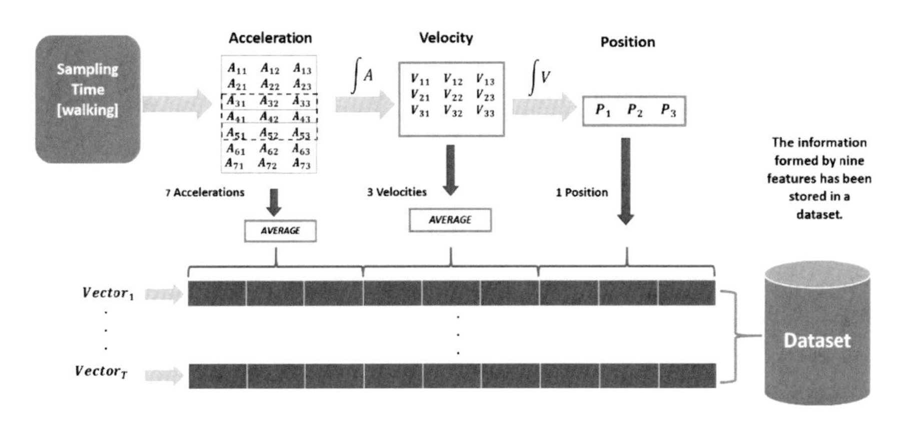

Fig. 2 Methodology used to construct the dataset

algorithm to find the optimal weights and delays. The objective function for the LIF model is shown in the following equation [10]:

$$f = \frac{1}{dist(AFR)} + \sum_{k=1}^{K} SDFR_k \tag{5}$$

where $dist(AFR)$ is the Euclidean distance between the average firing rate of the classes to be classified; the objective function for the SRM is the quadratic error of each individual, illustrated in the following equation [5]:

$$E_i = \sum_{k=1}^{T} (t_k - y_k)^2, \tag{6}$$

where T is the total of training vectors. As the stop criterion was used a maximum of 500 generations, reached this criterion, the best individual was used to perform the classification with the training set and test. Subsequently we took the next training set and test and a new population of individuals to find the weights and delays of the neural network. This process was performed for the entire dataset.

The previous process corresponds to one trial and we report the accuracy average of the ten folds and the execution time of each trial, executed in a computer with the following characteristics: processor Intel Core i7, CPU 2.00 GHz, Memory 8 GB, Operative System 64 bits. For each model, a total of 10 trials were performed. The values of metaheuristic parameters used in the experimentation are the following: $F = 0.8$ and $Cr = 0.9$. The classification process is summarized in Fig. 3.

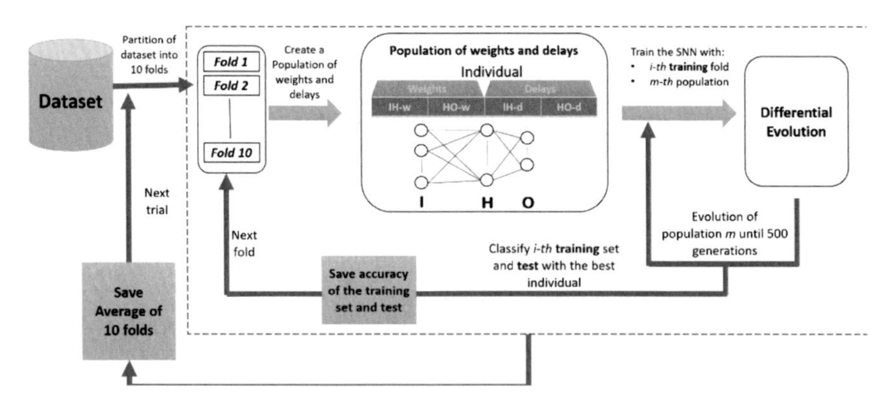

Fig. 3 Methodology followed for classification

Table 2 Results of classification with the SRM

Trial	SRM		
	Training (%)	Test (%)	Time (s)
1	94.55	92.59	473.48
2	93.31	98.21	483.89
3	95.16	92.59	507.18
4	94.64	91.53	550.53
5	93.95	96.29	520.57
6	93.31	92.85	528.26
7	95.36	96.29	556.10
8	91.70	94.64	496.94
9	94.73	92.85	625.32
10	91.70	94.64	475.72
Average	93.84	94.24	521.79
Mean	94.04%		
Std. Dev.	1.72		

4 Results

For the classification results we report: mean, standard deviation and runtime. The results with the SRM are shown in Table 2, where we obtained a classification mean of 94.04% for training set and test, a standard deviation of 1.72 and an average runtime of 521.79 s per trial.

For the LIF model, the results that were obtained from the experiments are shown in Table 3, we obtained a classification mean of 92.91% for training set and test, a standard deviation of 1.22 and with an average runtime of 193.93 s per trial.

Table 3 Results of classification with the LIF model

Trial	LIF model		
	Training (%)	Test (%)	Time (s)
1	94.95	94.44	144.90
2	94.53	94.64	142.91
3	93.75	94.44	144.90
4	94.53	98.21	145.12
5	94.15	92.59	146.48
6	93.31	92.85	145.28
7	93.75	96.29	145.67
8	93.92	94.64	145.46
9	96.35	94.64	633.40
10	94.73	94.64	145.25
Average	94.39	94.73	193.93
Mean	92.91%		
Std. Dev.	1.22		

From Table 2 and 3, we observe that the SRM obtained a higher accuracy than the LIF model, however, the individuals are evaluated faster in the LIF model.

5 Conclusions and Future Work

In this paper a new methodology for the collection of walking patterns, using accelerometers was presented. With this methodology it is shown that good results of classification efficiency using a SNN with the LIF and SRM neural models were obtained around 94%.

In future work new classes will be added to the dataset, corresponding to other variations of walking, obtained by the accelerometers. Also, this information will be send to a middleware to help in the detection of walking alterations on elderly people. The information can be useful to compare the performance of this proposal in relation to other classifiers.

Acknowledgements This work was partially supported by CONACYT and León Institute of Technology.

References

1. S. Chernbumroong, S. Cang, A. Atkins, H. Yu, Elderly activities recognition and classification for applications in assisted living. Expert Syst. Appl. **40**(5), 1662–1674 (2013)
2. L. Bao, S.S. Intille, in *Activity Recognition from User-Annotated Acceleration Data*, ed. by A. Ferscha, F. Mattern. Pervasive Computing. Pervasive 2004. Lecture Notes in Computer Science, vol 3001 (Springer, Berlin, Heidelberg, 2004)
3. U. Maurer, A. Smailagic, D.P. Siewiorek, M. Deisher, Activity recognition and monitoring using multiple sensors on different body positions, wearable and implantable body sensor networks, in *BSN 2006, International Workshop on* (2006), pp. 4–7
4. D. Sprute, K. Matthias, On-chip activity recognition in a smart home, in *12th International Conference on Intelligent Environments (IE)* (2016), pp. 95–102
5. M. Friedman, A. Kandel, *Introduction to Pattern Recognition—Statistical, Structural, Neural and Fuzzy Logic Approaches* (London, 1999), p. 345
6. G. Ou, Y.L. Murphey, Multi-class pattern classification using neural networks. Pattern Recognit. **40**(1), 4–18 (2007)
7. T.H. Oong, N. Ashidi, M. Isa, Adaptive evolutionary artificial neural networks for pattern classification. IEEE Trans. Neural Netw. **22**(11), 1823–1836 (2011)
8. S. Ghosh-dastidar, H. Adeli, Spiking neural networks. Int. J. Neural Sys. **19**, 295–308 (2009). https://dx.doi.org/10.1142/S0129065709002002
9. A. Belatreche, L.P. Maguire, Advances in design and application of spiking neural networks. Comput. A Fusion Found. Methodologies Appl. **11**, 239–248 (2007)
10. R.A. Vazquez, A. Cachón, Integrate and fire neurons and their application in pattern recognition. CCE, pp. 424–428 (2010)
11. V. Feoktistov, *Differential Evolution* (New York, 2006)

Partially-Connected Artificial Neural Networks Developed by Grammatical Evolution for Pattern Recognition Problems

Olga Quiroz-Ramírez, Andrés Espinal, Manuel Ornelas-Rodríguez, Alfonso Rojas-Domínguez, Daniela Sánchez, Héctor Puga-Soberanes, Martin Carpio, Luis Ernesto Mancilla Espinoza and Janet Ortíz-López

Abstract Evolutionary Artificial Neural Networks (EANNs) are a special case of Artificial Neural Networks (ANNs) for which Evolutionary Algorithms (EAs) are used to modify or create them. EANNs adapt their defining components ad hoc for solving a particular problem with little or no intervention of human expert. Grammatical Evolution (GE) is an EA that has been used to indirectly develop ANNs, among other design problems. This is achieved by means of three elements: a Context-Free Grammar (CFG) which includes the ANNs defining components, a search engine that drives the search process and a mapping process. The last component is a heuristic for transforming each GE's individual from its genotypic form into its phenotypic form (a functional ANN). Several heuristics have been proposed as mapping processes in the literature; each of them may transform a specific individual's genotypic form into a very different phenotypic form. In this paper, partially-connected ANNs are automatically developed by means of GE. A CFG is proposed to define the topologies, a Genetic Algorithm (GA) is the search engine and three mapping processes are tested for this task; six well-known pattern

O. Quiroz-Ramírez · M. Ornelas-Rodríguez · A. Rojas-Domínguez
H. Puga-Soberanes · M. Carpio · L. E. M. Espinoza
Tecnológico Nacional de México-Instituto Tecnológico de León, León, Mexico
e-mail: judithq21@gmail.com

M. Ornelas-Rodríguez
e-mail: manuel.ornelas@itleon.edu.mx

A. Espinal (✉)
División de Ciencias Económico Administrativas, Universidad de Guanajuato, Guanajuato, Mexico
e-mail: aespinal@ugto.mx

D. Sánchez
Tecnológico Nacional de México-Instituto Tecnológico de Tijuana, Tijuana, Mexico

J. Ortíz-López
Escuela Internacional de Doctorado, Universidad de Vigo, Vigo, Spain

O. Castillo et al. (eds.), *Fuzzy Logic Augmentation of Neural and Optimization Algorithms: Theoretical Aspects and Real Applications*, Studies in Computational Intelligence 749, https://doi.org/10.1007/978-3-319-71008-2_9

recognition benchmarks are used to statistically compare them. The aim of this work for using and comparing different mapping process is to analyze them for setting the basis of a generic framework to automatically create ANNs.

Keywords Evolutionary artificial neuronal networks · Grammatical evolution Mapping process · Pattern recognition

1 Introduction

Currently, the Artificial Neural Networks (ANNs) are powerful and successful tools that are frequently used to solve pattern recognition problems [1–4]. Despite their continuous evolution, ANNs still present issues related to the design of their architecture and their subsequent training, the aforementioned problems are the main drawbacks with respect to the use of ANNs [1, 5]. The design of ANNs includes parameters such as input features, a connectivity pattern and the number of neurons in the hidden and output layers; usually, they are defined by a human expert [1]. On the other hand, training is required to adjust the values of the connections between neurons, called weights. The best-known training method is the Back Propagation (BP) algorithm [1, 4, 6–8]. Although BP is a popular method, it possesses certain weaknesses such as a high possibility of being trapped in a local minimum [6, 9]; or being slow when the dimensionality of the data increases [10]. Several optimization techniques, known as Evolutionary Algorithms (EAs), have been proposed to address the problem of local minima in the training phase.; most of these are inspired by biological processes [6, 8]. One of the major advantages of these algorithms lies in their simultaneous search of the solution within a set of possible solutions, called a population of individuals [11]. Besides covering aspects of training, the EAs also contribute to the design of the network structure [8, 12]. The combination of EAs with ANN, known as Evolutionary Artificial Neural Networks (EANNs) represent a more efficient method due to its adaptability and its ability to avoid local minima [13, 14].

Most of the ANNs are designed as fully connected topologies. The main reason is that this simplifies the design of the networks; however, the resulting networks contain a high degree of redundancy [7, 15]. In order to avoid that redundancy, Partially Connected Neural Networks (PCNNs) are ANNs formed with a subset of connections of fully-connected ANNs but with equal or better performance than them.

Reducing the number of connections allow to improve the generalization and reduce the complexity of the network, as well as reducing the training and processing times [7].

In this paper, PCNNs are developed by means of an evolutionary methodology, known as Grammatical Evolution (GE); which has as its main components: a Context-Free Grammar (CFG) in Backus-Naur Form (BNF) related to the problem, a search engine that is usually an EA and a mapping process which is an heuristic that allows the indirect search of solutions. The GE is applied to design partially and weighted topologies of ANNs for solving pattern recognition problems, this is achieved by proposing a CFG in BNF to define the weighted topologies, using a Genetic Algorithm (GA) as search engine and for the mapping process are selected three of them reported in the stated of the art. Since the main aim of this paper is to set the basis of a generic framework to develop PCNNs, the selected mapping processes are used and tested through several experiments by using six well-known pattern recognition benchmarks. Finally, the performances obtained by the three mapping processes are statistically compared.

The rest of this paper is organized as follows: Sect. 2 presents the theoretical foundations used in this research. Section 3 shows the methodology proposed. The experiments and results are reported in Sect. 4. Our conclusions and future work are described in Sect. 5.

2 Background

2.1 Grammatical Evolution

The Grammatical Evolution (GE), proposed by Ryan and O'Neill in 1998 [16, 17], is an EA based on the combination of genetic algorithms with Context-Free Grammars (CFG) [1, 18, 19]. In the Backus Naur Form (BNF) [16]; which is a notation formed by a set of rules of production [18] and composed of terminal nodes and nonterminal nodes. The terminal nodes are those that appear in the language, i.e. they represent or express solutions, and the non-terminal nodes are elements that can be expanded in more non-terminal nodes or in terminal nodes [18]. Generally, grammars are expressed by the tuple: N, T, P, S where N defines non-terminal nodes, T represents terminal nodes, P indicates the set of production rules and S is the start symbol which belong to N [18]. To solve a problem with GE it is required to define a BNF grammar related with the problem, a search engine that drives the search process and a mapping process to obtain the functional form of solutions.

2.2 Mapping Process

In GE, a mapping process is a heuristic to transform an individual from its genotypic form into its phenotypic form. A mapping process requires an individual's chromosome, which is a variable numeric string, and a BNF grammar to derivate and obtain its phenotype [20]. There are several mapping process reported in the state of the art, but basically most of them employ a modulus-based derivation rule to process the chromosomes [20]. This process is illustrated in Fig. 1, where the codon value refers to an element of numeric string.

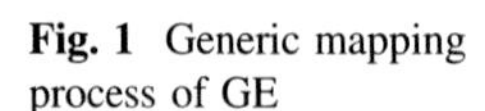

Fig. 1 Generic mapping process of GE

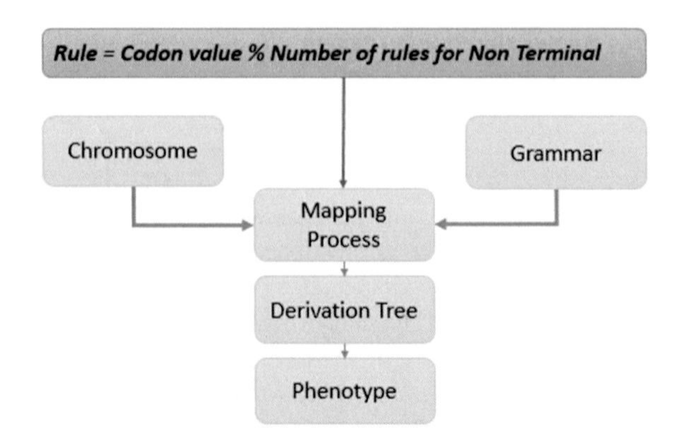

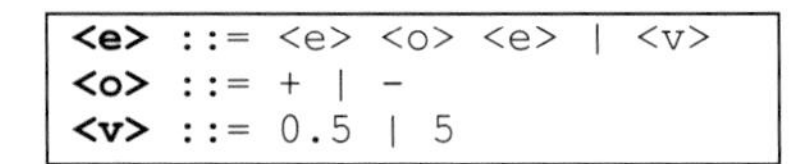

Fig. 2 Example of a BNF grammar [20]

In this work, three mapping processes were used, and these are discussed next. For explaining the mapping processes, it is used the same BNF grammar (see Fig. 2) and the same chromosome. It can be seen that the mapping processes produce different architectures. The first mapping is called Depth-First (DF), and it is the standard mapping in grammatical evolution. Beginning from a non-terminal symbol, this heuristic derives the leftmost non-terminal node along the codons until you find a terminal symbol, then moves to the right to derive the next node [20]. This process is illustrated in Fig. 3.

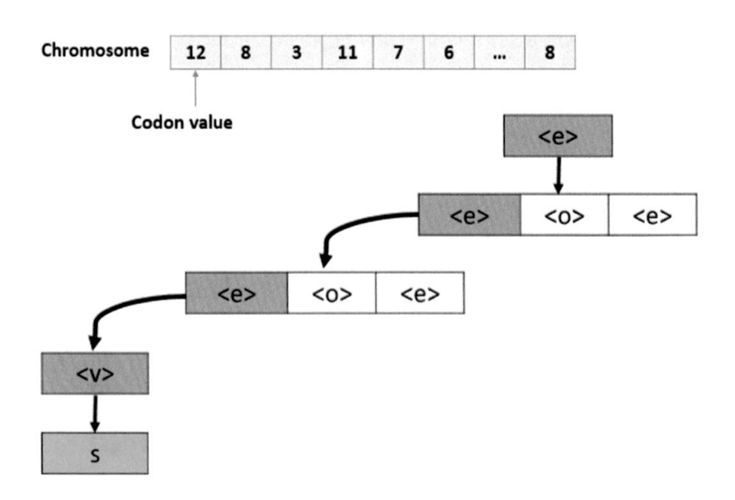

Fig. 3 Depth-first mapping process [20]

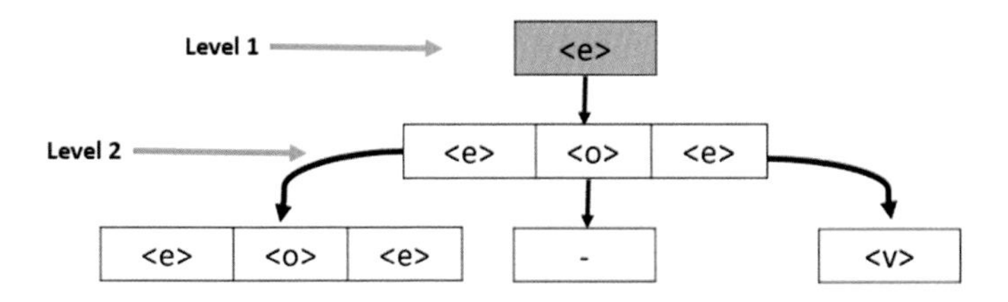

Fig. 4 Breadth-first mapping process [20]

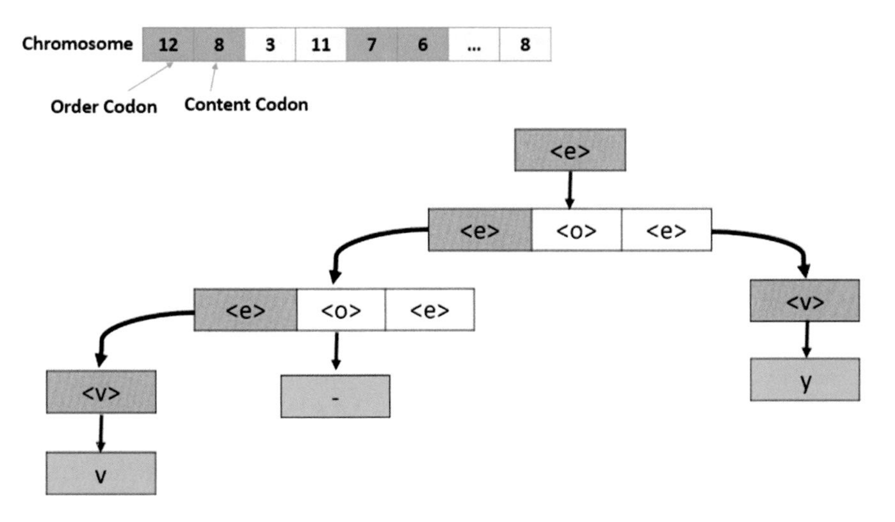

Fig. 5 Position invariant mapping process [20]

The second mapping process, called Breadth-First (BF), expands the derivation tree level by level in a left-to-right manner [20] illustrated in Fig. 4.

The third GE's mapping process, called Position Invariant mapping (PI), differs from the previous two in that each expansion of a non-terminal node requires two codons. Thus, the standard chromosome is sequentially divided into pairs of codon values. In each pair of codons, the first codon is used to choose which non-terminal node to expand, and the second codon indicates which production rule will be employed to expand that non-terminal node [20]. This process is illustrated in Fig. 5.

2.3 Search Engine

The search engine used in this work is the Genetic Algorithm (GA) [21–24]. This Algorithm is a well-known and widely used metaheuristic, it is a guided stochastic search technique, inspired by the mechanisms of natural selection, genetics, and evolution [7, 21, 25, 26]. The algorithm makes use of genetics operators: selection, crossover and mutation [7, 27]. GAs are techniques that assist RNAs in the search of optimizing the topology and find their ideal weights [7, 25, 26]. The GA is shown in Algorithm 1 [28]:

Algorithm 1. Genetic Algorithm

```
Begin
        t = 0;
          initialize p(t);
          while termination condition not satisfied do
          begin
                    t = t + 1;
                      select_repro C (t) from p(t-1);
                      recombine and mutate structures in C(t)
                      forming C'(t);
                      evaluate structures in C'(t);
                      select_replace p(t) from C'(t) and p(t-1);
          end
end
```

where p(t) is the current population at the generation t, C(t) represents the selection of individuals for mating and C′(t) are the new individual or child generated.

2.4 *Evolutionary Artificial Neural Networks*

The use of Artificial Neural Networks (ANNs) combined with Evolutionary Algorithms (EAs) is Knows as Evolutionary Artificial Neural Networks (EANNs) [1, 29]. The EANNs cover several ANN design parameters, such as architecture design, training, learning rule and others [29]. They are divided into three groups: in the first one the EAs focus exclusively on the adaptation of the weights of the network, in the second group, EAs cover only the design of the architecture, and in the third group EAs are used to cover both aspects (architecture and training) [1]. The architectures designed with these algorithms can be codified in two ways: with direct encoding and indirect encoding [8]. Direct encoding is an easy-to-implement type of coding in which all details of the architecture are compiled into the chromosome or genotype [1].

EANNs lead to significantly better intelligent systems than those empirically designed by an expert [30]. Previous work on EAs indicate that these are robust algorithms that allow the location of high quality solution areas, even when the search space is very large and complex [12]. This is what makes them suitable optimization techniques for the design and training of neural networks. The advantage of this combination of techniques is that the network architectures and learning rules can change according to different environments without human intervention [29]. There are several combinations of ANNs with EAs used in classification problems [4, 19, 31, 32], the three best EAs are: genetic algorithms, evolutionary strategies and evolutionary programming [12, 33].

3 Proposal

The proposed framework requires a training set from the pattern recognition problem to solve, a BNF grammar as indirect representation of topologies of neural networks, a fitness function, a search engine and a mapping process. This methodology is illustrated in Fig. 6.

The BNF grammar proposed in this work is shown in Fig. 7; the initial symbol is represented by the non-terminal: <network> and the non-terminal nodes are shown in bold face. The grammar returns the number of neurons in the hidden layer of a network, the connections between the neurons and their corresponding weights.

The general structure of a neural network designed by the methodology is shown in Fig. 8, where the continue lines represents the connections selected by the GE from the total of possible connections (dash lines).

An example of an expression and the corresponding topology generated by this proposal are provided in Fig. 9.

In the previous example we can identify the inputs represented by the symbol "i", hidden layer neurons "h" and outputs neurons "o" as well as the bias "i0" which can have connection with the hidden layer and output layer. In the expression, the first bracket indicates the number of neurons in the hidden layer separated by an underscore character, the connections and weights of the hidden layer with the input

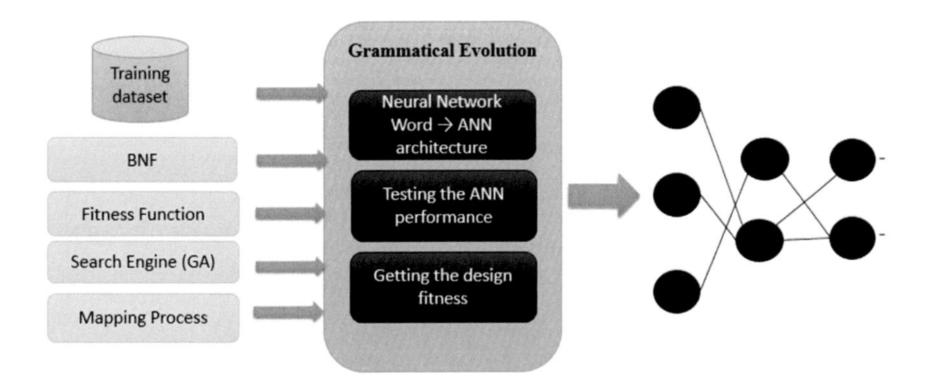

Fig. 6 Proposal metodology

```
<network>         ::= <hiddenNeurons> <outputLayer>
<hiddenNeurons>   ::= <hiddenNeuron> | <hiddenNeuron>_ <hiddenNeurons>
<outputLayer>     ::= i0 | <weight> _ i0 | <weight> _ i0 ,<weight>
<hiddenNeuron>    ::= <neuronInputs> / i0 | <weight> # <neuronOutputs>
<neuronInputs>    ::= <neuronInput> | <neuronInput> / <neuronInputs>
<neuronOutputs>   ::= <neuronOutput> | <neuronOutput>/<neuronOutputs>
<neuronInput>     ::= <inputNeuron> | <weight>
<neuronOutput>    ::= <outputNeuron> | <weight>
<inputNeuron>     ::= i1|i2|i3|i4
<outputNeuron>    ::= o1|o2|o3|
<weight>          ::= <sign> <digitList> . <digitList>
<sign>            ::= +|-
<digitList>       ::= <digit>|digit> <digitList>
<digit>           ::= 0|1|2|3|4|5|6|7|8|9
```

Fig. 7 Proposed BNF grammar

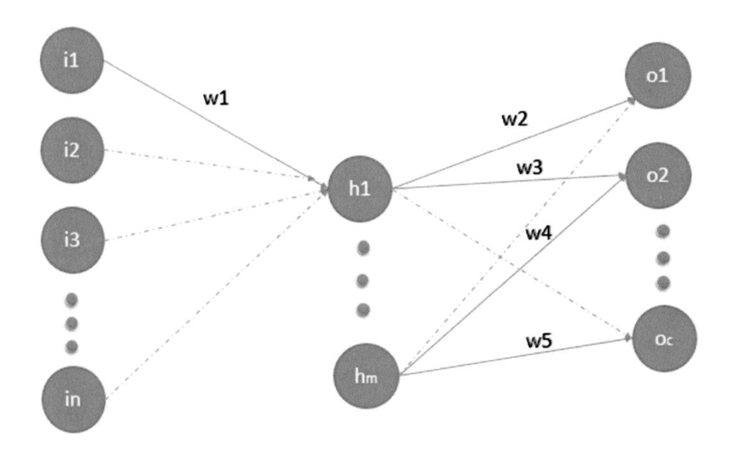

Fig. 8 General structure of an automatically designed neural network

and output layer; in this case, we have six neurons in the hidden layer. The second bracket represents the neurons in the output layer and their connections and weights with the hidden layer. Summary of the connections and weights in the expression are shown in the tables in Fig. 9.

The search engine is used to evolve solutions called genotypes (solutions are interpreted as coded by gray code) that will be treated by means of a BNF grammar designed for this research proposal and which through a mapping process (either depth first, breath first or position invariant) obtains the phenotypes, which represent the design of the ANN. Before carrying out the mapping process, the genotypes generated are transformed into binary coding and finally to integer coding. The integer coding is the genotype processed by the mapping process.

To measure the performance of an individual or solution, it was used a fitness function. In our case, the best solution is represented by the individual whose fitness function returns a minimum value. There are several fitness functions, which

Expression obtained

[i4,+5566.8/i0,-9390.0#o2,-679.02/o1,-67.459/o3,+8623.122_i2,+5.0099/i0, 9.99999# o2, +9.000 _i3,+2.4/i2,-3.89999/i0,-1.99#o2,-9.9/o3,+9.9944_i4,-285.0/i2,+77.8247/i0,+7.231#o2, 78113.54/ o1,+39.6_i2,-1.0/i0,-3.0#o3,+999.9_i3,-8.50/i0,+41.9#o3,-18.0/ o2,+9.99] [i0,-9.9_i0,-4.9_i0,-5.40]

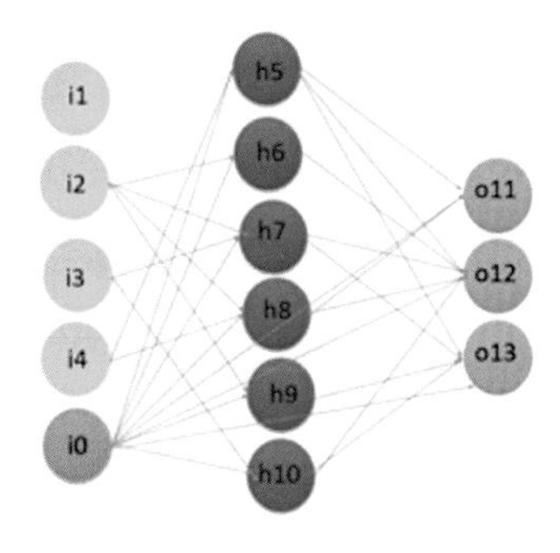

CONNECTIONS AND WEIGHTS						
neuron	h5	h6	h7	h8	h9	h10
i0	-9390	-9.99999	-1.99	7.231	-3	41.9
i1						
i2		5.0099	-3.89999	77.8247	-1	
i3			2.4			-8.5
i4	5566.8			-285		

CONNECTIONS AND WEIGHTS							
neuron	h5	h6	h7	h8	h9	h10	i0
o11	-67.459			39.6			-9.9
o12	-679.02	9	-9.9	-78113.54		9.99	-4.9
o13	8623.122		9.9944		999.9	-18	-5.4

Fig. 9 Expression and topology designed

evaluate the solutions with different metrics; in our case, we make use of the quadratic error. The procedure is repeated until it fulfills a stopping criterion based on the number of function calls. The desired result is to obtain the best network designed automatically for a particular problem.

4 Experiments and Results

This work focuses on the search for a generic framework for the design of PCNNs using a GE. The obtained PCNNs were tested with different classical benchmark datasets of pattern recognition problems from UCI Machine Learning Repository. The characteristics of the used datasets are shown in Table 1.

The parameters of GA used as search engine, were estimated experimentally. The parameter values are: *maximum number of function calls*: 1,000,000, *population size*: 100, *chromosome length*: 4000, *crossover rate*: random at a point, *mutation rate*: *5% of chromosome length, tournament size*: 5, *and elitism of 10%.*

Table 1 Characteristics of the used datasets

Datasets	Attributes	Classes	Instances
Iris-plant	4	3	150
Breast-cancer	10	2	683
Glass	10	6	214
Wine	13	3	178
Ionosphere	34	2	351
Balance scale	4	3	625

Table 2 Results of fitness values obtained with the designed PCNNs with the different mapping processes

		Min	Max	Average	Median	Standard deviation
Iris-plant	DF	2.73E−79	28.91	8.66	4.21	8.48
	BF	4.72E−61	13.03	12.79	13.03	10.58
	PI	3.61	33.33	17.38	16.67	7.24
Breast-cancer	DF	4	30.6	14.01	14.56	6.07
	BF	1.95	33.72	15.82	15.12	7.42
	PI	5.78	24.1	14.78	13.74	6.38
Glass	DF	31.11	78.38	54.44	53.54	9.8
	BF	24.7	58.52	49.52	51.21	10.08
	PI	40.7	67.62	53.69	51.87	4.83
Wine	DF	0.12	41.92	18.78	15.77	15.62
	BF	2.45	45.82	28.22	31.63	13.37
	PI	5.1	43.19	30.7	33.96	7.73
Ionosphere	DF	12	67.37	34.48	31.83	14.14
	BF	20.28	67.5	42.66	45.76	14.21
	PI	20.4	71.5	40.87	37.85	11.63
Balance scale	DF	54.82	177.65	121	125.93	36.2
	BF	37.85	172.83	113.25	112.48	40.37
	PI	41.45	166.77	106.93	111.19	38.51

To evaluate the performance of the methodology each dataset was randomly divided into two folders: one folder contains 50% of data for training and the other 50% of data for testing. The data of each folder were normalized obtaining from the training set the mean and the range by each dimension.

In order to provide statistical significance to our conclusions, 31 experiments were performed with each of the mapping processes and each of the databases. The statistical results of fitness function (quadratic error) are shown in Table 2.

The average and the maximum percentage of classification for the training set and testing set of the topologies obtained with this methodological proposal are shown in Table 3, where the highlighted values in bold represent the best results obtained for each dataset.

4.1 Statistical Significance of Results

The Friedman nonparametric statistical test was applied for giving statistical significance of results considering a comparison between the different mapping processes for the six classification problems. The aim of the test is to accept or reject a null hypothesis (Ho). The null and alternative hypotheses in this comparison are defined as follows.

Table 3 Classification performance of the designed PCNNs with the different mapping processes

	Training						Testing					
	DF		BF		PI		DF		BF		PI	
	Av. (%)	Max (%)	Av. (%)	Max (%)	Av. (%)	Max (%)	Av. (%)	Max (%)	Av. (%)	Max (%)	Av. (%)	Max (%)
Iris-plant	**83.83**	**100**	74.25	**100**	64.41	97	**80.58**	90	71.38	90	63.12	**94**
Breast-cancer	95.83	97	95.38	**98**	**95.87**	97	95.38	95	94.93	**96**	**95.64**	95
Glass	39.25	66	47.96	**70**	**48.22**	59	40.87	69	49.32	**70**	**49.87**	64
Wine	**75.77**	**100**	61.77	95	59.61	96	**67.7**	**85**	56.29	84	56.35	**93**
Ionosphere	**86.77**	**96**	79.29	92	83.06	93	**82.8**	**84**	75.51	**87**	79.45	85
Balance scale	60.62	84	60.93	**85**	**64.64**	84	59.51	84	61.38	83	**64.45**	83

Table 4 The Friedman test parameters

Columns (k)	Rows (n)	Significance level	Critical value
3	6	0.05	7

Table 5 Average ranking of the different mapping processes

Dataset	Mapping process		
	Depth-first	Breadth-first	Position-invariant
Iris-plant	1	2	3
Wine	1	2	3
Ionosphere	1	3	2
Glass	3	1	2
Balance scale	3	2	1
Breast cancer	2	3	1
Average	1.8333	2.1666	2.0

Table 6 Friedman test results

F	0.333
P value	0.846

Ho: *There are no differences in the fitness function medians of the different mapping processes*
Ha: *There are differences in the fitness function medians of the different mapping processes.*

Table 4 shows the Friedman test parameters.

In the Friedman test, the mapping processes are ranked according to the median obtained for each of them. The best mapping process gets rank one, the second best gets two, and so on. These ranks values are shown in Table 5.

The Friedman value (F) is computed from the medians of the Table 2. The results of the Friedman test are shown in Table 6.

From Table 6, the Friedman value is smaller than the critical value = 7 and the p-value is higher than the significance level, which indicate that not exist evidence to reject the null hypothesis.

5 Conclusions and Future Work

This paper presents a generic framework for the design of PCNNs. The partiality in the connections benefits in aspects of processing time in the design of the different topologies. The combination of ANNs and EAs such as GE allows through three elements: a CFG that includes the design components of ANNs, a search engine in this proposal called GA that drives the process of search and a mapping process, to obtain compact ANNs designed with good performances, as well as no need the knowledge of a human expert.

The Friedman test was applied to the obtained results in this work. From this, we conclude that not exist statistic evidence that indicate differences in the performance of the different mapping processes used in the PCNNs design.

In this proposal, the training process was omitted since the weight values were obtained by the GE process. We also want to underline the fact that in our network architectures some of the input neurons were not connected, which implicitly works as a feature selection, or dimensionality reduction mechanism. This can be seen as a desirable characteristic of our method.

As future work, the authors propose to present a methodology in which the grammar can define other neural network design parameters (selection of activation function, and number of hidden layers). Also the use of different metaheuristics to test the proposed grammar on the same problems and test with other datasets.

Acknowledgements We are grateful to the National Council for Science and Technology (CONACYT) of Mexico for the support provided by means of the Scholarship for Postgraduate Studies: **703036** (O. Quiroz) and Research Grant: **CÁTEDRAS-2598** (A. Rojas) as well as to the National Technology Institute of Mexico.

References

1. K. Soltanian, F.A. Tab, F.A. Zar, I. Tsoulos, Artificial neural networks generation using grammatical evolution, in *21st Iranian Conference on Electrical Engineering (ICEE)*, pp. 1–5 (2013)
2. C.M. Bishop, Neural networks for pattern recognition. J. Am. Stat. Assoc. **92**, 482 (1995)
3. G.P. Zhang, Neural networks for classification: a survey. IEEE Trans. Syst. Man Cybern. Part C Appl. Rev. **30**(4), 451–462 (2000)
4. B.A. Garro, H. Sossa, R.A. Vazquez, Design of artificial neural networks using a modified particle swarm optimization algorithm, in *2009 International Joint Conference on Neural Networks*, pp. 938–945 (2009)
5. F. Ahmadizar, K. Soltanian, F. Akhlaghiantab, I. Tsoulos, Engineering applications of artificial intelligence artificial neural network development by means of a novel combination of grammatical evolution and genetic algorithm. Eng. Appl. Artif. Intell. **39**, 1–13 (2015)
6. S. Kulluk, L. Ozbakir, A. Baykasoglu, Training neural networks with harmony search algorithms for classification problems. Eng. Appl. Artif. Intell. **25**(1), 11–19 (2012)
7. D. Elizondo, E. Fiesler, A survey of partially connected neural networks. Int. J. Neural Syst. **8**, 535–558 (1997)

8. E. Cantu-Paz, C. Kamath, An empirical comparison of combinations of evolutionary algorithms and neural networks for classification problems. IEEE Trans. Syst. Man Cybern. Part B Cybern. Publ. IEEE Syst. Man Cybern. Soc. **35**, 915–927 (2005)
9. D.J. Montana, L. Davis, Training feedforward neural networks using genetic algorithms, in *Proceedings of 11th International Joint Conference Artificial Intelligence, vol. 1*, vol. 89, pp. 762–767 (1989)
10. M. Gardner, S. Dorling, Artificial neural networks (the multilayer perceptron)—a review of applications in the atmospheric sciences. Atmos. Environ. **32**(14–15), 2627–2636 (1998)
11. D. Simon, *Evolutionary Algorithms Biologically-Inspired and Population-Based Approaches to Computer Intelligence* (Wiley, Hoboken, New Jersey, 2013)
12. J. Branke, Evolutionary algorithms for neural network design and training, in *Workshop on Genetic Algorithms and its Applications*, pp. 1–21 (1995)
13. S. Ding, H. Li, C. Su, J. Yu, Evolutionary artificial neural networks: a review. Artif. Intell. Rev. **39**(3) (2011)
14. X. Yao, Evolving artificial neural networks. Proc. IEEE **87**(9), 1423–1447 (1999)
15. L. Wang, Y. Zeng, T. Chen, Back propagation neural network with adaptive differential evolution algorithm for time series forecasting. Expert Syst. Appl. **42**(2), 855–863 (2015)
16. M. O'Neill, C. Ryan, *Grammatical Evolution: Evolutionary Automatic Programming in a Arbitrary Language*, vol. 4 (Springer, US, 2003)
17. M. O'Neill, C. Ryan, Grammatical evolution. IEEE Trans. Evol. Comput. **5**(4), 349–358 (2001)
18. C. Ryan, J.J. Collins, M. Neill, *Grammatical Evolution: Evolving Programs for an Arbitrary Language* (Springer, Berlin Heidelberg, 1998)
19. I. Tsoulos, D. Gavrilis, E. Glavas, Neural network construction and training using grammatical evolution. Neurocomputing **72**(1–3), 269–277 (2008)
20. D. Fagan, "Analysing the genotype-phenotype map in grammatical evolution," for the degree of Ph.D. at the School of Computer Science and Informatics College of Science (2013)
21. T. Bäck, *1996—Back—Evolutionary Algorithms in Theory And Practice.pdf* (Oxford University Press, Inc., 1996)
22. M. Mitchell, L.D. Davis, Handbook of genetic algorithms. Artif. Intell. **100**(1–2), 325–330 (1998)
23. J.P.M. De Sa, *Pattern Recognition: Concepts Methods and Applications* (Springer, 2001)
24. J.H. Holland, *Adaptation in Natural and Artificial Systems: An introductory Analysis with Applications to Biology, Control and Artificial Intelligence* (*MIT Press*, 1975), p. 183
25. J.D. Schaffer, L.J. Eshelman, Combinations of genetic algorithms and neural networks: a survey of the state of the art, in *International Workshop on Combinations of Genetic Algorithms Neural Networks, 1992, COGANN-92 June 6, 1992, Balt. Maryland/Cat. No. 92Th0435-8 E-b.*, pp. 1–37 (1992)
26. J. Arifovic, R. Gencay, Using genetic algorithms to select architecture of a feedforward artificial neural network. Phys. A Stat. Mech. Appl. **289**(3–4), 574–594 (2001)
27. P.E. Valencia, Optimización Mediante Algoritmos Genéticos, *Anales del Instituto de Ingenieros de Chile*, vol. 109, no. 2, pp. 83–92 (1997)
28. T. Bäck, D.B. Fogel, Z. Michalewicz, *Evolutionary Computation 1: Basic Algorithms and Operators,* 1st edn. (CRC Press, 2000)
29. X. Yao, Evolutionary artificial neural networks. Int. J. Neural Syst. **4**, 203–222 (1993)
30. X. Yaot, A review of evolutionary artificial neural networks. Common. Sci. Ind. Res. Organ. **8**, 539–567 (1993)
31. B.A. Garro, H. Sossa, R.A. Vazquez, Design of artificial neural networks using differential evolution algorithm, in *Proceedings of 17th International Conference Neural Information Processing Models and Applications*, vol. Part II, pp. 201–208 (2010)
32. B.A. Garro, R.A. Vázquez, Swarm optimization algorithms. Comput. Intell. Neurosci. **2015**, 20 (2015)
33. D. Whitley, An overview of evolutionary algorithms: practical issues and common pitfalls. Inf. Softw. Technol. **43**, 817–831 (2001)

Part III
Metaheuristics: Theory and Applications

Bio-inspired Metaheuristics for Hyper-parameter Tuning of Support Vector Machine Classifiers

Adán Godínez-Bautista, Luis Carlos Padierna,
Alfonso Rojas-Domínguez, Héctor Puga and Martín Carpio

Abstract Support Vector Machines (SVMs) are machine learning models with many diverse applications. The performance of these models depends on a set of assigned hyper-parameters. The task of hyper-parameter tuning has been performed by metaheuristics methods and recent studies have shown that the effectiveness of these methods is statistically equivalent. In this work we compare four bio-inspired metaheuristics (Bat Algorithm, Firefly Algorithm, Particle Swarm Optimization Algorithm and Social Emotional Optimization Algorithm) to test the hypothesis that the efficiency among these differs while the effectiveness remains. Experimental results on several classification problems indicate that there exist bio-inspired algorithms with higher efficiency, in terms of the required number of SVM evaluations to find optimal hyper-parameters. Based on these results the Bat Algorithm is recommended for SVM hyper-parameter tuning.

Keywords Bio-inspired metaheuristics · Hyper-parameter optimization
Support vector machines · Pattern classification

A. Godínez-Bautista · L. C. Padierna · A. Rojas-Domínguez (✉) · H. Puga · M. Carpio
Tecnológico Nacional de México - Instituto Tecnológico de León,
Av. Tecnológico s/n, Fracc. Julián de Obregón, 37290 León, Gto, Mexico
e-mail: alfonso.rojas@gmail.com

A. Godínez-Bautista
e-mail: gb.adan88@gmail.com

L. C. Padierna
e-mail: luiscarlos.padierna@itleon.edu.mx

H. Puga
e-mail: pugahector@yahoo.com

M. Carpio
e-mail: juanmartin.carpio@itleon.edu.mx

O. Castillo et al. (eds.), *Fuzzy Logic Augmentation of Neural and Optimization Algorithms: Theoretical Aspects and Real Applications*, Studies in Computational Intelligence 749, https://doi.org/10.1007/978-3-319-71008-2_10

1 Introduction

Since their creation by Vapnik in 1995 [1], SVMs have been widely studied and applied to solve classification and regression problems [2, 3]. For a classification problem, an SVM constructs a model that separates two sets of labeled data into defined classes, to predict whether a new sample belongs to one class or another. Initially SVMs were designed as a linear classifier, but because most problems are not linearly separable, a penalty factor C and a kernel function were added [4].

The factor C allows the creation of a soft margin that penalizes errors in the classification. Obtaining a good performance in classification depends mostly on the values used for factor C and kernel function parameters, called hyper-parameters. Inappropriate values of hyper-parameters lead to poor results in classification, therefore, different metaheuristic algorithms have been used to find the most appropriate values [5, 6]. Some of these algorithms are based on the behavior of biological organisms, such as Bat Algorithm (BA) [7], Firefly Algorithm (FA) [8], Particle Swarm Optimization (PSO) [9] and Social Emotional Optimization Algorithm (SEOA) [10].

The bio-inspired algorithms are all capable of finding hyper-parameter values to produce near optimal performance of SVMs. However, in terms of efficiency, some algorithms can find these values with fewer SVM evaluations, i.e., function calls. This leads to the hypothesis that the bio-inspired algorithms are statistically equivalent in terms of effectiveness, but in terms of efficiency they differ. In this paper we present an analysis of bio-inspired algorithms to verify this hypothesis, measuring the effectiveness by the classification Accuracy and Percentage of Support Vectors (PSV); while the difference in efficiency is measured in terms of Percentage of Function Calls (PFC).

The rest of this paper is organized as follows: Sect. 2 provides a brief explanation of SVMs and bio-inspired algorithms. Section 3 describes the methodology and datasets used in the experiments. The results and their discussion are presented in Sect. 4. Finally in Sect. 5 the conclusions and the future work are provided.

2 Theoretical Background

In this section we describe basic concepts of SVMs for classification of two-class problems (a more detailed exposition of these concepts can be found in [11]) as well as description and pseudo-code of bio-inspired algorithms.

2.1 *Support Vector Machines*

Given a classification problem with n features and l training points represented by $D = \{(\mathbf{x}_1, y_1), \ldots, (\mathbf{x}_l, y_l)\}$, where $\mathbf{x} \in \mathbb{R}^n$ and $y \in \{1, -1\}$ is the label of each class, the aim of an SVM is to find a hyper-plane defined as $\mathbf{w}^{\mathrm{T}}\mathbf{x}+b=0$, that separates the training points into two regions, according to their respective classes.

For a linearly separable problem, an optimal hyper-plane is obtained solving the following quadratic optimization problem:

$$\begin{gathered}\min \tfrac{1}{2}\|\mathbf{w}\|^2 \\ s.t.\ y_i(\mathbf{w}^{\mathrm{T}}\mathbf{x}+b)\geq 1, \quad i=1,\ldots,l\end{gathered} \tag{1}$$

When the problem is not linearly separable there are points that violate the constraint of the problem (1). Slack variables ξ_i are introduced to relax the constraint and to achieve separation of the points, while allowing some misclassifications. To avoid making slack variables too large, a penalty coefficient C is added, resulting in the following equation:

$$\begin{gathered}\min \tfrac{1}{2}\|\mathbf{w}\|^2 + C\sum_{i=1}^{l}\xi_i \\ s.t.\ y_i(\mathbf{w}^{\mathrm{T}}\mathbf{x}+b)\geq 1-\xi_i,\ \xi_i\geq 0, \quad i=1,\ldots,l,\end{gathered} \tag{2}$$

Solving the problem (2) is equivalent to solving its dual problem (3), where Lagrange multipliers α_i are introduced:

$$\begin{gathered}\mathrm{L}(\mathbf{w},b,\boldsymbol{\xi},\boldsymbol{\alpha}) = \tfrac{1}{2}\|\mathbf{w}\|^2 + C\sum_{i=1}^{l}\xi_i - \sum_{i=1}^{l}\alpha_i(y_i(\mathbf{w}^{\mathrm{T}}\mathbf{x}_i+b)-1+\xi_i), \\ s.t.\ \alpha_i\geq 0\end{gathered} \tag{3}$$

Introducing the Karush-Kuhn-Tucker conditions, the solution to the problem (3) can be found by solving the following dual problem:

$$\begin{gathered}\max L(\boldsymbol{\alpha}) = -\tfrac{1}{2}\sum_{i=1}^{l}\sum_{j=1}^{l} y_i y_j \alpha_i \alpha_j K(\mathbf{x},\mathbf{x}') + \sum_{j=1}^{l}\alpha_j \\ s.t.\ \sum_{i=1}^{l} y_i\alpha_i = 0, \\ C-\alpha_i = 0,\ \alpha_i\geq 0, \quad i=1,\ldots,l\end{gathered} \tag{4}$$

Another way to deal with problems that are not linearly separable is to use a kernel function, which maps the input data to a higher dimensional feature space where linear classification is feasible [12]. A function $K(\mathbf{x},\mathbf{x}')$ defined on $\mathbb{R}^n\times\mathbb{R}^n$ is a kernel function if there exists a map Φ to the Hilbert space, $\Phi:\mathbb{R}^n\rightarrow\mathbb{H}$ such that $\mathrm{K}(\mathbf{x},\mathbf{x}')=(\Phi(\mathbf{x})\cdot\Phi(\mathbf{x}'))$.

To obtain the best solution of a problem, different kernel functions have been proposed by researchers, such as the Linear kernel, $K(\mathbf{x},\mathbf{x}')=\mathbf{x}^{\mathrm{T}}\mathbf{x}'$, the Radial Basis Function (RBF) kernel, $K(\mathbf{x},\mathbf{x}')=e^{-\gamma\|\mathbf{x}-\mathbf{x}'\|^2}$ and the Polynomial kernel, $K(\mathbf{x},\mathbf{x}')=(a(\mathbf{x}^{\mathrm{T}}\mathbf{x}')+b)^d$. Here C,γ,a,b and d are kernel parameters that affect the classification accuracy rate. Several techniques have been used to find the best values for these parameters. The next section describes some of these techniques.

2.2 Bio-inspired Algorithms

Bio-inspired algorithms are stochastic optimization methods based on the behavior of biological entities such as birds, fish, humans, etc., in search for food, defense mechanisms, attraction between species or other social behaviors. In this work four bio-inspired algorithms: BA, FA, PSO and SEOA are studied. For all these algorithms, maximization of the fitness function is considered.

2.2.1 Bat Algorithm (BA)

BA is based on echolocation. This behavior allows bats to differentiate between obstacles and preys. The idealization of this behavior can be summarized as follows: initially bats fly randomly from position $\mathbf{x}_i$ with velocity $\mathbf{v}_i$, a fixed frequency f and with a loudness A_0. In search for their prey, bats adjust their frequency, loudness and pulse rate r, depending on the distance to their objective. A frequency adjustment is performed to control the dynamic behavior of a swarm of bats. The search for the best solution continues until a stopping criterion is reached. The general process of BA is explained in Algorithm 1 [7], where f_{min} and f_{max} are the minimum and maximum frequencies. $\beta \in [0, 1]$ is a random number and $\mathbf{x}^*$ is the best solution in generation t.

Algorithm 1: BA

1. Objective function $f(\mathbf{x})$, $\mathbf{x} = (x_1, \ldots, x_n)^{\mathrm{T}}$
2. Initialize the bat population $\mathbf{x}_i$ and $\mathbf{v}_i$,$(i = 1,2, \ldots, l)$.
3. Define pulse frequency f_i at $\mathbf{x}_i$
4. Initialize pulse rates r_i and the loudness A_i
5. WHILE (t < Max of generations)
6. Generate new solutions by adjusting frequency, and updating velocities and locations/solutions:
7. $f_i = f_{min} + (f_{max} - f_{min})\beta$
8. $\mathbf{v}_i^t = \mathbf{v}_i^{t-1} + (\mathbf{x}^* - \mathbf{x}_i^{t-1})f_i$
9. $\mathbf{x}_i^t = \mathbf{x}_i^{t-1} + \mathbf{v}_i^t$
10. If$(rand > r_i)$
11. Randomly generate a local solution, $\mathbf{x}_{new}$ around the best solution (exploitation)
12. End If
13. Evaluate the new solution (f_{new})
14. If$(rand < A_i \text{AND } f(\mathbf{x}_i) < f(\mathbf{x}_{new}))$
15. Accept the new solution
16. Increase r_i and reduce A_i
17. End If
18. Rank the bats and find the current best, $\mathbf{x}^*$
19. End WHILE

2.2.2 Firefly Algorithm (FA)

FA is based on the flashing light generated by fireflies, whose fundamental function is to attract their mating partners and potential prey. They also use it as a protection mechanism to keep predators away. The intensity of light I between one firefly and another is inversely proportional to the distance s between them. In addition I can be affected by the absorption of air, ω. For FA the following three rules are applied:

1. All fireflies are unisex, which implies that they can attract each other regardless of sex.
2. If at any moment there's no firefly brighter than the others, then they will move randomly.
3. The brightness of a firefly is determined by an objective function.

Based on this behavior, the FA can be summarized in Algorithm 2 [8], where $\beta_0 e^{-\omega s_{ij}^2}$ models the attraction between fireflies and β_0 is the attractiveness when the distance $s = 0$ (in this paper we set $\beta_0 = 0.3$); $\eta\epsilon_i$ performs a randomization with $\eta \in [0, 1]$ and $\in_i$ is a uniformly distributed vector (in this work $\eta = 0$ and $\omega = 0$ were used in order to improve convergence of the FA algorithm).

Algorithm 2: FA

1. Objective function $f(\mathbf{x}), \mathbf{x} = (x_1, \ldots, x_n)^{\mathrm{T}}$
2. Initialize population of fireflies $\mathbf{x}_i (i = 1,2, \ldots, l)$
3. Light intensity I_i at $\mathbf{x}_i$ is determined by $f(\mathbf{x})$
4. Define light absorption coefficient ω
5. WHILE (t < Max of generations)
6. For $i = 1 : l$(for all fireflies)
7. Move firefly i towards $\mathbf{x}^*$ by means of
8. $\mathbf{x}_i = \mathbf{x}_i + \beta_0 e^{-\omega s_{i*}^2}(\mathbf{x}^* - \mathbf{x}_i) + \eta\epsilon_i$
9. End For
10. Evaluate new solutions and update light intensity
11. Rank the fireflies and find the current best, $\mathbf{x}^*$
12. End WHILE

2.2.3 Social Emotional Optimization Algorithm (SEOA)

SEOA is based on the decision making of humans in society. The basic idea is that humans perform certain activities according to their emotions, and these activities are evaluated as more- or less-successful. Each individual represents a virtual person and his behavior is evaluated according to his emotional index in each generation. A state value is assigned by society to each individual, which determines whether his behavior is correct or not. If the behavior of a certain individual is considered correct, his emotional index increases, which will guide his next behavior for a better social evaluation; otherwise, his emotional index decreases.

The emotional index (*EI*) of all individuals is initialized as 1 (the maximum emotional index). The SEOA is summarized in Algorithm 3 [10]; where k_i are parameters to control the emotion changing rate, ς_i are random numbers with uniform distribution in [0,1] and Th_1 and Th_2 are thresholds that restrict the different behavior. The values used in this work for these parameters are: $Th_1 = 0.8, Th_2 = 0.95, k_1 = k_2 = k_3 = 0.35$.

Algorithm 3: SEOA

1. Objective function $f(\mathbf{x}), \mathbf{x} = (x_1, \ldots, x_n)^{\mathrm{T}}$
2. Initialize population of individuals $\mathbf{x}_i$,$\mathbf{x}_i^*$ and EI_iat 1, $(i = 1,2, \ldots, l)$
3. WHILE (t < Max of generations)
4. For $i = 1 : l$ (all individuals)
5. Calculate fitness value
6. If $(f(\mathbf{x}_i) > f(\mathbf{x}_i^*))$
7. Update $\mathbf{x}_i^*$ and Increase EI_i to 1
8. Else
9. Decrease EI_i
10. End If
11. End For
12. Rank the individuals and find the current best, $\mathbf{x}^*$
13. For $i = 1 : l$
14. If ($t = 1$),update individual $\mathbf{x}_i$ according to:
$$\mathbf{x}_i^{t+1} = \mathbf{x}_i^t - k_1\varsigma_1 \sum_{j=1}^{l} (\mathbf{x}_j^1 - \mathbf{x}_i^1)$$
15. Else
16. If $(EI_i < Th_1)$, update according to:
17. $\mathbf{x}_i^{t+1} = \mathbf{x}_i^t + k_3\varsigma_3(\mathbf{x}_i^* - \mathbf{x}_i^t) + k_2\varsigma_2(\mathbf{x}^* - \mathbf{x}_i^t)$
18. Else If $(Th_1 \leq EI_i \text{AND} EI_i < Th_2)$, update according to:
19. $$\mathbf{x}_i^{t+1} = \mathbf{x}_i^t + k_3\varsigma_3(\mathbf{x}_i^* - \mathbf{x}_i^t) + k_2\varsigma_2(\mathbf{x}^* - \mathbf{x}_i^t) - k_1\varsigma_1 \sum_{j=1}^{l} (\mathbf{x}_j^1 - \mathbf{x}_i^1)$$
20. Else
21. $$\mathbf{x}_i^{t+1} = \mathbf{x}_i^t + k_3\varsigma_3(\mathbf{x}_i^* - \mathbf{x}_i^t) - k_1\varsigma_1 \sum_{j=1}^{l} (\mathbf{x}_j^1 - \mathbf{x}_i^1)$$
22. End If
23. End If
24. End For
25. End WHILE

2.2.4 Particle Swarm Optimization (PSO)

PSO is based on the behavior of swarms in nature, such as bird flocking or fish schooling, where a group of organisms move collectively (in search of food). Each organism is called a particle; it has a position $\mathbf{x}_i$ and moves with a velocity $\mathbf{v}_i$ through the search space. The velocity is adjusted in each generation and the position is evaluated by an objective function. Each particle moves towards the best location that has been found in its history (local best), and at the same time all particles are attracted to the best location that has been found during the search (global best). The implementation of PSO is presented in Algorithm 4 [13]; where $Rand(0, \phi)$ generates a uniformly distributed vector in $(0, \phi)$, the operator $\times$ indicates a component-by-component vector multiplication and τ is termed the inertia weight. In this work $\phi_1 = \phi_2 = 2.05$ and $\tau = 0.7$ were used.

Algorithm 4: PSO

1. Objective function $f(\mathbf{x}), \mathbf{x} = (x_1, \ldots, x_n)^{\mathrm{T}}$
2. Initialize population of particles $\mathbf{x}_i, \mathbf{x}_i^*$ and $\mathbf{v}_i$, $(i = 1,2, \ldots, l)$
3. WHILE (t < Max of generations)
4. For $i = 1 : l$ (all particles)
5. Calculate fitness value
6. If $(f(\mathbf{x}_i) > f(\mathbf{x}_i^*))$
7. Update $\mathbf{x}_i^*$
8. End If
9. End For
10. Rank the particles and find the current best, $\mathbf{x}^*$
11. For $i = 1 : l$
12. Calculate new velocity and update particle position:
13. $\mathbf{v}_i^t = \tau \mathbf{v}_i^{t-1} + Rand(0, \phi_1) \times (\mathbf{x}_i^* - \mathbf{x}_i) + Rand(0, \phi_2) \times (\mathbf{x}^* - \mathbf{x}_i)$
14. $\mathbf{x}_i^t = \mathbf{x}_i^{t-1} + \mathbf{v}_i$
15. End For
16. End WHILE

3 Experimental Methodology

This section describes the classification problems and configuration setting used in this work. In order to obtain a general conclusion, SVM classifiers with Linear, RBF and Polynomial kernels were used. The experiments were performed on nine datasets (two-class problems) from the University of California at Irvine (UCI) Machine Learning Repository. General information about these datasets is provided in Table 1.

Table 1 Description of classification problems used in the experiments

No.	Dataset	No. of instances	No. of features
1	Breast	683	10
2	Chronic	400	25
3	Diabetic	1151	20
4	Fertility	100	10
5	Ionosphere	351	34
6	Liver	345	6
7	Monks-2	169	6
8	Parkinson	1040	27
9	Sonar	208	60

Table 2 Ranges used for hyper parameters in experimentation

Hyper-parameter	Min	Max
Penalty factor (C)	0.5	32
Gamma (γ)	2^{-10}	4
Scale factor (a)	2^{-10}	4
Offset term (b)	1	1
Polynomial degree (d)	1	6

For each dataset in Table 1, the following process was performed: first the data was divided into test data and training data under a 10 fold cross-validation scheme. For each fold a population of 50 random individuals was generated. Each individual was formed by the hyper-parameters: C (used by all kernels), γ (used by the RBF kernel), and a, b and d (used by the polynomial kernel) depending on the kernel employed. The search ranges for each hyper-parameter are presented in Table 2.

Next, the hyper-parameter tuning of a SVM classifier was carried out with a specific metaheuristic and starting from said population. Classification accuracy on the corresponding test data was used as the objective function and the optimization process was terminated according to two criteria: that a minimum standard deviation of 0.07 within a subset of the best 25 individuals was found, or that a maximum of 10 generations were completed.

The process was repeated for a different metaheuristic but using the same initial population until all four metaheuristics considered had been evaluated. Then this evaluation was performed using the next data fold and a different initial population, until all the 10 folds had been employed. A single experimental trial was completed when all datasets had been used. The process of one trial is summarized in Fig. 1. A total of 35 such experimental trials were performed in order to provide statistical support to our conclusions.

Notice that, since the degree of a polynomial is a discrete value, a roulette method was used to update this hyper-parameter. That is, the degree of each individual for the next generation was assigned probabilistically, based on probabilities computed from the best individuals of the current generation.

To measure the efficiency of the different metaheuristics three quantities were employed: (1) the classification accuracy of the best SVM classifier produced by

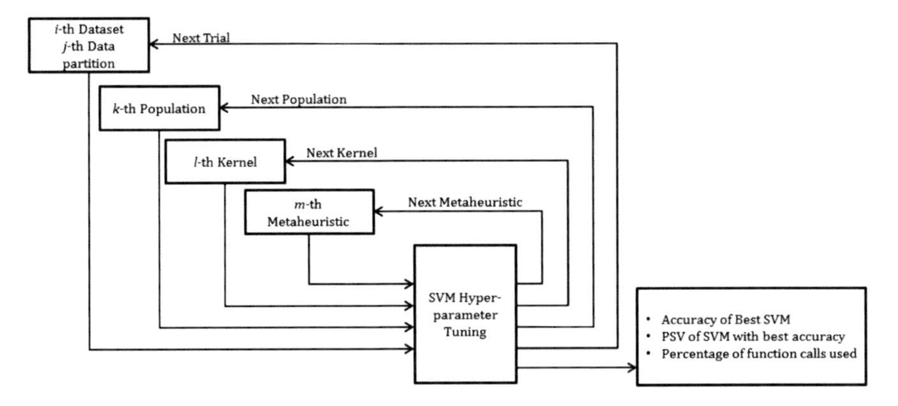

Fig. 1 Diagram of experiments for parameters determination for SVM

each algorithm, per data fold; (2) the PSV required by said SVM (the smaller the PSV, the better the classifier); and (3) the number of function calls that the metaheuristic required to optimize the hyper parameters of the SVM. Thus, for each dataset, kernel and metaheuristic, the average accuracy, average PSV and a PFC over the 35 trials is reported.

4 Results

The results obtained from the experimental methodology are summarized in this section. In Tables 3, 5 and 8, the best result per performance index and dataset is shown in bold unless the difference between the top result and the rest is not significant; in Tables 4, 6 and 7, the values in bold correspond to the best average rank per performance index and to p-values that indicate statistically significant differences (at significance level of 0.05). Table 3 shows the results on the considered datasets using SVM classifiers with linear kernel. It shows that the four metaheuristics obtained very similar average accuracy and PSV on most datasets (breast, chronic, diabetic, fertility, liver, monks-2 and Parkinson). The FA obtains slightly lower accuracy on the ionosphere and sonar datasets, but with a lower (i.e. better) PSV. Regarding the PFC, a difference is observed in favor of BA, which requires significantly lower PFC than the other metaheuristics on 8 of the 9 datasets. The exception is the dataset Monks-2, where the four metaheuristics converged quickly, satisfying the stopping criterion of a minimum standard deviation in the first generation. In contrast, the SEOA needed more function calls on 5 of the 9 datasets (chronic, diabetic, fertility, liver and sonar), and particularly on liver and diabetic.

In order to evaluate whether there are statistically significant differences between the metaheuristics in terms of accuracy, PSV and function calls, non-parametric

Table 3 SVM performance results with Linear kernel

Dataset	Accuracy				PSV				PFC			
	BA	FA	PSO	SEOA	BA	FA	PSO	SEOA	BA	FA	PSO	SEOA
Breast	96.90	96.88	96.90	96.90	7.75	7.71	7.75	7.77	**34.29**	36.57	41.43	40.57
Chronic	98.35	98.33	98.34	98.36	6.55	6.40	6.57	6.70	**54.86**	64.86	70.00	79.71
Diabetic	73.83	73.84	73.84	73.83	62.13	62.19	62.19	62.18	**34.00**	63.43	59.71	96.00
Fertility	88.14	88.14	88.14	88.14	49.75	49.75	49.75	49.75	**16.00**	17.43	16.29	20.86
Ionosphere	89.00	88.95	89.04	89.01	25.02	24.06	25.02	24.97	**74.57**	90.57	95.43	89.43
Liver	69.69	69.68	69.70	69.71	74.79	74.74	74.79	74.84	**44.00**	63.43	62.00	90.29
Monks-2	62.15	62.15	62.15	62.15	91.16	91.16	91.16	91.16	**10.00**	**10.00**	**10.00**	**10.00**
Parkinson	65.13	65.14	65.14	65.11	77.38	77.35	77.39	77.40	**51.71**	69.43	68.57	62.00
Sonar	78.46	77.69	78.83	78.83	48.59	46.25	49.38	49.73	**93.43**	98.57	98.00	100.00

Table 4 Average ranks and p-values of statistical tests on the linear kernel

Metaheuristic	Friedman			Aligned Friedman			Quade		
	Acc.	PSV	PFC	Acc.	PSV	PFC	Acc.	PSV	PFC
BA	2.722	2.111	**1.166**	20.61	16.22	**6.388**	2.777	2.100	**1.033**
FA	3.111	**1.666**	2.833	25.33	**8.888**	19.27	3.455	**1.433**	2.855
PSO	**2.000**	3.000	2.722	**13.38**	22.77	21.27	1.900	3.144	2.722
SEOA	2.166	3.222	3.277	14.66	26.11	27.05	**1.866**	3.322	3.388
p-value	0.237	**0.033**	**0.003**	0.065	0.081	0.072	**0.002**	**3.1E−4**	**1.1E−5**

statistical tests were performed on the results presented above, including the Friedman, Aligned Friedman and Quade tests with a significance value of 5% (0.05). Table 4 shows the p-values and average ranks obtained from evaluating the results corresponding to SVMs with linear kernel. After post hoc tests, significant differences were found between:

- SEOA and FA on accuracy, according to the Quade test.
- FA versus SEOA and PSO on PSV, according to Friedman and Quade tests.
- BA and the rest of the methods on PFC, under the Friedman and Quade tests.

In general, there is no statistically significant difference between the BA and the rest of the metaheuristics with respect to the accuracy and PSV indexes. However, BA is significantly better than the other metaheuristics in terms of function calls required. In other words, BA possesses a superior convergence capability.

Table 5 shows the results corresponding to experiments of tuning SVM classifiers with RBF kernel. In this case, all the metaheuristics considered obtained very similar accuracy on most datasets, particularly on chronic, diabetic, fertility and monks-2 datasets. FA produced slightly lower accuracy on breast, sonar and liver datasets. SEOA produced the lowest accuracy on the Parkinson dataset, but only slightly below the rest of the algorithms. With respect to the PSV, the results vary between the algorithms; however, FA stands out by producing the highest (worst) PSV on the chronic, ionosphere and sonar datasets.

Considering the PFC, BA offers a remarkable advantage over the other algorithms on all datasets with the exception of the diabetic and liver. On these, all the metaheuristics consumed all the function calls allocated, corresponding to a maximum of 10 generations. A point to be highlighted is that, with the exception of the fertility dataset, the four metaheuristics used all the allocated function calls in the majority of the experimental trials. Table 6 shows the p-values and average ranks obtained from evaluating the results corresponding to SVMs with RBF kernel. After post hoc tests, significant differences were found between:

Table 5 SVM performance results with RBF kernel

Dataset	Accuracy				PSV				PFC			
	BA	FA	PSO	SEOA	BA	FA	PSO	SEOA	BA	FA	PSO	SEOA
Breast	97.18	96.98	97.23	97.19	11.45	13.20	11.93	12.18	**94.57**	96.29	97.43	97.43
Chronic	99.61	99.41	99.61	99.58	12.14	19.27	11.54	10.70	**97.43**	100.00	100.00	100.00
Diabetic	74.49	74.09	74.39	74.34	64.31	64.35	64.60	65.36	**100.00**	**100.00**	**100.00**	**100.00**
Fertility	89.19	89.19	89.19	89.20	94.25	94.25	93.49	92.56	**33.14**	38.86	65.71	46.86
Ionosphere	95.73	95.54	95.81	95.60	38.34	49.29	39.58	43.06	**98.86**	100.00	100.00	99.71
Liver	74.03	73.89	74.01	73.94	70.78	70.11	71.79	69.98	**100.00**	**100.00**	**100.00**	**100.00**
Monks-2	82.27	82.23	82.22	82.00	74.26	76.66	74.17	73.97	**94.29**	100.00	100.00	98.00
Parkinson	72.10	72.08	72.11	71.95	72.27	71.90	73.57	72.50	**99.71**	100.00	100.00	100.00
Sonar	90.28	89.70	90.33	90.01	69.13	78.26	68.57	66.80	**99.71**	100.00	100.00	100.00

Table 6 Average ranks and p-values of statistical tests on the RBF kernel

Metaheuristic	Friedman			Aligned Friedman			Quade		
	Acc.	PSV	PFC	Acc.	PSV	PFC	Acc.	PSV	PFC
BA	1.833	2.277	**1.333**	10.72	**14.61**	**7.000**	1.733	2.411	**1.100**
FA	3.555	3.166	2.777	28.11	26.94	19.44	3.644	3.566	2.755
PSO	**1.722**	2.555	3.166	**10.61**	17.11	25.83	**1.577**	2.333	3.388
SEOA	2.888	**2.000**	2.722	24.55	15.33	21.72	3.044	**1.688**	2.755
p-value	**0.005**	1.405	**0.015**	0.074	0.072	0.072	**1.6E−5**	**0.002**	**3.7E−5**

- PSO versus FA and SEOA on accuracy, under the Friedman and Quade tests.
- SEOA against FA on PSV, according to the Quade test.

There is no significant difference between the BA and the rest of the metaheuristics with respect to accuracy or PSV. However, BA is significantly better than the other algorithms in terms of function calls.

The results corresponding to SVM classifiers with polynomial kernel are shown in Table 7. Once again, the average accuracy obtained by the four metaheuristics on the breast, chronic, diabetic, fertility, Parkinson and sonar datasets are all very similar. This time FA achieves a slightly higher performance on the liver dataset; however, on the ionosphere and monks-2 datasets, produces lower accuracy.

The results in Table 7 regarding PSV index vary between algorithms, but it can be observed that this time the FA produced better results (lower PSV) than the rest on the ionosphere and monks-2datasets. In terms of function calls, FA obtains better results on breast, chronic, diabetic, liver, monks-2 and Parkinson. BA produces better performance on ionosphere and liver; and finally, PSO and SEOA are better on sonar and fertility, respectively.

Finally, Table 8 shows results of non-parametric tests for the SVMs with polynomial kernel. The post hoc tests indicate that significant differences were found between:

- No significant differences on accuracy.
- FA against SEOA and PSO on PSV with the Friedman and Quade tests.
- FA and SEOA on Function Calls with the Quade test.

In general:

- FA reaches better PSV and Function Calls indexes with a degradation in accuracy. The BA algorithm is slightly better in accuracy but stands in the second place on PSV and Function Calls.
- Although the BA and FA metaheuristics obtained the two first places, for the polynomial kernel a trade-off between efficiency and effectiveness should be considered for choosing the appropriate algorithm.

Table 7 SVM performance results with polynomial kernel

Dataset	Accuracy				PSV				PFC			
	BA	FA	PSO	SEOA	BA	FA	PSO	SEOA	BA	FA	PSO	SEOA
Breast	97.17	97.01	97.13	97.15	11.61	10.67	11.66	11.04	66.57	**43.43**	56.57	69.43
Chronic	99.49	99.46	99.48	99.49	9.12	9.36	9.27	9.25	25.71	**25.14**	25.71	39.71
Diabetic	76.73	76.79	76.78	76.75	53.72	53.62	53.65	53.71	94.86	**85.14**	100.00	96.57
Fertility	88.15	88.14	88.14	88.14	47.55	48.02	47.97	47.54	41.43	38.29	37.71	**22.86**
Ionosphere	92.42	91.38	92.32	92.24	29.00	23.54	29.62	29.40	**98.00**	100.00	100.00	100.00
Liver	72.94	73.02	72.99	72.97	70.02	69.69	70.05	70.50	**96.00**	**96.00**	100.00	99.71
Monks-2	71.78	71.52	72.07	72.04	54.91	52.62	55.82	55.79	86.29	**77.43**	90.29	95.14
Parkinson	69.40	69.45	69.40	69.37	63.96	62.74	64.15	64.82	100.00	**99.43**	100.00	100.00
Sonar	87.62	87.62	87.59	87.72	57.20	57.02	57.33	57.41	33.43	31.14	**30.57**	54.57

Table 8 Average ranks and p-values of statistical tests on the polynomial kernel

Metaheuristic	Friedman			Aligned Friedman			Quade		
	Acc.	PSV	PFC	Acc.	PSV	PFC	Acc.	PSV	PFC
BA	**2.277**	2.222	2.444	20.50	18.55	18.00	**2.200**	2.133	2.677
FA	2.777	**1.666**	**1.611**	23.00	**09.11**	**11.16**	3.044	**1.400**	**1.611**
PSO	2.444	3.222	2.722	**15.11**	24.00	19.38	2.311	3.488	2.422
SEOA	2.500	2.888	3.222	15.38	22.33	25.44	2.444	2.977	3.288
p-value	0.213	**0.050**	0.061	0.068	0.075	0.067	0.100	**1.6E-4**	**0.007**

5 Conclusions and Future Work

The main conclusion derived from the statistical analysis carried out on the three kernels is that the BA metaheuristic is equally effective and more efficient that the FA, PSO and SEOA methods. This result is more evident for the RBF kernel, which is considered the benchmark kernel for SVM implementations. On the other hand, the BA algorithm reduces to the PSO when a particular configuration of its parameters (loudness and pulse rate) are given. Further, it requires less parameters than the SEOA method. Due to these reasons, we propose the use of the BA algorithm for SVM hyper-parameter tuning and recommend adding PFC as a basic measure of efficiency.

This conclusion allows us to suggest as future work a comparison between BA and other metaheuristic methods such as Estimation of Distribution Algorithms in terms of their efficiency and effectiveness, as well as an analysis of their algorithmic complexity.

Acknowledgements This work was partially supported by the National Council of Science and Technology (CONACYT) of Mexico [grant numbers: 375524 (Luis C. Padierna), CATEDRAS-2598 (A. Rojas), 416761 (Adán Godínez)].

References

1. V.N. Vapnik, *Statistical Learning Theory* (New York, 1998)
2. M. Heydari, M. Teimouri, Z. Heshmati, S.M. Alavinia, Comparison of various classification algorithms in the diagnosis of type 2 diabetes in Iran. Int. J. Diab. Dev. Countries (2015)
3. R. Langone, C. Alzate, B. De Ketelaere, J. Vlasselaer, W. Meert, J.A.K. Suykens, Engineering applications of artificial intelligence LS-SVM based spectral clustering and regression for predicting maintenance of industrial machines. Eng. Appl. Artif. Intell. **37**, 268–278 (2015)
4. V.N. Vapnik, An overview of statistical learning theory. IEEE Trans. Neural Netw. **10**(5), 988–999 (1999)
5. L.C. Padierna, A. Rojas, Hyper-parameter tuning for support vector machines by estimation of distribution algorithms, pp. 787–800 (2017)
6. C.L. Huang, C.J. Wang, A GA-based feature selection and parameters optimization for support vector machines. Expert Syst. Appl. **31**(2), 231–240 (2006)

7. X. Yang, A new metaheuristic bat-inspired algorithm, pp. 1–10 (2010)
8. X. Yang, L. Press, *Nature-Inspired Metaheuristic Algorithms Second Edition*
9. J. Kennedy, R. Eberhart, Particle swarm optimization, pp. 1942–1948 (1995)
10. Z. Cui, Y. Xu, J. Zeng, Social emotional optimization algorithm with random emotional selection strategy. Theory New Appl. Swarm Intell (2012)
11. C.Z. Naiyang Deng, Y. Tian, *Support Vector Machines* (CRC Press, 2013)
12. J. Shawe-Taylor, N. Cristianini, *Kernel Methods for Pattern Analysis* (Cambridge, 2004)
13. R. Poli, J. Kennedy, T. Blackwell, Particle swarm optimization an overview (2007)

Galactic Swarm Optimization with Adaptation of Parameters Using Fuzzy Logic for the Optimization of Mathematical Functions

Emer Bernal, Oscar Castillo, José Soria and Fevrier Valdez

Abstract In this paper the Galactic Swarm Optimization (GSO) algorithm with the use of fuzzy systems for the adaptation of the parameters in the GSO algorithm is proposed. This algorithm is inspired by the movement of stars, galaxies and superclusters of galaxies under the force of gravity. The GSO algorithm uses multiple cycles of exploration and exploitation phases to achieve a balance between exploring new solutions and exploiting existing solutions. In this work different fuzzy systems were designed for the dynamic adaptation of the c3 and c4 parameters to measure the operation of the algorithm with 7 mathematical functions with different number of dimensions. A statistical comparison was made between the different variants to test the performance of the method applied to optimization problems.

Keywords Galactic swarm optimization · GSO · Fuzzy system
Adaptation of parameters · Mathematical function

1 Introduction

In this paper we aim at solving a problem that frequently manifests itself, which is the optimization of multimodal functions by a modification of a metaheuristic called Galactic Swarm Optimization (GSO) presented by Muthiah-Nakarajan and Noel. This method increases the efficiency of the algorithm by providing multiple cycles of exploration and exploitation, thus increasing the chances of accurately finding a global minimum [12].

The GSO algorithm takes as inspiration the movement of stars within galaxies and the movement of cumulus and superclusters of galaxies. The initial population

---

E. Bernal · O. Castillo (✉) · J. Soria · F. Valdez
Tijuana Institute of Technology, Tijuana, BC, Mexico
e-mail: ocastillo@tectijuana.mx

O. Castillo et al. (eds.), *Fuzzy Logic Augmentation of Neural and Optimization Algorithms: Theoretical Aspects and Real Applications*, Studies in Computational Intelligence 749, https://doi.org/10.1007/978-3-319-71008-2_11

is divided into subpopulations and each individual solution is attracted to the best solutions. However, on a larger scale, full galaxies appear as point masses and are attracted to other galaxies.

We propose the optimization of the parameters of the galactic swarm optimization algorithm using fuzzy logic. Two different fuzzy systems were designed to control the c3 and c4 parameters and to measure the performance of our proposed fuzzy galactic swarm optimization versus the original GSO algorithm.

Zadeh proposed fuzzy logic and fuzzy set theory. Fuzzy systems are built based on fuzzy if-then rules formed through knowledge and heuristics based on human knowledge [10]. Fuzzy logic has been recognized as an effective tool for information management in rule-based systems because of their tolerance to imprecision, ambiguity and lack of information [4, 6].

Swarm intelligence techniques have gained popularity in recent decades because of its ability to find partially optimal solutions for combinatorial optimization problems. These have been applied in various areas such as engineering, economics and industry. These problems benefit from the use of swarm intelligence techniques, are usually very difficult to solve accurately since there is no precise algorithm to solve them [1, 2, 13–15].

Our research was based on the paper called of the galactic swarm optimization originally presented by Muthiah-Nakarajan and Noel [12]. The notion of parameter adaptation was inspired on the works published by Melin et al. in optimal design of fuzzy classification systems using PSO with adaptive dynamic parameters through fuzzy logic [11] and Bernal et al. [3] in Imperialist Competitive Algorithm with Dynamic Parameter Adaptation Applied to the Optimization of Mathematical Functions.

The manuscript is organized as show below: in Sect. 2 a description of the Galactic swarm optimization is shown. In Sect. 3 we present the mathematical functions used to measure performances. In Sect. 4 the methodology of our proposal for the adaptation of parameters is described. In Sect. 5 the experimental results is presented so we can to observe the GSO algorithm behavior when implementing the fuzzy systems. In Sect. 6 the conclusions obtained after the study of the galactic swarm optimization utilizing mathematical functions are described.

2 Galactic Swarm Optimization

The GSO algorithm imitates the movement of stars, galaxies and superclusters of galaxies in the universe. Stars are not distributed uniformly in the universe, but clustered into galaxies, which in turn are not uniformly distributed. On a large enough scale individual galaxies appear as point masses [12].

The attraction of a set of stars in a galaxy and a set of galaxies to a bigger set of galaxies is emulated within the GSO algorithm using the PSO algorithm. Firstly all individuals or solutions from each of the subpopulations are influenced by the best solutions in each subpopulation. Then each subpopulation is represented by the best solution in each of the subpopulations and treated as superswarm. The superswarm is composed with the best solutions of each subpopulation. In this way every one of the individuals or solutions will be attracted towards the global best solution.

In the GSO algorithm the swarm is represented by means of an X set containing $x_j^{(i)}$ elements consisting of M partitions called subswarms x_i all of them of size N. All elements of the set X are initialized in the search space $[x_{min}, x_{max}]^D$.

The GSO algorithm consists of two levels of grouping for its implementation through the updating of all subswarms in level one and then it is passed to the formation and update of the superswarms in level two [5].

Level 1

Instead of having a single swarm to explore towards a particular direction, when they have several swarms they have a synergistic effect that results in a better exploration.

Each subswarm explores the search space independently on its own. This process begins by calculating the velocity and position of the particles said expressions to update the velocity and position are [5, 12]:

$$v_j^{(i)} \leftarrow \omega_1 v^i + c_1 r_1 \left(p_j^{(i)} - x_j^{(i)}\right) + c_2 r_2 \left(g^{(i)} - x_j^{(i)}\right) \tag{1}$$

$$x_j^{(i)} \leftarrow x_j^{(i)} + v_j^{(i)} \tag{2}$$

where $v_j^{(i)}$ is the velocity of the particle, $p_j^{(i)}$ is the best solution found until that moment, $g^{(i)}$ is the best global solution, $x_j^{(i)}$ is the position of current particle, c_1 and c_2 are the acceleration constants that give the direction to the best local solutions and global, ω_1 is inertial weight, and r_1 and r_2 are random numbers between 0 and 1.

Level 2

The global best solutions take part in the next stage of clustering to form the superswarm. A new Superswarm *Y* is created through the compilation of the global best solutions of the subswarms x_i.

$$\begin{aligned} y^{(i)} &\in Y : i = 1, 2, \ldots, M \\ y^{(i)} &= g_{(i)}. \end{aligned} \tag{3}$$

The superswarm utilizes the global best solution already calculated by the subswarms and therefore takes advantage of the already calculated information. The best global solutions of subswarms have influence on the super swarm, but there is no feedback effect or flow of information from the super swarm to the subswarms to conserve the diversity of solutions. Avoiding feedback helps the GSO metaheuristic to keep very diverse sub-swarms and a constant global search capability.

The velocity and position of the super swarm in the second clustering level are updated according to the equations shown below [5]:

$$v^{(i)} \leftarrow \omega_2 v^i + c_3 r_3 \left(p^{(i)} - y^{(i)}\right) + c_4 r_4 \left(g - y^{(i)}\right) \tag{4}$$

$$y^{(i)} \leftarrow y^{(i)} + v^{(i)} \tag{5}$$

where $v^{(i)}$ the velocity is associated with Yi, $p^{(i)}$ is the best personal solution, ω_2 is weight of inertia, r_3 and r_4 are random numbers similar to those presented in level 1. At this level **g** serves as the best global and is not updated unless a better solution is found. The super swarm focuses on the best global solutions of the sub swarms, thus improving exploitation.

3 Mathematical Functions

In the area of algorithms for optimization it is common to use mathematical functions to test new methods and these are also used in this work: that is a modification to GSO, where the adaptation of the c3 and c4 parameters will be performed.

In the present work we utilized 7 mathematical functions to measure the performance of the galactic swarm optimization with adaptation of parameters. In the Table 1 are listed with the search space in which they are generally evaluated [8, 11].

Table 1 Test functions

Function	
$Rosenbroke(x) = \sum_{i=1}^{n-1} \left[100\left(x_i + x_i^2\right)^2 + \left(x_{i-1}\right)^2\right]$ $Search\ Space\ x_j \in [-5, 10]\ and\ f(x^*) = 0$	(6)
$Sum\ Square(x) = \sum_{i=1}^{n} ix_i^2$ $Search\ Space\ x_j \in [-10, 10]\ and\ f(x^*) = 0$	(7)
$Zakharov(x) = \sum_{i=1}^{n} x_i^2 + \left(\sum_{i=1}^{n} 0.5ix_i\right)^2 + \left(\sum_{i=1}^{n} 0.5ix_i\right)^4$ $Search\ Space\ x_j \in [-5, 10]\ and\ f(x^*) = 0$	(8)
$Shubert(x) = \left(\sum_{i=1}^{5} i\cos((i+1)x_1 + i)\right)\left(\sum_{i=1}^{5} i\cos((i+1)x_2 + i)\right)$ $Search\ Space\ x_j \in [-10, 10]\ and\ f(x^*) = -186.7309$	(9)
$Baele(x) = (1.5 - x_1 + x_1x_2)^2 + \left(2.25 - x_1 + x_1x_2^2\right)^2 + \left(2.625 - x_1 + x_1x_2^3\right)^2$ $Search\ Space\ x_j \in [-4.5, 4.5]\ and\ f(x^*) = 0$	(10)
$Booth(x) = (x_1 + 2x_2 - 7)^2 + (2x_1 + x_2 - 5)^2$ $Search\ Space\ x_j \in [-10, 10]\ and\ f(x^*) = 0$	(11)
$Dixon{-}Price(x) = (x_1 - 1)^2 + \sum_{i=2}^{n} i\left(2x_i^2 + x_{i-1}\right)^2$ $Search\ Space\ x_j \in [-10, 10]\ and\ f(x^*) = 0$	(12)

4 Proposed Methodology

The galactic swarm optimization (GSO) is a strong search technique utilized for solving optimization problems. In the present work the modification to the GSO called Fuzzy Galactic Swarm Optimization (FGSO) using fuzzy logic for the optimization of benchmark functions is proposed.

The fuzzy logic and fuzzy sets allow what is now known as approximate reasoning. With fuzzy sets, an element may belong to a set with a certain degree of certainty. Fuzzy logic allows reasoning with these uncertain or vague facts to infer new facts, with a certain degree of certainty associated with each fact. Fuzzy logic and fuzzy sets allow us to model common sense or facts as they do in real life [5, 7].

The correct values of the parameters for the fuzzy systems help the algorithm to find best solutions. The objective of the research is to find the optimal values of the parameters for improve the performance of the GSO algorithm. The following shows the fuzzy systems designed to adapt the parameters of the GSO algorithm.

Figure 1. Shows the design of a fuzzy system Mamdani type with the iteration input that is grainy in three triangular membership functions labeled as low, medium and high, the output variables of the fuzzy system are c_3 and c_4 granulated same way in three triangular membership functions labeled as was mentioned in the input variable. Figure 2. Shows the design of a fuzzy system Mamdani type with the input variables iteration and diversity that are grainy in three triangular membership functions labeled as low, medium and high, the output variables of the fuzzy system are c_3 and c_4 granulated in five triangular membership functions labeled as low, medium low, medium, medium high and high.

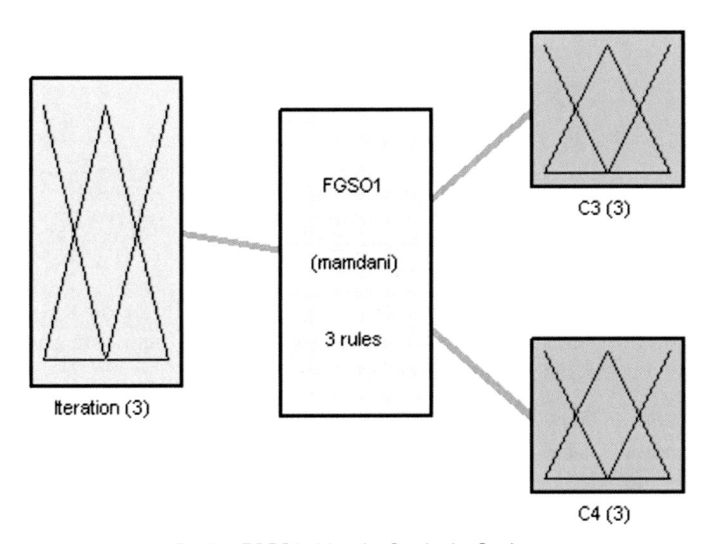

Fig. 1 Fuzzy system FGSO1

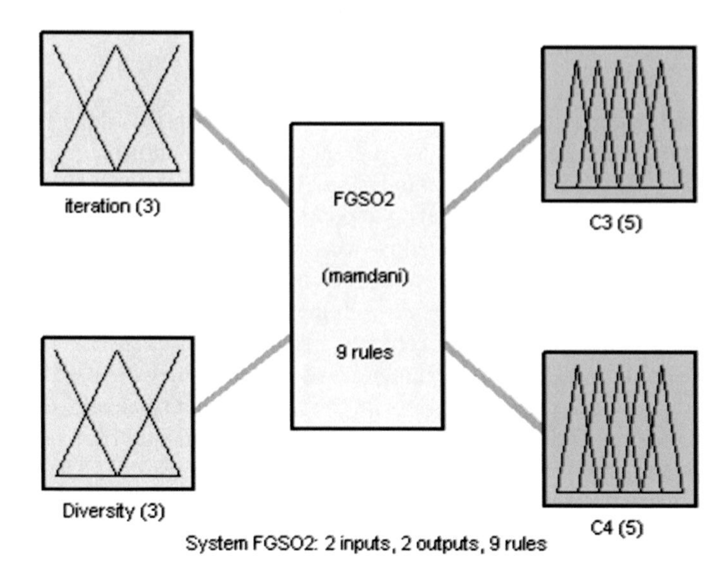

Fig. 2 Fuzzy system FGSO2

The input variables are in the range of 0–1, the iteration variable is a percentage of the total number of iterations, the diversity is the degree of dispersion of the particles, the output variables are in the range of 0–3, where c_3 is the cognitive factor that tell us the importance of the best previous position of the particle, c_4 is the social factor that tell us the importance of the best global position of the swarm.

The values of the input variables are obtained by the expressions shown below [3, 11]:

$$Iterations = \frac{Current\ Iterations}{Maximum\ of\ Iterations} \tag{13}$$

$$Diversity\,(s(t)) = \frac{1}{n_s}\sum_{j=1}^{n_s}\sqrt{\sum_{j=1}^{n_s}\left(x_{ij}(t)\right)^2 - \left(\bar{x}_j(x)\right)^2} \tag{14}$$

After of design of the fuzzy systems, we design their fuzzy rules with the idea that c_3 is on the increase and c_4 is on decrease, as shown below, respectively.

Rule set for the fuzzy system FGSO1:

1. If (Iteration is Low) then (c_3 is Low) (c_4 is High)
2. If (Iteration is Medium) then (c_3 is Medium) (c_4 is Medium)
3. If (Iteration is High) then (c_3 is High) (c_4 is Low)

Rule set for the fuzzy system FGSO2:

1. If (iteration is Low) and (Diversity is Low) then (c_3 is Low) and (c_4 is High)
2. If (iteration is Low) and (Diversity is Medium) then (c_3 is MediumLow) and (c_4 is MediumHigh)
3. If (iteration is Low) and (Diversity is High) then (c_3 is Medium) and (c_4 is High)
4. If (iteration is Medium) and (Diversity is Low) then (c_3 is MediumLow) and (c_4 is MediumHigh)
5. If (iteration is Medium) and (Diversity is Medium) then (c_3 is Medium) and (c_4 is Medium)
6. If (iteration is Medium) and (Diversity is High) then (c_3 is MediumHigh) and (c_4 is MediumLow)
7. If (iteration is High) and (Diversity is Low) then (c_3 is MediumHigh) and (c_4 is MediumLow)
8. If (iteration is High) and (Diversity is Medium) then (c_3 is MediumHigh) and (c_4 is Medium)
9. If (iteration is High) and (Diversity is High) then (c_3 is High) and (c_4 is Low)

5 Experimental Results

The experiments performed with the galactic swarm optimization algorithm (GSO) and the proposal fuzzy galactic swarm optimization (FGSO), we used 7 benchmark functions with 10, 30 and 50 dimensions for to adapt the c_3 and c4 parameters. The results obtained by the GSO algorithm and our proposal are shown in separate tables by the number of dimensions, respectively. Each table contains the average obtained after 30 executions.

Values for the parameters used with the GSO algorithm with fixed parameters and the proposed FGSO with parameter adaptation.

Table 2 shows the values for the parameters used during the experiments with the GSO algorithm and FGSO algorithm using 10, 30 and 50 dimensions.

Table 2 Values of the parameters for GSO and FGSO

Parameters	10 Dimensions	30 Dimensions	50 Dimensions
	Value	Value	Value
Population	5	5	5
Subpopulation	10	20	20
Iteration 1	100	150	250
Iteration 2	1000	1500	1500
No epochs	5	5	9

Table 3 Results for 10 dimensions

Function	GSO	FGSO1	FGSO2
Rosenbrock	6.67E−07	7.51E−12	3.1502E−08
Sumsquare	0	0	0
Zakharov	0	0	0
Shubert	−186.730909	−186.730909	−186.730909
Baele	3.00E−21	1.91E−24	2.057E−23
Booth	0	0	0
Dixon-Price	0.07725316	0	0

Table 4 Results for 30 dimensions

Function	GSO	FGSO1	FGSO2
Rosenbrock	4.91E−07	3.21E−08	0
Sumsquare	0	0	0
Zakharov	0	0	0
Shubert	−186.730909	−186.730909	−186.730909
Baele	0	0	0
Booth	0	0	0
Dixon-Price	0.66666668	0	0

Table 5 Results for 50 dimensions

Function	GSO	FGSO1	FGSO2
Rosenbrock	4.40E−09	3.69E−12	0
Sumsquare	0	0	0
Zakharov	0	0	0
Shubert	−186.730909	−186.730909	−186.730909
Baele	0	0	0
Booth	0	0	0
Dixon-Price	0.66666667	0	0

Table 3 shows the results after the execution of 30 times the GSO algorithm and our proposal of adaptation of the c_3 and c_4 parameters and we can observe the mean of the experiments for all the functions with 10 dimensions.

Table 4 shows the results after the execution of 30 times the GSO algorithm and our proposal of adaptation of the c_3 and c_4 parameters and we can observe the mean of the experiments for all the functions with 30 dimensions.

Table 5 shows the results after the execution of 30 times the GSO algorithm and our proposal of adaptation of the c_3 and c_4 parameters and we can observe the mean of the experiments for all the functions with 50 dimensions.

6 Conclusions

In this paper several fuzzy systems were developed for the adaptation of parameters in the galactic swarm optimization are presented. We performed a comparison between these methods and the result of the experiments shows that our proposal can achieve an improvement against the original algorithm in some cases where the original algorithm fails to reach zero, based on the results shown in Tables 3, 4 and 5.

In the analysis of the results of each of our proposals, we can conclude that we have made improvements to the galactic swarm optimization, since our proposals, on average, are able to improve in some cases the original algorithm.

Acknowledgements We would like to express our gratitude to the CONACYT and Tijuana Institute of Technology for the facilities and resources granted for the development of this research.

References

1. E. Atashpaz-Gargari, F. Hashemzadeh, R. Rajabioun, C. Lucas, Colonial competitive algorithm: a novel approach for PID controller design in MIMO distillation column process. Int. J. Intell. Comput. Cybern. **1**, 337–355 (2008)
2. E. Bernal, O. Castillo, J. Soria, Imperialist competitive algorithm applied to the optimization of mathematical functions: a parameter variation study, in *Design of Intelligent Systems Based on Fuzzy Logic, Neural Networks and Nature-Inspired Optimization*, vol. 601 (Springer International Publishing, 2015), pp. 219–232
3. E. Bernal, O. Castillo, J. Soria, F. Valdez, Imperialist competitive algorithm with dynamic parameter adaptation using fuzzy logic applied to the optimization of mathematical functions. Algorithms **10**(1), 18 (2017a)
4. E. Bernal, O. Castillo, J. Soria, A fuzzy logic approach for dynamic adaptation of parameters in galactic swarm optimization, in *Annual Conference of the North American Fuzzy Information Processing Society (NAFIPS), IEEE* (2017b)
5. E. Bernal, O. Castillo, J. Soria, Fuzzy logic for dynamic adaptation in the imperialist competitive algorithm, in *IEEE Symposium Series on Computational Intelligence (SSCI), IEEE* (2017c)
6. J. Cepeda-Negrete, R.E. Sanchez-Yanez, Automatic selection of color constancy algorithms for dark image enhancement by fuzzy rule-based reasoning. Appl. Soft Comput. **28**, 1–10 (2015)
7. A.P. Engelbrecht, *Computational intelligence* (Wiley, Pretoria, South Africa, 2007)
8. A.R. Hedar, Test functions for unconstrained global optimization [online], Egypt, Assiut University. Available: http://www-optima.amp.i.kyoto-u.ac.jp/member/student/hedar/Hedar_files/TestGO.htm
9. B.S. Khehra, A.P.S. Pharwaha, M. Kaushal, Fuzzy 2-partition entropy threshold selection based on Big Bang-Big Crunch Optimization algorithm. Egypt. Inf. J. **16**(1), 133–150 (2015)
10. M.J. Mahmoodabadi, H. Jahanshahi, Multi-objective optimized fuzzy-PID controllers for fourth order nonlinear systems. Eng. Sci. Technol. Int. J. **18**, 1084–1098 (2016)
11. P. Melin, F. Olivas, O. Castillo, F. Valdez, J. Soria, M. Valdez, Optimal design of fuzzy classification systems using PSO with dynamic parameter adaptation through fuzzy logic. Expert Syst. Appl. **40**(8), 3196–3206 (2013)

12. V. Muthiah-Nakarajan, M.M. Noel, Galactic swarm optimization: a new global optimization metaheuristic inspired by galactic motion. Appl. Soft Comput. **38**, 771–787 (2016)
13. A. Sombra, F. Valdez, P. Melin, O. Castillo, A new gravitational search algorithm using fuzzy logic to parameter adaptation, in *IEEE Congress on Evolutionary Computation*, Cancun, México (2013), pp. 1068–1074
14. F. Valdez, P. Melin, O. Castillo, Evolutionary method combining particle swarm optimization and genetic algorithms using fuzzy logic for decision making, in *IEEE International Conference on Fuzzy Systems* (2009), pp. 2114–2119
15. F. Valdez, P. Melin, O. Castillo, An improved evolutionary method with fuzzy logic for combining particle swarm optimization and genetic algorithms. Appl. Soft Comput. **11**(2), 2625–2632 (2011)

Protein Folding Problem in the Case of Peptides Solved by Hybrid Simulated Annealing Algorithms

Anylu Melo-Vega, Juan Frausto-Solís, Guadalupe Castilla-Valdez, Ernesto Liñán-García, Juan Javier González-Barbosa and David Terán-Villanueva

Abstract Protein Folding Problem (PFP) is a computational challenge with many implications in bioinformatics and computer science. This problem consists in determining the biological and functional three-dimensional structure of the atoms of the amino acid sequence of the protein, which is named Native Structure (NS). Whereas there are a huge number of possible structures this problem is classified as NP-Hard. For PFP, hybrid methods based on Simulated Annealing (SA) have been applied to different kinds of proteins with high-quality solutions. Nevertheless, at the time of presenting this work, they are not enough review of successful algorithms in the group of proteins called peptides. In addition, how successful are the algorithms applied to this kind of proteins has not been previously published. In this paper, we present the main variants of these methods, their applications, and the main characteristics of those algorithms that have made them successful in that area.

Keywords Simulated annealing · Protein folding problem · Peptides Metropolis

A. Melo-Vega · J. Frausto-Solís (✉) · G. Castilla-Valdez
TecNM - Instituto Tecnológico de Cd. Madero, Ciudad Madero, Mexico
e-mail: juan.frausto@itcm.edu.mx

A. Melo-Vega
e-mail: anylumelovega@gmail.com

G. Castilla-Valdez
e-mail: gpe_cas@yahoo.com.mx

E. Liñán-García · J. J. González-Barbosa · D. Terán-Villanueva
Universidad Autónoma de Coahuila, Saltillo, Mexico
e-mail: ernesto_linan_garcia@hotmail.com

J. J. González-Barbosa
e-mail: jjgonzalezbarbosa@itcm.edu.mx

D. Terán-Villanueva
e-mail: david_teran01@yahoo.com.mx

O. Castillo et al. (eds.), *Fuzzy Logic Augmentation of Neural and Optimization Algorithms: Theoretical Aspects and Real Applications*, Studies in Computational Intelligence 749, https://doi.org/10.1007/978-3-319-71008-2_12

1 Introduction

Proteins are molecules that perform vital functions in living beings. They are responsible for transferring small and important molecules called enzymes which are catalysts of biological functions and many other functions. Proteins are composed of monomer units called amino acids. The ordering of amino acids dictates how the protein folds and take the final three-dimensional structure called the Native Structure (NS). The NS of a protein is associated to the functionality of the protein. The folding of proteins as a natural phenomenon happens extremely fast, so it is difficult to analyze in detail and find its exact NS structure just as nature does. Because, even for a small chain of amino acids the number of possible solutions is in the order of 10^{11} [1].

To predict the three-dimensional structure of proteins two approaches have been followed: Ab initio and Template-based. Ab initio is a Latin term meaning "*from the beginning*", this method focuses on finding the NS from the amino acid sequence using as only information the primary structure of the amino acids sequence, and guide the search through the energy values of the different structures. Ab Initio models all the energies involved in the real process of folding and finds the structure with lowest energy [2]. Template-based also initiates with the primary structure of the amino acids sequence, afterwards predicts the secondary structure and then searches in the Protein Data Bank (PDB) for fragments or structures similar to the target sequence. The Template-based approach is based on the ability to identify suitable templates that aligns the sequence to some known template [3].

The Protein Folding Problem (PFP) is a challenge that scientists have been facing since 1960 [4]. PFP is an NP-Complete optimization problem [5], each instance of PFP even the smallest ones called peptides has a several large number of possible conformations or states. Levinthal [6] was the first in pointing out the conformational space of proteins, noting that proteins rapidly converge in microseconds to their NS even with the vast conformational space. This dynamics of protein folding given in the nature is known as Levinthal's paradox, since if a protein explores all its possible conformations to find its functional folding, this would take even more time than the age of the universe. Because the proteins fold to their NS in a matter of microseconds, it is clear that they do not explore all possible conformations [3]. Therefore, the Levinthal's postulate say that if the physical mechanism of protein folding was emulated by algorithms that focused on obtaining the NS of the amino acid sequences, then they could correctly solve PFP [6].

This paper is organized as follows: in Sect. 2, the proteins and peptides are described; Sect. 3 shows the research of Hybrid Simulated Annealing Algorithms for PFP, and finally, in Sect. 4, the conclusions are presented.

2 Proteins and Peptides

A protein, is a polypeptide chain of amino acids linked together in a well-defined sequence. There are twenty amino acids and its conformation has the following structure: a central atom of Carbon (Cα) bonded to a carboxylic acid group (COOH), a side chain (R), an atom of Hydrogen (H), and an amino group (NH2). The group R is in fact the distinctive feature of the peptide, because this side chain is the only difference between the twenty types of amino acids [1].

With the sequence given, the peptide begins the folding going through different states until reaching their final three-dimensional structure. The primary structure is a linear polymer with an amino acids sequence. The second structure begins when the peptide twist and fold to form α helices and β sheet. The α helix is a spiral conformation (helix) in which every backbone NH group donates a hydrogen bond to the backbone COOH group of the amino acid located at three or four residues earlier along the protein sequence. An alpha helix is wound just tightly enough that the atoms that are interacting form bonds at very stable, low-energy angles. Unlike the α helix structure, β sheets are multiple strands of polypeptides connected all together through hydrogen bonds, which are arranged in the form of sheets. Peptides and proteins fold from a primary to a tertiary three-dimensional structure. This final three-dimensional structure thermodynamically stable is just the Native Structure (NS) that we mentioned earlier. In this structure, peptides and proteins are completely fold and they are capable to execute their biochemistry functions. Moreover, some tertiary structures could be combined in a new structure to form a larger molecule. This new structure is called “quaternary structure” which also reaches its functional folded when the Gibbs free energy is minimum. In Fig. 1, the four structures of the proteins folded process are shown [7].

2.1 Complexity of Protein Folding

How fast peptides and proteins fold, leads to a mystery known as Levinthal’s Paradox which develop the next question: How a protein can reach its final structure between the many possible conformations so quickly and that conformation has the minimum energy? [5, 6]. Suppose a protein of 200 amino acids with only three states the number of configurations will be 3200. In this case, with a very high rate of 10^{13} s, the time consuming for a very powerful computer can be longer than the age of the universe. However, is well known than the time for the folding process of nature is only some seconds or even lower.

The literature about complexity of PFP shows a more compelling restatement of that paradox. *On the Complexity of Protein Folding* [5] shows the reduction of PFP to the Hamilton cycle problem known as NP-complete. Starts by showing that PFP for a set of sequences is NP-complete, then proceeds to establish the result for a single sequence using some variants of the planar Hamilton Cycle problem.

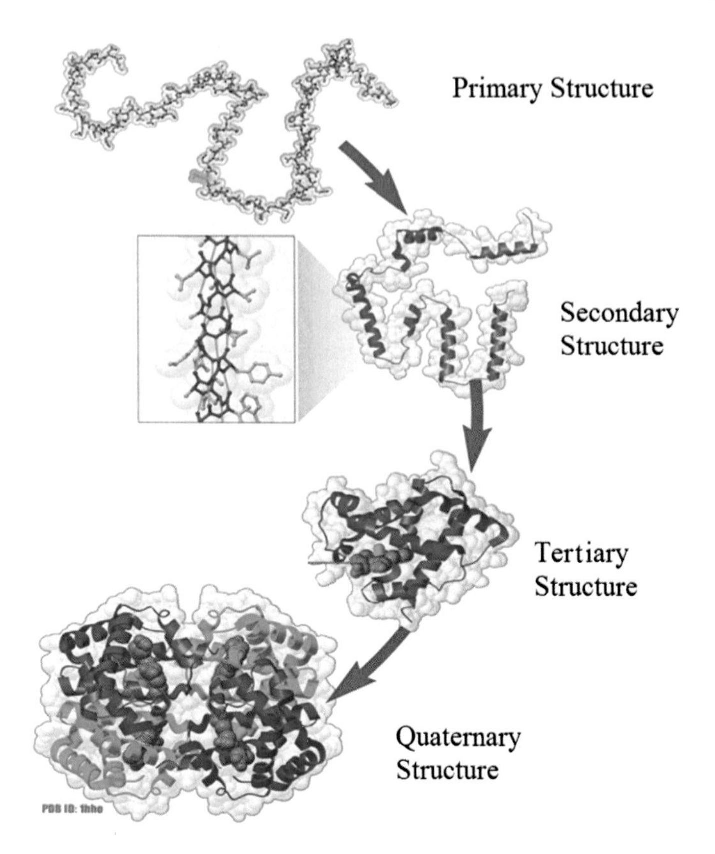

Fig. 1 Primary, secondary, tertiary and quaternary protein structures of protein 1HHO

2.2 *Dihedral Angles and Ramachandran Plots*

In PFP all the bond angles are essential to solve the problem. The geometry of the amino acid residues is determined by using small molecule crystal structures [8]. The dihedral angle ω in the peptide is very close to 180°, nevertheless a slight variation of this value in real structures may occur. But, the remaining dihedral angles are the variability in protein folding.

There are four types of torsion angles Phi (φ), Psi (ψ), Omega (ω) and Chi (χ) where:

- φ is the dihedral angle between the amino group and the alpha carbon.
- ψ is the dihedral angle between the alpha carbon and the carboxyl group.
- ω is the angle between two amino acids in a sequence.
- χ is the angle of the side chains.

As is mentioned above, the ω torsion angle within the protein backbone is essentially flat and fixed to 180°, the torsion angles φ and ψ are responsible for

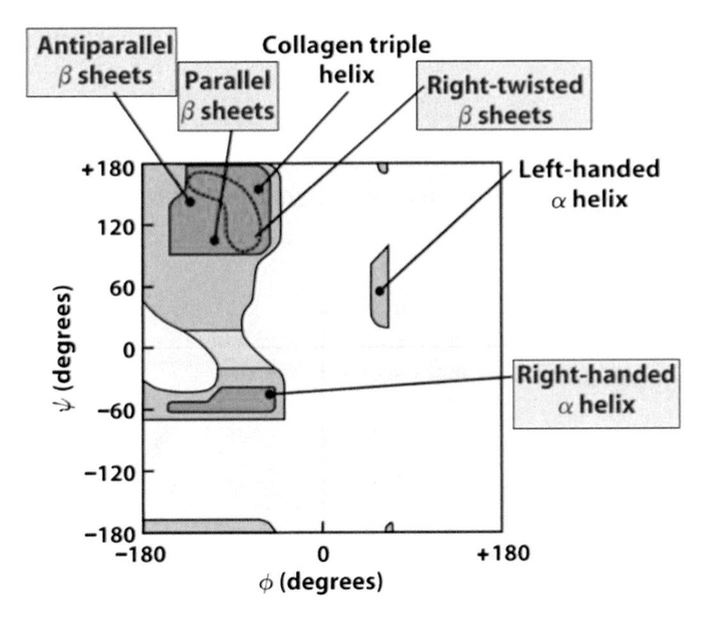

Fig. 2 Ramachandran plot [9]

providing the flexibility required for the polypeptide backbone to adopt a certain fold; they are known as the Ramachandran [9] angles. The rotation of the protein chain can be described as rotation of the peptide bond planes relative to each other.

Ramachandran plot indicates which values, or conformations, of the ψ and φ angles are possible for an amino-acid residue in a peptide or protein. The most favorable combination of ψ and φ values are represented by the darkest shaded areas in this plot. Figure 2 shows a Ramachandran plot.

2.3 *Applications of Peptides*

In computational biology, predict a protein's three-dimensional structure from amino acid sequence is the major goal [4]. Studies on peptides are a good way for exploring general domains of locally ordered structures and understand its behavior [10]. The research on peptides could help in:

(a) Replacing the expensive and slow biological experiments by faster computer simulations in the search for new drugs [4];
(b) Designing proteins and peptides for therapeutic applications like Cancer, HIV, Alzheimer's disease;

(c) Design or redesign of enzymes and biocatalysts; and
(d) Manufacturing therapeutics antibodies [3].

3 Hybrid Simulated Annealing Algorithms for PFP Peptides

The most common heuristic methods using for solving PFP are, Genetic Algorithms, Simulated Annealing, Neural Network and Tabu Search. SA provides excellent solutions [10] in a short execution time.

3.1 Critical Assessment of Techniques for Protein Structure Prediction

The Critical Assessment of protein Structure Prediction (CASP) [11] aim is establishing the current state of the art in protein structure prediction, identifying what progress has been made, and highlighting where future effort may be most productively focused.

CASP is an international proteins structure prediction contest, in which the amino acid sequences of unknown structures are given to the participants, then they provided a structure model. These models are compared with the newly determined experimental structures. Finally, results of CASP are published. In nowadays, have been twelve CASP, these competition takes place every two years since 1994. In CASP XI almost 60,000 models from 207 modeling groups for the prediction of 100 targets were collected [12].

The evaluation of models performance is divided into a number of categories: Models based on homologous templates and free models without detectable homologous templates. In Table 1 is described the best servers and its characteristics; is a top ranking of the best in resolving the PFP.

3.2 Hybrid Simulated Annealing Algorithms for Peptides

With the current knowledge of folding codes, "*Levinthal's paradox seems to now have a viable hypothesis: A protein can fold quickly and solve its big global optimization puzzle by piecewise solutions of a smaller component puzzle*" [4]. Current Computer Algorithms predict three-dimensional structures for small proteins with notable precision. Since 80s several groups of investigators began using Monte Carlo in the case of peptide Met Enkephalin [10] which is a good test case

Table 1 Top ranking of best public CASP-certified protein structure prediction servers

Author	Structure prediction servers	Model	Approach
Zhang [13]	I-TASSER	Pro-file-profile threading alignment (PPA) and Iterative Template Based Model	Monte Carlo-Simulated Annealing
Kim et al. [14]	ROBETTA	Ginzu (Homology Model and Novo Model)	Multiple sequence alignment (MSA)
Söding et al. [15]	HHpred	Ab Initio Model with FASTA or PSI-BLAST format	HMM (Hidden Markov Model)
Zhou et al. [16]	METATASER	SPARKS, SP^3, PROSPECTOR_3 Threading methods and 3D-Jury approach	Parallel Hyperbolic Monte Carlo
Wang et al. [17]	MULTICOM	Template Based Model	Hhsearch or Hidden Markov Model
Lundström et al. [18]	Pcons	Homology Model	Neural networks
Karplus [19]	SAM-T08	Homology Model	Hidden Markov Model
Ginalski et al. [20]	3D-Jury	Ab Initio Model	CAFASP—Consensus groups
Jones [21]	THREADER	Threading Method	Double dynamic Programming
Källberg et al. [22]	RaptorX	Template Based Model	Conditional (Markov) Random Fields (CRFs)

showing that even a small peptide involves a lot of efforts, because for this peptide with five amino acid sequence is expected to contain 10^{11} local minima [23].

3.3 Simulated Annealing

The Simulated Annealing (SA) algorithm was proposed in 1983 [24]. This algorithm emulates the process of tempering metals until its melting point. SA is tuned to a high temperature T_0, and when the process advances, this temperature is gradually decreased until reaching a final temperature T_f which is very close to zero. In order to decrease the temperature, SA uses different cooling functions, with the lineal geometrical function $T_{k+1} = \alpha T_k$ among the most commonly used. For PFP, the value of alpha recommended is in the range $0.70 \leq \alpha < 1.0$. The values of T_0, T_f and α are experimentally set. SA has a method that produces neighboring solutions, called Metropolis Cycle (MC). MC generates solutions based on the current solution, using Boltzmann probability distribution as acceptance parameter. In PFP, SA or Genetic algorithms tend to perform better than

Table 2 Hybrid simulated annealing algorithms for peptides

Author	Title	Model	Hybrid approach	Proteins	Max amino acids
Agostini et al. [27]	Generalized simulated annealing applied to protein folding studies	Grid Model	Simulated Annealing and Generalized Statistical Mechanics	18-alanine	Small peptide
Morales et al. [10]	Applications of simulated annealing to the multiple-minima problem in small peptides	Ab Initio	Simulated Annealing technique that includes Anti-correlations	Met-Enkephalin Leu-Enkephalin N-acetyl-N′-methyl-aspartic acid TRH(+) N-acetyl-N′-methyl-glycineamide, N-acetyl-N′-methyl-alanineamide	5
Zhan et al. [28]	Conformational study of Met-enkephalin based on the ECEPP force fields	Ab Initio	Basin Paving Method and Monte Carlo	Met-Enkephalin	5
Fengbin et al. [29]	Protein folding study based on parallel group annealing algorithms	Ab Initio	Simulated annealing and message passing interface (MPI)	Met-Enkephalin	5
Frausto-Solis and Román [30]	Analytically tuned simulated annealing applied to the protein folding problem	Ab Initio	Markov Chain and Simulated Annealing	Met Enkephalin C-peptide	13
Sakae et al. [31]	Combination of genetic crossover and replica-exchange method for conformational search of protein systems	Ab Initio	Simulated Annealing and Genetic Algorithm	Trp-cage	20
Okamoto et al. [32]	Tackling the multiple-minima problem in protein folding by Monte Carlo simulated annealing and generalized-ensemble algorithms	Ab Initio	Monte Carlo Simulated Annealing and Multicanonical Algorithms	C-peptide Fragment of BPTI (16–36)	20

(continued)

Table 2 (continued)

Author	Title	Model	Hybrid approach	Proteins	Max amino acids
Hiroyasu et al. [33]	Energy minimization of protein tertiary structure by parallel simulated annealing using genetic crossover	Ab Initio	Parallel Simulated Annealing with an operator of Genetic Algorithm (GA),	Met-Enkephalin C-peptide Parathyroid Hormone Fragment	34
Hao et al. [34]	Multiple simulated annealing-molecular dynamics (MSA-MD) for conformational space search of peptide and miniprotein	Ab Initio	Simulated Annealing-Molecular Dynamics (SA-MD)	ALPHA1 Trp-cage PolyAla 1UAO 1E0Q 1ERD 1GAB	40
Frausto-Solis et al. [35]	Golden ratio simulated annealing for protein folding problem	Ab Initio	Golden Ratio and Simulated annealing	Met-Enkephaline C-Peptide 1EOG 1ENH 1BDD	60
Frausto-Solis et al. [36]	Chaotic multiquenching annealing applied to the protein folding problem	Ab Initio	Simulated annealing, Multi Quenching annealing and Chaotic Local Search	Met-Enkephalin Proinsulin T0549 T0335 T0281	90
Frausto-Solis et al. [26]	Multiphase simulated annealing based on Boltzmann and Bose-Einstein distribution applied to protein folding problem	Ab Initio	Multiphase Simulated Annealing Algorithm using Boltzmann and Bose-Einstein distributions (MPSABBE)	Met-Enkephalin Proinsulin T0549 T0335 T0281	90

others algorithms [25]. Table 2 has a compilation of Hybrid Simulated Annealing Algorithms for peptides or tested in peptides.

Table 2 contains in columns one and two, the authors and titles of the analyzed works respectively, while columns three and four have the models and hybrid solution methods correspondingly. Finally, in the fifth column the test peptides and the maximum number of amino acids included in the studied peptides chains are listed.

As can be observed in Table 2, algorithms with a Simulated Annealing (SA) approach have been applied over time in the successful hybrid methods that have addressed not only for small proteins or peptides but also it was applied in larger proteins. This metaheuristic has been hybridized with different stochastic strategies that work with a single solution and not often with methods based on populations of solutions.

4 Conclusions

In this work a review of successful PFP algorithms for proteins with an emphasis in peptides is presented. A compilation of the methods and successful models used in the CASP is shown; as we can see in CASP is common to use Homology and Threading strategies from most of the servers, and only a few Ab Initio Models are reported. From the presented review can be noticed that hybrid simulated annealing algorithms for peptides and proteins are commonly used in homology and threading strategies. For Ab initio strategy, algorithms using simulated annealing are the most widely used algorithm. In addition, the tendency is to use new performance metrics of algorithms; most of these metrics looks for a good alignment between the best solution or even the NS of solved proteins and not only lower energy values reached by the algorithms. However, there is not an agreement about which is the best metric to use for PFP problems.

Acknowledgements The authors would like to acknowledge to CONACYT, TECNM and PRODEP. An special acknowledgement to the Laboratorio Nacional de Tecnologías de la Informaciőn del Instituto Tecnológico de Ciudad Madero for the access to the cluster. This work has been partially supported by CONACYT Project 254498.

References

1. J.S. Richardson, in *The Anatomy and Taxonomy of Protein Structure* (1981), pp. 167–339
2. D. Osguthorpe, Ab initio protein folding. Curr. Opin. Struct. Biol. **10**(2), 146–152 (2000)
3. G.A. Khoury, J. Smadbeck, C.A. Kieslich, C.A. Floudas, Protein folding and de novo protein design for biotechnological applications. Trends Biotechnol. **32**(2), 99–109 (2014)
4. K.A. Dill, S.B. Ozkan, M.S. Shell, T.R. Weikl, The protein folding problem. Annu. Rev. Biophys. **37**, 289–316 (2008)

5. P. Crescenzi, D. Goldman, C. Papadimitriou, A. Piccolboni, M. Yannakakis, On the complexity of protein folding. J. Comput. Biol. **5**(3), 423–465 (1998)
6. C. Levinthal, Are there pathways for protein folding? J. Chim. Phys. Physico-Chimie Biol. **65**, 44–45 (1968)
7. C.-I. Brändén, J. Tooze, in *Introduction to Protein Structure* (Garland Pub, 1999)
8. F.A. Momany, R.F. McGuire, A.W. Burgess, H.A. Scheraga, Energy parameters in polypeptides. VII. Geometric parameters, partial atomic charges, nonbonded interactions, hydrogen bond interactions, and intrinsic torsional potentials for the naturally occurring amino acids. J. Phys. Chem. **79**(22), 2361–2381 (1975)
9. G.N. Ramachandran, C. Ramakrishnan, V. Sasisekharan, Stereochemistry of polypeptide chain configurations. J. Mol. Biol. **7**(1), 95–99 (1963)
10. L.B. Morales, R. Garduño-Juárez, D. Romero, Applications of simulated annealing to the multiple-minima problem in small peptides. J. Biomol. Struct. Dyn. **8**(4), 721–735 (1991)
11. Protein Structure Prediction Center. [Online]. Available: http://www.predictioncenter.org/index.cgi. Accessed 18 Mar 2017
12. J. Moult, K. Fidelis, A. Kryshtafovych, T. Schwede, A. Tramontano, Critical assessment of methods of protein structure prediction: Progress and new directions in round XI. Proteins Struct. Funct. Bioinf. **84**(S1), 4–14 (2016)
13. Y. Zhang, I-TASSER server for protein 3D structure prediction. BMC Bioinf
14. D.E. Kim, D. Chivian, D. Baker, Protein structure prediction and analysis using the Robetta server. Nucleic Acids Res. **32**(Web Server), W526–W531 (2004)
15. J. Soding, A. Biegert, A.N. Lupas, The HHpred interactive server for protein homology detection and structure prediction. Nucleic Acids Res. **33**(Web Server), W244–W248 (2005)
16. H. Zhou, S.B. Pandit, S.Y. Lee, J. Borreguero, H. Chen, L. Wroblewska, J. Skolnick, Analysis of TASSER-based CASP7 protein structure prediction results. Proteins Struct. Funct. Bioinf. **69**(S8), 90–97 (2007)
17. Z. Wang, J. Eickholt, J. Cheng, MULTICOM: a multi-level combination approach to protein structure prediction and its assessments in CASP8. Bioinformatics **26**(7), 882–888 (2010)
18. J. Lundström, L. Rychlewski, J. Bujnicki, A. Elofsson, Pcons: a neural-network-based consensus predictor that improves fold recognition. Protein Sci. **10**(11), 2354–2362 (2008)
19. K. Karplus, SAM-T08, HMM-based protein structure prediction. Nucleic Acids Res. **37**(Web Server), W492–W497 (2009)
20. K. Ginalski, A. Elofsson, D. Fischer, L. Rychlewski, 3D-Jury: a simple approach to improve protein structure predictions. Bioinformatics **19**(8), 1015–1018 (2003)
21. D. Jones, THREADER: protein sequence threading by double dynamic programming (1998), pp. 285–311
22. M. Källberg, H. Wang, S. Wang, J. Peng, Z. Wang, H. Lu, J. Xu, Template-based protein structure modeling using the RaptorX web server. Nat. Protoc. **7**(8), 1511–1522 (2012)
23. A. Nayeem, J. Vila, H.A. Scheraga, A comparative study of the simulated-annealing and monte carlo-with-minimization approaches to the minimum-energy structures of polypeptides: [Met]-Enkephalin. J. Comput. Chem. **12**(5), 594–605 (1991)
24. S. Kirkpatrick, C.D. Gelatt, M.P. Vecchi, Optimization by simulated annealing. Sci. New Ser. **220**(4598), 671–680 (1983)
25. H. Zhou, J. Skolnick, J. Skolnick, V.S. Pande, M.B. Swindells, J.M. Thornton, J.J. Ward, K.M.S. Misura, D. Baker, Ab initio protein structure prediction using chunk-TASSER. Biophys. J. **93**(5), 1510–1518 (2007)
26. J. Frausto-Solis, E. Liñán-García, J.P. Sánchez-Hernández, J.J. González-Barbosa, C. González-Flores, G. Castilla-Valdez, Multiphase simulated annealing based on Boltzmann and Bose-Einstein distribution applied to protein folding problem. Adv. Bioinf. **2016**, 7357123 (2016)
27. F.P. Agostini, D.D.O. Soares-Pinto, M.A. Moret, C. Osthoff, P.G. Pascutti, Generalized simulated annealing applied to protein folding studies. J. Comput. Chem. **27**(11), 1142–1155 (2006)

28. L. Zhan, J.Z.Y. Chen, W.-K. Liu, Conformational study of Met-enkephalin based on the ECEPP force fields. Biophys. J. **91**(7), 2399–2404 (2006)
29. P. Fengbin, Z. Huilin, W. Yanjie, F. Shengzhong, Y. Zhixiang, Protein folding study based on parallel group annealing algorithms **4**(5), 26–34 (2013)
30. J. Frausto-Solis, E. Román, Analytically tuned simulated annealing applied to the protein folding problem, in *International Conference on Computational Science, 2007* (2007), pp. 370–377
31. Y. Sakae, T. Hiroyasu, M. Miki, K. Ishii, Y. Okamoto, Combination of genetic crossover and replica-exchange method for conformational search of protein systems (2015)
32. Y. Okamoto, Tackling the multiple-minima problem in protein folding by monte carlo simulated annealing and generalized-ensemble algorithms. Int. J. Mod. Phys. C **10**(8), 1571–1582 (1999)
33. T. Hiroyasu, M. Miki, S. Ogura, K. Aoi, T. Yoshida, Y. Okamoto, J. Dongarra, Energy minimization of protein tertiary structure by parallel simulated annealing using genetic crossover, in *2002 Genetic and Evolutionary Computation Conference (GECCO 2002) Workshop Program* (2002), pp. 49–51
34. G.-F. Hao, W.-F. Xu, S.-G. Yang, G.-F. Yang, Multiple simulated annealing-molecular dynamics (MSA-MD) for conformational space search of peptide and miniprotein. Sci. Rep. **5**, 15568 (2015)
35. J. Frausto-Solis, J.P. Sánchez-Hernández, M. Sánchez-Pérez, E.L.L. García, Golden ratio simulated annealing for protein folding problem. Int. J. Comput. Methods **12**(6), 1550037 (2015)
36. J. Frausto-Solis, E. Liñáan-García, M. Sanchez-Perez, J.P. Sanchez-Hernandez, Chaotic multiquenching annealing applied to the protein folding problem. Sci. World J. **2014** (2014)

Optimization of the Parameters of Smoothed Particle Hydrodynamics Method, Using Evolutionary Algorithms

Juan de Anda-Suárez, Martín Carpio, Solai Jeyakumar, Héctor José Puga-Soberanes, Juan F. Mosiño and Laura Cruz-Reyes

Abstract Smooth particle hydrodynamics (SPH) is a mesh free numerical method for solving hydrodynamical equations. For its functioning, the method uses; one integer-domain parameter (the total number of particles) and three real domain parameters (smoothing parameters and artificial viscosity). For a given problem (geometry and initial conditions) these parameters can be tuned to reduce the computational cost and improve the accuracy of the solutions. Optimized values of the SPH parameters using the evolutionary algorithms, Differential Evolution (DE) and Boltzmann Univariate Marginal Distribution Algorithm (BUMDA) are obtained for different Sod shock tube test problems. Comparison of the numerical solution of the physical variables with that of the exact solution shows that this optimization strategy can be used to make an initial guess of the SPH parameters based on the initial conditions of the simulation domain. The performance of the two algorithms are statistically compared.

Keywords Evolutionary algorithms · Differential Evolution · Smooth Particle Hydrodynamics · Riemann solver · Parameters tunning · Optimization of Algorithms

---

J. de Anda-Suárez · M. Carpio (✉) · H. J. Puga-Soberanes · J. F. Mosiño
Tecnológico Nacional de México- Instituto Tecnológico de León, León, Gto, Mexico
e-mail: juanmartin.carpio@itleon.edu.mx

J. de Anda-Suárez
e-mail: Juandanda2@gmail.com

S. Jeyakumar
Departamento de Astronomía, Universidad de Guanajuato, Guanajuato, Gto, Mexico

L. Cruz-Reyes
Tecnológico Nacional de México- Instituto Tecnológico de Ciudad Madero, Ciudad Madero, Tamaulipas, Mexico

O. Castillo et al. (eds.), *Fuzzy Logic Augmentation of Neural and Optimization Algorithms: Theoretical Aspects and Real Applications*, Studies in Computational Intelligence 749, https://doi.org/10.1007/978-3-319-71008-2_13

1 Introduction

Smooth Particle Hydrodynamics (SPH) is a method used to numerically solve fluid flow problems. Though it was proposed for hydrodinamical simulations of fluid dynamics in astronomical objects, nowadays this method is applied to several branches of physics, engineering as well as computer graphics and animation [1, 2]. The main idea behind the SPH is the discretization of continuous space into discrete particles and these particles approximate the physical variables of the fluid. The fluid dynamical effects are fed back during the update of the particle position and velocities. The SPH method is intrinsically mesh-free, therefore it is well suited for simulations of free surface fluid problems and ablation of material from solids in engineering problems [3]. A new hybrid technique that incorporates aspects of mesh based Godunov scheme has been shown to improve the accuracy of the solutions. This method uses solution to the Riemann problem of the interface between particles and their neighbours thus capturing the detailed physics of the local interaction [4].

The SPH is also used in the field of computer graphics such as animation, video games and medical simulators. These applications require solution to the fluid flow problems in real time, in order to find a better approximation to natural phenomena [5–8]. In order to track the dynamics of the fluid in real time, several authors have explored different methodologies such as different kernels, real-time particle reduction and adaptive time discretization strategies [9].

In addition, the SPH method is also used for different problems in engineering involving free surface flows such as sloshing of liquids in tanks and sea breakwater [10, 11]. In addition, SPH method is also used to simulate problems involving interaction of solid bodies and fluids [12–14].

In order to approximate the fluid dynamical problem in a domain with ‘N’ particles, the SPH method requires a kernel and tunable parameters associated with it [15]. Secondly, it is necessary to introduce a new parameter, artificial viscosity, in order to simulate problems that involve shocks [16]. These parameters influence the computational cost and the accuracy of the solutions. Therefore several authors have attempted to find or fix optimum values for these parameters, in order to reduce the computational cost and the accuracy of the solutions. For the first time, Gingold and Monaghan [15], used a methodology to determine a suitable kernel and the smoothing length associated with it. In their study, they estimate that the smoothing length parameter can be determined by

$$h \propto \frac{1}{N^{\frac{1}{d}}}$$

where N is the number of particles and d is the dimension. Thus the smoothing length parameter is proportional to the value of the density and is given by

$$h \propto \frac{1}{\rho^{\frac{1}{d}}}$$

Based on their study, they found that the smoothing length can fixed for all particles to a constant value in the range $0.15 \leq h \leq 0.5$ [16].

However, Steinmetz and Muller [17], studied the problem of smoothing length and their results contradicted with the classical method for determining it using the mean density [17]. Using a heuristic method, Springel [18] showed a method to determinate the smooth parameter based on the behavior of its neighborhood. However, the heuristics need a fixed value of neighbors to interact with [18]. In order to further optimize the smoothing length, Hongfu and Weiran [19], used a new equation for estimating the density field from particles, in which it is possible to vary the smoothing length. This approach deviates from the basic methodology of SPH where it is not possible to vary the smoothing length. This strategy permits further exploration of optimization techniques in SPH simulations [19].
In the case of problems in engineering such as shape optimization and in computer graphics rendering optimizing the computational cost, while keeping the solution accuracy is the most important requirement. However, all of the studies mentioned above focussed on finding optimized values of the parameters of the SPH using heuristic methods. However, the use of SPH in areas of engineering, computer graphics and other fields require optimization that cannot be handled with heuristic methods alone. Therefore meta-heuristic methods are tied to optimize the SPH simulations. The Stochastic evolutionary algorithms such as Differential Evolution and its variants are very popular in the optimization of fluid problems. For example optimization of airfoils for aircrafts and turbines is successfully achieved using DE and mesh based CFD solvers [20]. Techniques involving hybrids of two optimization algorithms are also used to achieve fast convergence of Airfoil problem [21].

Based on these previous studies it is possible to achieve computationally cost effective, real time simulations of fluid flows using SPH by optimizing the SPH solver parameters. In this work we attempt to optimize the SPH based fluid simulations using evolutionary algorithms, for a given initial conditions of the fluid state variables and the characteristics of the immersed solid body surface by varying the SPH parameters dynamically during the simulations as well as using fixed values, treating all the parameters at once. We present a comparative study between two different families of evolutionary algorithms for tuning SPH parameters. We tested the algorithms DE and BUMDA using the six standard shock tube problems, where the known exact solution permits us to evaluate the performance of the results of optimization.

2 Smoothed Particle Hydrodynamics

The SPH is a mesh free method to solve the Euler equations in Lagrangian coordinates. In the SPH methodology, all the primitive variables of the fluids, density, velocity, and pressure, are interpolated from particles using a kernel function.

The basic flow of the algorithm of SPH is summarized in the following points:

- The fluid is approximated by discrete particles with mass, m. The fluid state variables at a point in space r is obtained by the contribution of each particle in the region of interaction.
- The region of interaction is defined by a kernel that tries to approximate the Dirac delta function. In general, a Gaussian function is used as the kernel function.
- The solution to the Euler equations are approximated as sum of contributions from finite number of individual particles, weighted by a kernel or the derivative of the kernel.
- New positions of the particles are obtained using the fluid forces calculated above.
- The above steps are repeated until the time when the solution is sought.

Mathematically the discretization of the fluid variables is described in the following equations Eqs. (1)–(5).

$$A(r)_i = \sum_j m_j \frac{A(r)_j}{\rho_j} W\left(\left|r_i - r_j\right|, h\right) \tag{1}$$

where $A(r)_i$ represents the primitive variable representing the fluid state variables, m_j is the mass of the particles, ρ_j is the density of fluid, W and h are the smoothing kernel and smoothing parameters respectively.

Based on the above equation, the density can be represented by Eq. (14).

$$\rho_j = \sum_j m_j W\left(\left|r_i - r_j\right|, h\right) \tag{2}$$

The momentum equation can be converted as

$$\frac{dv_i}{dt} = -\sum_j m_j \left(\frac{p_j}{\rho_j^2} + \frac{p_i}{\rho_i^2}\right) \nabla_i W\left(\left|r_i - r_j\right|, h\right) \tag{3}$$

where v_i represents the velocity of particles and p_j is the pressure of the fluid associated with the particle. In order to treat the shock problems the momentum equations are modified to incorporate an artificial viscosity between the particles Eq. (4).

$$\frac{dv_i}{dt} = -\sum_j m_j \left(\frac{p_j}{\rho_j^2} + \frac{p_i}{\rho_i^2} + \Pi_{ij} \right) \nabla_i W(|r_i - r_j|, h) \tag{4}$$

where Π_{ij} is giving by Eq. (5)

$$\Pi_{ij} = \begin{cases} \frac{-\alpha \bar{c}_{ij} \mu_{ij} + \beta \mu_{ij}^2}{\bar{\rho}} & v_{ij r_{ij}} < 0; \\ 0 & v_{ij r_{ij}} > 0; \end{cases} \tag{5}$$

The parameters of the SPH formalism are the total number of particles, the mass of the particles, the smoothing length parameter and the artificial viscosity parameters.

2.1 Parameters of SPH Algorithm and the Computational Domain

In a CFD simulation, the initial continuous fluid variables are discretized as a mesh. The surface of the solid body is immersed inside the computational mesh. In such a scenario the resolution of the mesh may or may not be sufficient to simulate the fine scale details for the given initial conditions, for example, density or pressure jumps or near the sharp corners of the solid bodies. In the case of simulations using SPH, the initial distribution of particles and their masses (m_p) are again further approximates the mesh based discretization. In order to improve the accuracy, the total number of particles (*N*) used for the simulation has to be increased, thus adding further computational cost. Therefore in order to capture all the physical effects of the fluid flow, several parameters need to be tuned.

In order to illustrate the problem, a Sod Shock tube problem is simulated for various parameters as shown in Table 1. The results of the simulation runs are shown in Fig. 1. The effect of variation of the parameters on the final solution can be seen in the figure. The cross symbols represent the exact solution to the problem. The solution to case 3, shown as squares, is a closer match to the theoretical solution. The other two cases the solution is not acceptable. However, such a wrong result was expected since very few particles are being used in these simulations.

As can be seen from the example problem above, some parameters produce unacceptable solutions. Then the fundamental question is what are the optimal

Table 1 The parameters of the Sod shock tube test

Parameter	N	m_p	h	α	β
Case 1	10	0.021	0.5	1	1
Case 2	50	0.021	0.23	0.5	1
Case 3	350	0.021	0.1	0.05	1.2
Case 4	500	0.021	0.04	0.5	1.2

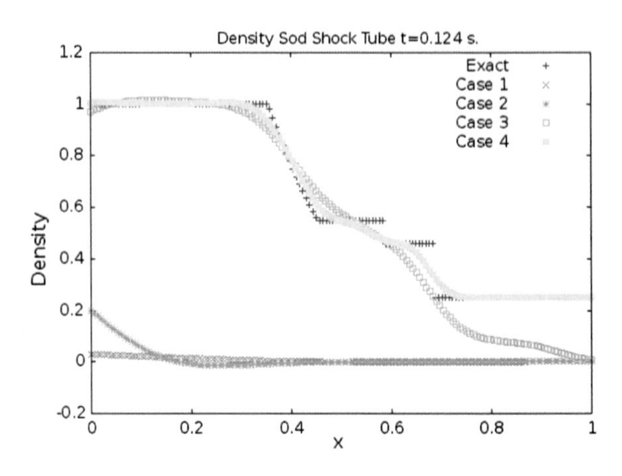

Fig. 1 Comparison between the exact solution and the SPH results

Table 2 Parameters of the SPH methodology

Parameter	Description	Domain
N	Total number of particles	$(0, n_{max}]$
m_p	Mass of each particle	$(0, m_{max}]$
h	Smoothing length parameter	$(0, L_{max}]$

parameters to use, for a given initial conditions and geometries, in order to obtain a correct simulation. So the problem of finding the parameters of an acceptable solution, in computational science terminology is NP-Complete [22, 23]. Therefore the problem that has been raised in this work is a candidate for the use of evolutionary algorithms that belong to the family of the metaheuristics.

3 Optimization Using Evolutionary Algorithms

Evolutionary Algorithm is a stochastic method used in optimization where the next set of trial parameters are chosen based on concepts of species evolution, such as mutation and crossover, the optimization is achieved through the fitness or the objective function and an engine (algorithm) of optimization, such as the well-known Genetic Algorithm. In this work as engines of optimization, the Differential Evolution (DE) and Boltzmann Marginal Univariate Distribution Algorithm (BUMDA) are used.

3.1 Fitness Function

The important component of the evolutionary algorithm is the definition of the fitness function. In the case of the test problems where an exact solution is available, the fitness function can make use of the known solution. For Sod shock tube problems, the exact solution to the Riemann problems are well known [24].

Therefore, the fitness function is defined as the sum of the squares of differences between the SPH solution and that of the exact one, mathematically represented as

$$f_{fitness} = \frac{1}{M}\frac{1}{N}\sum_{i}^{N}\sum_{j}^{M}\frac{\left[\mathrm{S}_{\mathrm{Exact}}(r_i, A_j, t) - \mathrm{S}_{\mathrm{SPH}}(\mathrm{p}, r_i, A_j, t)\right]^2}{\left[\mathrm{S}_{\mathrm{Exact}}(r_i, A_j, t) + \mathrm{S}_{\mathrm{SPH}}(\mathrm{p}, r_i, A_j, t)\right]^2} \tag{6}$$

$S_{Exact}(r_i, A_j, t)$ is the exact solution in the space-time, $S_{SPH}(p, r_i, A_j, t)$ represent the SPH solution and p are the parameters that are present in the SPH method. Here r_i are the interpolated mesh points A_j are the physical variables. These parameters are given in Table 2.

The parameter n_{max} is allowed to vary no more than five thousand particles because of the computational cost for each of the elements in the population is very high. The maximum mass m_{max} of each particle is not more than 1.0, because particles with very large masses do not make sense. The parameters α and β are related to the artificial viscosity and are bounded by the ranges given in Gingold and Monaghan [15].

4 Differential Evolution

The DE is part of algorithms known as metaheuristics. Although DE is good for finding the global optimal value, it does not guarantee that an optimal solution can be found in all situations. Price and Storm [25], proposed the DE algorithm while seeking to solve problems of global optimization, for which the derivative of the fitness function is not necessarily known or it is impossible to obtain. Since its first implementation many modifications have been made to the fundamental algorithm [25].

The DE algorithm utilizes elements (members) of a uniformly distributed population in the search parameter space. New elements are generated based on the existing elements as described below:

- **Initialization**: The initialization of the population is from a uniform distribution of elements in the search space, given by Eq. (14)

$$x_{ij0} = ran(0, 1)(b_{j,U} - b_{j,L}) + b_{j,L} \tag{7}$$

- **Mutation**: Once the population is created, the elements of the population are combined in a vectorial way allowing mutations in the newly generated elements. These mutations have the purpose that in the best of cases allows us to reach the optimum value. The mutation process is given by Eq. (14)

$$V_{ig} = x_{i0,g} + F(x_{i1,g} - x_{i,g}) \tag{8}$$

- **Crossover**: To carry out the crossing process two elements of the population are used and a new element of the population is generated taking a random part of the elements that make up the parents of the new vector, the mathematical formulation is given by Eq. (14)

$$u_{ij} = \begin{cases} V_{ij,g} & \textit{if } ran(0,1) \leq Cr \\ x_{ij,g} & \text{otherwise.} \end{cases} \tag{9}$$

  The crossover probability, $Cr \in [0, 1]$, is a user-defined value that controls the fraction of parameter values that are copied from the mutant.
- **Selection**: The procedure for selection is based on the criteria of fitness. If the test element is better than the element in the population then it is replaced by the new one. This is achieved using Eq. (14).

$$x_{i,g+1} = \begin{cases} u_{i,g} & \textit{if } f(u_{i,g}) \leq f(x_{i,g}) \\ x_{i,g} & \text{otherwise} \end{cases} \tag{10}$$

The DE algorithm has been applied to different problems of science and engineering. As we mentioned in the introduction it is also applied to problems of fluid flows. A flow chart of the DE algorithm is shown below:

Algorithm

1. Initialize a random population of N with (j) individuals, d-dimensional vectors:

$$X_j = (x_{i1}, x_{i2}, \ldots, x_{id}), \quad j = 1, 2, \ldots, n. \tag{11}$$

2. Calculate the fitness of the initial population.
3. Using mutation operator, for each vector $X_j(t)$:

$$V_j(t+1) = X_m(t) + F \cdot (X_k(t) - X_l(t)).$$

   where:

   - $(k, l, m) \in 1, \ldots, N$: are mutually different, randomly selected indices and $(k, l, m) \neq j$.
   - X_m: is the base vector.
   - $X_k(t) - X_l(t)$: is the difference vector.
   - $F \in (1, 2)$: is the scaling factor.

4. Calculate a trial element [25] $U_j(t+1) = (u_{j1}, u_{j2}, \ldots, u_{jd})$ using equation:

$$U_{jn}(t+1) = \begin{cases} V_{jn}, & if(rand \leq CR) \vee (j = rnbr(i)); \\ X_{jn}, & if(rand > CR) \wedge (j \neq rnbr(i)). \end{cases}$$

   where:

   - $n = 1, 2, \ldots, d$
   - $rand \in (0, 1)$: is a uniform random number.
   - $CR \in (0, 1)$: is the user-specified crossover constant.
   - $rnbr(i)$: is a randomly chosen index selected from the range $(1, 2, \ldots, n)$.

5. A solution to the fluid test problem is obtained using an SPH code.
6. Selection: The resulting trial element replaces its predecessor, if it has higher fitness, otherwise the predecessor survives unchanged into the next iteration (Eq. 6), then a new population is created, and the process is repeated.

$$U_{jn}(t+1) = \begin{cases} U_{jn}(t) & If\ fit(U_{jn}(t)) > fit(x_{jn}) \\ x_{nj} & Otherwise \end{cases}$$

5 Boltzmann Univariate Marginal Distribution Algorithm

The BUMDA is a type of evolutionary algorithm that belongs to the family known as Estimation of Distribution Algorithms (EDA), this algorithm uses Boltzmann distribution and does not have any parameter apart from the initial distribution. Based on the comparative study using segmentation problems of Coronary Artery with genetic algorithms and univariate marginal distribution algorithm, it is found to perform better than the other algorithms [26].

The main idea in the algorithm of BUMDA is the approximation of the Boltzmann function as a normal distribution with parameters (μ, σ) [26]. The approximation to the mean and variance are given by Eqs. (12) and (13).

$$\mu \approx \frac{\sum_i^n x_i g(x_i)}{\sum_i^n g(x_i)} \tag{12}$$

$$v \approx \frac{\sum_i^n g(x_i)(x_i - \mu_i)^2}{1 + \sum_i^n g(x_i)} \tag{13}$$

where $g(x)$ represents the fitness function of the problem under optimization.

This algorithm does not have any parameters apart from the size of the population and is known to converge to an optimum value [26]. The steps of the algorithm are given below:

Algorithm

1. Initialize a random population of N with (j) individuals, d-dimensional vectors:
2. Evaluate the fitness and truncate the population [26]
3. Compute mean and variance as follows:

$$\mu = \frac{\sum_i^n x_i \acute{g}(x_i)}{\sum_i^n \acute{g}(x_i)}$$

$$Error = \|Exact(x,t) - SPH(x,t,p)\|$$

$$v = \frac{\sum_i^n \acute{g}(x_i)(x_i - \mu_i)^2}{1 + \sum_i^n \acute{g}(x_i)}$$

4. Generate random numbers with normal distribution using μ and v.
5. Take new population until convergence is readied.

6 Methodology of Optimization

The fundamental part of the fitness function is the exact solution to the test problems. Therefore a list of test problems (Sod shock tubes) for which exact solutions can be calculated are selected. In Table 3, the parameters of the test problems are given. The exact solutions were calculated based on the approach presented in Toro [24].

In Fig. 2, the methodology used to optimize the test problems is shown. First, we start by defining the test problem and obtain the exact solution. For a selected set of parameters of SPH, a numerical solution to the test problems are obtained. The evolutionary algorithm is used with the SPH solutions of the test problems, to start generating optimized parameters. Depending on test criteria on the fitness function, the final parameters are obtained and entered in SPH to generate the optimized solution.

This technique is tested with one-dimensional (spatial) fluid problems, as a first experiment. For the experimentation, 31 experiments were generated each with a different seed. The experiments were iterated until a relative error of 1×10^{-4}, or a maximum of 1000 iterations are reached. For both algorithms, the population size was taken as 100. The parameters of the DE are fixed as F = 1.5 and CR = 0.5.

Table 3 Parameters of the Sod shock tube test problems

Test	ρ_l	ρ_r	u_l	u_r	p_l	p_r
1	1.0	0.25	0.0	0.0	1.0	0.1795
2	1.0	0.25	0.7	0.0	1.0	0.1795
3	1.0	0.25	0.5	0.7	1.0	0.1795
4	1.0	0.2	0.5	−0.7	1.0	0.1795
5	1.0	0.2	−0.5	−0.7	1.0	0.1795
6	1.0	0.2	0.0	0.0	10.0	0.1795

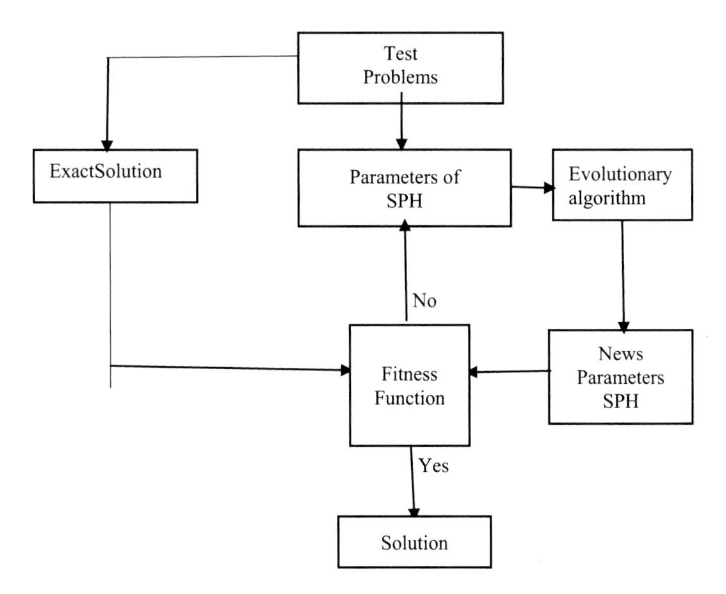

Fig. 2 Diagram of methodology of optimization using the evolutionary algorithm

In order to measure the error between the exact solution and the optimized SPH solution, Eq. (14) is used.

$$\text{Error} = \frac{1}{N}\sum_{i}^{N}\frac{|\text{S}_{\text{Exact}}(r_i, A_j, t) - \text{S}_{\text{SPH}}(r_i, A_j, t)|}{|\text{S}_{\text{Exact}}(r_i, A_j, t) + \text{S}_{\text{SPH}}(r_i, A_j, t)|} \tag{14}$$

Here p represents the best parameters obtained using the evolutionary algorithm.

7 Experimental Results

In Table 4 we present the statistical results of the 31 experiments, we calculated the error between the exact solution and its approximation as given above, but for different physical variables. These results are also plotted in Figs. 3, 4 and 5.

As can be seen in Fig. 4 and Table 4, in general the best results for the physical variable are obtained by the BUMDA algorithm. The DE algorithm may need a large population size to converge to an optimum value.

The pressure, density and velocity are plotted in Figs. 3 and 4 respectively.

From these figures, it can be seen that the BUMDA algorithm obtained best parameters as compared to the DE algorithm.

Table 4 Statistical results of the 31 experiments for mean of the physical variables

Test	BUMDA			DE		
	μ_ρ	μ_v	μ_P	μ_ρ	μ_v	μ_P
1	0.07928	1.483469	0.10546	0.36143	2.524051	0.61427
2	0.385441	1.096785	0.652479	0.65332	4.819627	1.12232
3	0.069689	0.559656	0.099817	0.289513	1.340455	0.44833
4	0.225399	2.383026	0.441639	0.584547	4.122395	1.02367
5	0.088645	0.701525	0.127884	0.261608	1.846571	0.39843
6	5.347165	250.4526	480.0676	23.70632	363.2583	1273.40

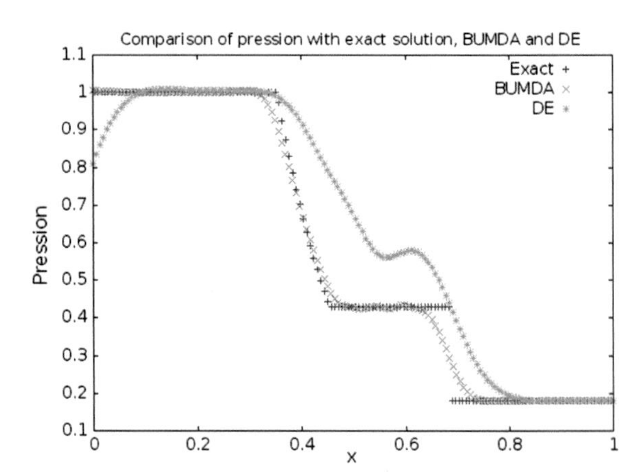

Fig. 3 Results of the evolutionary algorithms for the pressure of the fluid

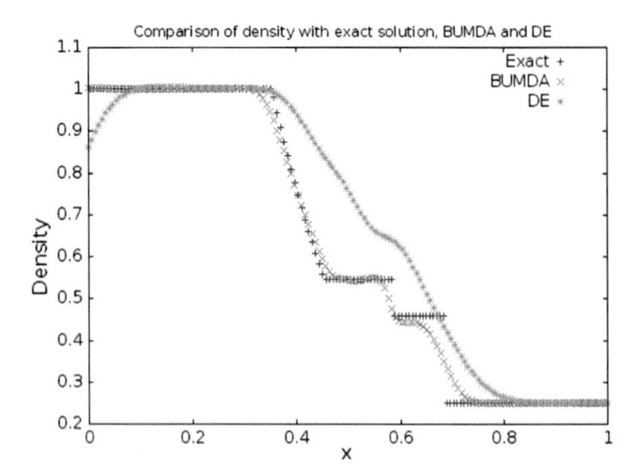

Fig. 4 Results of the evolutionary algorithms for the density of the fluid

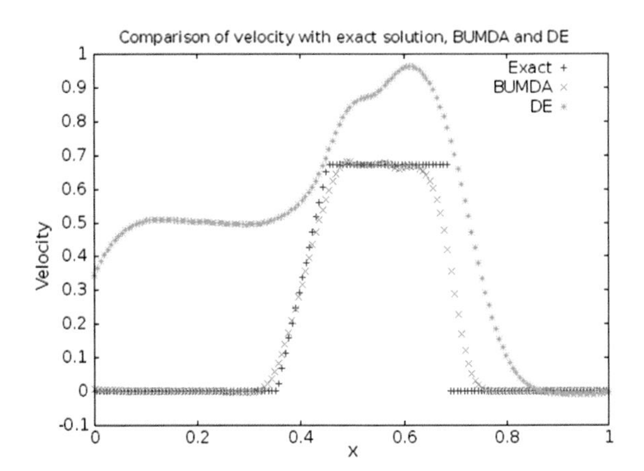

Fig. 5 Results of the evolutionary algorithms for the velocity of the fluid

7.1 Statistical Significance of the Results

In order to compare the performance of the algorithms, we use the Wilcoxon test with the following hypothesis:

$$\begin{aligned} H_0 &: Median_{differences} = 0 \\ H_1 &: Median_{differences} \neq 0 \end{aligned} \quad (15)$$

The null hypothesis is that the samples are the same, and the test returns a ***W_value***. For a chosen $Critical_{value}$, if $W_{value} \leq Critical_{value}$, it will be taken to indicate that the samples are drawn from different distributions.

The Wilcoxon Rank-Sum test was applied to the optimization experiments, using the median values from Table 4, for each of the physical variables, with a significance value of 0.05. From this statistical test, we infer that the BUMDA algorithm performs better than the DE.

8 Conclusions

Based on this study we conclude that it is possible to use evolutionary algorithms for parameter tuning in the SPH methodology. In particular, the BUMDA algorithm is more efficient. On the basis of the results, it can be seen that the DE algorithm may not obtain an optimum value for a fixed population size. Therefore Evolutionary Algorithms can be utilized to extend the optimization in real time.

References

1. J.J. Monaghan, *Smoothed Particle Hydrodynamics, Area*, vol. 30 (1992), pp. 543–574
2. N. Xiao, C. Leiting, X. Tao, Real-time incompressible fluid simulation on the GPU. Int. J. Comput. Games Technol. **2015**, 12 (2015)
3. A.W. Alshaer, B.D. Rogers, L. Li, Smoothed particle hydrodynamics (SPH) modelling of transient heat transfer in pulsed laser ablation of Al and associated free-surface problems. Comput. Mater. Sci. **127**, 161–179 (2017)
4. S.I. Inutsuka, Reformulation of smooth particle hydrodynamics with Riemann solver. J. Comput. Phys. **179**, 238–267 (2002)
5. M. Muller, D. Charypar, M. Gross, Particle-based fluid simulation for interactive applications, in *Proceedings of the 2003 ACM SIGGRAPH/Eurographics Symposium on Computer Animation* (2003), pp. 154–159
6. G. Fourtakas, B.D. Rogers, Modelling multi-phase liquid-sediment scour and resuspension induced by rapid flows using smoothed particle hydrodynamics (SPH) accelerated with a graphics processing unit (GPU). Adv. Water Resour. **92**, 186–199 (2016)
7. I. Markus, O. Jens, S. Barbara, K. Andreas, T. Matthias, *SPH Fluids in Computer Graphics* (The Eurographics Association, 2014)
8. H. Chen, Z. Jian, S. Hanqiu, W. Enhua, Parallel-optimizing SPH fluid simulation for realistic VR environments. Comput. Anim. Virtual Worlds **26**, 43–54 (2015)
9. R.A. Gingold, J.J. Monagahan, Kernel estimates as a basis for general particle methods in hydrodynamics. J. Comput. Phys. **46**, 429–453 (1982)
10. A. Barreiro, A.J.C. Crespo, J.M. Domínguez, M. Gómez-Gesteira, Smoothed particle hydrodynamics for coastal engineering problems. Comput. Struct. **120**, 96–106 (2013)
11. A. Corrado, A.J.C. Crespoc, J.M. Domínguez, M. Gómez-Gesteira, S. Tomohiro, V. Toon, Applicability of smoothed particle hydrodynamics for estimation of sea wave impact on coastal structures. Coast. Eng. **96**, 1–12 (2015)
12. E.L. Fasanella, K.E. Jackson, Impact testing and simulation of a crashworthy composite fuselage section with energy-absorbing seats and dummies. J. Am. Helicopter Soc. **49**, 140–148 (2004)
13. D. Guibert, M. de Leffe, G. Oger, J.-C. Piccinali, Efficient parallelisation of 3D SPH schemes, in *7th International SPHERIC Workshop*, Prato, Italy, SPHERIC (2012), pp. 259–265
14. M.S. Shadloo, M. Zainali, M. Yildiz, Improved solid boundary treatment method for the solution of flow over an airfoil and square obstacle by SPH method, in *5th International SPHERIC Workshop*, Manchester, UK, SPHERIC (2010), pp. 37–41
15. R.A. Gingold, J.J. Monaghan, Smoothed particle hydrodynamics - theory and application to non-spherical stars. Mon. Not. R. Astron. Soc. **181**, 375–389 (1977)
16. J.J. Monaghan, R.A. Gingold, Shock simulation by the particle method SPH. J. Comput. Phys. **52**, 374–389 (1983)
17. M. Steinmetz, E. Mueller, Hydrodynamical cosmology: galaxy formation in a cosmological context, metallicities and metallicity gradients. Astron. Ges. Abstract Ser. **281** (1992)
18. V. Springel, The cosmological simulation code GADGET-2, mnras, vol. 364 (2005), pp. 1105–1134
19. Q. Hongfu, G. Weiran, A new SPH equation including variable smoothing lengths aspects and its implementation. Comput. Mech. ISBN: 9783540759997_143 (2009)
20. J.C. Travis, H.D. Brian, X.H. Zhen, P.B. Wang, Aerodynamic shape optimization of a vertical-Axis wind turbine using differential evolution. ISRN Renew. Energy **2012**, 16 (2012)
21. T. Rogalsky, R.W. Derksen, Hybridization of differential evolution for aerodynamic design, in *Proceedings of the 8th Annual Conference of the Computational Fluid Dynamics Society of Canada* (2000), pp. 729–736
22. K.W. Wagner, The complexity of combinatorial problems with succinct input representation. Acta Inf. **23**, 325–356 (1986)

23. J.K. Lenstra, A.H.G. Rinnooy, P. Brucker, Complexity of machine scheduling problems. Ann. Discret. Math. **1**, 343–362 (1977)
24. E.F. Toro, *Riemann Solvers and Numerical Methods for Fluid Dynamics: A Practical Introduction* (Springer, ISBN. 3540616764, 1997)
25. R. Storn, K. Price, *Differential Evolution - A Simple and Efficient Adaptive Scheme for Global Optimization Over Continuous Spaces* (1995)
26. S.I. Valdez, A. Hernández, S. Botello, A Boltzmann based estimation of distribution algorithm. Inf. Sci. **236**, 126–137 (2013)

Symbolic Regression by Means of Grammatical Evolution with Estimation Distribution Algorithms as Search Engine

M. A. Sotelo-Figueroa, Arturo Hernández-Aguirre, Andrés Espinal, J. A. Soria-Alcaraz and Janet Ortiz-López

Abstract Grammatical Evolution (GE) is a Grammar-based form of Genetic Programming (GP) and it has been used to evolve programs or rules. The GE uses a population of linear genotypic strings and it is transformed by mapping process, those string are evolved using a search engine like the Genetic Algorithm (GA), Differential Evolution (DE), Particle Swarm Optimization (PSO), among others. One of the big trouble of these algorithms is the parameter tuning. In this paper is proposed an Estimation Distribution Algorithm (EDA) as search engine using the Symbolic Regression as a benchmark, due to the few parameters used by the EDA. The results were compared against the obtained by DE as search engine using the Friedman nonparametric test.

Keywords Grammatical Evolution · Estimation Distribution Algorithm
Symbolic Regression

1 Introduction

The Symbolic Regression (SR) is the process to obtain a representative expression from available data, it is used when we want to know which was the expression who generated the data. There are many mathematical methods to obtain the

M. A. Sotelo-Figueroa (✉) · A. Espinal · J. A. Soria-Alcaraz
División de Ciencias Económico Administrativas, Departamento de Estudios Organizacionales, Universidad de Guanajuato, Fraccionamiento I El Establo, 36250 Guanajuato, GTO, Mexico
e-mail: masotelof@gmail.com

A. Hernández-Aguirre
Departamento de Ciencias de La Computación, Centro de Investigación En Matemáticas, Jalisco S/N, 36240 Guanajuato, GTO, Mexico

J. Ortiz-López
Escuela Internacional de Doctorado, Universidad de Vigo, Róa do Conde de Torrecedeira 105, 36208 Vigo, Pontevedra, Spain

O. Castillo et al. (eds.), *Fuzzy Logic Augmentation of Neural and Optimization Algorithms: Theoretical Aspects and Real Applications*, Studies in Computational Intelligence 749, https://doi.org/10.1007/978-3-319-71008-2_14

expression, however, those methods try to find a linear equation. The SR tries to modeling the problem through an equation or expression finding the combination of variables, symbols, and coefficients.

The Grammatical Evolution (GE) [18] is a grammatical-base form of the Genetic Programming (GP) [11] that allow evolving programs or expressions. The GE is composed of a grammar, a problem instance, and the search engine. Several metaheuristics have been tested as a search engine like Differential Evolution (DE) [13, 14, 19], Genetic Algorithms (GA) [5], Particle Swarm Optimization (PSO) [21, 20], etc. However, is necessary to apply a strategy to obtain the values of they parameters and by this way to obtain good results with GE, this is known as a parameter tuning process [6].

In the present paper was applied an Estimation Distribution Algorithm (EDA) [12] as a search engine, because those algorithms have few parameters, the instances set used were the SR because is a well-known instance set used with the GP and GE, and it was proposed a grammar based on the state-of-art to generate equations. The results obtained was compared using the Friedman non-parametric test.

2 Grammatical Evolution

The Grammatical Evolution (GE) [18] is a grammar-based form of the Genetic Programming (GP) [11]. Both GE and GP use the concept of genotype and phenotype but in comparison with the GP that uses a genotype based on tree structures, the GE uses a linear genotype and combines it with a formal grammar.

The GE is composed of three main components, as seen in Fig. 1.

- **Problem Instance**: it allow to apply the proposed solution, with the problem instances is necessary to use a fitness function to realize the optimization process.
- **Grammar**: it allows to map between the genotype and the phenotype, this grammar is designed based on the problem instance.
- **Search Engine**: it manages the optimization and makes the changes over the genotype based on the quality of the phenotype applied to the instances problem (Fig. 2).

The GE is based on GA [5], however, is possible to use other metaheuristics as a search engine like the Differential Evolution (DE) [13, 15, 19], Particle Swarm Optimization (PSO) [21, 20], Evolutive Strategies (ES) [7] among others. However, those metaheuristics need to used a fine tuning process to obtain the value of the parameters used by each one.

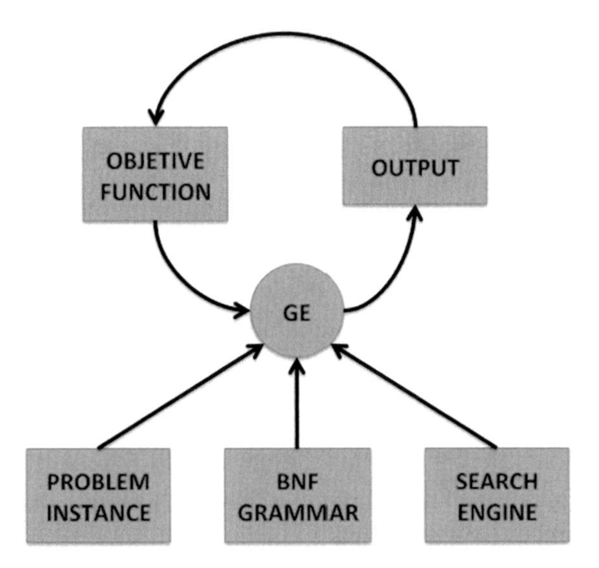

Fig. 1 The methodology proposed in [20] includes the main GE components. It can see the GE process in which the output is the result to mapping a genotype using the grammar and this output is evaluated in the objective function to obtain the fitness of each genotype. The values obtained by the objective function are used by the search engine to guide the search

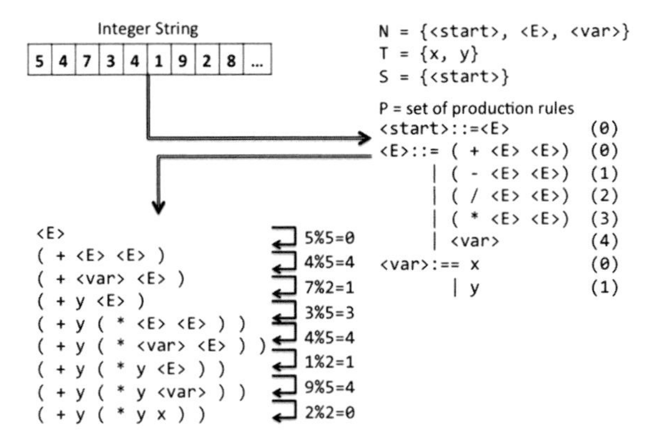

Fig. 2 An example of a transformation from genotype to phenotype using a BNF Grammar. The genotype is represented by an integer string and it can see the components of a BNF grammar including the set of productions rules used to map the genotype to phenotype. The process starts with the start symbol and the more left item is mapping using the current codon or item from the genotype using the corresponding rule

2.1 Estimation Distribution Algorithm

The Estimation Distribution Algorithms (EDA) [8, 12] are metaheuristics who builds an explicit probabilistic model with candidate solutions and uses the sampling to generates the new solutions.

The EDA starts to generate the solutions with uniform distribution and by each generation construct a probabilistic model to generate the solutions of the next generation, as seen in the Algorithm 2.1.

Algorithm 1 Estimation Distribution Algorithm

Require: n population size, m selected individuals
1: $l \leftarrow 0$
2: $D_l \leftarrow$ Create n individuals, initial population.
3: **while** stopping criterion not met **do**
4: $D_l^{Se} \leftarrow Select\,(D_l, m)$, select m individuals from the actual population according to the selection method.
5: $p_{l+1}(x) = p\left(x|D_l^{Se}\right)$, estimate the probability distribution using the selected individuals.
6: $D_{l+1} \leftarrow Sample\,(p_{l+1}, n)$, sample n individuals using the probability distribution.
7: **end while**

The EDA works construct probabilistic models and is possible to use different models depending on the problem, like the Univariate Marginal Distribution Algorithm (UMDA) [14], Probabilistic Based Incremental Learning (PBIL) [2], Mutual Information Maximization for Input Clustering (MIMIC) [4], Bivariate Marginal Distribution Algorithm (BMDA) [16], among others.

The UMDA uses the joint probability distribution to estimate the probability distribution through the product of independent univariate marginal distributions, Eq. 1. The distribution is estimated from the marginal frequencies, Eq. 3.

$$p_{l+1} = (x) = p(x|D_l^{Se}) = \prod_{i=1}^{n} p_{l+1}(x_i) \tag{1}$$

$$p_{l+1} = \frac{\sum_{j=1}^{N} \delta_j(X_i = x_i|D_l^{Se})}{N} \tag{2}$$

where

$$\delta_j(X_i = x_i|D_l^{Se}) = \begin{cases} 1 & \textit{if in the case } j\text{th } \textit{of } D_l^{Se}, X_i = x_i \\ 0 & \textit{otherwise} \end{cases} \tag{3}$$

3 Symbolic Regression

The Symbolic Regression (SR) [1, 3, 9] is the process to obtain a representative expression from available data, it is used when we want to know which was the expression who generated the data. It's also considered as a process to identify an exact mathematical expression from experimental data [17, 22].

Figure 3 shows the available data, those data can be obtained without knowing the function equation who generate them. Figure 4 shows an Equation $X^3 + X^2 + X$ who generates the available data and lets to know the unknown points.

4 Experiments

The proposal is compared with GE with DE and the Artificial Bee Colony Programming (ABCP) [9]. The parameters used by each metaheuristic are shown in Table 1.

The 10 functions used are shown in Table 2, those functions were taken from [9]. To evolve the expressions was designed a grammar, Grammar 1, who was based on the state of art. To discern the quality of those expressions was use Eq. 4 as a fitness function.

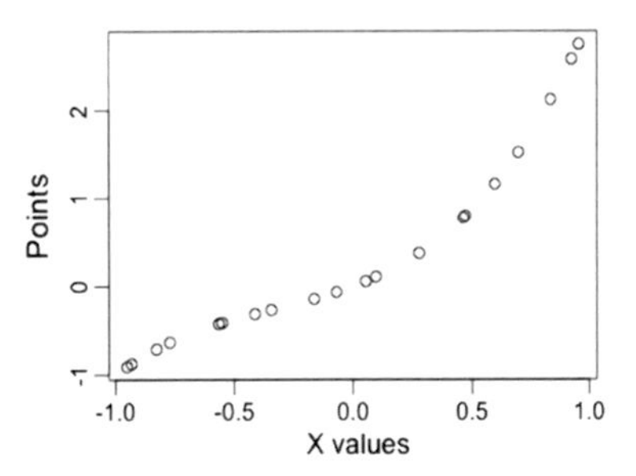

Fig. 3 Available data

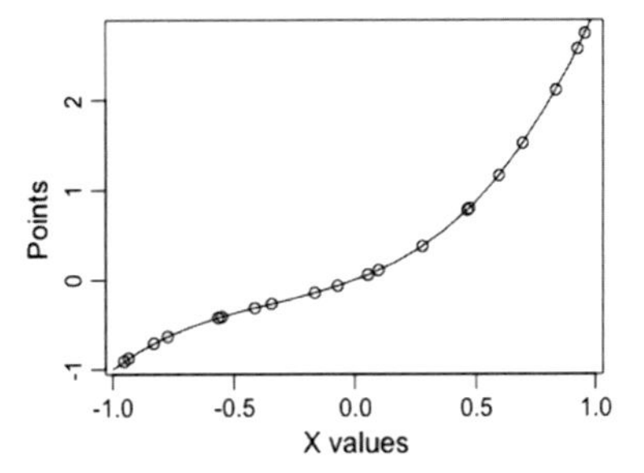

Fig. 4 Representative expression

Table 1 Parameters used by each metaheuristic

Parameter	Algorithm		
	GE-UMDA	GE-DE	ABCP
Population size	1000	100	500
F	–	0.8	–
Cr	–	0.9	–
Limit	–	–	500
Function calls	250,000		

Table 2 Symbolic regression functions used as instances set [10]

Function	Fit cases
$F_1 = X^3 + X^2 + X$	20 random points $\subseteq[-1, 1]$
$F_2 = X^4 + X^3 + X^2 + X$	
$F_3 = X^5 + X^4 + X^3 + X^2 + X$	
$F_4 = X^6 + X^5 + X^4 + X^3 + X^2 + X$	
$F_5 = \sin(x^2)\cos(x) - 1$	
$F_6 = \sin(x) + \sin(x + x^2)$	
$F_7 = \log(x + 1) + \log(x^2 + 1)$	20 random points $\subseteq[0, 2]$
$F_8 = \sqrt{x}$	20 random points $\subseteq[0, 4]$
$F_9 = \sin(x) + \sin(y^2)$	2*100 random points $\subseteq[-1, 1]$, $X\subseteq[-1, 1]$, $Y\subseteq[-1, 1]$
$F_{10} = 2\sin(x)\cos(y)$	

$$
\begin{aligned}
\langle \text{inicio} \rangle &\models \langle \text{expr} \rangle \\
\langle \text{expr} \rangle &\models (\langle \text{expr} \rangle \langle \text{op} \rangle \langle \text{expr} \rangle) | \langle \text{pre-op} \rangle (\langle \text{expr} \rangle) | \langle \text{var} \rangle \\
\langle \text{op} \rangle &\models + | - | / | * \\
\langle \text{pre-op} \rangle &\models \sin | \cos | \exp | \log \\
\langle \text{var} \rangle &\models X | Y | 1
\end{aligned}
$$

Grammar 1 Grammar proposed to Symbolic Regression problem

$$
Fitness = \sum_{j=1}^{N} \left\| g_j - t_j \right\| \tag{4}
$$

For each function was performed 33 individual experiments and was chosen the median to compared the results. Those results were compared with the obtained by GE with DE and ABCP. The comparison was implemented using the Friedman non-parametric test to discern if exists differences among them and select the best algorithm as a search engine.

Table 3 Results obtained for each metaheuristic at instance set

Instance	GE-EDA	GE-DE	ABCP
F1	0.00000000	27.80000E−18	0.010000
F2	1.110000E−16	0.146795	0.050000
F3	0.00000000	0.136496	0.070000
F4	6.110000E−16	0.044394	0.100000
F5	0.229721	0.097149	0.050000
F6	0.156171	0.149126	0.020000
F7	0.169668	0.044259	0.060000
F8	0.294106	0.001773	0.100000
F9	0.00000000	0.00000000	0.470000
F10	40.70000E−16	0.475674	1.060000

5 Results

Table 3 shows the results obtained for each instance. The results shown are the median from the 33 experiments performed. With this results was performed the Friedman non-parametric test.

The value obtained with the Friedman non-parametric test is 0.65 and a p-value 0.75. The p-value is bigger than 0.1 and it's no possible to execute a post hoc procedure to determine which metaheuristic gave better results.

6 Conclusions

In the present paper was proposed an Estimation Distribution Algorithm (in this case the Univariate Marginal Distribution Algorithm) as the Search Engine of Grammatical Evolution. The new approach removes the fine tuning process to select the parameters. The results obtained by the new approach was compared against those obtained by Differential Evolution and the Artificial Bee Colony Programming using the Friedman non-parametric test. The test shows there not exists statistical information to discern about the performances of the metaheuristics to solve the Symbolic Regression Problem. With those results is possible to study the use of another kind of Estimation Distribution Algorithms looking dependence among codons, trying to build a more effective probabilistic model with the candidate solutions information.

Acknowledgements The authors want to thank to *Universidad de Guanajuato* (UG) for the support to this research.

References

1. D.A. Augusto, H.J.C. Barbosa, Symbolic regression via genetic programming, in *Proceedings of the VI Brazilian Symposium on Neural Networks (SBRN '00)* (IEEE Computer Society, Washington, DC, USA, 2000), p. 173. http://dl.acm.org/citation.cfm?id=827249.827526
2. S. Baluja, Population-based incremental learning: a method for integrating genetic search based function optimization and competitive learning. Tech. rep., Pittsburgh, PA, USA (1994)
3. P. Barmpalexis, K. Kachrimanis, A. Tsakonas, E. Georgarakis, Symbolic regression via genetic programming in the optimization of a controlled release pharmaceutical formulation. Chemom. Intell. Lab. Syst. **107**(1), 75–82 (2011). http://www.sciencedirect.com/science/article/pii/S0169743911000153
4. J.S. De Bonet, C.L. Isbell, P. Viola, Mimic: finding optima by estimating probability densities, in *Advances in Neural Information Processing Systems*, vol. 9 (1997)
5. I. Dempsey, M. O'Neill, A. Brabazon, Foundations in grammatical, in *Foundations in Grammatical Evolution for Dynamic Environments*, vol. 194 (Springer, New York, 2009)
6. A. Eiben, S. Smit, Parameter tuning for configuring and analyzing evolutionary algorithms. Swarm Evol. Comput. **1**(1), 19–31 (2011)
7. A. Espinal, M. Carpio, M. Ornelas, H. Puga, P. Melín, M. Sotelo-Figueroa, *Evolutionary Indirect Design of Feed-Forward Spiking Neural Networks*, vol. 101 (Springer International Publishing, 2015), p. 89. http://dx.doi.org/10.1007/978-3-319-17747-2\s\do5(7)
8. M. Hauschild, M. Pelikan, An introduction and survey of estimation of distribution algorithms. Swarm Evol. Comput. **1**(3), 111–128 (2011). http://www.sciencedirect.com/science/article/pii/S2210650211000435
9. D. Karaboga, C. Ozturk, N. Karaboga, B. Gorkemli, Artificial bee colony programming for symbolic regression. Inf. Sci. **209**, 1–15 (2012). http://www.sciencedirect.com/science/article/pii/S0020025512003295
10. T.M. Khoshgoftaar, N. Seliya, Y. Liu, Genetic programming-based decision trees for software quality classification, in *Proceedings of the 15th IEEE International Conference on Tools with Artificial Intelligence*, pp. 374–383 (Nov 2003)
11. J.R. Koza, R. Poli, Genetic programming, in *Search Methodologies: Introductory Tutorials in Optimization and Decision Support Techniques*, ed. by E.K. Burke, G. Kendall (Kluwer, Boston, 2005), pp. 127–164
12. P. Larrañaga, J.A. Lozano, *Estimation of Distribution Algorithms*, vol. 2 (Springer, US, 2002)
13. A. Moraglio, S. Silva, Geometric differential evolution on the space of genetic programs, in *Genetic Programming*, vol. 6021, Lecture notes in computer science, ed. by A. Esparcia-Alcázar, A. Ekárt, S. Silva, S. Dignum, A. Uyar (Springer, Berlin, 2010), pp. 171–183
14. H. Mühlenbein, The equation for response to selection and its use for prediction. Evol. Comput. **5**(3), 303–346 (1997). https://doi.org/10.1162/evco.1997.5.3.303
15. M., O'Neill., A, Brabazon, Grammatical differential evolution, in *International Conference on Artificial Intelligence (ICAI '06)* (CSEA Press, Las Vegas, Nevada, 2006)
16. M. Pelikan, H. Muehlenbein, *The Bivariate Marginal Distribution Algorithm* (Springer, London, 1999), pp. 521–535. http://dx.doi.org/10.1007/978-1-4471-0819-1\s\do5(3)9
17. R. Riolo, T. Soule, B. Worzel (eds.), *Genetic Programming Theory and Practice IV*, No. 1 (Springer, US, 2007)
18. C. Ryan, J.J. Collins, M. O'Neill: Grammatical evolution: evolving programs for an arbitrary language, in *Proceedings of the First European Workshop on Genetic Programming*. Lecture notes in computer science, vol. 1391 (Springer, Berlin, 1998), pp. 83–95
19. M. Sotelo-Figueroa, H. Puga Soberanes, J. Martin Carpio, H. Fraire Huacuja, L. Cruz Reyes, J. Soria-Alcaraz, Evolving and reusing bin packing heuristic through grammatical differential evolution, in *2013 World Congress on Nature and Biologically Inspired Computing (NaBIC)*, pp. 92–98 (2013)

20. Sotelo-Figueroa, M.A., Puga Soberanes, H.J., Carpio, J.M., Fraire Huacuja, H.J., Cruz Reyes, L., Soria-Alcaraz, J.A.: Improving the bin packing heuristic through grammatical evolution based on swarm intelligence. Mathematical Problems in Engineering 2014 (2014)
21. J. Togelius, R.D. Nardi, A. Moraglio, Geometric pso + gp = particle swarm programming, in *IEEE Congress on Evolutionary Computation*, pp. 3594–3600 (2008)
22. P. Widera, J.M. Garibaldi, N. Krasnogor, Gp challenge: evolving energy function for protein structure prediction. Genet. Program Evolvable Mach. **11**(1), 61–88 (2010). https://doi.org/10.1007/s10710-009-9087-0

Part IV
Fuzzy Control

Impact Study of the Footprint of Uncertainty in Control Applications Based on Interval Type-2 Fuzzy Logic Controllers

Emanuel Ontiveros, Patricia Melin and Oscar Castillo

Abstract Fuzzy control is one of the most important applications of Fuzzy Logic, and with the emergence of Type-2 Fuzzy Logic, now Type-2 Fuzzy Logic Controllers provide the possibility of consider the uncertainty in the controller design, and these results are very useful for noisy environments and with multiple uncertainty sources. However, it is important to identify the relationship between the FOU and noise robustness, observing the behavior of the IT2 FLC with different FOUs in different uncertainty context. The main goal of this paper is to evaluate the impact of the Footprint of Uncertainty (FOU) in the performance of an Interval Type-2 Fuzzy Logic Controllers (IT2 FLC). The experiments considered two plants, evaluating the performance of the same IT2 FLC by changing only the FOU, evaluated in with different noise levels, this in order to find the controller behavior by the variation of the FOU. In addition, we propose to use a heuristic optimization method based on the behavior knowledge and by adjusting the FOU.

Keywords Fuzzy control · Interval Type-2 Fuzzy Logic · Footprint of uncertainty

1 Introduction

Nowadays, Type-1 Fuzzy logic controllers (T1 FLC) [1], have been successfully applied in many kinds of problems. However, with the emergence of Interval Type-2 Fuzzy logic controllers (IT2 FLC), that consider the uncertainty in their design, it is interesting to evaluate the performance of both in similar perturbation conditions.

The capability of IT2-FLC to consider the uncertainty in their design takes importance in real applications, this is because real world problems have many uncertainty sources, and this uncertainty sources can affect the control performance or stability of controllers in many applications.

E. Ontiveros · P. Melin · O. Castillo (✉)
Tijuana Institute of Technology, Tijuana, Mexico
e-mail: ocastillo@tectijuana.mx

O. Castillo et al. (eds.), *Fuzzy Logic Augmentation of Neural and Optimization Algorithms: Theoretical Aspects and Real Applications*, Studies in Computational Intelligence 749, https://doi.org/10.1007/978-3-319-71008-2_15

However, this IT2-FLC capability have some disadvantages, for example, the computational cost, because it demands approximately twice the computational cost than T1-FLCs, on the other hand, the degree of uncertainty considered in the IT2-FLC's design is not directly proportional to the controller performance.

The relevance of this study relates to understanding the effect of the variation of the footprint of uncertainty (FOU) on the controller performance.

For understanding the IT2 FLC behavior by varying the FOU, we propose to use a heuristic optimization method [2] to improve the controller performance, based in the variation of the FOU.

The contribution of this study is to contribute to better understanding of IT2 FLCs, finding the relation between the FOU and controller performance in order to use this relation for controller optimization.

The organization of this work is the following, in Sect. 2 we are presenting the background of Fuzzy Logic and Fuzzy Logic Controllers, Sect. 3 describes the experiments, firstly the setup and then the experiments results. We propose to use two benchmark problems, the D.C. Motor speed controller and the Water Tank Level control, both controlled by a non optimized IT2 FLC. Also this section reports on the implementation results of the optimization algorithm proposed by changing only the FOU. Finally, Sect. 4 presents the conclusions and future works.

2 Fuzzy Logic and Fuzzy Logic Controllers

In this section, we briefly describe basic concepts of Fuzzy Logic and specifically one of the most popular application areas, Fuzzy Control.

2.1 T1 Fuzzy Logic Controllers

Fuzzy Sets were introduced by Zadeh in 1965 [3], and propose the use of continuous membership degrees described by mathematical equations known as membership functions (MF), these MFs are associated with concepts and by special operations can be expressed in rules form in order to make decisions. The mathematical form of Fuzzy Sets is expressed following Eq. (1).

$$A = \{(x, \mu_A(x)) | x \in X\} \tag{1}$$

where $\mu_A(x) : X \rightarrow [0, 1]$ is the membership degree of x to the set A.

In 1974 Mamdani [4] implemented successfully a controller based on Fuzzy Logic, in other words, a Fuzzy Logic Controller (FLC), and betters results than classical controllers where achieved with easy controller design.

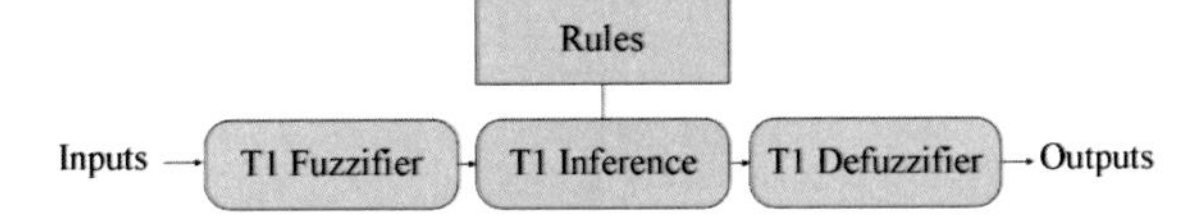

Fig. 1 Mamdani T1 FLC

Figure 1 illustrates a Mamdani Fuzzy Logic Controller.

Today there exist variations of FLCs for many control applications [1, 5–7], and these achieve obtain betters results than classical controllers.

2.2 *IT2 Fuzzy Logic Controllers*

Type-2 Fuzzy logic was proposed as an extension of T1 FL, it is based on the statement "Words can mean different things to different people" [8].

In order to consider the uncertainty in their design, Interval Type-2 Membership functions (IT2 MF) are composed by two T1 membership functions (T1 MF), the upper membership function and lower membership function. The area between the upper and lower T1 MF is called the Footprint of Uncertainty (FOU) [9]. The mathematical expression of the IT2 MF can be expressed as in Eq. (2).

$$\tilde{A} = \{((x, u), 1) | \forall x \in X, \forall u \in J_x \subseteq [0, 1]\} \tag{2}$$

The graphical representation of an IT2 MF is illustrated in Fig. 2.

On the other hand, the implementation of the Mamdani FLC with IT2 FL is very similar to the T1 FL implementation, the main difference is that two separate blocks of the fuzzifier and inference are used, for the upper and lower fuzzy sets, and a new block called Type-Reduction [10] is also used. The schematic of the IT2 FLC is illustrated in Fig. 3.

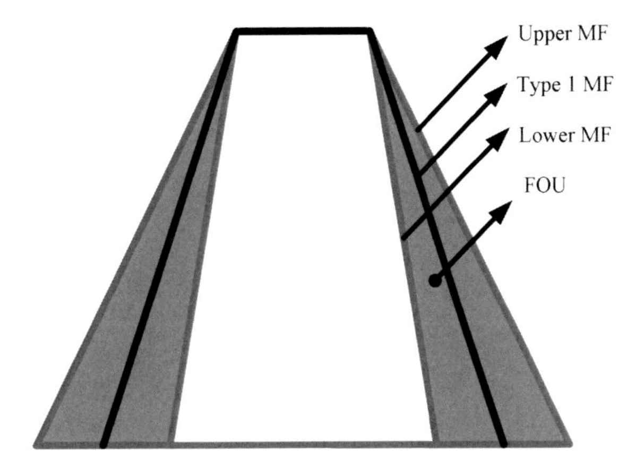

Fig. 2 IT2 MF

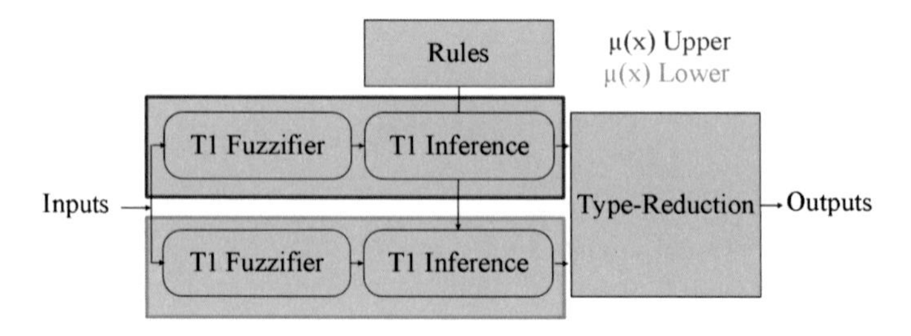

Fig. 3 Mamdani IT2 FIS

The IT2 FLC has been studied extensively because of its capabilities, and some examples of these studies are [11, 12]. Several approaches of IT 2 FLC have been recently reported, for example in [12–18].

3 Experiments and Results

This section starts with the experimental setup in order to understand better the experiments realized, secondly, each benchmark problem is presented, the corresponding controller setup and the experiments results. Finally based on these results a controller behavior with respect to the FOU's variation is proposed.

Finally, the experimental results of the implementation of a heuristic optimization method are reported.

3.1 Experimental Setup

In order to find the behavior of the IT2 FLC by variation of the FOU, the experiments were proposed with the setup illustrated in Fig. 4.

The variation of the FOU is proposed based on a T1 Trapezoidal MF [3], and we propose the design of an IT2 Trapezoidal MF expressed in Eq. (3).

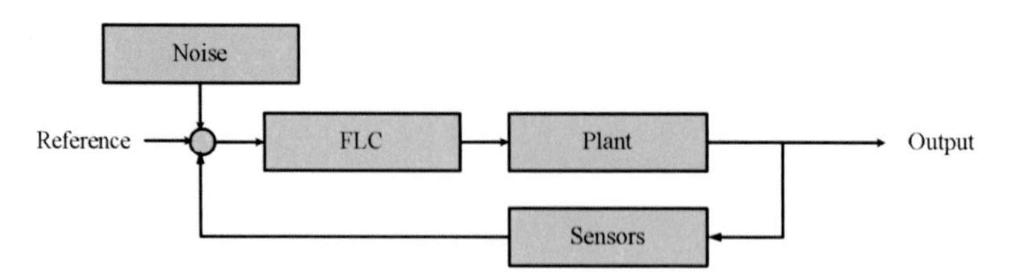

Fig. 4 Experimental setup

$$IT2MF = \begin{cases} trapmf\left(x, \left[a - \frac{u(d-a)}{2}, b, c, d + \frac{u(d-a)}{2}\right]\right) \\ trapmf\left(x, \left[a + \frac{u(d-a)}{2}, b, c, d - \frac{u(d-a)}{2}\right]\right) \end{cases} \quad (3)$$

where u is a proposed variable to change the FOU of the IT2MF, and the design considers the upper and lower membership functions. The graphical representation of Eq. (3) is illustrated in Fig. 5.

As can be observed, this approach proposes to consider only symmetric MFs, in this study.

On the other hand, the experiments also consider evaluating the IT2 FLC behavior in noisy environments, and this noise was introduced by the Random Number block of Simulink illustrated in Fig. 6.

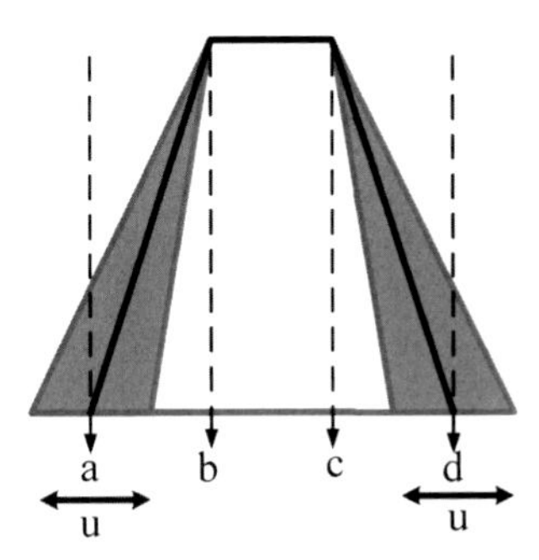

Fig. 5 IT2 TrapMF

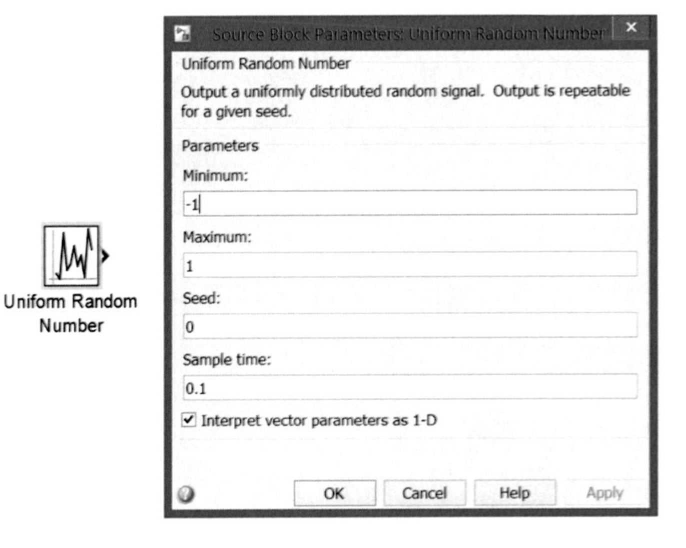

Fig. 6 Uniform random number block

The *Minimum* and *Maximum* parameters are proposed in Eq. (4).

$$\begin{aligned} Minim &= -n(SignalMagnitude) \\ Maxim &= n(SignalMagnitude) \end{aligned} \tag{4}$$

where $n \in (0, 1]$ allows changing the noise proportionally in any input.

Next, is realized the performance evaluation of IT 2 FLVs by changing the parameters n and u.

Remembering that u is directly related to the FOU and n is related to the noise magnitude.

The performance metrics reported in this work are the Integral Absolute Error (IAE) and Integral Time-weighted Absolute Error (ITAE).

3.2 Plant 1: D.C. Motor Speed Control

The D.C. motor speed control (Fig. 7) is a common benchmark problem in many works [17], in its different versions, but is a stable second order plant in which the goal is reducing the error in the reference speed.

In the present work, we propose that the IT2 FLC works with two inputs (Error and Error change) and with one output (Delta voltage). This structure is illustrated in Fig. 8.

As previously shown the IT2 FLC output is accumulated, and this controller is the equivalent of a PID controller.

The IT2 FLC fuzzy sets are illustrated in Fig. 9.

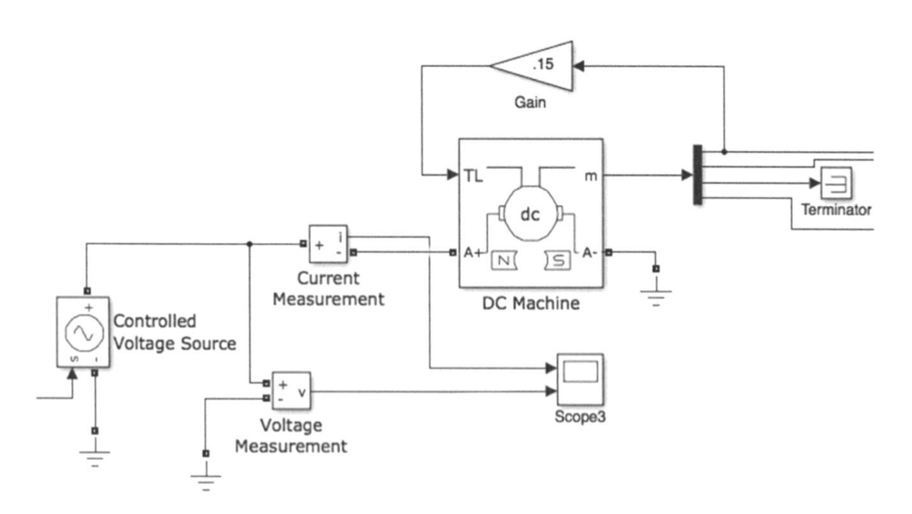

Fig. 7 D.C. motor speed control graphical representation

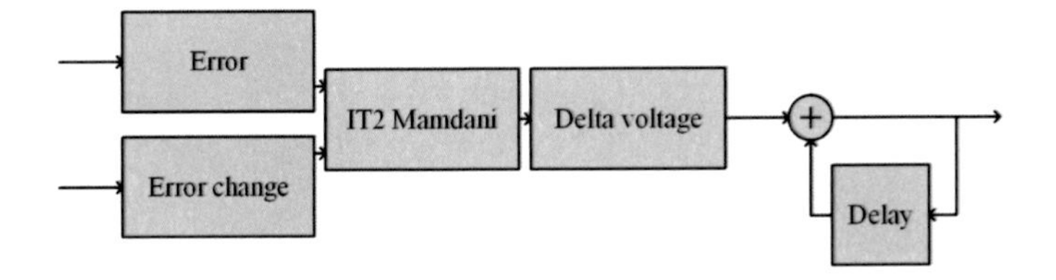

Fig. 8 D.C. motor speed IT2 FLC

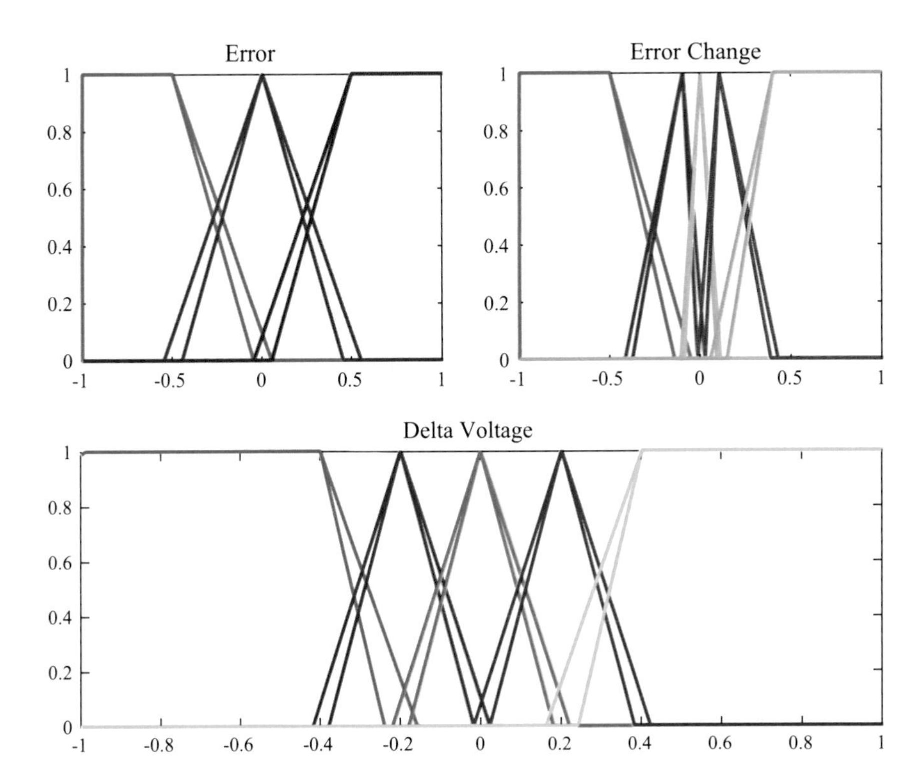

Fig. 9 D.C. motor speed controller fuzzy sets

In the experiments, we proposed to use nine values of u and n, realizing a total de 81 different performance evaluations. Figure 10 illustrates the different control surfaces obtained by changing the parameter u.

As we can observe, most of the control surfaces are very similar, but when the parameter u overpass 0.3 the control surface changes considerately.

Tables 1 and 2 report the IT2 FLC's IAE and ITAE respectively by changing u and n, in order to find the control behavior of the IT2 FLC.

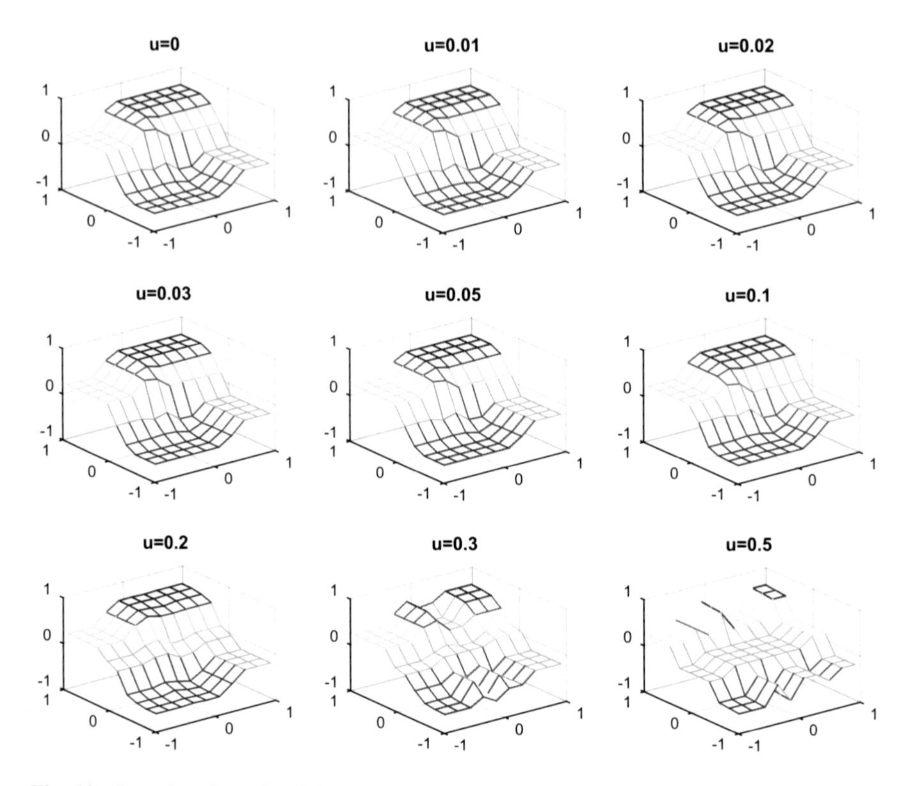

Fig. 10 Control surfaces for different u parameter values

Table 1 D.C. motor speed controller IAE by changing parameters u and n

n	u								
	0	0.01	0.02	0.03	0.05	0.10	0.20	0.30	0.50
0	0.11080	**0.11070**	0.11070	0.11350	0.11970	0.11980	0.11350	0.12070	0.12180
0.01	0.11160	0.11130	**0.11090**	**0.11090**	0.11100	0.11140	0.11330	0.12070	0.12180
0.02	0.11270	0.11220	0.11150	**0.11110**	**0.11100**	0.11140	0.11330	0.12070	0.12180
0.03	0.11370	0.11320	0.11240	0.11170	**0.11110**	0.11140	0.11330	0.12070	0.12940
0.05	0.11530	0.11500	0.11400	0.11350	0.11180	**0.11140**	0.11360	0.11520	0.12940
0.1	0.11680	0.11690	0.11720	0.11730	0.11710	0.11420	**0.11390**	0.11620	0.11830
0.2	0.11970	0.11970	0.11970	0.11980	0.12030	0.11890	**0.11750**	0.12150	0.12400
0.3	**0.13130**	0.13140	0.13160	0.13270	0.13400	0.13460	0.13270	0.13650	0.13430
0.5	0.16130	0.16120	0.16100	0.16030	**0.15980**	**0.15980**	0.16070	0.16280	0.16710

Bold values mean "best values" of the results presented in the Table

Figure 11 illustrates the control performance behavior for four different rows of Table 1, with different values of noise by changing the parameter n.

Figure 12 illustrates the control performance behavior for four different rows of Table 2, with different noise values by changing the n parameter.

Table 2 D.C. motor speed controller ITAE by changing u and n parameters

n	*u*								
	0	0.01	0.02	0.03	0.05	0.10	0.20	0.30	0.50
0	**0.01530**	**0.01530**	**0.01530**	0.01600	0.01780	0.01780	0.01570	0.01630	0.01650
0.01	0.01550	0.01540	**0.01530**	**0.01530**	**0.01530**	0.01540	0.01560	0.01630	0.01650
0.02	0.01580	0.01570	0.01550	0.01540	**0.01530**	0.01540	0.01560	0.01630	0.01650
0.03	0.01600	0.01590	0.01570	0.01550	**0.01540**	**0.01540**	0.01560	0.01630	0.01760
0.05	0.01640	0.01630	0.01600	0.01590	0.01550	**0.01540**	0.01570	0.01580	0.01760
0.1	0.01650	0.01650	0.01660	0.01670	0.01670	0.01600	**0.01560**	0.01590	0.01610
0.2	0.01690	0.01690	0.01690	0.01690	0.01710	0.01670	**0.01610**	0.01650	0.01690
0.3	0.01840	0.01840	0.01840	0.01870	0.01900	0.01910	**0.01810**	0.01850	0.01790
0.5	0.02230	0.02230	0.02220	0.02200	**0.02190**	**0.02190**	0.02200	0.02220	0.02260

Bold values mean "best values" of the results presented in the Table

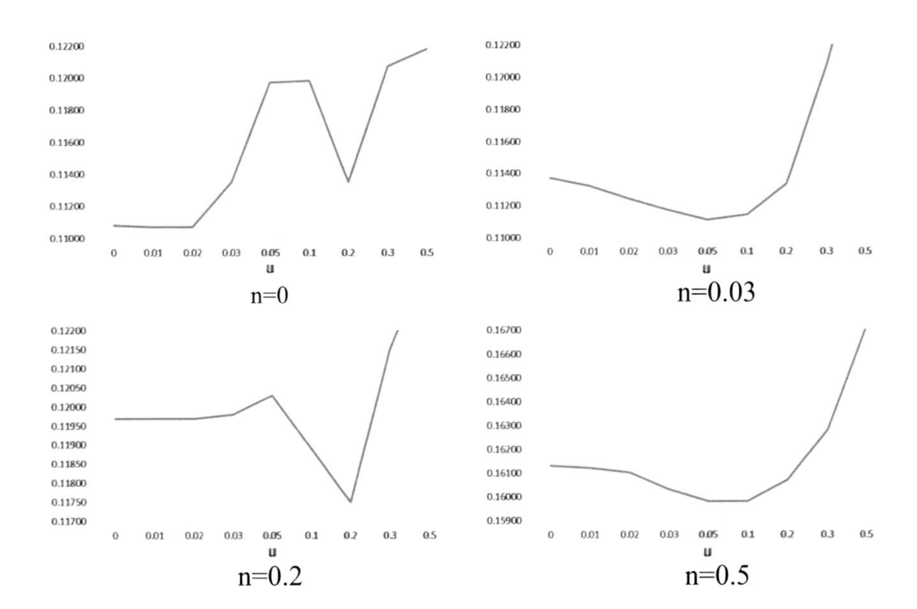

Fig. 11 IAE behavior by changing the *u* parameter and with different levels of noise

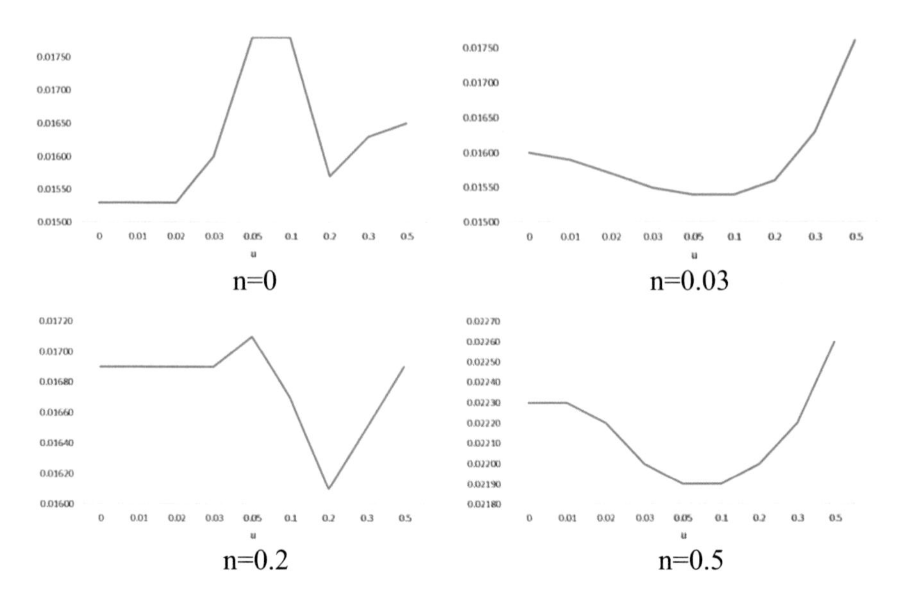

Fig. 12 ITAE behavior by changing the *u* parameter and with different noise levels

3.3 *Plant 3: Water Tank Level Controller*

Water level control (Fig. 13) is also a benchmark problem used to test the performance of different controllers and optimization methods [19].

The controller of this plant has two inputs, *level* and *rate*, and with one output, valve, this is a IT2 Mamdani FLC, and is illustrated in Fig. 14.

The membership functions of the fuzzy logic controller are illustrated in Fig. 15.

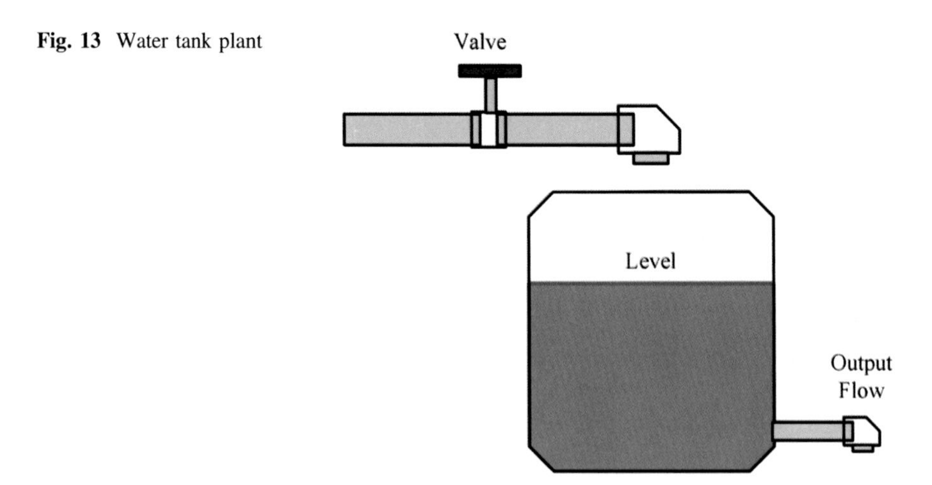

Fig. 13 Water tank plant

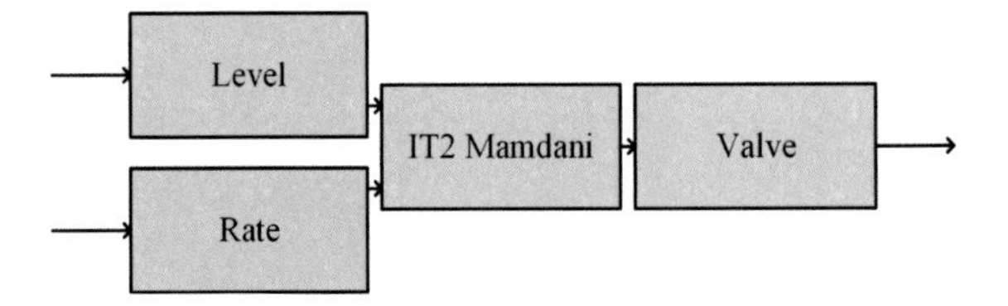

Fig. 14 Water tank level IT2 FLC

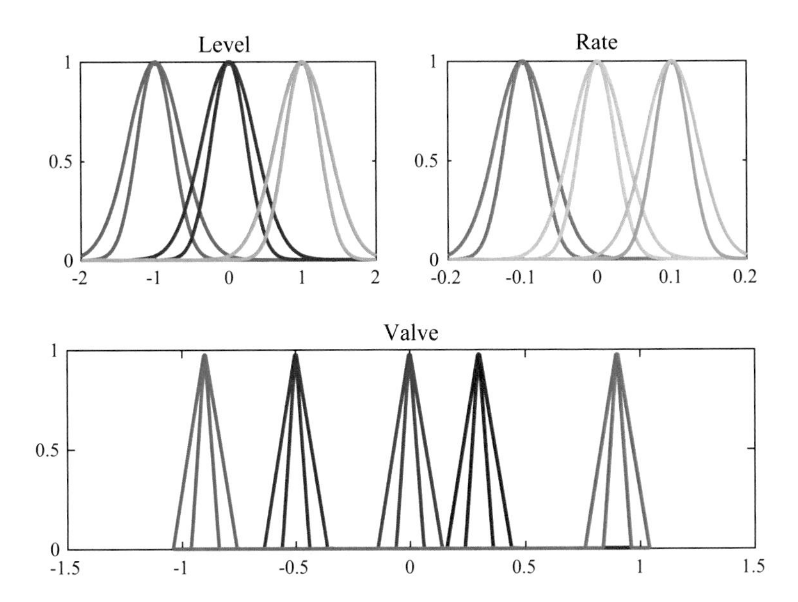

Fig. 15 MFs of water tank level IT2 FLC

The experiments with this plant consist in evaluating nine IT2 FLCs with different FOUs by changing the u parameter, and the control surfaces of these controllers are illustrated in Fig. 16.

Tables 3 and 4 report the performance results obtained by the combination of different values of, the n and u parameters, and this performance results are reported by IAE and ITAE respectively.

Figure 17 illustrates four of the rows of Table 3 and it represents the IAE's behavior in four different noise environments.

Figure 18 illustrates four of the rows of Table 4 and it represents the ITAE behavior in four different noise environment.

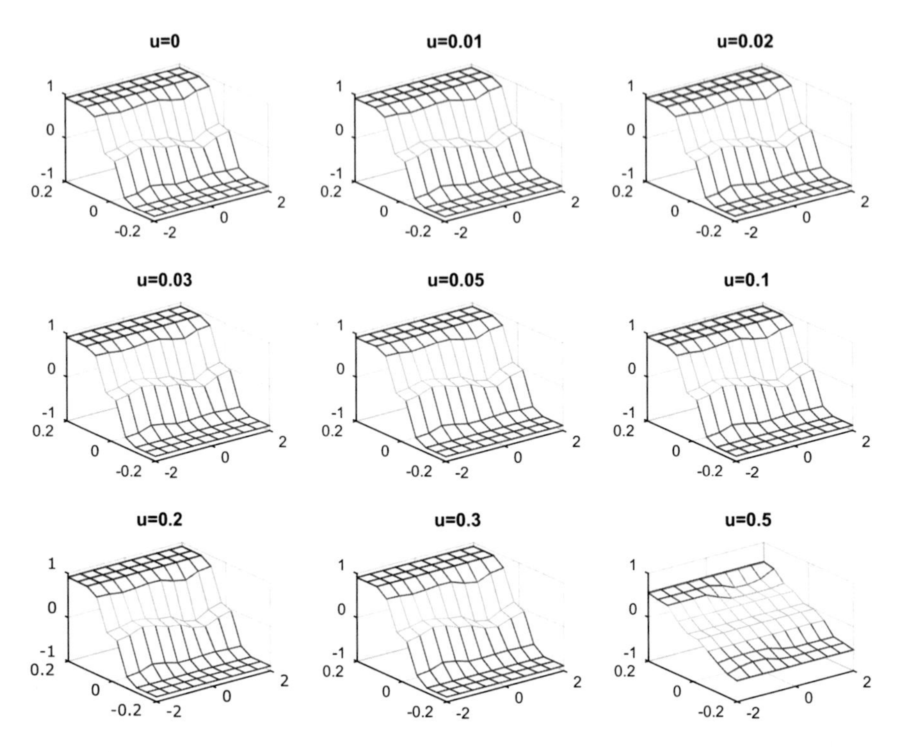

Fig. 16 Control surfaces by changing the *u* parameter

Table 3 IT2 FLC IAE by changing parameters u and n

n	*u*								
	0	1	2	3	5	10	20	30	50
0	2.9359	2.9353	2.9345	2.9328	2.9287	**2.9249**	2.9541	2.9581	2.9436
0.01	2.9327	2.9323	2.9312	2.9295	2.9256	**2.9232**	2.9574	2.9544	2.9385
0.02	2.9291	2.9286	2.9274	2.9258	2.9220	**2.9214**	2.9595	2.9495	2.9348
0.03	2.9248	2.9245	2.9233	2.9221	**2.9184**	2.9199	2.9599	2.9433	2.9324
0.05	2.9146	2.9142	2.9132	2.9118	**2.9095**	2.9144	2.9550	2.9277	2.9316
0.1	2.8742	**2.8741**	2.8744	2.8750	2.8771	2.8840	2.9027	2.8800	2.9536
0.2	**2.8260**	2.8261	2.8264	2.8269	2.8280	2.8316	2.8624	2.9362	3.0578
0.3	**3.0488**	3.0491	3.0497	3.0506	3.0488	3.0532	3.0748	3.1273	3.2323
0.5	**3.3175**	3.3187	3.3206	3.3247	3.3357	3.3608	3.4576	3.5812	3.7666

Bold values mean "best values" of the results presented in the Table

Table 4 IT2 FLC ITAE by changing u and n parameters

n	u								
	0.000	1.000	2.000	3.000	5.000	10.000	20.000	30.000	50.000
0	7.7526	**7.7502**	7.7491	7.7552	7.7886	8.0082	9.4798	10.4000	11.0204
0.01	7.6656	7.6624	**7.6616**	7.6645	7.6950	7.9290	9.4480	10.3081	10.9234
0.02	7.5764	7.5734	**7.5717**	7.5733	7.6038	7.8554	9.4077	10.2051	10.8338
0.03	7.4852	7.4805	7.4796	**7.4782**	7.5138	7.7900	9.3554	10.0892	10.7492
0.05	7.2947	7.2912	**7.2898**	7.2933	7.3440	7.6701	9.2121	9.8286	10.6086
0.1	**6.8661**	6.8692	6.8929	6.9337	7.0589	7.3943	8.5501	9.1199	10.4427
0.2	**7.1059**	7.1093	7.1275	7.1555	7.2310	7.4153	8.1792	9.3113	10.8711
0.3	**9.6494**	9.6558	9.6695	9.6934	9.7260	9.8834	10.5422	11.3978	12.8726
0.5	**12.7049**	12.7207	12.7559	12.8230	13.0110	13.4330	14.9673	16.7778	19.5106

Bold values mean "best values" of the results presented in the Table

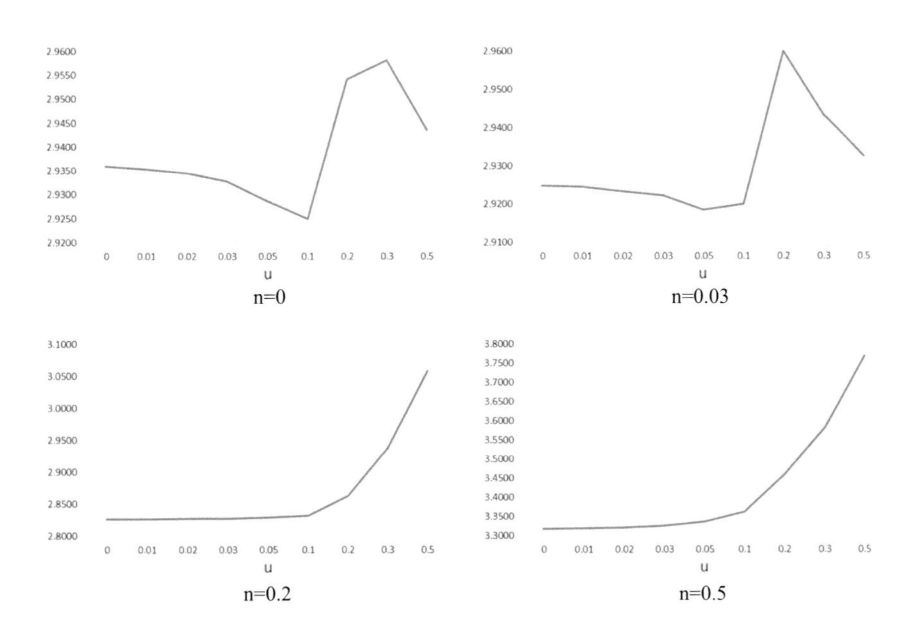

Fig. 17 IAE behavior by changing u parameter and with different noise levels

3.4 *Heuristic Optimization*

Based on the results reported in Tables 1, 2, 3 and 4, we propose a performance behavior by changing the parameter u of the controller and is illustrated in Fig. 19.

With the proposed behavior we implemented a heuristic algorithm called Perturb and Observe (P&O), and this algorithm is used in solar energy in the Maximum Power Point Tracking (MPPT) [2, 20].

This algorithm basically effectuates two performance evaluations with different FOU levels, then calculate the delta value, expressed in Eq. (5) and decides if

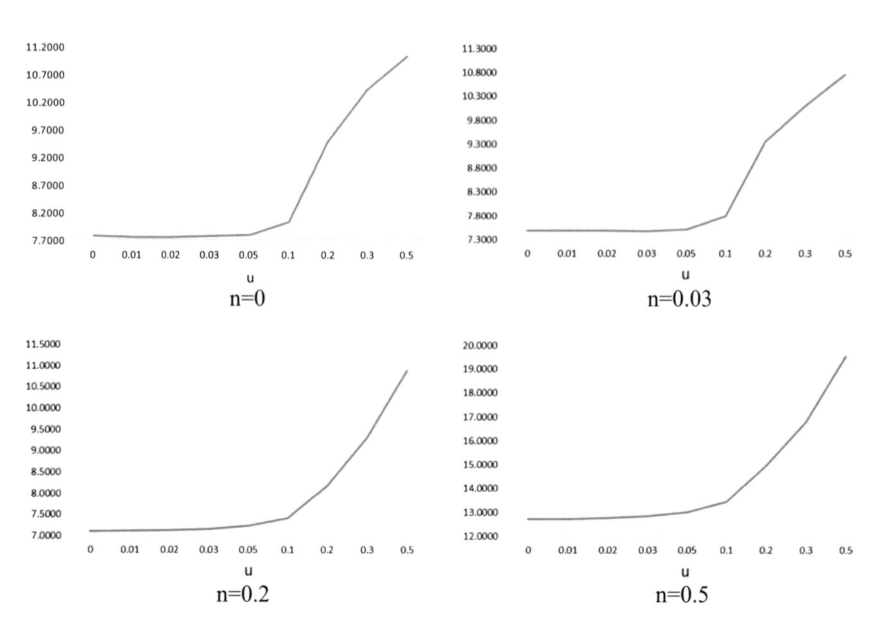

Fig. 18 ITAE behavior by changing u parameter and with different noise levels

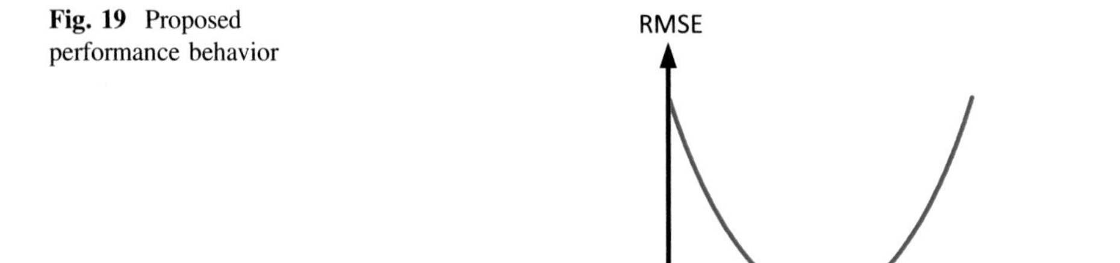

Fig. 19 Proposed performance behavior

increase or decrease the u parameter. This decision is only possible if the controller behavior is known in advance. The algorithm flowchart is illustrated in Fig. 20.

$$Delta = RMSE_i - RMSE_{i-1} \tag{5}$$

Next we show an implementation of P&O to optimize a D.C. Motor speed IT2 FLC, and the results of this implementation are illustrated in Fig. 21.

As can be observed, the P&O algorithm achieve the optimization of the IT2 FLC, and this is possible because the behavioral proposed is consistent with the controller behavior by changing the u parameter, i.e. by changing the IT2 FLC FOU.

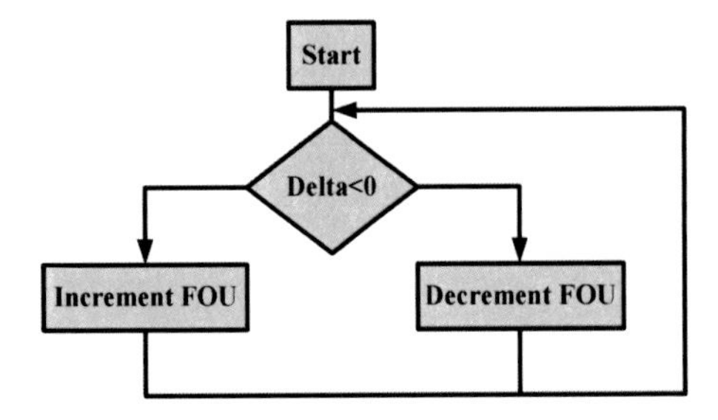

Fig. 20 P&O flowchart

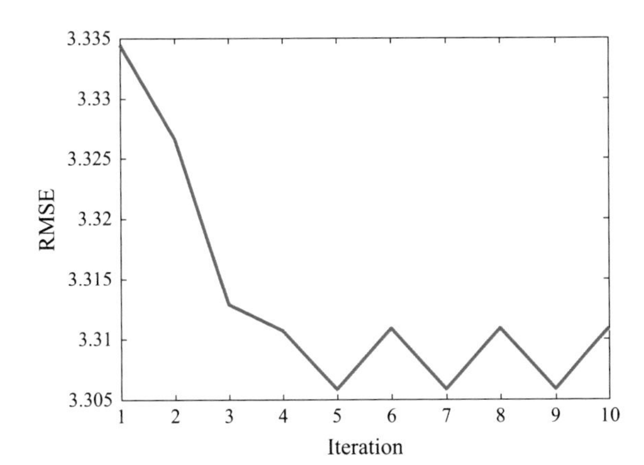

Fig. 21 D.C. motor speed IT2 FLC

4 Conclusions

As a conclusion, the experiments realized in the present work allow obtaining a better understanding of the IT2 FLC performance. Based on this, an IT2 FLCs behavior was proposed by changing the FOU; this behavior can be appreciated in Fig. 19 and shows a performance improvement by increasing the FOU but with a critical point where it begins to decrease the performance of the controller. This behavior can be appreciated in the control surfaces illustrated in Figs. 10 and 16, where when the parameter u surpasses the critical point the control surface is appreciated with significant changes. Based on the proposed behavior, the optimization algorithm P&O was successfully implement, achieving to optimization of the IT2 FLC's control performance by changing only the FOU, that is proportional at parameter u. As future work we propose to optimize the IT2 FLC by changing parameter *u,* but with another optimization method, for example, Fuzzy Logic [21]. On the other hand, the controllers studied in the present work are not the optimal but is interesting to implement our optimization method based on the FOU, to optimized controllers by MF's parameters, like in [22, 23]. In addition, other forms or types of fuzzy logic could be used like in [24–30].

References

1. C.C. Lee, Fuzzy logic in control systems: fuzzy logic controller. I. IEEE Trans. Syst. Man Cybern. **20**(2), 404–418 (1990)
2. J. Ahmed, Z. Salam, An improved perturb and observe (P&O) maximum power point tracking (MPPT) algorithm for higher efficiency. Appl. Energy **150**, 97–108 (2015)
3. L.A. Zadeh, Fuzzy logic = computing with words. IEEE Trans. Fuzzy Syst. **4**(2), 103–111 (1996)
4. E.H. Mamdani, Application of fuzzy algorithms for control of simple dynamic plant. Proc. Inst. Electr. Eng. **121**(12), 1585–1588 (1974)
5. H.O. Wang, K. Tanaka, M.F. Griffin, An approach to fuzzy control of nonlinear systems: stability and design issues. IEEE Trans. Fuzzy Syst. **4**(1), 14–23 (1996)
6. M.S. Masmoudi, N. Krichen, M. Masmoudi, N. Derbel, Fuzzy logic controllers design for omnidirectional mobile robot navigation. Appl. Soft Comput. **49**, 901–919 (2016)
7. J. Liu, W. Zhang, X. Chu, Y. Liu, Fuzzy logic controller for energy savings in a smart LED lighting system considering lighting comfort and daylight. Energy Build. **127**, 95–104 (2016)
8. J.M. Mendel, R.I.B. John, Type-2 fuzzy sets made simple. IEEE Trans. Fuzzy Syst. **10**(2), 117–127 (2002)
9. J.M. Mendel, R.I. John, F. Liu, Interval type-2 fuzzy logic systems made simple. IEEE Trans. Fuzzy Syst. **14**(6), 808–821 (2006)
10. A.D. Torshizi, M.H.F. Zarandi, H. Zakeri, On type-reduction of type-2 fuzzy sets: a review. Appl. Soft Comput. **27**, 614–627 (2015)
11. H. Hagras, Type-2 FLCs: a new generation of fuzzy controllers. IEEE Comput. Intell. Mag. **2** (1), 30–43 (2007)
12. O. Castillo, P. Melin, A review on interval type-2 fuzzy logic applications in intelligent control. Inf. Sci. **279**, 615–631 (2014)
13. M.E. Abdelaal, H.M. Emara, A. Bahgat, Interval type 2 fuzzy sliding mode control with application to inverted pendulum on a cart. IEEE Int. Conf. Ind. Technol. (ICIT) **2013**, 100–105 (2013)
14. U. Farooq, J. Gu, J. Luo, On the interval type-2 fuzzy logic control of ball and plate system. IEEE Int. Conf. Robot. Biomimetics (ROBIO) **2013**, 2250–2256 (2013)
15. R. Sepúlveda, O. Montiel, O. Castillo, P. Melin, Embedding a high speed interval type-2 fuzzy controller for a real plant into an FPGA. Appl. Soft Comput. **12**(3), 988–998 (2012)
16. H. Sahu, R. Ayyagari, Interval fuzzy type-II controller for the level control of a three tank system. IFAC-PapersOnLine **49**(1), 561–566 (2016)
17. I.F. Davoudkhani, M. Akbari, Adaptive speed control of brushless DC (BLDC) motor based on interval type-2 fuzzy logic, in *2016 24th Iranian Conference on Electrical Engineering (ICEE)* (2016), pp. 1119–1124
18. M.A. Sanchez, O. Castillo, J.R. Castro, Generalized type-2 fuzzy systems for controlling a mobile robot and a performance comparison with interval type-2 and type-1 fuzzy systems. Expert Syst. Appl. **42**(14), 5904–5914 (2015)
19. O. Castillo, L. Amador-Angulo, J.R. Castro, M. Garcia-Valdez, A comparative study of type-1 fuzzy logic systems, interval type-2 fuzzy logic systems and generalized type-2 fuzzy logic systems in control problems. Inf. Sci. **354**, 257–274 (2016)
20. J. Gosumbonggot, Maximum power point tracking method using perturb and observe algorithm for small scale DC voltage converter. Proc. Comput. Sci. **86**, 421–424 (2016)
21. H. Bounechba, A. Bouzid, K. Nabti, H. Benalla, Comparison of Perturb & Observe and fuzzy logic in maximum power point tracker for PV systems. Energy Proc. **50**, 677–684 (2014)
22. A. Sombra, F. Valdez, P. Melin, O. Castillo, A new gravitational search algorithm using fuzzy logic to parameter adaptation, in *IEEE Congress on Evolutionary Computation* (Cancun, Mexico 2013), pp. 1068–1074

23. F. Valdez, P. Melin, O. Castillo, Evolutionary method combining particle swarm optimization and genetic algorithms using fuzzy logic for decision making, in *IEEE International Conference on Fuzzy Systems* (2009), pp. 2114–2119
24. G.M. Mendez, O. Castillo, Interval type-2 TSK fuzzy logic systems using hybrid learning algorithm, in *The 14th IEEE International Conference on Fuzzy Systems, 2005 FUZZ'05*, pp. 230–235
25. O. Castillo, P. Melin, Design of intelligent systems with interval type-2 fuzzy logic, in *Type-2 Fuzzy Logic: Theory and Applications* (2008), pp. 53–76
26. O. Castillo, P. Melin, E. Ramírez, J. Soria, Hybrid intelligent system for cardiac arrhythmia classification with Fuzzy K-Nearest Neighbors and neural networks combined with a fuzzy system. Expert Syst. Appl. **39**(3), 2947–2955
27. L. Aguilar, P. Melin, O. Castillo, Intelligent control of a stepping motor drive using a hybrid neuro-fuzzy ANFIS approach. Appl. Soft Comput. **3**(3), 209–219
28. P. Melin, O. Castillo, in *Modelling, Simulation and Control of Non-linear Dynamical Systems: An Intelligent Approach Using Soft Computing and Fractal Theory* (CRC Press, 2001)
29. P. Melin, C.I. Gonzalez, J.R. Castro, O. Mendoza, O. Castillo, Edge-detection method for image processing based on generalized type-2 fuzzy logic. IEEE Trans. Fuzzy Syst. **22**(6), 1515–1525
30. P. Melin, O. Castillo, Intelligent control of complex electrochemical systems with a neuro-fuzzy-genetic approach. IEEE Trans. Ind. Electron. **48**(5), 951–955

Optimization of Membership Function Parameters for Fuzzy Controllers of an Autonomous Mobile Robot Using the Firefly Algorithm

Marylu L. Lagunes, Oscar Castillo and Jose Soria

Abstract Bio-Inspired Algorithms are effective for solving optimization problems, in particular for finding the appropriate parameter values for the membership functions used in fuzzy control. Fuzzy controllers are widely used in engineering, industrial, and medical solutions and other fields. Fuzzy models help to represent informal, unstructured abstract knowledge into formal mathematical models. In this paper the firefly algorithm is used to optimize fuzzy controllers for autonomous mobile robots. In this work optimization of parameters of the membership functions in fuzzy control systems allows a better performance of the actuators that are controlling an autonomous robot. This article will explain the proposed methodology for the optimization of parameters of membership functions of a tracking fuzzy controller for a mobile autonomous robot using the firefly algorithm.

Keywords Firefly algorithm · Membership functions · Optimization problems Fuzzy system

1 Introduction

This article describes the optimization of the membership function parameters of a fuzzy controller for an autonomous mobile robot. Currently, bio-inspired algorithms are being used extensively [1–5], to solve optimization problems. These algorithms are making a great contribution to the scientific community, because with their application many real problems are solved or improve results. Industry, in particular, is taking advantage of fuzzy logic [6, 7], because fuzzy mathematics makes mathematical models by capturing human knowledge, through linguistic variables, which is a representative form of how humans describe and specify knowledge [8]. These fuzzy models [9] are used to build fuzzy controllers that are very useful today for the creation of tools that we use day to day. For example, the intelligent washer

M. L. Lagunes · O. Castillo (✉) · J. Soria
Tijuana Institute of Technology, Tijuana, BC, Mexico
e-mail: ocastillo@tectijuana.mx

O. Castillo et al. (eds.), *Fuzzy Logic Augmentation of Neural and Optimization Algorithms: Theoretical Aspects and Real Applications*, Studies in Computational Intelligence 749, https://doi.org/10.1007/978-3-319-71008-2_16

has a controller that uses fuzzy logic, which helps know how much water, soap, fabric softener, should we use in each wash cycle. Meta-heuristics [10–12] are also taking on much relevance since they are better at solving problems than traditional methods [13, 14], and these are usually used to solve optimization problems.

The most popular meta-heuristics to solve optimization problems are those inspired by nature. Such as the Genetic Algorithm (GA), which is inspired on biological evolution [15]. The representation of solutions is based on a phenotype, which can be transformed by selection, Crossover, Mutation, and replacement. The ant colony optimization (ACO) algorithm is inspired by swarm intelligence, particularly in the ant colony which is looking for the shortest path between its anthill and its food source [16, 17]. The Particle Swarm Optimization (PSO) algorithm is a recent addition to the list of global search methods. It is a stochastic optimization technique based on the population developed by [18], inspired by the social behavior of the birds, the schooling of fish and the theory of the swarm.

The rest of the paper is organized as follows. Section 2 describes the bio-inspired algorithm to be used for optimization [19–21]. Section 3 explains the development of the solution of the problem, and Sect. 4 we are presenting the results obtained.

2 Firefly Algorithm

The firefly algorithm has been increasingly used, for example in [22–25], where we can observe their ability to solve problems in different areas of research.

This Algorithm was first presented in [26] at the University of Cambridge, and is based on the intermittent patterns and behavior of fireflies.

The FA is based on three idealized rules [26]:

1. Fireflies are unisex so that one firefly will be attracted to other fireflies regardless of their sex.
2. The attractiveness is proportional to the brightness, and decreases when the distance increases between two fireflies. Thus for any two flashing fireflies, the less brighter one will move towards the brighter one. If there is no brighter one than a particular firefly, then it will move randomly.
3. The brightness of a firefly is determined by the landscape of the objective function.

2.1 Equation of Attractiveness

As the attractiveness of a firefly is proportional to the luminous intensity observed by the adjacent fireflies, we can now define the variation of attractiveness β with the distance r and is given by:

$$\beta = \beta_0 e^{-\gamma r^2} \tag{1}$$

where the initial attractiveness at r = 0 is represented by β_0, γ determines the variation of attractiveness (absorption coefficient) by increasing the distance between the fireflies, and r is the distance between each two fireflies.

2.2 Equation of Movement

The movement of a firefly i attracted by another more attractive firefly (brighter) j is determined by:

$$x_i^{t+1} = x_i^t + \beta_0 e^{-\gamma r_{ij}^2} \left(x_j^t - x_i^t \right) + \alpha_t \varepsilon_i^t \tag{2}$$

The current position is represented by the term x_i^t the second term is determining the initial attraction of the firefly β_0 where the exploitation is, γ is the absorption coefficient, r is the Euclidean distance between the positions of the firefly i and the firefly j. The last term handles the exploration, where α is the parameter that controls how much randomness the firefly is allowed to have in its movement and is a vector containing random numbers drawn from a Gaussian distribution or uniform distribution at time t.

2.3 Equation of Distance

The variable r is the distance between two fireflies i and j, which is the Cartesian distance, between the corresponding vectors

$$r_{ij} = \sqrt{\sum_{k=1}^{d} \left(x_{i,k} - x_{j,k} \right)^2}, \tag{3}$$

where $x_{i,k}$ is the *k*th component of the firefly spatial coordinate x_i firefly, r_{ij} is the Euclidean distance between two fireflies i and j.

3 Fuzzy Controller

In related work on optimization the usefulness of meta-heuristics is described and in this case the bio-inspired algorithms in nature [27–30]. For this reason in this work the fuzzy controller is optimized with the firefly algorithm inspired by the behavior of blinking fireflies.

This fuzzy controller is of Mamdani type, it has two inputs, the first is the linear velocity error (ev) and the second is the angular velocity error (ew), the first output is torque 1 (t1), and finally the second output is torque 2 (t2). The linguistic variables are: negative (N), zero (Z) and positive (P) for all inputs and outputs.

This fuzzy controller has 9 if-then rules, which are shown below [31]:

1. If (ev is N) and (ew is N) then (T1 is N) (T2 is N) (1)
2. If (ev is N) and (ew is Z) then (T1 is N) (T2 is Z) (1)
3. If (ev is N) and (ew is P) then (T1 is N) (T2 is P) (1)
4. If (ev is Z) and (ew is N) then (T1 is Z) (T2 is N) (1)
5. If (ev is Z) and (ew is Z) then (T1 is Z) (T2 is Z) (1)
6. If (ev is Z) and (ew is P) then (T1 is Z) (T2 is P) (1)
7. If (ev is P) and (ew is N) then (T1 is P) (T2 is N) (1)
8. If (ev is P) and (ew is Z) then (T1 is P) (T2 is Z) (1)
9. If (ev is P) and (ew is P) then (T1 is P) (T2 is P) (1)

These rules of the fuzzy system describe a relation between the linguistic variables based on the membership functions, taking into account the inputs and outputs.

4 Optimization of Parameters of the Membership Functions

For the development of this methodology the following steps were carried out in the firefly algorithm:

1. A dimension of 40 for the position of the points to be optimized
2. Lower and upper limit of −1 to 1 is used as rank in membership functions
3. The controller fuzzy inference system was established as an objective function.

For the fuzzy controller only generated a function that received the data (the fireflies) for the optimization of the parameters of the membership functions.

The methodology to optimize the parameters of the membership functions with the firefly algorithm, then evaluate the fuzzy controller in the plant and the results obtained in the desired path, as shown in Fig. 1.

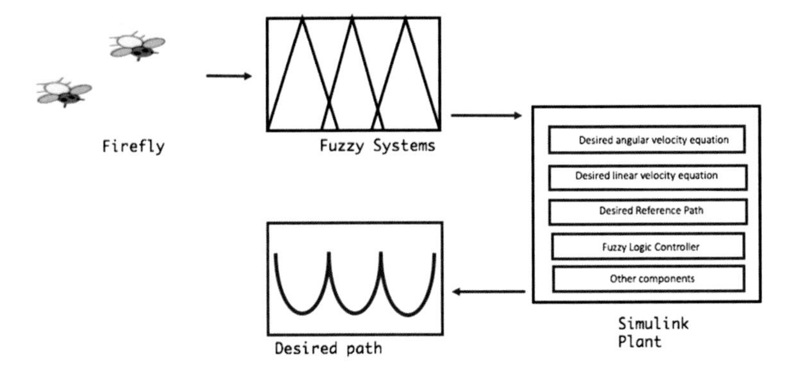

Fig. 1 Methodology for optimization of the parameters of the membership functions

5 Simulation

We performed 30 experiments using the proposed methodology for the optimization of the membership parameters, obtaining a minimum error of 0.0048.

In Table 1 the results obtained from the methodology for the optimization of a fuzzy inference system for a controller of a mobile autonomous robot using the firefly algorithm are observed.

6 Conclusions

In this work, a methodology was proposed for the optimization of parameters of membership functions, of an autonomous mobile controller using the firefly algorithm.

The objective of the methodology is described, for which it is explained and described that it is necessary to lower the minimum error that gives the simulation of the robot, once understood the objective and the methodology to be performed, the necessary coding of linking of the bio-inspired algorithm data with the fuzzy controller for its optimization.

We performed 30 experiments to observe the efficacy of the proposed methodology, using the mean square error as a metric.

As future work we would like to implement other optimization methods, like in [32, 33], or hybrid approaches for designing controllers, like in [34, 35].

Table 1 Fuzzy controller for trajectory tracking

FireFly						
Experiments	FireFlies	Iterations	Alpha	Beta	Gamma	Error
1	30	1000	0.5	1	0.1	0.082291808
2	30	1000	0.6	1	0.1	0.073915385
3	35	550	0.7	0.4	0.2	0.004817189
4	30	500	0.6	1	0.1	0.075920923
5	30	100	0.8	1	0.1	0.627
6	25	100	0.8	1	0.1	0.67
7	25	100	0.7	1	0.1	0.26
8	25	100	0.6	1	0.1	0.6754
9	25	100	0.5	1	0.1	0.67
10	40	800	0.8	0.2	0.1	0.0772
11	38	850	0.7	0.5	0.1	0.1119
12	45	850	0.9	0.5	0.1	0.0739
13	28	600	0.9	0.3	0.1	0.08229181
14	45	750	0.9	0.4	0.1	0.110954
15	35	1000	0.8	0.4	0.1	0.0739
16	35	550	0.3	0.3	0.2	0.2818
17	35	550	0.7	0.4	0.2	0.11095
18	50	800	0.9	0.5	0.1	0.1952
19	40	300	0.8	0.3	0.1	0.073915
20	45	600	0.9	0.9	0.1	0.3279
21	50	780	0.7	0.3	0.1	0.077152
22	33	900	0.5	0.3	0.1	0.07391539
23	32	2000	0.8	0.1	0.1	0.073915
24	42	600	0.9	0.2	0.1	0.077152391
25	42	800	0.8	0.5	0.1	0.7715239
26	50	680	0.6	0.6	0.1	0.0759
27	20	600	0.9	0.3	0.1	**0.0048**
28	20	800	0.8	0.5	0.1	0.0772
29	33	700	1	0.5	0.1	0.077152
30	33	500	0.6	0.2	0.1	0.073915
		Average				0.212358196

The bold value indicate "best value" of the results presented in the table

References

1. D. Karaboga, B. Basturk, A powerful and efficient algorithm for numerical function optimization: artificial bee colony (ABC) algorithm. J. Glob. Optim. **39**(3), 459–471 (2007)
2. I. Fister, I. Fister Jr, J. Brest, V. Zumer, Memetic artificial bee colony algorithm for large-scale global optimization, in *IEEE Congress on Evolutionary Computing*, pp. 1–8 (2012)
3. X. Yang, S. Deb, Cucko search via levy flights, in *World Congress on Nature & Biologically Inspired Computing*, pp. 210–214 (2009)

4. X. Yang, A new metaheuristic bat-inspired algorithm. Stud. Comput. Intell. **284**, 65–74 (2010)
5. A. Gandomi, A. Alavi, Krill herd: a new bio-inspired optimization algorithm. Commun. Nonlinear Sci. Numer. Simul. **17**(12), 4831–4845 (2012)
6. G. Aluja, A. Kaufmann, *Operational management techniques for the treatment of uncertainty* (Hispano Europea, Barcelona, 1987)
7. A. Kaufmann, J. Gil Aluja, Theory of expertons and fuzzy logic, in *Fuzzy Sets and Systems*, España, Milladoiro, pp. 295–304 (1986)
8. L.A. Zadeh, Fuzzy logic. Computer **21**(4), 83–93 (1988)
9. L. Zadeh, "Fuzzy sets. Inf Control **8**, 338–353 (1965). Department of Electrical Engineering and Electronics Research Laboratory
10. B. Gonzalez, F. Valdez, P. Melin, A gravitational search algorithm using type-2 fuzzy logic for parameter adaptation, in *Nature-Inspired Design of Hybrid Intelligent Systems* (Springer, Tijuana, Mexico, 2017), pp. 127–138
11. L. Rodriguez, O. Castillo, J. Soria, A study of parameters of the grey wolf optimizer algorithm for dynamic adaptation with fuzzy logic, in *Nature-Inspired Design of Hybrid Intelligent Systems* (Springer, Tijuana, Mexico, 2017), pp. 371–390
12. E. Mendez, O. Castillo, J. Soria, A. Sandollah, Fuzzy dynamic adaptation of parameters in the water cycle algorithm, in *Nature-Inspired Design of Hybrid Intelligent Systems* (Springer, Tijuana, Mexico, 2017), pp. 297–311
13. M. Crepinsek, M. Mernik, S. Liu, Analysis of exploration and exploitation in evolutionary algorithms by ancestry trees. Int. J. Innovative Comput. Appl. **3**(1), 11–19 (2011)
14. K. Tashkova, J. Silc, N. Atanasova, S. Dzeroski, Parameter estimation in a nonlinear dynamic model of an aquatic ecosystem with meta-heuristic optimization. Ecol Model **226**, 36–61 (2012)
15. D. Goldberg, *Genetic Algorithms in Search* (1989)
16. M. Dorigo, C. Blum, Ant colony optimization theory: a survey. Theoret. Comput. Sci. **334**, 243–278 (2005)
17. P. Korosec, J. Silc, B. Filipic, The differential ant-stigmergy algorithm. Inf. Sci., pp. 82–97 (2012)
18. J. Kennedy, R. Eberthart, The particle swarm optimization: social adaptation in information processing, in *New Ideas in Optimization,* pp. 379–387 (1999)
19. B. Jakimovski, B. Meyer, E. Maehle, Firefly flashing synchronization as inspiration for self-synchronization of walking robot gait patterns using a decentralized robot control architecture, in *Architecture of Computing Systems,* pp. 61–72 (2010)
20. S. Severin, J. Rossmann, A comparison of different metaheuristic algorithms for optimizing blended PTP movements for industrial robots, in *Intelligent Robotics and Applications,* pp. 321–330 (2012)
21. A. Chatterjee, G. Mahanti, A. Chatterjee, Design of a fully digital controlled reconfigurable switched beam concentric ring array antenna using firefly and particle swarm optimization algorithm. Prog. Electromagn. Res., pp. 113–131 (2012)
22. Y. Zhang, L. Wu, A novel method for rigid image registration based on firefly algorithm. Int. J. Res. Rev. Soft Intell. Comput. **2**(2), 141–146 (2012)
23. B. Basu, G. Mahanti, Firefly and artificial beescolony algorithm for synthesis of scanned and broadside linear array antenna. Prog. Electromagn. Res., pp. 169–190 (2011)
24. G. Giannakouris, V. Vassiliadis, G. Dounias, Experimental study on a hybrid nature-inspired algorithm for financial portfolio optimization, in *Artificial Intelligence: Theories, Models and Applications,* pp. 101–111 (2010)
25. A. Santos, H. Campos Velho, E. Luz, S. Freitas, G. Grell, M. Gan, Firefly optimization to determine the precipitation field on South America, Inverse Problems in Science and Engineering, pp. 1–16 (2013)
26. X. Yang, Firefly algorithm, in *Nature-Inspired Metaheuristic Algorithms,* pp. 79–90 (2008)

27. M. Sanchez, O. Castillo, J. Castro, Generalized Type-2 Fuzzy Systems for controlling a mobile robot and a performance comparison with interval type-2 and type-1 fuzzy systems. Expert Syst. Appl. **42**, 5904–5914 (2015)
28. O. Castillo, P. Melin, O. Montiel, R. Sepulveda, W. Pedrycz, *Theoretical Advances and Applications of Fuzzy Logic and Soft Computing* (Springer, Tijuana, BC, 2007)
29. R. Martinez, O. Castillo, L.T. Aguilar, Optimization of interval type-2 fuzzy logic controllers for a perturbed autonomous wheeled mobile robot using genetic algorithms. Inf. Sci. **179**(13), 2158–2174 (2009)
30. C. Soto, F. Valdez,O. Castillo, A review of dynamic parameter adaptation methods for the firefly algorithm, in *Nature-Inspired Design of Hybrid Intelligent Systems* (Springer, Tijuana, BC, 2007), pp. 285–295
31. L. Astudillo, P. Melin, O. Castillo, *Chemical Optimization Algorithm for Fuzzy Controller Design, Tijuana* (Springer, Mexico, 2014)
32. A. Sombra, F. Valdez, P. Melin, O. Castillo, A new gravitational search algorithm using fuzzy logic to parameter adaptation, in *IEEE Congress on Evolutionary Computation, Cancun, México*, pp. 1068–1074 (2013)
33. F. Valdez, P. Melin, O. Castillo, Evolutionary method combining particle swarm optimization and genetic algorithms using fuzzy logic for decision making, in *IEEE International Conference on Fuzzy Systems,* pp. 2114–2119 (2009)
34. L. Aguilar, P. Melin, O. Castillo, Intelligent control of a stepping motor drive using a hybrid neuro-fuzzy ANFIS approach. Appl. Soft Comput. 3(3), 209–219
35. P. Melin, O. Castillo, Intelligent control of complex electrochemical systems with a neuro-fuzzy-genetic approach, IEEE Trans. Ind. Electron. **48**(5), 951–955

Parallel Bacterial Potential Field Algorithm for Path Planning in Mobile Robots: A GPU Implementation

Ulises Orozco-Rosas, Oscar Montiel and Roberto Sepúlveda

Abstract Path planning is a fundamental task in autonomous mobile robot navigation and one of the most computationally intensive tasks. In this work, a parallel version of the bacterial potential field (BPF) method for path planning in mobile robots is presented. The BPF is a hybrid algorithm, which makes use of a bacterial evolutionary algorithm (BEA) with the artificial potential field (APF) method, to take advantage of intelligent and classical methods. The parallel bacterial potential field (parallel-BPF) algorithm is implemented on a graphics processing unit (GPU) to speed up the path planning computation in mobile robot navigation. Simulation results to validate the analysis and implementation are provided; the experiments were specially designed to show the effectiveness and the efficiency of the parallel-BPF algorithm.

Keywords Bacterial potential field · Path planning · Mobile robots GPU

1 Introduction

In this work, the parallel bacterial potential field (parallel-BPF) algorithm is an extension of the bacterial potential field (BPF) method presented in [1]. The parallel-BPF algorithm is implemented on a graphics processing unit (GPU) to speed up the path planning computation. The path planning problem is one of the

U. Orozco-Rosas (✉) · O. Montiel · R. Sepúlveda
Centro de Investigación y Desarrollo de Tecnología Digital (IPN-CITEDI), Instituto Politécnico Nacional, Av. Instituto Politécnico Nacional no. 1310, Nueva Tijuana, 22435 Tijuana, BC, Mexico
e-mail: uorozco@citedi.mx

O. Montiel
e-mail: oross@ipn.mx

R. Sepúlveda
e-mail: rsepulvedac@ipn.mx

O. Castillo et al. (eds.), *Fuzzy Logic Augmentation of Neural and Optimization Algorithms: Theoretical Aspects and Real Applications*, Studies in Computational Intelligence 749, https://doi.org/10.1007/978-3-319-71008-2_17

most computationally intensive tasks in mobile robot navigation, heterogeneous computing helps to gain performance. By means of GPUs is possible to process data-intensive tasks (evaluation process) efficiently.

The parallel-BPF algorithm is employed to ensure a feasible, smooth, and safe path for autonomous mobile robot navigation. The parallel-BPF employs concepts from the artificial potential field (APF) method and from the bacterial evolutionary algorithms (BEA) to obtain an enhanced flexible path planner.

Path planning methods can be classified in two big categories: global and local path planning. Global path planning methods evaluate all the information available in order to generate an obstacle-free path between the start position and the goal position. The generated path must satisfy certain constraints, such as the dynamic constraints of the mobile robot and environment constraints. Local path planning algorithms take into account the surrounding space of the mobile robot and are able to respond to unpredicted factors, such as unmapped obstacles [2].

Previous research and academic works have addressed the path planning problem in mobile robots from the global or the local perspective, and very often complement each other. In [3] a global path planning algorithm is presented; the algorithm considers different parameters to maximize the path quality: the maximum slope allowed by the robot and the heading changes during the path. These constraints allow discarding infeasible paths while minimizing the heading changes. The artificial potential field (APF) method is commonly employed for local path planning, and combinations or improvements of the APF method are employed to complement the global and local path planning. In [4] a path planning system is presented for rovers based on an improved version of the Fast Marching (FM) method. Scalar and vector properties are considered when computing the potential field which is the basis of the proposed technique. In [5] a path planning algorithm based on the artificial potential field method was proposed. The algorithm is employed for the path planning problem of the multiple unmanned aerial vehicles (UAV) formation in a known environment. In [6] a set-point control for a mobile robot with obstacle detection and avoidance using navigation function is presented. The controller is based on the navigation potential function, the navigation potential aggregates information of the robot position and orientation but also the repelling potentials of obstacles.

In this work, the main contribution is a parallel version of the BPF algorithm for global path planning implemented on a GPU to speed up the path planning computation in mobile robot navigation. The remaining of this paper is organized as follows. Section 2 briefly summarizes the fundamentals of the BPF method based on the bacterial evolutionary algorithms (BEA) and the artificial potential field (APF) method. Section 3 details the parallel-BPF algorithm for path planning. Section 4 describes the experiments and simulation results. Finally, the conclusions are presented in Sect. 5.

2 Bacterial Potential Field

In this section, a brief description of the main components of the parallel bacterial potential field (parallel-BPF) algorithm is presented. The parallel-BPF algorithm combines the artificial potential field (APF) method and mathematical programming, using a metaheuristic based on a bacterial evolutionary algorithm (BEA) as the global optimization method, and parallel computing techniques, to solve efficiently the path planning problem in mobile robots.

2.1 *Bacterial Evolutionary Algorithms*

In 1999, Nawa and Furuhashi proposed the bacterial evolutionary algorithm (BEA) for fuzzy rule base extraction [7]. The BEA was based on the pseudo-bacterial genetic algorithm (PBGA) proposed by Nawa, et al., in 1997 [8]. The bacterial evolutionary algorithm (BEA) introduces two operations inspired by the microbial evolution phenomenon. The bacterial mutation operation which labor is to optimize the chromosome of a single bacterium, and the gene transfer operation to provide the transfer of information between the bacteria in the population [9].

The convergence to the global optimal of the bacterial evolutionary algorithms in comparison with the genetic algorithms is faster, which is important since the computation process cost of combinations is omitted [10].

A bacteriological approach is more an adaptive approach than an optimization approach as with genetic algorithms. The bacteriological approach aims at mutating the initial population to adapt it to a particular environment. The adaptation is based on small changes in the individuals. The individuals within the population are called bacteria and correspond to atomic units. Unlike the genetic model, the bacteria cannot be divided. The crossover operation is not used. A bacterium can only be reproduced and altered to improve the population [11].

Bacterial mutation operator

The bacterial mutation operator is inspired by the biological model of bacterial cells. This operator makes that the bacterial evolutionary algorithm mimics the phenomenon of microbial evolution [12]. Exploration of different regions in the search space that has not been covered by the current population, it is a necessary task to find the global optimum. This task is achieved by adding new information to the bacteria, information that is generated randomly by the bacterial mutation operator. The bacterial mutation operator is applied to all the bacteria in the population, one by one. The following steps are employed to perform the bacterial mutation operator [13]:

1. A number (N_{clon}) of clones are created (copies of the bacteria from the population).

2. In each clone, one segment with length l_{bm} is randomly selected, this segment will be operated by the mutation. The mutation operator is applied to all clones except for one, just one clone is left unmutated.
3. After mutation, the clones are evaluated using the fitness function. The mutated clones with the best evaluation will transfer the undergone mutation segment to the other clones.
4. The steps two and three (mutation of the clones, selection of the best clone, and transfer the segment subjected to mutation) are repeated until each segment of the bacterium has been once subjected to mutation.
5. The best bacterium is maintained, and the clones are removed.

Gene transfer operator

In the bacterial evolutionary algorithms, the gene transfer operator is introduced to establish the relationship among the individuals from the population. The following steps are employed to perform the gene transfer operator [13]:

1. The bacterial population is sorted and divided into two parts. One part of the population is called superior half, and it contains the bacteria with the best fitness value. The second part is called inferior part, and it contains the bacteria with the worst fitness value.
2. Two bacteria are randomly chosen, one from the superior half and one from the inferior half. The bacterium from the superior half is called the source bacterium, and the bacterium from the inferior half is called the destination bacterium.
3. A segment of length l_{gt} from the source bacterium is randomly chosen and this segment is used to overwrite a random segment of the destination bacterium, if the source segment is not already in the destination bacterium.
4. Steps one to three are repeated N_{inf} number of times, where N_{inf} is the number of infections per generation.

2.2 *Artificial Potential Field*

The main idea of the artificial potential field (APF) method is to establish an attractive potential field around the goal point, as well as to establish a repulsive potential field around obstacles. The two potential fields, attractive and repulsive form the total potential field [14].

The APF method searches the falling direction of the potential field to find a collision-free path, which is built from the start to the goal position [15]. The APF method is basically operated by the gradient descent method, which is directed toward minimizing the total potential field, see (1), in each position of the mobile robot. The goal position represents an attractive pole and the obstacles represent repulsive forces in the surface where the mobile robot will perform its movement.

The total potential field, $U(q)$, it is the sum of the attractive potential field, $U_{att}(q)$, and the repulsive potential field, $U_{rep}(q)$, and it is described by (1),

$$U(q) = U_{att}(q) + U_{rep}(q). \tag{1}$$

The attractive potential field is described by (2),

$$U_{att}(q) = \frac{1}{2} k_a \left(q - q_f\right)^2, \tag{2}$$

where q represents the robot position vector in a workspace of two dimensions. The vector q_f represents the goal coordinates, and k_a is a positive scalar-constant that represents the proportional gain of the attractive potential field function. The expression $\left(q - q_f\right)$ is the distance between the robot position and the goal position.

The repulsive potential field has a limited range of influence that condition prevents that the movement of the mobile robot will be affected by distant obstacles. The repulsive potential field is described by (3),

$$U_{rep}(q) = \begin{cases} \frac{1}{2} k_r \left(\frac{1}{\rho} - \frac{1}{\rho_0}\right)^2 & \text{if} \rho \le \rho_0 \\ 0 & \text{if} \rho > \rho_0 \end{cases} \tag{3}$$

where ρ_0 represents the limit distance of influence of the potential field, ρ is the shortest distance to the obstacle, and k_r is a positive scalar-constant that represents the proportional gain of the repulsive potential field function.

The generalized force $F(q)$ is obtained by the negative gradient of the total potential field, $U(q)$, and it is described by (4),

$$F(q) = -\nabla U(q) \tag{4}$$

The APF method requires the appropriate values of the proportional gains (k_a, and k_r) to make the mobile robot to successfully reach the goal position. In that sense, in this work a BEA is employed to search for the appropriate values of the proportional gains.

3 Parallel-BPF Algorithm for Path Planning

In this section, the parallel bacterial potential field (parallel-BPF) algorithm is described. The parallel-BPF algorithm for path planning is a high-performance hybrid metaheuristic that combines the artificial potential field (APF) method with the bacterial evolutionary algorithms (BEA). To obtain a flexible path planning algorithm for mobile robots; the parallel-BPF algorithm uses parallel computing in the evaluation process for taking advantage of the graphics processing unit (GPU).

3.1 Algorithm

In this work, the parallel-BPF algorithm is an extension of the bacterial potential field (BPF) method presented in [1]. Figure 1 shows a simplified version of the parallel-BPF algorithm. The vertebral column of the parallel-BPF algorithm is the use of the bacterial potential field (BPF) method and parallel computing for searching dynamically the optimal values of the proportional gains (k_a, and k_r) required in (2) and (3), respectively for optimal path planning. Hence, the parallel-BPF algorithm achieves the task of path planning generation, with the particular characteristic that it provides an optimal or nearly optimal reachable set of configurations (path) if it exists. Allowing the mobile robot navigation without being trapped in local minima, making the parallel-BPF algorithm suitable for path planning in complex environments.

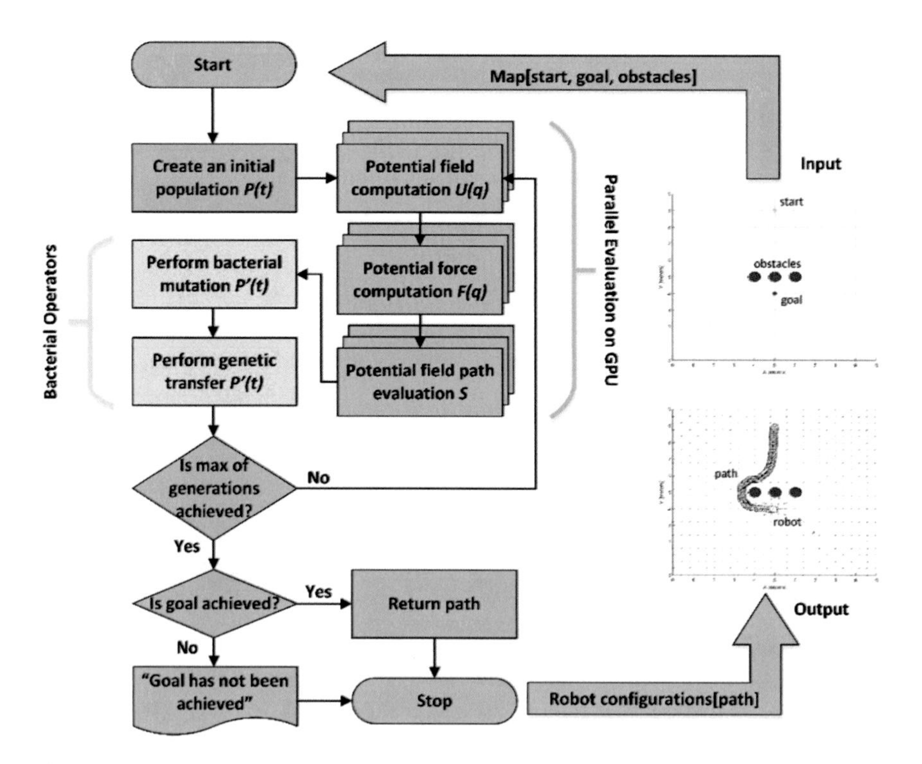

Fig. 1 Parallel-BPF algorithm for path planning

Figure 1 shows the parallel-BPF algorithm and its parallel evaluation process implemented on GPU (green blocks). The parallel-BPF algorithm for path planning uses the start, goal, and obstacles positions as input parameters to give a sequence of robot configurations, in other words, a sequence of objective points that form the path that the mobile robot must attain for a safe and efficient mobile robot navigation.

The process starts with a creation of a random population of bacteria, $P(t)$, where each bacterium (possible solution) is codified with the values of the proportional gains, attraction, k_a, and repulsion, k_a, required to generate a feasible path. For the parallel evaluation on GPU, where each bacterium is evaluated. First, the potential field, $U(q)$, computation is performed employing (1). Then the generalized potential force, $F(q)$, is computed with (4). In that sense, the generalized potential force is used to drive the mobile robot. Last, for the parallel evaluation on GPU, the potential field path, S, is computed with (5)

$$S = \sum_{i=0}^{n-1} \|q_{i+1} - q_i\| \tag{5}$$

After the parallel evaluation, the bacterial operators are applied to evolve the bacteria in the population, $P(t)$. In concrete terms, the parallel-BPF algorithm employs the single population master slave approach, in which, the master node executes the bacterial evolutionary algorithm (BEA) operators (bacterial mutation and genetic transfer) working on a single population, $P'(t)$, on CPU, and the fitness evaluation of the bacteria in the population, $P'(t)$, is divided in $P_i(t)$ subpopulations distributed among several slave threads on the GPU.

All the process (the parallel evaluation in the GPU and the bacterial operators) is repeated over and over again until the maximum number of generations has been achieved. If the resultant path successfully achieves the goal in a safe and efficient manner, the path is returned as a result, otherwise, a message of "goal has not been achieved" is displayed.

3.2 Implementation

For the GPU implementation, we have chosen Matlab-CUDA as a platform to implement the parallel-BPF algorithm. We used CUDA kernel integration in the Matlab application, a kernel (code written in CUDA) called from Matlab application is executed on the GPU to accelerate the evaluation process (potential field computation, $U(q)$, potential force computation, $F(q)$, and potential field path evaluation, S) of the parallel-BPF algorithm.

For the CUDA kernel integration, a kernel is a programming code written in CUDA (compute unified device architecture) for its execution on the GPU. CUDA kernels are functions that can run on a large number of threads. Parallelism arises from each thread independently running the same program (potential field function, potential force function, and potential field path evaluation function) on different data (bacterium composed by the attraction and repulsion proportional gains).

The parallel-BPF evaluation CUDAKernel requires the creation of an executable kernel CU or PTX (parallel thread execution) file for running CUDA kernels on the GPU from a Matlab application. The kernel is a CUDAKernel object, which can operate on Matlab gpuArray variables. In a general form, for the use of CUDA kernels in a Matlab application. First, use a compiled PTX code to create the CUDAKernel object, which contains the GPU executable code. Then, set the properties on the CUDAKernel object to control its execution on the GPU. Finally, call feval on the CUDAKernel with the required inputs, to run the kernel on the GPU.

4 Experiments and Simulation Results

In this section, the experiments, the simulation results, and the performance of the parallel-BPF algorithm for path planning are presented. The experimental framework consists of four test problems (i.e., instances). The four problems were selected to demonstrate the relevance of the parallel-BPF algorithm implemented on a GPU.

For uniformity among the different experiments, we have employed the same parameter values in all the experiments. These parameters have provided the best results for path planning using the parallel-BPF algorithm implemented on the GPU.

1. Each bacterium contains two genes (one gene for the attraction proportional gain, k_a, and one for the repulsion proportional gain, k_r). Each gene is codified with eight bits.
2. The proportional gains are constrained, $\{0 < k_a, k_r \leq 10\}$.
3. The population size, N_{pop}, it was varied from 32 to 1024 bacteria for the performance evaluation.
4. Number of clones per bacterium $N_{clon} = 4$.
5. Bacterial mutation rate of 40%.
6. Gene transfer rate of 15%.
7. Maximum number of generations $N_{gen} = 10$.

The results of the experiments were achieved using a personal computer equipped with the Intel i7-5820K CPU@3.30 GHz, and a NVIDIA GeForce GTX TITAN X. The computer runs with the Ubuntu Trusty distribution of Linux, the NVIDIA CUDA 7.5 version, and Matlab R2015A installed.

4.1 Experiments

Considering that the aim of this work is to demonstrate the effectiveness and the efficiency of the parallel-BPF algorithm to solve path planning problems, four test problems were carefully selected. In all the experiments, to obtain the statistical results, thirty independent tests were performed.

Experiment 1
The aim of this experiment is to demonstrate the ability of the parallel-BPF algorithm to escape from a "U-shaped" obstacle. In real-world scenarios, U-shaped obstacles contain geometrical characteristics that cannot be captured entirely by the sensors of the mobile robot. In such situation, the mobile robot is incapable of escaping from the trap.

In this experiment, the test instance has been labeled as Map01. The instance Map01 is a two-dimensional map, divided by a grid of 10 by 10 m, where the start position of the mobile robot is at coordinates (5.00, 8.00), and the goal position is at coordinates (5.00, 2.00); the scenario is composed by five obstacles forming a U-shaped trap, as it shown in Fig. 2a. The center position coordinates for the five obstacles in the U-shaped trap are (5.00, 4.50), (3.50, 4.50), (3.50, 6.00), (6.50, 4.50), and (6.50, 6.00). Each obstacle in the U-shaped trap has a radius of 0.50 m.

Table 1 shows the statistical results for the instance Map01 based on the total number of evaluations, in other words, the total number of bacteria that were evaluated in one independent test. The total number of bacteria that were evaluated is obtained by the product of the population size by the number of clones per bacterium by the maximum number of generations, (i.e., Total number of evaluations $= N_{pop} * N_{clon} * N_{gen}$). Among the results for the Map01, we can observe how the parallel-BPF algorithm improves the path distance (minimize, the shortest path distance is the best) for the mobile robot to travel from the start position to the goal position. Figure 2b shows the best path founded. The best path has a length of 9.315 m, the path was generated by the best proportional gains ($k_a = 0.784$, and $k_r = 4.627$) obtained from a total number of 40,960 bacteria that were evaluated. The worst distance founded for the instance Map01 is 11.787 m with a total number of 1280 bacteria that were evaluated.

Experiment 2
The target of this experiment is to test the parallel-BPF algorithm in a more complex scenario populated with several obstacles.

In this experiment, the test instance has been labeled as Map02. The instance Map02 is a two-dimensional map, divided by a grid of 10 by 10 meters, were the start position of the mobile robot is at coordinates (5.00, 9.00), and the goal position is at coordinates (5.00, 1.00); the scenario is composed of three blocks of three obstacles, as it shown in Fig. 2c. The center position coordinates for the first block of obstacles are (2.00, 7.50), (3.00, 7.50), and (4.00, 7.50). The center position coordinates for the second block of obstacles are (4.00, 5.00), (5.00, 5.00), and (6.00, 5.00). The center position coordinates for the third block of obstacles are

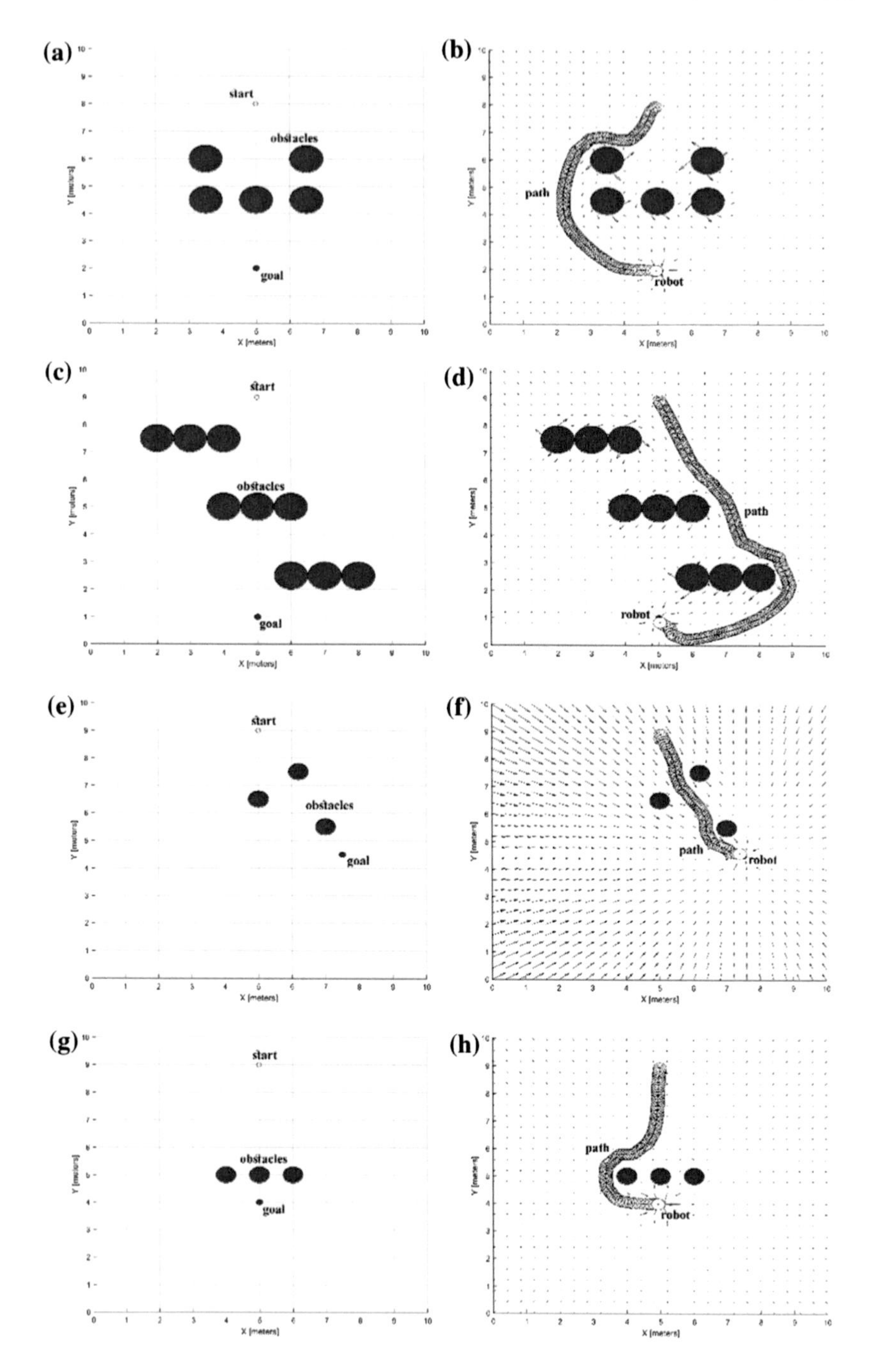

Fig. 2 **a** Map01.**b** Resultant path for Map01.**c** Map02.**d** Resultant path for Map02.**e** Map03. **f** Resultant path for Map03. **g** Map04. **h** Resultant path for Map04

Table 1 Statistical results for path planning using the parallel-BPF algorithm in the different test instances

Instance	Total number of evaluations	Best (m)	Mean (m)	Worst (m)	Std. dev. (s)
Map01	1280	9.670	10.163	11.787	0.582
	2560	9.406	9.776	10.965	0.451
	5120	9.386	9.487	9.702	0.099
	10,240	9.345	9.393	9.551	0.058
	20,480	9.330	9.376	9.418	0.032
	40,960	9.315	9.359	9.393	0.022
Map02	1280	13.267	13.475	13.712	0.134
	2560	13.263	13.451	14.041	0.217
	5120	13.244	13.304	13.401	0.048
	10,240	13.207	13.280	13.394	0.064
	20,480	13.203	13.241	13.299	0.027
	40,960	13.199	13.235	13.262	0.022
Map03	1280	5.339	5.358	5.392	0.017
	2560	5.333	5.341	5.347	0.004
	5120	5.332	5.337	5.340	0.002
	10,240	5.331	5.334	5.339	0.002
	20,480	5.329	5.334	5.338	0.002
	40,960	5.327	5.331	5.335	0.002
Map04	1280	7.101	7.206	7.348	0.073
	2560	7.024	7.099	7.192	0.048
	5120	6.999	7.052	7.129	0.046
	10,240	6.999	7.025	7.090	0.026
	20,480	6.989	6.992	7.068	0.024
	40,960	6.984	6.990	7.043	0.017

(6.00, 2.50), (7.00, 2.50), and (8.00, 2.50). Each obstacle in the three blocks has a radius of 0.50 m.

Table 1 shows the statistical results for the instance Map02 based on the total number of bacteria that were evaluated. Through the results for the Map02, we can observe how the parallel-BPF algorithm minimizes the path distance for the mobile robot to travel from the start position to the goal position. Figure 2d shows the best path founded. The shortest path has a length of 13.199 m, the path was generated by the best proportional gains ($k_a = 0.784$, and $k_r = 4.000$) obtained from a total number of 40,960 bacteria that were evaluated. The worst distance founded for the instance Map02 is 13.712 m with a total number of 1280 bacteria that were evaluated.

Experiment 3

The objective of this experiment is to demonstrate the ability of the parallel-BPF algorithm to escape from a local minima situation. Local minima can be caused by either one obstacle or a combination of them [16].

In this experiment, the test instance has been labeled as Map03. The instance Map03 is a two-dimensional map, divided by a grid of 10 by 10 m, were the start position of the mobile robot is at coordinates $(5.00, 9.00)$, and the goal position is at coordinates $(7.50, 4.50)$; the scenario is composed by three obstacles, as it shown in Fig. 2e. The center position coordinates for the obstacles are $(5.00, 6.50)$, $(6.20, 7.50)$, and $(7.00, 5.50)$. Each obstacle in the scenario has a radius of 0.30 m.

Table 1 shows the statistical results for the instance Map03 based on the total number of evaluations. To obtain the results thirty independent tests for each total number of evaluations were carried out. From the results for the Map03, we can observe how the parallel-BPF algorithm obtains the shortest path distance for the mobile robot to travel from the start position to the goal position. Figure 2f shows the best path founded. The best path has a length of 5.327 m, the path was generated by the best proportional gains ($k_a = 3.843$, and $k_r = 3.529$) obtained from a total number of $40, 960$ bacteria that were evaluated. The worst distance founded for the instance Map03 is 5.392 m with a total number of 1280 bacteria that were evaluated.

Experiment 4

The aim of this experiment is to demonstrate the ability of the parallel-BPF algorithm with the problem called goals non-reachable with obstacles nearby (GNRON), where the repulsive force generated by an obstacle close to the goal generates a force higher than the attractive force, as a result, the mobile robot cannot converge to the correct point, which is the goal [17].

In this experiment, the test instance has been labeled as Map04. The instance Map04 is a two-dimensional map, divided by a grid of 10 by 10 m, were the start position of the mobile robot is at coordinates $(5.00, 9.00)$, and the goal position is at coordinates $(5.00, 4.00)$; the scenario is composed by a barrier of three obstacles, as it shown in Fig. 2g. The center position coordinates for the obstacles are $(5.00, 5.00)$, $(4.00, 5.00)$, and $(6.00, 5.00)$. Each obstacle has a radius of 0.30 m.

Table 1 shows the statistical results for the instance Map04 based on the total number of evaluations. To obtain the statistical results thirty independent tests for each total number of evaluations were carried out. Among the results for the Map04, we can observe how the parallel-BPF algorithm improves the path distance for the mobile robot to travel from the start position to the goal position. Figure 2h shows the best path founded. The best path has a length of 6.984 meters, the path was generated by the best proportional gains ($k_a = 0.627$, and $k_r = 1.647$) obtained from a total number of $40, 960$ bacteria that were evaluated. The worst distance founded for the instance Map04 is 7.348 m with a total number of 1280 bacteria that were evaluated.

4.2 Performance

Table 2 shows the performance of the parallel-BPF algorithm based on the total number of evaluations. To evaluate the performance of the parallel implementation of the BPF algorithm on GPU versus the sequential implementation on CPU of the BPF algorithm, we carried out independently thirty tests for each total number of evaluations. The total number of bacteria that were evaluated is obtained by the product of the population size by the number of clones per bacterium by the maximum number of generations, (i.e., Total number of evaluations $= N_{pop} * N_{clon} * N_{gen}$).

Among the results, we can observe how the speedup is increased on the parallel-GPU implementation when the total number of evaluations is incremented. For a total number of $40,960$ bacteria that were evaluated in one independent test, we have more than the half of computation time reduced for the parallel-GPU implementation of the BPF algorithm in all the instances (Map01, Map02, Map03,

Table 2 Average time, standard deviation and speedup of each population size evaluated in the sequential form on the CPU and in the parallel form on the GPU

	Total number of evaluations	Sequential—CPU		Parallel—GPU		Speedup
		Mean (s)	Std. dev. (s)	Mean (s)	Std. dev. (s)	
Map01	1280	5.167	1.387	51.596	15.980	**0.100**
	2560	11.058	3.079	70.052	1.913	**0.158**
	5120	26.060	6.266	72.355	0.265	**0.360**
	10,240	42.697	8.016	72.732	0.175	**0.587**
	20,480	93.685	14.215	76.162	0.610	**1.230**
	40,960	185.886	7.409	90.394	2.091	**2.056**
Map02	1280	4.990	2.152	63.217	12.287	**0.079**
	2560	12.741	3.664	66.363	11.916	**0.192**
	5120	23.697	6.101	71.716	2.336	**0.330**
	10,240	47.343	5.251	73.288	0.208	**0.646**
	20,480	100.830	8.458	77.202	0.744	**1.306**
	40,960	200.153	12.125	92.273	2.177	**2.169**
Map03	1280	3.307	0.551	23.602	12.379	**0.140**
	2560	6.273	0.507	39.486	11.488	**0.159**
	5120	13.017	0.724	56.167	12.278	**0.232**
	10,240	26.705	1.330	66.611	3.746	**0.401**
	20,480	51.336	1.209	73.077	1.764	**0.702**
	40,960	105.553	2.841	81.796	0.357	**1.290**
Map04	1280	5.588	0.738	61.500	15.284	**0.091**
	2560	13.462	2.136	71.024	0.169	**0.190**
	5120	28.675	3.651	72.075	0.069	**0.398**
	10,240	59.229	3.170	72.272	0.151	**0.820**
	20,480	117.702	3.960	78.674	0.742	**1.496**
	40,960	239.986	8.154	100.191	1.366	**2.395**

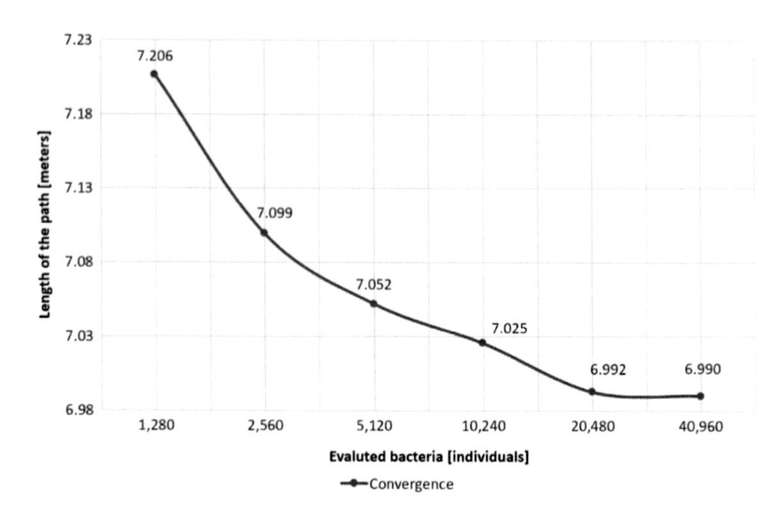

Fig. 3 Convergence

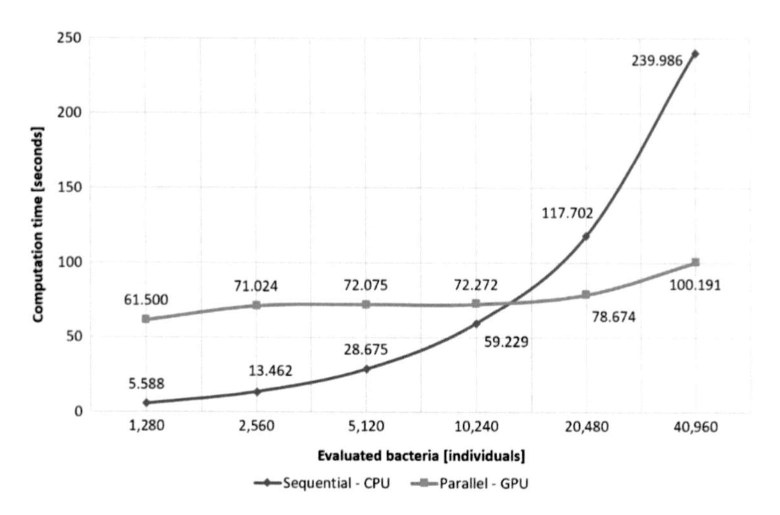

Fig. 4 Performance

and Map04). The best speed up for the parallel-GPU implementation is 2.395×, that was achieved in the instance of Map04. Through the performance results, we can conclude that the increment of the number of evaluations demonstrates that the parallel-BPF algorithm on GPU implementation outperforms the sequential-CPU implementation.

The best performance results are shown by the results on instance Map04, in that sense, the convergence plot (in red) for the Map 04 is shown in Fig. 3. As well as the average computation time plot (in green) for the parallel implementation on GPU of the BPF algorithm and the average computation time plot (in blue) for the sequential implementation on CPU of the BPF algorithm are shown in Fig. 4.

5 Conclusions

In this work, we have presented the parallel-BPF algorithm implemented on GPU as an extension of the bacterial potential field (BPF) method for path planning in mobile robots. The parallel-BPF algorithm has demonstrated its capability to perform the path planning in different test environments with a different number of bacteria, that were evaluated in the optimization process.

The performance results show that the parallel-BPF algorithm implemented on GPU accelerates the evaluation process by a factor of $2.4\times$ for the bigger number of bacteria ($40,960$ individuals) evaluated in the whole optimization process.

The performance results show that a small number of bacteria (1280 individuals) employed in the whole optimization process could guide the parallel-BPF algorithm to reasonable solutions (paths), but with lower quality in terms of path length. In the other hand, employing a large number of bacteria ($40,960$ individuals) in the whole process, the results are improved in terms of path length, but the parallel-BPF algorithm spent more computation time to find best solutions.

Making a compromise between solution quality and computation time, we have found that the best performance of the parallel-BPF algorithm on GPU is obtained when the number of bacteria evaluated in the whole optimization process is $15,000$ individuals or bigger.

Furthermore, the path planning results demonstrate the efficiency of the parallel-BPF algorithm to solve the path planning problem in all the different environments tested.

Acknowledgements We thank Instituto Politécnico Nacional (IPN), the Comisión de Operación y Fomento de Actividades Academicas of IPN (COFAA), and the Mexican National Council of Science and Technology (CONACYT) for supporting our research activities.

References

1. O. Montiel, U. Orozco-Rosas, R. Sepúlveda, Path planning for mobile robots using Bacterial Potential Field for avoiding static and dynamic obstacles. Expert Syst. Appl. **42**(12), 5177–5191 (2015)
2. M. Candeloro, A.M. Lekkas, A.J. Sørensen, A Voronoi-diagram-based dynamic path-planning system for underactuated marine vessels. Control Eng. Pract. **61**, 41–54 (2017)
3. P. Muñoz, M.D. R-Moreno, B. Castaño, 3Dana: a path planning algorithm for surface robotics. Eng. Appl. Artif. Intell. **60**, 175–192 (2017)
4. S. Garrido, L. Moreno, F. Martín, D. Álvarez, Fast marching subjected to a vector field-path planning method for mars rovers. Expert Syst. Appl. **78**, 334–346 (2017)
5. Y. Chen, J. Yu, X. Su, G. Luo, Path planning for multi-UAV formation. J. Intell. Rob. Syst. **77**(1), 229–246 (2015)
6. W. Kowalczyk, M. Przybyla, K. Kozlowski, Set-point control of mobile robot with obstacle detection and avoidance using navigation function—experimental verification. J. Intell. Rob. Syst. **85**(3), 539–552 (2017)

7. N.E. Nawa, T. Furuhashi, Fuzzy system parameters discovery by bacterial evolutionary algorithm. IEEE Trans. Fuzzy Syst. **7**(5), 608–616 (1999)
8. N.E. Nawa, T. Hashiyama, T. Furuhashi, Y. Uchikawa, A study on fuzzy rules discovery using pseudo-bacterial genetic algorithm with adaptive operator. Paper presented at the IEEE international conference on evolutionary computation, Indianapolis, IN, 13–16 Apr 1997 (1997)
9. J. Botzheim, L.T. Kóczy, Model identification by bacterial optimization. National Scientific Research Fund, Budapest, Hungary. https://pdfs.semanticscholar.org/b717/7925a514452b03d9802a7a6390b1910db30a.pdf. Accessed 05 Apr 2017 (2004)
10. M.F. Rad, F. Akbari, A.J. Bakht, Implementation of common genetic and bacteriological algorithms in optimization testing data in mutation testing. Paper presented at the 2010 international conference on computational intelligence and software engineering, Wuhan, 10–12 Dec 2010 (2010)
11. B. Baudry, F. Fleurey, J.M. Jézéquel, Y.L. Traon, From genetic to bacteriological algorithms for mutation-based testing. Softw. Test. Verif. Reliab. **15**(2), 73–96 (2005)
12. J. Botzheim, L. Gál, L.T. Kóczy, Fuzzy rule base model identification by bacterial memetic algorithms, in *Recent Advances in Decision Making*, vol. 222, ed. by E. Rakus-Andersson, R. R. Yager, N. Ichalkarange, L.C. Jain (Springer, Berlin, 2009), pp. 21–43
13. J. Botzheim, Y. Toda, N. Kubota, Path planning for mobile robots by bacterial memetic algorithm. Paper presented at the 2011 IEEE workshop on robotic intelligence in informationally structured space, Paris, 11–15 Apr 2011 (2011)
14. O. Kathib, Real-time obstacle avoidance for manipulators and mobile robots. Int. J. Robot. Res. **5**(1), 90–98 (1986)
15. Q. Zhang, D. Chen, T. Chen, An obstacle avoidance method of soccer robot based on evolutionary artificial potential field. Procedia **16**, 1792–1798 (2012). Paper presented at the 2012 international conference on future energy, environment, and materials
16. V. Sezer, M. Gokasan, A novel obstacle avoidance: "Follow the Gap Method". Robot. Auton. Syst. **60**(9), 1123–1134 (2012)
17. S.S. Ge, Y.J. Cui, New potential functions for mobile robot path planning. IEEE Trans. Robot. Autom. **16**(5), 615–620 (2000)

Path Following Fuzzy System for a Nonholonomic Mobile Robot Based on Frontal Camera Information

Yoshio Rubio, Kenia Picos, Ulises Orozco-Rosas, Carlos Sepúlveda, Enrique Ballinas, Oscar Montiel, Oscar Castillo and Roberto Sepúlveda

Abstract This work proposes a fuzzy approach for path following of a nonholonomic mobile robot, based on the information of a frontal camera. The proposed methodology is divided in three stages. The first stage gets the image of the frontal camera and processes the image to detect and isolate the desired path to follow and eliminate non-useful information. The second stage estimates the orientation for different sections of the path to follow. Finally, in the last stage, a fuzzy system is designed and simulated to control the steering direction of the mobile robot. We show the design, simulations, and experiments using the fuzzy controller. The results are evaluated and discussed in terms of quantitative metrics.

Y. Rubio (✉) · K. Picos · U. Orozco-Rosas · C. Sepúlveda
E. Ballinas · O. Montiel · R. Sepúlveda
Instituto Politécnico Nacional, Centro de Investigación y Desarrollo de Tecnología Digital (IPN-CITEDI), Av. Instituto Politécnico Nacional No. 1310, Nueva Tijuana, 22435 Tijuana, B.C., Mexico
e-mail: rrubio@citedi.mx

K. Picos
e-mail: kpicos@citedi.mx

U. Orozco-Rosas
e-mail: uorozco@citedi.mx

C. Sepúlveda
e-mail: csepulveda@citedi.mx

E. Ballinas
e-mail: lballinas@citedi.mx

O. Montiel
e-mail: oross@ipn.mx

R. Sepúlveda
e-mail: rsepulvedac@ipn.mx

O. Castillo
Tijuana Institute of Technology, Calzada Tecnológico S/N, Tomas Aquino, 22414 Tijuana, B.C., Mexico
e-mail: ocastillo@tectijuana.mx

O. Castillo et al. (eds.), *Fuzzy Logic Augmentation of Neural and Optimization Algorithms: Theoretical Aspects and Real Applications*, Studies in Computational Intelligence 749, https://doi.org/10.1007/978-3-319-71008-2_18

Keywords Autonomous mobile robot · Fuzzy controller · Image processing
Path following · Line detection

1 Introduction

In recent years, numerous autonomous mobile robotic systems have emerged like mobile-wheeled robots, emergency service robots, mobile robotic manipulators, unmanned aerial vehicles (UAV), and autonomous underwater vehicles (AUV), just to mention some examples. These autonomous mobile robots are replacing human beings on board in military missions. In common applications, there is an increasing demand for autonomous mobile robots in various fields of application, such as material transport, cleaning, monitoring, and guiding people, among others. An important task for mobile robotic systems is the path following to accomplish its duties or mission. In this sense, the path following problem has become one of the most challenging tasks in the robotic community.

Path following consists of controlling and directing the mobile robot, in this work a car-like robot that follows a specific and predefined path with respect to an imposed performance criterion, where it needs strategies to complete the execution of the path following. Thus, the actions must be determined and adapted to the path changes. For this aim, the car-like robot uses a fuzzy control system to provide suitable input velocities to drive the car-like robot over the desired path. Fuzzylogic controllers can provide a formal methodology to represent and implement human heuristic knowledge on how to control a system subject to uncertainties, which makes it become a popular approach to attain artificial intelligence functionalities [1].

Previous research and academic works have used fuzzylogic as a powerful soft computing technique to control complex and non-linear systems based on human expert knowledge [2]. In [3], an adaptive fuzzy logic controller was designed to keep track of a mobile robot on the desired smoothed path, by transmitting the appropriate right and left velocities using wireless communication. Xiang et al., present a three-dimensional path following control for a underactuated autonomous underwater vehicle, the heuristic adaptive fuzzy algorithm is based on the guidance and feedback linearization Proportional-Integral-Derivative (PID) controller to account for the nonlinear dynamics and system uncertainties, including inaccurate parameters and time-varying environmental disturbances [1]. In [4], a study focused on the Takagi-Sugeno fuzzy-model-based control design is proposed for turning and forward motion of a differentially-driven wheeled mobile robot, where the position and posture of the mobile robot are estimated by visual odometry. Montiel et al., proposed a fuzzy system for obstacle avoidance and autonomous navigation based on measurements made with a global positioning system (GPS) to estimate the position and orientation of the mobile robot [5]. In [2], an approach to design a fuzzy logic PI controller (Fuzzy-PI) was presented to overcome the drawbacks of the PI control when it is applied to omnidirectional robot navigation.

In this work, the main contribution is a fuzzy control system design for path following to drive a nonholonomic mobile robot with Ackerman architecture (car-like robot) based on the frontal camera information. The remaining of this paper is organized as follows. Section 2 briefly summarizes the theoretical background based on fuzzy inference systems and vision systems. Section 3 details the nonholonomic car-like robot model. Section 4 presents the methodology employed to design the fuzzy control system for path following. Section 5 describes the results and experiments. Finally, the conclusions are presented in Sect. 6.

2 Theoretical Background

In this section, we provide the minimal theoretical background on fuzzy and vision systems needed to understand the design of the path following system focused on controlling an Ackerman scaled robotic car, for which mathematical models and simulation were developed.

2.1 Fuzzy Inference Systems

Fuzzy logic is based on the concept of perception; it gives the opportunity to obtain the approximate reasoning, emulating the human brain mechanism from facts described using natural language with uncertainty. It allows handling vague information.

A Fuzzy inference system (FIS) or a fuzzy system, can implement a non-linear mapping of its input to the output space. This mapping is achieved through the IF—THEN fuzzy rules, where each rule describes the local behavior of the mapping. As is shown in Fig. 1, a fuzzy system is composed of a knowledge base that comprises the information given by the process operator in form of linguistic control rules. A fuzzification interface has the effect of transforming crisp data into fuzzy sets. An inference system, that uses them in conjunction with the knowledge base to make the inference by means of an inference mechanism. The last stage is the

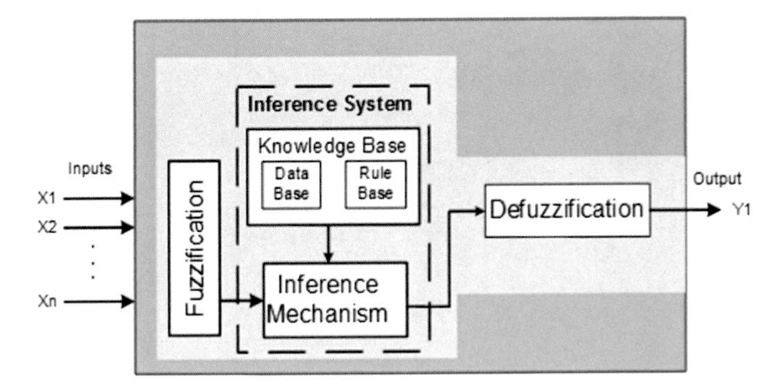

Fig. 1 A fuzzy system

defuzzification interface that translates the fuzzy control action so obtained to a real control action using a defuzzification method [6].

These systems are capable of using human knowledge, which is crucial in control problems where is difficult or sometimes impossible to build accurate mathematical models.

2.2 Vision-Based Navigation

Path planning, lane-based navigation, and obstacle avoidance are the most important tasks for the decision-making process in a robot visual system. Lane detection plays an important role in robot navigation. The main objective of lane detection research is to assure safety in driver assistance. A basic process of lane detection uses information from the camera for the visibility of roads by the boundary recognition of the lane marks. This boundary yields the estimation of the geometry of the floor, the localization and orientation of the lane marks. In the state-of-the-art, lane detection techniques estimate the markings of the road using a simple Canny filter to compute edges from the input scene. This provides a region of interest of white marking on a black surface. Despite this simplicity, several issues can be presented in real-world scenes. For instance, different marks, such as solid, dashed and curved lines, need to be considered [7–9]. Another difficulty is the quality of the image captured by the camera, that is, noisy frames, motion distortion, poor lighting, clutter, contrasting shadows, and occlusions. For this kind of applications, image preprocessing algorithms are needed in order to improve the quality of the input scene [10–12]. Other consideration includes the geometrical distortion due to the perspective projection which is given by the camera placed over the mobile robot, so a geometrical adjustment is computed using the position of the camera [13, 14], see Fig. 2.

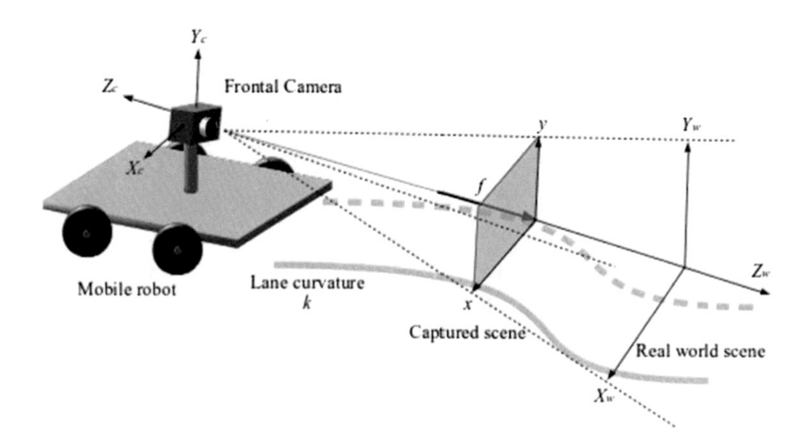

Fig. 2 Samples of real world captured scenes from frontal camera

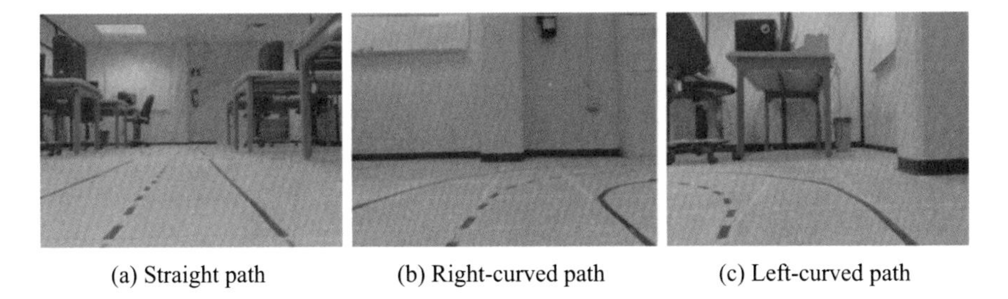

(a) Straight path (b) Right-curved path (c) Left-curved path

Fig. 3 Samples of real-world captured scenes from frontal camera

Generally, the visual system is composed of a frontal camera that captures the light intensities from the scene transforming it into a discrete signal. Considering the geometrical model representation shown in Fig. 3, the camera sensor returns an intensity value from a pixel coordinate $r(x, y)$, where the range values are used to estimate the position of the point in the real world, computed for each 3D coordinate as [15], where f is the focal length of the camera, see (1), (2) and (3). A sample of an input visualization from the captured scene is shown in Fig. 3. Observe that different marks can appear as Fig. 3a a straight path formed by solid and dashed lines to distinguish right and left lanes, respectively; Fig. 3b a path curved from right formed by different lines; and Fig. 3c a left-curved path with a missing lane.

$$X = Z\frac{x}{f}, \tag{1}$$

$$Y = Z\frac{y}{f}, \tag{2}$$

and

$$Z = r(x, y)\frac{f}{\sqrt{f^2 + x^2 + y^2}}, \tag{3}$$

As seen in Fig. 3, real-world scenes appear approximately half of the image with the observed roads and background information. Due to the background area does not provide significant information for lane recognition, is commonly cropped and discarded. This procedure assumes the continuous lane markings without significant changes. So, the region of interest (ROI) of the input scene is just the segmented image which contains the lane markings.

Considering that Fig. 4a is the input scene captured from the frontal camera, thus Fig. 4b is the segmented ROI only with the visualization of the roads. Note that the information of the roads is geometrically distorted with a perspective projection. To solve this issue, a 4-points projection mapping can be implemented using [16]. Figure 4c shows the resulting mapped image, where each row of the image

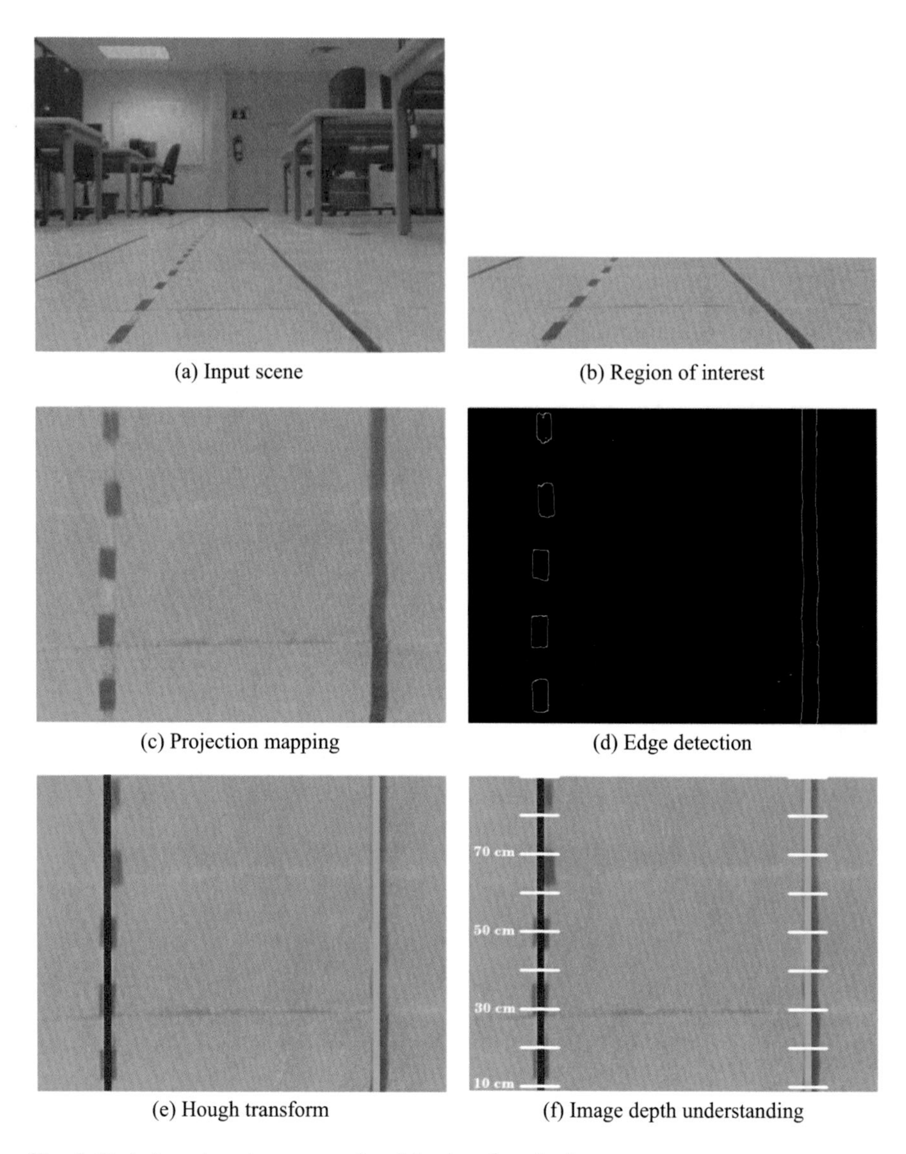

(a) Input scene (b) Region of interest

(c) Projection mapping (d) Edge detection

(e) Hough transform (f) Image depth understanding

Fig. 4 Basic lane detection process for vision-based navigation

corresponds to a linear transformation of depth information shown in the input scene. An edge detection process is carried on with the Canny algorithm, which reduces the amount of data to process [17]. The Canny algorithm extracts features of edges in the input image. As shown in Fig. 4d the information of the scene is simplified by only the intensity changes given by the boundary, in which the lane mark is located [18].

In addition to this feature extraction, a Hough transform is computed in order to fill unconnected lines and to classify different lines by a voting procedure using an accumulating threshold [18]. In addition, Hough transform can be useful also to detect and locate the lane markings [19]. Figure 4e shows the lane marking detection. As we can see, the procedure is able to classify lanes by detecting groups of different types of marks. For instance, red and green lines are used to distinguish left and right lanes, respectively.

An inverse projection mapping can be calculated using the physical dimensions of the real-world scene. The distance d (measured in cm) from robot to the last point perceived of the scene, is related to the number of rows of the image ROI. A depth retrieval from a segment, d_s, can be computed using (4)

$$ds = d\left(\frac{Ns}{Nt}\right) \quad (4)$$

where Ns is the number of rows of the segment of interest, and Nt is the total number of rows in the image ROI. Figure 4f shows the image depth understanding in terms of distance from robot origin given by a real-world scene. Monocular depth retrieval is crucial in the decision-making process in path planning. Thus, the accuracy of the vision system plays an important role in mobile robot navigation.

3 Nonholonomic Car-Like Robot

From a kinematic point of view, robots have two categories: omnidirectional and constrained driven robots. The first approach enables the movement of the robot in any direction (of their plane of reference), the robot has as many controllable degrees of freedom as the total amount of their degrees of freedom. This type of systems is called fully actuated. The second approach has less controllable degrees of freedom than the total number of degrees of freedom that the system has, these robots are called under-actuated [20].

The above can be explained with the analysis of the kinematics constraints and depending on how these constraints can be expressed, the robotic system is holonomic or nonholonomic.

A system is holonomic if all the constraints of the system are expressed without any derivatives, in the form $f(\boldsymbol{q}, t) = 0$, where $\boldsymbol{q}$ are the constraints of the system. If the constraints of the system have derivatives in the expression in the form $f(\boldsymbol{q}, \dot{\boldsymbol{q}}, \ddot{\boldsymbol{q}}, t) = 0$, the system is called nonholonomic. The equations that describe nonholonomic systems are non-integrable, which means that the total degrees of freedom are greater that the controllable ones [20, 21].

Wheeled robots move on a two-dimensional plane, granting them three degrees of freedom: x-axis, y-axis, and orientation with respect to some of the axis. Car-like Robots are a special type of wheeled robot that has a steered-wheel which imposes some constraints on the linear velocity along specific directions. These limitations

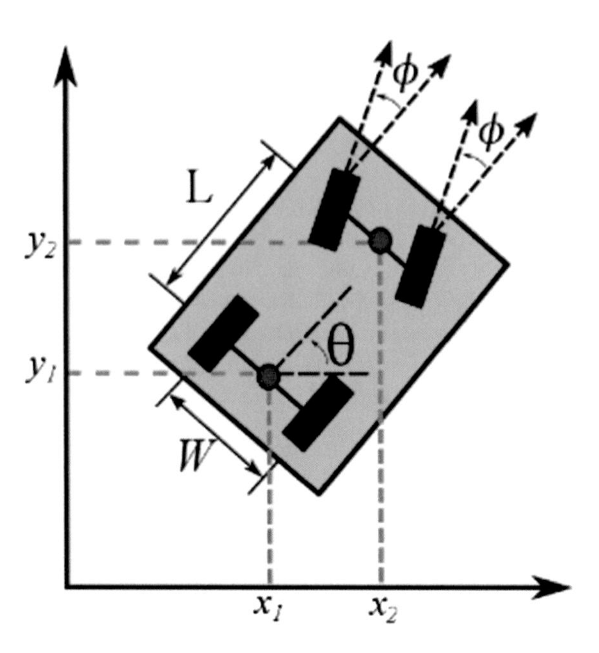

Fig. 5 Kinematic scheme of a rear driven front-steered car-like robot

are based on the fact that the kinematics constraints have derivative relationships, this classifies a car-like robot as a nonholonomic system.

The position of car-like robots, Fig. 5, is described by a point in a two coordinate system $P = (x, y)$, which is located in the middle of the rear axle, and an orientation angle with respect to a fixed reference system (which in most cases is the x-axis) denoted by θ. At any given point P, the system has associated an instantaneous linear velocity v, and an angular velocity for the steering wheels represented by ω.

Since the system is nonholonomic, there exist some limitations in the degrees of freedom of the robot, these constraints are imposed by the steering angle and represented by ϕ that limits the position and orientation [22–24]. Considering that the robot is front-steered, the nonholonomic restrictions of the system are determined by (5) and (6).

$$\dot{y}_1 \cos\theta - \dot{x}_1 \sin\theta = 0 \tag{5}$$

$$\dot{y}_2 \cos\{\theta + \phi\} - \dot{x}_2 \sin\{\theta + \phi\} = 0 \tag{6}$$

where $x_2 = x_1 + L\cos\theta$, and $y_2 = y_1 + L\sin\theta$. Having the position, orientation, and steering angle, any given state of the robot can be represented by (7).

$$q = \begin{bmatrix} x \\ y \\ \theta \\ \phi \end{bmatrix} \tag{7}$$

To calculate the relation of the steering angle with the position and orientation, it is needed to obtain the instantaneous curvature of the robotic system. Considering a model in which the mass and velocities are not large enough to induce slippage [25], the instantaneous curvature k is affected only by two parameters: the steering angle of the wheels and the length of the length of the robot. The curvature is calculated using (8),

$$k = \frac{\tan \phi}{L} \tag{8}$$

The rate of change for the steering angle can be calculated with the first order differential equation $\dot{\phi} = \omega$, in which the only dependence is given by the angular velocity of the steering wheels [24, 26, 27]. Taking into account the linear velocity and the instantaneous curvature is possible to obtain the kinematics equations with (9),

$$\dot{q} = \begin{bmatrix} \dot{x} \\ \dot{y} \\ \dot{\theta} \\ \dot{\phi} \end{bmatrix} = \begin{bmatrix} \cos \theta & 0 \\ \sin \theta & 0 \\ \frac{\tan \phi}{L} & 0 \\ 0 & 1 \end{bmatrix} \begin{bmatrix} v \\ \omega \end{bmatrix} \tag{9}$$

With these set of equations, and considering fixed velocities, the only control variable for the system is the steering angle.

$$\begin{cases} |v| \leq v_{max} \\ |\phi| \leq \phi_{max} \\ |\omega| \leq \omega_{\max} \end{cases} \tag{10}$$

Finally, the last considerations of the models are the physical constraints of the robot, they usually are given in the form of (10).

4 Methodology

To control a robot in a path following problem, a feedback stage is needed. In this work, feedback based on frontal camera information is used. The main difference between using this approach and another type of sensor is the look-ahead distance given by the characteristics of the camera and its position. This look-ahead distance makes it harder to control the position of the robot and keep it on the desired track. To solve this problem, we proposed a fuzzy system designed to control a car-like robot with the information given by the camera.

The goal of the fuzzy system is to keep the middle of the rear axle with the less possible error with the desired path. To achieve this, some feedback is needed, which in this case is given by the processing stage of the frontal camera.

The feedback provided is the difference between the actual position P with components $(x_P,\ y_P)$, and the desired position D with components $(x_D,\ y_D)$, and this will be the first input of the fuzzy system.

With the difference of the component in both axis and trigonometry, an angle can be calculated. This value, $u = \tan^{-1}((x_D - x_P)/(y_D - y_P))$, is the instantaneous angular displacement that the robot has to assume to reduce the deviation of the robot with the desired path. Another important information that can be calculated is the change of rate of u at each discrete time step, which is the angular velocity of the steering angle, this change is calculated with $\dot{u} = u_t - u_{t-1}$. Using the information above, it is evident that both u and $\dot{u}$ are the main candidates for being the input of the fuzzy system.

The output of the fuzzy system is going to be the only control variable of the system, which is the steering angle ϕ. For this problem, a Mamdani-type fuzzy inference system is a good selection for its simplicity. The inputs and outputs of the proposed fuzzy system can be seen in Fig. 6.

The next step is to choose the number of membership functions (MFs) for the inputs and outputs. For both inputs u and $\dot{u}$, five MFs are proposed: NB (far left), N (left), Z (center), P (right), and PB (far right). For the output of the system, a similar approach was taken: BD (far left), D (left), Z (center), I (right), BI (far right).

According to the human expert, the rule matrix is shown in Table 1.

The ranges of the input and output variables depend on the physical characteristics of the robot. In this specific case, the car-like robot used is in a 1:10 scale car, see Fig. 7. The height of the camera H is 18 cm of the ground, the total length of the car l_1 is 0.43 m, the distance from axels l_2 is 0.26 m, the width of the car W is 0.2 m, the look-ahead of the camera is 0.46 m, and the maximum steering angle is 38°.

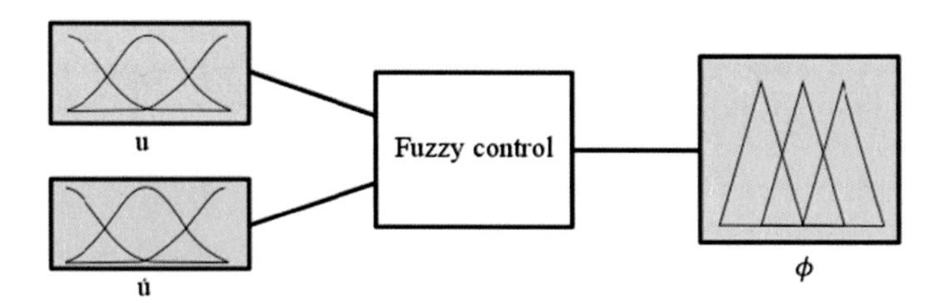

Fig. 6 Inputs and outputs of the proposed fuzzy system

Table 1 Rules for the fuzzy controller

u\û	NB	N	Z	P	PB
NB	BI	BI	BI	BI	BI
N	BI	I	I	Z	D
Z	BI	I	Z	D	BD
P	I	Z	D	D	BD
PB	BD	BD	BD	BD	BD

Fig. 7 Experimental platform

With these values and experimental information of the vehicle, the input MFs were defined, see Figs. 8 and 9. The ranges of each input are within the ranges of the maximum steering angle. For u, the center, far right and far left MFs have bigger area than the left and right MFs; this is to keep the robot most of the time in the center of the path and to reduce wobbling of the steering wheels.

For input $\dot{u}$, the distribution and area for the center, left and right membership functions are almost the same, this is because in the case of the change of rate of u, there exist physical limitations in the response time of the actuator that controls the steering. With this in mind, the value of $\omega_{\max}$ was set to 2.4 rad/s.

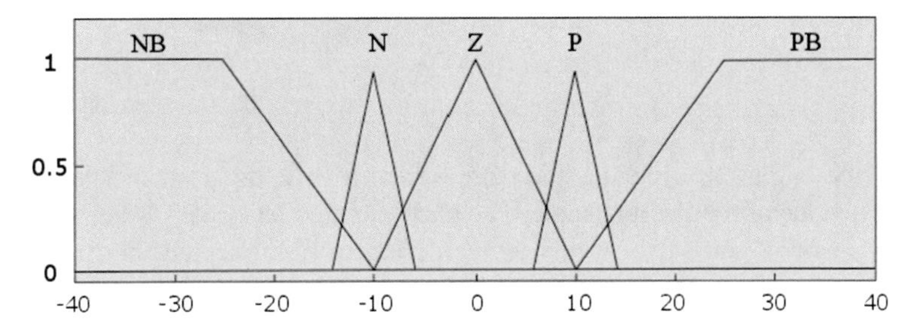

Fig. 8 First input membership function, angular displacement u to correct the actual position

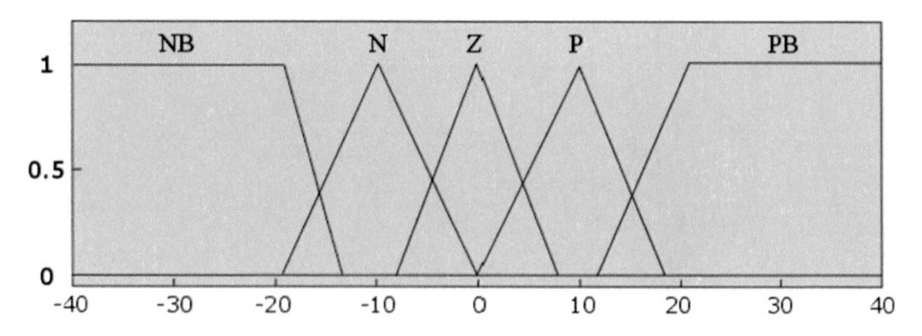

Fig. 9 Second input membership function $\dot{u}$

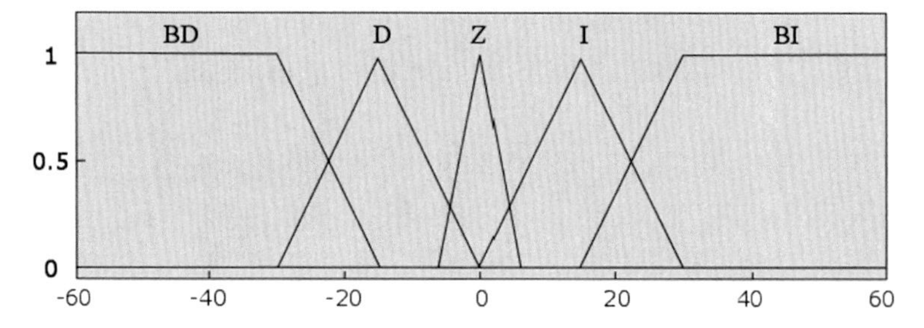

Fig. 10 Output membership function, steering angle ϕ

Fig. 11 Fuzzy system surface, err-axis is the input u, cerr-axis is the input $\dot{u}$, and increment-axis corresponds to the value of the steering angle

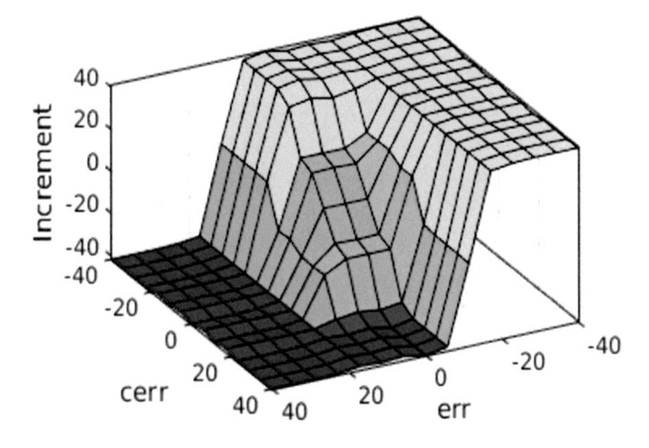

For the output, see Fig. 10, the range of the MFs is bigger than the maximum physical value of the steering angle. The maximum output of the system is restricted by the shape of the MFs as well as the defuzzification method, in this case, the centroid type. This makes that the maximum output of the fuzzy and ϕ_{max} be the same. The surface of the fuzzy inference system shows that the maximum output of the system matches with ϕ_{max}, see Fig. 11.

5 Experiments

For this stage, a simulation with the same parameters as the experimental platform was used, both the simulation and the control stage were implemented in Matlab. The circle and the sinusoidal paths are commonly used to evaluate path following algorithms, see Fig. 12a, b, whereas the paths showed in Fig. 12c, d are common tasks that a car has to do in normal conditions, such as the change between lanes or avoid an object. The last path, Fig. 12e, is an arbitrary curved path designed to test the maximum steering angle of the model.

To compare the performance of the fuzzy controller, a PD controller was developed, and it was tuned using the first path as an input, see Fig. 13, the constants used for the PD controller are 0.1 for K_P and 0.8 for K_D. For the simulations, the maximum linear velocity stablish was 0.66 m/s based on the physical limitations of the robot.

The goal of the fuzzy controller is to obtain the minimum amount of error since there is a physical limitation on how much error can have the system in the x-axis. This limitation comes from the camera lens, if the path the error is too big in the x-axis of the camera, the path might not be detected by the lens of the camera, and the system will be unstable.

The maximum error in the x-axis is half the distance of the width of the car, around 0.1 m. Considering this information, the maximum error given by the difference of the Euclidean distance between the desired path and the one given by the controller is established in 0.1 m.

6 Results

For every path, the fuzzy inference system and the PD controllers were implemented. The path obtained using each controller was compared with the reference path. With this information the main square error (MSE), the root means square error (RMSE), and the maximum square root error (ME) were obtained. The information of the different errors is collected in Tables 2 and 3.

Considering the five paths used for testing the algorithms, the fuzzy inference system fulfilled the goal in all the paths, see Table 2. The PD system fulfilled the goals only in four paths (paths 2, 3 and 4), but fail in path 1, as is shown in Table 1.

Examining the ME goal for the controller, although the fuzzy controller has a bigger maximum square root error in four of the paths, this controller meets the requirements for the ME goal established, that was 0.1 m. On the other hand, the PD controller fails to meet the maximum error requirement established for path 1.

Since the PD controller failed to meet the ME requirements for path 1, the K_P value was adjusted ($K_P = 2$), the results are detailed in Table 4. As the data show, an increment in the proportional constant give better results in the ME for path 1,

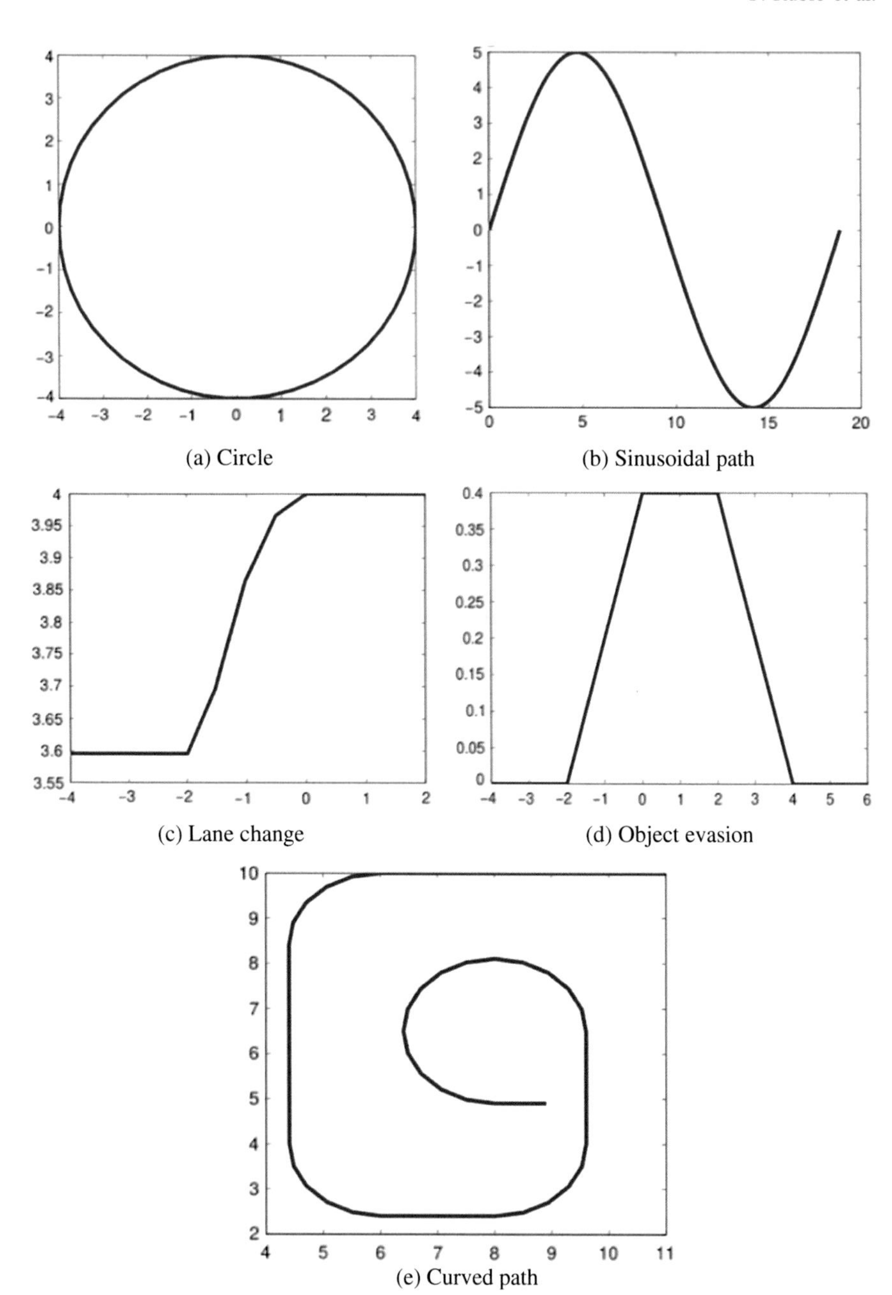

(a) Circle

(b) Sinusoidal path

(c) Lane change

(d) Object evasion

(e) Curved path

Fig. 12 Different paths used for the experiments

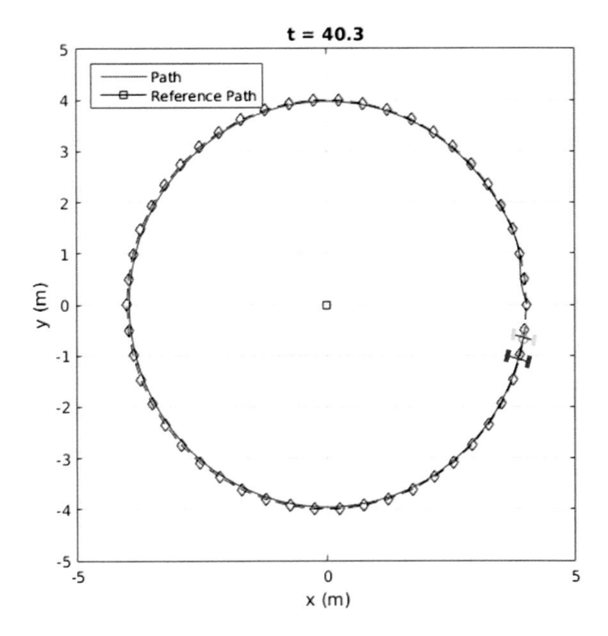

Fig. 13 Path following simulation, the time is measured in seconds

Table 2 Fuzzy controller error (in meters) for every path

Path	MSE	RMSE	ME
1	*0.0015*	*0.0381*	*0.0911*
2	4.38×10^{-4}	0.0209	0.0362
3	9.65×10^{-5}	0.0098	0.0246
4	8.04×10^{-5}	0.0090	0.0311
5	0.0019	0.0432	0.0698

Path with the highest path in italics

Table 3 PD controller error (in meters) for every path

Path	MSE	RMSE	ME
1	0.002	0.0444	**0.1151**
2	1.94×10^{-4}	0.0139	0.0250
3	1.02×10^{-4}	0.0101	0.0227
4	6.16×10^{-5}	0.0078	0.0214
5	0.0012	0.0348	0.0635

Error higher than the desired goal in bold

Table 4 PD controller error (in meters) for every path

Path	MSE	RMSE	ME
1	0.0012	0.0353	0.1030
2	4.9644×10^{-4}	0.0223	0.0471
3	1.0202×10^{-4}	0.0101	0.0247
4	8.6254×10^{-5}	0.0093	0.0258
5	0.0022	0.0472	0.0766

but the ME for the other paths get higher that the fuzzy controller output. This means that a higher value of K_P, would get a better performance for path 1, but a worse performance for the other paths.

7 Conclusions

The present work details the design, simulation and experimental tests of a fuzzy system controller for a nonholonomic car-like robot for path following based on the information of a frontal camera. Simulation results, using Matlab have shown that the proposed fuzzy inference system meet all the ME goals for the five paths used in the experiments; the PD system failed in one. Although, the PD system have smaller MSE and ME on four of the paths, modifying its characteristics to achieve the most important goal (ME less than 0.1 m) in all the paths, it gives worst results than the fuzzy system. Thus, the proposed fuzzy system controller has shown its adaptation to the dynamics of the car-like robot.

Therefore, we conclude that the fuzzy controller fulfill with all the design criteria, even then it was tuned by the knowledge expertise, so it can be expected to improve its performance by tuning the fuzzy controller using an automatic tuning method, such as an evolutionary algorithm to perform this task.

Future work will be focused on the implementation of the proposed fuzzy system controller on the real-world four-wheeled model vehicle (scale 1:10), with the purpose of performing experimental tests in real-world robotic systems to validate the proposed fuzzy system controller in complex path following problems, which certainly is the focus of any real-world path following system. In addition, type-2 fuzzy logic could be used, like in [28, 29, 33] or hybrid approaches like in [30–32, 34].

Acknowledgements We thank Instituto Politecnico Nacional (IPN), to the Comisión de Fomento y Apoyo Académico del IPN (COFAA), and the Mexican National Council of Science and Technology (CONACYT) for supporting our research activities.

References

1. X. Xiang, C. Yu, Q. Zhang, Robust fuzzy 3D path following for autonomous underwater vehicle subject to uncertainties. Comput. Oper. Res. (in press) (2016)
2. M.S. Masmoudi, N. Krichen, M. Masmoudi, N. Derbel, Fuzzy logic controllers design for omnidirectional mobile robot navigation. Appl. Soft Comput. **49**, 901–919 (2016)
3. A. Bakdi, A. Hentout, B. Boutami, A. Maoudj, O. Hachour, B. Bouzouia, Optimal path planning and execution for mobile robots using genetic algorithm and adaptive fuzzy-logic control. Robot. Auton. Syst. **89**, 95–109 (2017)

4. C.-H. Sun, Y.-J. Chen, Y.-T. Wang, S.-K. Huang, Sequentially switched fuzzy-model-based control for wheeled mobile robot with visual odometry. Appl. Math. Model. (in press) (2016)
5. O. Montiel, R. Sepúlveda, I. Murcio, U. Orozco-Rosas, Geo-navigation for a mobile robot and obstacle avoidance using fuzzy controllers, in *Recent Advances on Hybrid Approaches for Designing Intelligent Systems*. Studies in Computational Intelligence 547, ed. by O. Castillo, P. Melin, W. Pedrycz, J. Kacprzyk. (Springer, 2014), pp. 647–669
6. J.-S.R. Jang, C.-T. Sun, E. Mizutani, Neuro-fuzzy and soft computing: a computational approach to learning and machine intelligence Prentice Hall. MATLAB curriculum series (1997)
7. M. Aly, Real time detection of lane markers in urban streets, in *2008 IEEE Intelligent vehicles symposium* (Eindhoven, The Netherlands, 2008), 4–6 June 2008, pp. 7–12
8. A. Borkar, M. Hayes, T. Smith, Robust lane detection and tracking with RANSAC and Kalman filter. Paper presented at the 2009 16th IEEE International Conference on Image Processing (ICIP), Cairo, 7–10 Nov 2009, pp. 3261–3264 (2009)
9. G. Antonelli, S. Chiaverini, Experiments of fuzzy lane following for mobile robots. Paper presented at the Proceedings of American Control Conference, Boston, 30 June–2 July 2004, pp. 1079–1084 vol. 2 (2004)
10. C. Coutinho, F. Melicio, Road lane detection for autonomous robot guidance. Paper presented at the 2014 IEEE International Conference on Autonomous Robot Systems and Competitions, Espinho, 14–15 May, pp. 297–302 (2014)
11. A.M. Kumar, P. Simon, Review of lane detection and tracking algorithms in advanced driver assistance system. Int. J. Comput. Sci. Inform. Technol. **7**(4), 65–78 (2015)
12. D. Cáceres-Hernández, L. Kurnianggoro, A. Filomenko, K.H. Jo, Real-time lane region detection using a combination of geometrical and image features. Sensor **16**(1935), 1–19 (2016)
13. A. Lopez, C. Cañero, J. Serrat, J. Saludes, F. Lumbreras, T. Graf, Detection of lane markings based on ridgeness and RANSAC, in *Proceedings 2005 IEEE Intelligent Transportation Systems*, (Vienna, 2005), 16 September 2005, pp. 254–259
14. D. Scaramuzza, 1-Point-RANSAC structure from motion for vehicle-mounted cameras by exploiting non-holonomic constraints. Int. J. Comput. Vis. **95**(1), 74–85 (2011)
15. F. Mufti, R. Mahony, J. Heinzmann, Robust estimation of planar surfaces using spatio-temporal RANSAC for applications in autonomous vehicle navigation. Robot. Auton. Syst. **60**(1), 16–28 (2012)
16. P. Lu, Q. Liu, J. Guo, Camera Calibration Implementation Based on Zhang Zhengyou Plane Method, in *Proceedings of the 2015 Chinese Intelligent Systems Conference*. Lecture Notes in Electrical Engineering Springer, Berlin, Heidelberg, 2016, ed. by Y. Jia, J. Du, H. Li, W. Zhang
17. W.K. Pratt, *Digital Image Processing: PIKS Scientific Inside*, 4th edn. (Wiley, USA, 2007)
18. C.Y. Low, H. Zamzuri, S.A. Mazlan, Simple robust road lane detection algorithm. Paper presented at the 2014 5th International Conference on Intelligent and Advanced Systems (ICIAS), Kuala Lumpur, 3–5 June 2014, pp. 1–4 (2014)
19. D.H. Ballard, Generalizing the Hough transform to detect arbitrary shapes. Pattern Recogn. **13** (2), 111–122 (1981)
20. A. Prasad, B. Sharma, J. Vanualailai, A solution to the motion planning and control problem of a car-like robot via a single-layer perceptron. Robotica **32**(6), 935–952 (2014)
21. B. Li, Z. Shao, Simultaneous dynamic optimization: a trajectory planning method for nonholonomic car-like robots. Adv. Eng. Softw. **87**, 30–42 (2015)
22. E. Masehian, H. Kakahaji, NRR: a nonholonomic random replanner for navigation of car-like robots in unknown environments. Robotica **32**(7), 1101–1123 (2014)
23. J. Ni, J. Hu, Dynamic control of autonomous vehicle at driving limits and experiment on an autonomous formula racing car. Mech. Syst. Signal Process. **90**, 154–174 (2017)

24. T.H. Nguyen, D.H. Kim, C.H. Lee, H.K. Kim, S.B. Kim, Mobile robot localization and path planning in a picking robot system using kinect camera in partially known environment, in *AETA 2016: Recent Advances in Electrical Engineering and Related Sciences. AETA 2016.* Lecture Notes in Electrical Engineering, vol. 415, ed. by V. Duy, T. Dao, S. Kim, N. Tien, I. Zelinka (Springer, Cham, 2017), pp. 686–701
25. U. Rosolia, S. De Bruyne, A.G. Alleyne, Autonomous vehicle control: a nonconvex approach for obstacle avoidence. IEEE Trans. Control Syst. Technol. **25**(2), 469–484 (2017)
26. A.V. Pesterev, A linearizing feedback for stabilizing a car-like robot following a curvilinear path. J. Comput. Syst. Sci. Int. **52**(5), 819–830 (2013)
27. N. Ghita, M. Kloetzer, Trajectory planning for a car-like robot by environment abstraction. Robot. Auton. Syst. **60**(4), 609–619 (2012)
28. G.M. Mendez, O. Castillo, http://scholar.google.com.mx/citations?view_op=view_citation&hl=en&user=1C8gb8IAAAAJ&cstart=40&citation_for_view=1C8gb8IAAAAJ:qxL8FJ1GzNcC, in *The 14th IEEE International Conference on Interval Type-2 TSK Fuzzy Logic Systems Using Hybrid Learning Algorithm, Fuzzy Systems, 2005. FUZZ'05*, pp. 230–235
29. O. Castillo, P. Melin, Design of Intelligent Systems with Interval Type-2 Fuzzy Logic, in *Type-2 Fuzzy Logic: Theory and Applications*, pp. 53–76
30. O. Castillo, P. Melin, E. Ramírez, J. Soria, Hybrid intelligent system for cardiac arrhythmia classification with Fuzzy K-Nearest Neighbors and neural networks combined with a fuzzy system, Expert Syst. Appl. **39**(3), 2947–2955
31. L. Aguilar, P. Melin, O. Castillo, Intelligent control of a stepping motor drive using a hybrid neuro-fuzzy ANFIS approach, Appl. Soft Comput. **3**(3), 209–219
32. P. Melin, O. Castillo, Modelling, simulation and control of non-linear dynamical systems: an intelligent approach using soft computing and fractal theory (CRC Press, 2001)
33. P. Melin, C.I. Gonzalez, J.R. Castro, O. Mendoza, O. Castillo, Edge-detection method for image processing based on generalized type-2 fuzzy logic, IEEE Trans. Fuzzy Syst. **22**(6), 1515–1525
34. P. Melin, O. Castillo, Intelligent control of complex electrochemical systems with a neuro-fuzzy-genetic approach, IEEE Trans. Ind. Electron. **48**(5), 951–955

Design and Implementation of a Fuzzy Path Optimization System for Omnidirectional Autonomous Mobile Robot Control in Real-Time

Felizardo Cuevas and Oscar Castillo

Abstract The main goal is to maintain a specified position for a robot using fuzzy logic to control its behavior. In this paper we propose a novel approach for fuzzy control of an omnidirectional mobile platform to pursuit complicated trajectories in order to complete the whole robot route without losing the main trajectory. Simulation result show the advantages of the proposed approach.

Keywords Mobile robots · MLR (Multiple linear regression) · Experimental numerical model · FS (Fuzzy systems) · FLC (Fuzzy logic controller) Kinematic

1 Introduction

Fuzzy logic theory is a powerful soft computing technique to control complex and non-linear systems based on human expert knowledge.

A system based on fuzzy logic usually divides its work into three tasks. The transformation of numerical values into fuzzy logic values; The inference engine that employs the rules, and the block converting the values of the fuzzy logic into numeric values [1].

Fuzzy Logic is a type of logic that recognizes more than simple true and false values. With fuzzy logic, propositions can be represented with degrees of truth or falsity. By means of Fuzzy Logic, notions can be formulated mathematically, such as a little hot or very cold, so that they are processed by computers and quantify human vague expressions, such as "Very Cold" or "Hot Water". Thus, it is an attempt to apply the human way of thinking to computer programming. It also allows to quantify imprecise descriptions that are used in the language and gradual transitions in appliances such as going from dirty air to clean air in an air conditioner, or allowing to adjust environmental cleaning cycles through sensors. The use

F. Cuevas · O. Castillo (✉)
Tijuana Institute of Technology, Tijuana, BC, Mexico
e-mail: ocastillo@tectijuana.mx

O. Castillo et al. (eds.), *Fuzzy Logic Augmentation of Neural and Optimization Algorithms: Theoretical Aspects and Real Applications*, Studies in Computational Intelligence 749, https://doi.org/10.1007/978-3-319-71008-2_19

of Fuzzy Logic to process true partial values has been of great help in engineering [2, 3].

Currently, omnidirectional platforms are becoming increasingly popular in both the academic and industrial fields, e.g. mobile robots. For this reason, it is interesting to perform research work in this area.

This paper is organized as follows. Section 2 describes kinematic modeling of the omnidirectional robot with 3 wheels. Section 3 details the fuzzy logic controller architecture and simulation, results obtained using an engineering program and the simulation environments of the robotic system. Section 4 presents an Experimental Numerical Model of the Mobile Robot. Section 5 presents some final comments and conclusions.

2 Kinematic Model

Kinematics is applied in the study of robot movement in function of its geometry. Among the immediate applications of kinematics are the possibility of using it as a starting mathematical model for controller design, simulation of the kinematic behavior of the platform, or to establish equations for odometer calculations. Normally, the following limitations for the construction of the kinematic model are analyzed [4, 5]:

- The robot moves on a flat surface.
- There are no flexible elements in the robot structure (including wheels).
- The wheels have one or no steering axis, so that the latter is always perpendicular to the ground.
- Friction is not taken into consideration.

3 Kinematic Model of the Omnidirectional Robot

Table 1 compiles the positions, orientations, and distribution of the three omnidirectional wheels as a function of the robot radius with respect to the center of the robotic system [4, 5].

Table 1 Omni-directional triangular platform kinematic initial parameters configuration

	Wheel 1	Wheel 2	Wheel 3
α_i	180°	60°	–60°
β_i	0°	0°	0°
γ_i	0°	0°	0°
δ_i	(0, 0, 0)	(0, 0, 0)	(0, 0, 0)
λ_i	$(-L, 0, 0)$	$(\frac{L}{2}, \frac{L\sqrt{3}}{2}, 0)$	$(\frac{L}{2}, \frac{-L\sqrt{3}}{2}, 0)$

3.1 Kinematic Behavior

The kinematic behavior is established on the principle that the wheels in contact with the ground behave as a planar joint of three degrees of freedom [4, 6].

Wheels behave as a planar joint as described below [5]. Assuming that the wheel is a rigid element, it is always in contact with the ground in a single point and it serves as the origin for the reference system shown in Fig. 1. Direction v determines the normal direction of the wheel, v represents the side's slippage and w_z is the rotational speed when the robot turns. In the case of a conventional wheel the component v_x is null, but there are other wheels providing a different behavior as presented in Fig. 2.

The omnidirectional wheel is defined as a standard wheel, which is endowed with a crown of rollers, whose axes of rotation are perpendicular to the normal direction of advance. Thus, by applying a lateral force, the rollers rotate on themselves and allow the component Vx to be non-zero, and therefore, the non-holomicity constraint is eliminated [5, 6].

3.2 Kinematic Structure

Figure 3 presents the localization and the orientation of the omni-directional robot shown as a vector $(x, y, \emptyset)^T$. The globalvelocity of the robot is shown with the vector $(\dot{x}, \dot{y}, \dot{\phi})^T$ and the angular velocities of each wheel are represented by the vector $(\dot{\theta}_1, \dot{\theta}_2, \dot{\theta}_3)^T$.

Based on the model described in [7], the platform kinematic model is given by:

$$\begin{bmatrix} \dot{\theta}_1 \\ \dot{\theta}_2 \\ \dot{\theta}_3 \end{bmatrix} = \frac{1}{r} \cdot \begin{bmatrix} V_1 \\ V_2 \\ V_3 \end{bmatrix} = \frac{1}{r} \cdot \overbrace{\begin{bmatrix} -\sin(\phi + \varphi_1) & \cos(\phi + \varphi_1) & R \\ -\sin(\phi + \varphi_2) & \cos(\phi + \varphi_2) & R \\ -\sin(\phi + \varphi_3) & \cos(\phi + \varphi_3) & R \end{bmatrix}}^{P(\phi)} \cdot \begin{bmatrix} \dot{x} \\ \dot{y} \\ \dot{\phi} \end{bmatrix} \quad (1)$$

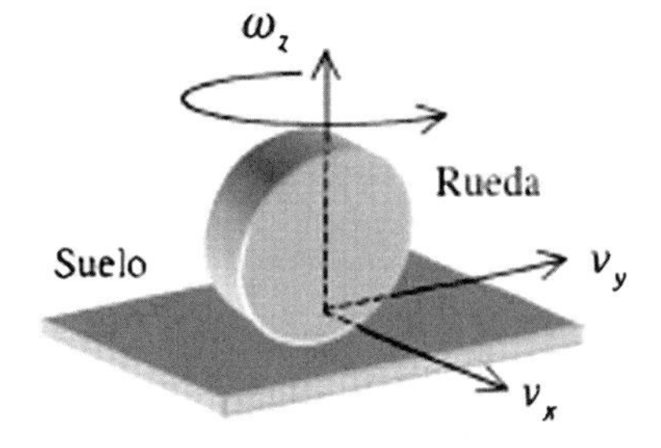

Fig. 1 The wheel contact with the floor [4]

Fig. 2 The omni-directional wheels [6]

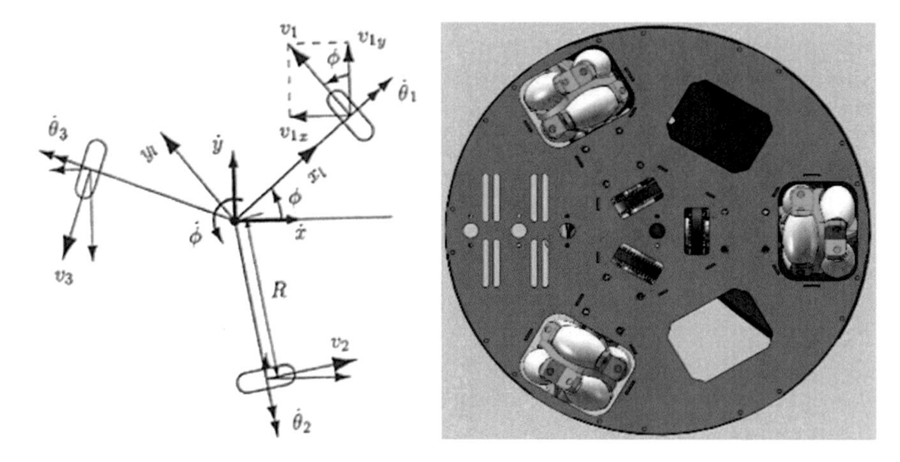

Fig. 3 Kinematic diagram of the robot

With:

θi	Angular speed of the wheel i
V_i	Linear speed of the wheel i
r	Radius of the wheel
R	Distance between a wheel and the center of the robot
ϕ	Angular speed of the robot
φ_i	Angular location of the wheel i
$P(\phi)$	Transformation matrix between the angular speeds of the wheels and the global velocity vector $(\dot{x}, \dot{y}, \dot{\phi})^T$

4 Fuzzy Control

The fuzzy control system is in charge of receiving the signals from the sensors, processing them, interpreting them and then making a decision based on the results, and thus giving an output that later by means of the actuators will be transformed into an actual physical result.

4.1 Fuzzy Logic

A system based on fuzzy logic usually divides its work into three blocks [1].

1. FUZZIFICATION BLOCK: In this block each input data is assigned a degree of membership to each of the fuzzy sets. The inputs to this block are concrete values of the variable to be analyzed and the output data are the degrees of belonging to the studied sets.
2. INFERENCE BLOCK: This block relates the fuzzy sets (input and output) representing the rules that define the system.
3. DEFUZZIFICATION BLOCK: In this block from the fuzzy sets coming from the inference a concrete numeric result is obtained by the application of mathematical methods of defuzzification.

As it is presented in Figs. 4 and 5, The fuzzification Interface, is responsible for transforming the numerical value of the input variable into a fuzzy set. For its part, the knowledge base characterizes the rules and manipulation of data in the fuzzy controller, whereas the Rule Base determines The control action to take. As for the Inference Mechanism, it simulates human decision making based on fuzzy concepts and infers control actions. The defuzzification process consists of converting the fuzzy control action to an equivalent numerical value.

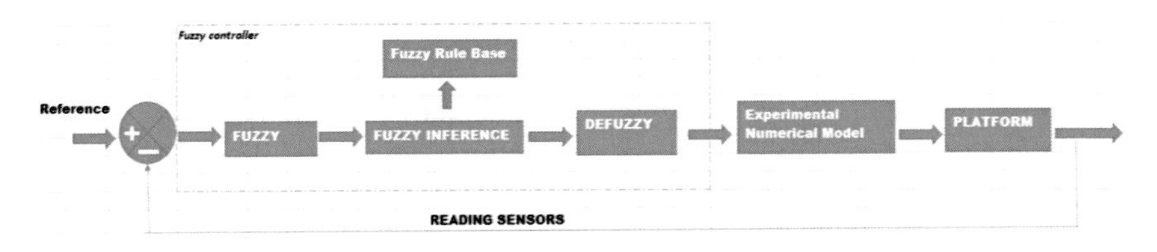

Fig. 4 Fuzzy control FeedbackLoop

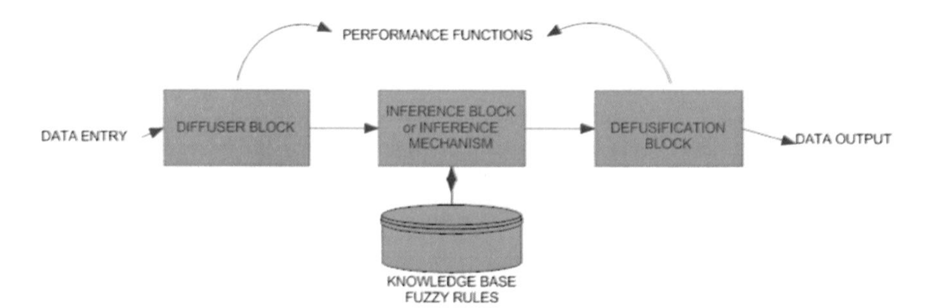

Fig. 5 Fuzzy controller structure

4.2 *Fuzzy System Design*

To have a clearer idea of what is the design of a fuzzysystem (FS) with the help of a computational program, the system was previously designed analytically.

The selected fuzzy system has six inputs, a system based on "if-then" rules, and two outputs (this is presented in Fig. 6).

The main purpose of robot navigation is to reach and maintain the target path. Various parameters could affect navigation steps such as desired position and angle values, distances traveled, curvature of the path as well as linear and angular velocities. We began by designing a fuzzy navigation controller to help the robotic platform maintain its central position in the path with a desired final angle (see Figs. 7, 8 and 9). To improve robotic platform navigation performance, a Mamdani type fuzzy controller was developed that combines the advantages of conventional PI with fuzzy logic theory and an experimental numerical model that allows us to calculate the distance from the trajectory center [8]. The Mamdani model is

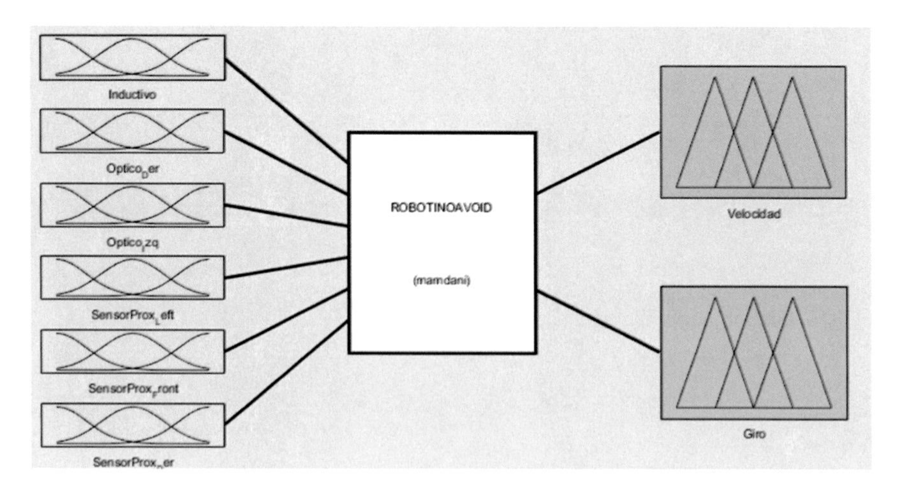

Fig. 6 Fuzzy system

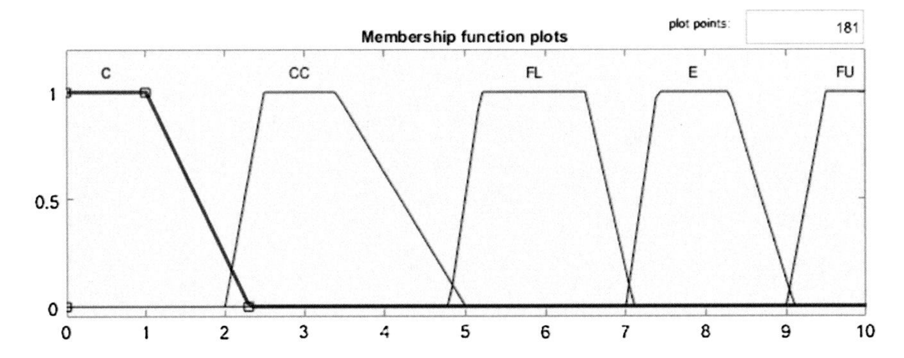

Fig. 7 Membership functions of the inductive sensor input

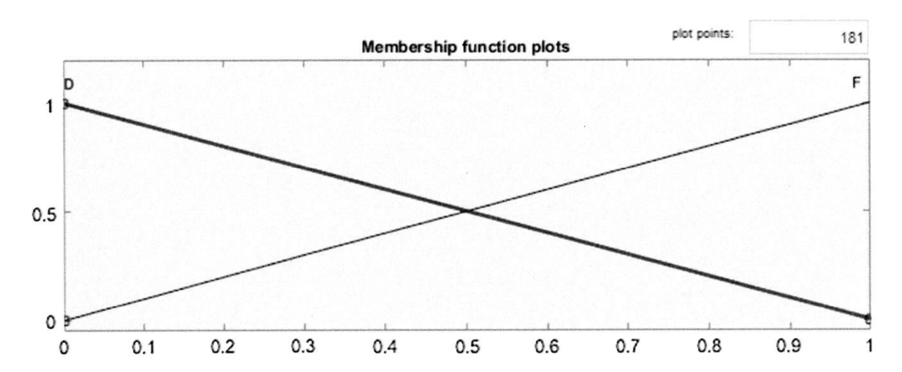

Fig. 8 Membership functions of the left and right input sensors

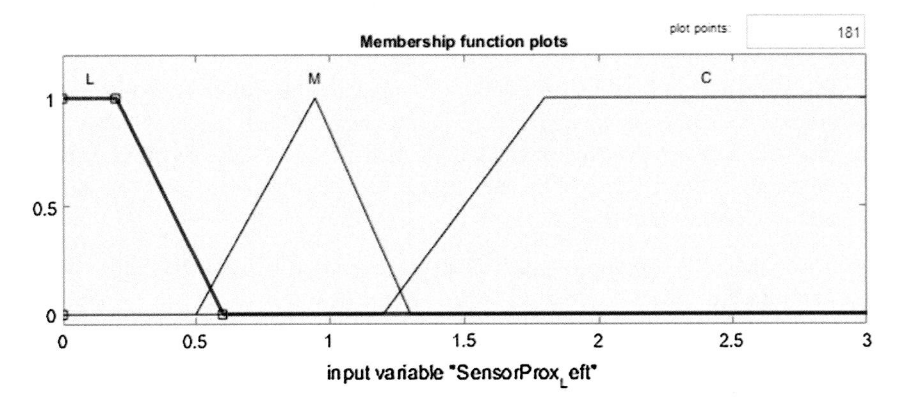

Fig. 9 Membership functions of the left, front and right sensors input proximity

Table 2 Fuzzy rules that define the behavior of the system

	Fuzzy rules							
Antecedent							Consequent	
	Inductive	OSleft	OSright	SPleft	SPcenter	SPright	Speed	Angle
1	C	'...	F	C	C	C	A	MRCR
2	C	F	'...	C	C	C	A	MRR
3	CC	D	F	C	C	L	ML	MRCR
4	CC	F	'...	C	C	L	ML	MRR
5	FL	F	'...	C	L	C	M	MRR
6	FL	'...	F	C	L	C	M	RCR
7	E	'...	F	C	L	L	L	MRCR
8	E	F	'...	C	L	L	L	MRR
9	FU	F	'...	L	C	C	ML	RR
10	FU	'...	F	L	C	C	ML	RCR
11	CC	F	F	L	C	L	MR	D
12	C	F	F	L	C	L	MR	D
13	FU	F	'...	L	L	C	ML	MRR
14	FU	'...	F	L	L	L	ML	MRR
15	'...	'...	'...	C	C	C	L	MRR
16	'...	'...	'...	C	C	L	L	R
17	'...	'...	'...	C	L	C	M	D
18	'...	'...	'...	C	L	L	L	R
19	'...	'...	'...	L	C	C	L	CR
20	'...	'...	'...	L	C	L	L	MRR
21	'...	'...	'...	L	L	C	L	CR
22	'...	'...	'...	L	L	L	A	D

expressive and interpretable, the consequence of the rules is fuzzy and more simple and intuitive [1].

Based on human knowledge, a set of 22 linguistic rules summarized on Table 2 of inference could be defined. These rules are obtained from the knowledge of the relationship between the experimental mathematical model and its variation and the variation of the input values of the inductive sensor. Each case in Table 2 contains a logical rule and should read as:

- If (inductive is Ai) and (OptDer is Bj) and (OptIzq is Ci) then (Speed is Xij) (TurnRate is Yij)

where Ai, Bj, Ci, Xij, Yij are respectively the membership functions of Inductive, OptDer, OptIzq, Speed and TurnRate respectively.

4.3 Obstacle Avoidance

The main goal of this work is the design of an Obstacle avoidance controller based on the Mamdani type fuzzy controller that allows the omnidirectional mobile robot to avoid fixed obstacles in its path making use of IR sensors (this is presented in Fig. 10).

Each case in Table 2 contains a logical rule and should read as:

- If (inductive is Di) and (OptDer is Ej) and (OptIzq is Fi) then (Speed is Xij) (TurnRate is Yij)

where Di, Ej, Fi, Xij, Yij are respectively the fuzzy membership functions of SensorProx_Left, SensorProx_Front, SensorProx_Right, Speed and TurnRate.

5 Experimental Numerical Model

In this paper an experimental numerical model is used to optimize the distance that we move away from the center of the trajectory, making use of the outputs of a Fuzzy Logic Controller and we consider Triangular and trapezoidal MFs.

In the design of the proposed experimental numerical model multiple linear regression (MLR) is used, because the proposed system is a function that will depend on two other variables, Acceleration and type of rotation, and the obtained function is the distance that the robotic platform moved away from the center the trajectory.

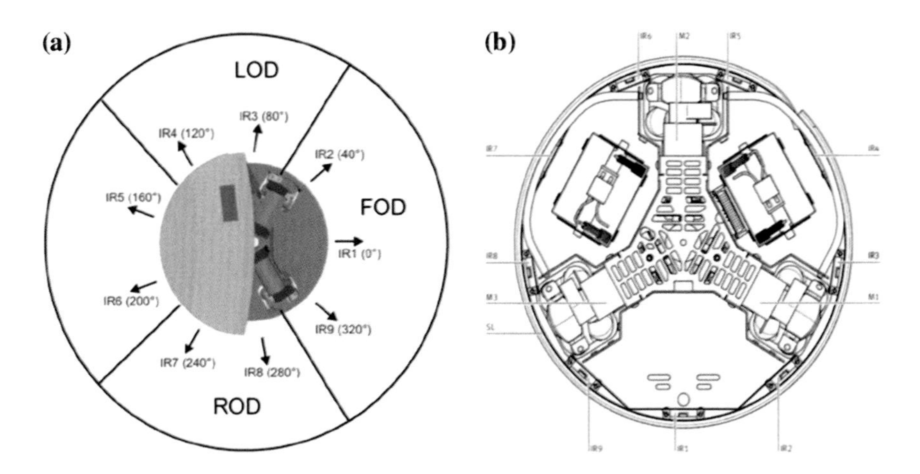

Fig. 10 IR sensors position (**a**) and angle (**b**) in robotic platform [16, 17]

This distance is a function of the voltage emitted by the inductive sensor, since when the platform is in the center of the line the sensor emits 2.2 V, which will increase as we are moving towards the edge of the trajectory until the 9.97–9 V.

5.1 *Experimental Numerical Model Design*

The distance from the center to the edge of the path is 25 mms. To obtain the real voltage emitted by the inductive sensor in function of the voltage, we perform a calculation in the form of two blocks:

BLOCK 1 = Voltage emitted by inductive sensor—Average voltage in inductive sensor (5 V).
BLOCK 2 = BLOCK 1. (Maximum Distance of the path (50 mms)/Maximum voltage that the inductive sensor can emit (approx. 9.97–9 V).

5.2 *Multiple Linear Regression*

For the development of the model it is necessary to record a database, which consists of the voltage emitted by the sensor, the speed and turn rate of the robot when it is working. Once having obtain the distance, we proceed to work with the multiple Linear regression (MLR) method.

To obtain the best values for the coefficients, this is done by means of the sum of the squares of the residues:

$$S = \sum_{i=1}^{n} (y_i - a_0 - a_1 x_{1i} - a_2 x_{2i})^2 \tag{2}$$

With the coefficients that give us the minimum sum of the squares of the residues which are obtained by deriving the partial derivatives to zero, we will form a 3 × 3 matrix to obtain the system that will describe the behavior of the system when moving away from the path.

$$\begin{bmatrix} n & \sum x_{1i} & \sum x_{2i} \\ \sum x_{1i} & \sum x_{1i}^2 & \sum x_{1i} x_{2i} \\ \sum x_{2i} & \sum x_{1i} x_{2i} & \sum x_{2i}^2 \end{bmatrix} \begin{bmatrix} a_0 \\ a_1 \\ a_2 \end{bmatrix} = \begin{bmatrix} \sum y_i \\ \sum y_i x_{1i} \\ \sum y_i x_{2i} \end{bmatrix} \tag{3}$$

Function that describes the operation of the system

$$Y = a_0 + a_1 x_1 + a_2 x_2 \quad (4)$$

where:

Y Distance from center
a_0 Independent value
$a_1 x_1$ Angle
$a_2 x_2$ Velocity

6 Conclusions

The proposed approach implements a fuzzy controller that produces the values of speed and angle of rotation in the trajectory tracking, making use of an autonomous mobile omnidirectional robot, in order for the robot to traverse the path without losing it and in the shortest time.

In the route that the robot performs with the fuzzy controller it is possible to observe how the values of speed and rotation are changing besides having the conditioning of the signal of the inductive sensor to calculate the distance that is away from the center of the trajectory.

In all the tests carried out, the effectiveness of the FIS controller was demonstrated when implementing it in a mobile omnidirectional robot, and its use gives an advantage in the efficiency of the route, smoothness in the route and correct decision of the acceleration and braking according to the Type of curve we are traversing.

The solutions reported in this paper are evidently different, confirming that Fuzzy Systems can be obtained from human control expertise, and in this case, each solution can be a representation of the knowledge of experts. As future work, we would like to consider type-2 fuzzy logic in control, like in [9–11], or hybrid approaches, like in [12–15].

References

1. Zadeh LotfiA, Fuzzy sets. Inf. Control **8**, 338–353 (1965)
2. D. Garcia-Sillas, E. Gorrostieta-Hurtado, J.E. Vargas, J. Rodríguez-Reséndiz, S. Tovar, Kinematics modeling and simulation of an autonomous omni-directional mobile robot. Ingeniería e Investigation **35**(2), 74–79 (2015)
3. R.J. Wai, Y.C. Chen, Design of automatic fuzzy control for parallel DC-DC converters. Eng. Lett. **17**(2), 83–92 (2009)
4. R. ParhiDayal, B.B.V.L. Deepak, Kinematic model of three wheeled mobile robot. J. Mech. Eng. Res. **3**(9), 307–318 (2011)

5. Y. Zhao, S-L. BeMent, Kinematics, dynamicsand control of wheeled mobile robot, in *Proceedings of the 1992 IEEE International Conference on robotics and Automation*, pp. 91–96, nice (France), 1992
6. J.B. Song, K.S. Byun, Design and control of an omnidirectional mobile robot with steerable omnidirectional wheels, in *Mobile Robots, Moving Intelligence*, ed. by J. Bushli (ARS/pIV, Germany, 2016), pp. 576. ISBN: 3-86611-284-X
7. T.A. Baede Motion control of an omnidirectional mobile robot (TraineeshipReport DCT, Vol. 2006.084) (2006)
8. S.M. Mohamed, K. Najla, M. Mohamed, D. Nabil, Fuzzy logic controllers design for omnidirectional mobile robot navegation. Original Res. Art. Appl. Soft Comput. **49**, 901–919 (2016)
9. G.M. Mendez, O. Castillo, Interval type-2 TSK fuzzy logic systems using hybrid learning algorithm, fuzzy systems, in *The 14th IEEE International Conference on FUZZ'05*, 2005, pp. 230–235
10. O. Castillo, P. Melin, Design of intelligent systems with interval type-2 fuzzy logic, type-2 fuzzy logic: theory and applications, pp. 53–76
11. P. Melin, C.I. Gonzalez, J.R. Castro, O. Mendoza, O. Castillo, Edge-detection method for image processing based on generalized type-2 fuzzy logic. IEEE Trans. Fuzzy Syst. **22**(6), 1515–1525
12. O. Castillo, P. Melin, E. Ramírez, J. Soria, Hybrid intelligent system for cardiac arrhythmia classification with fuzzy K-nearest neighbors and neural networks combined with a fuzzy system. Exp. Syst. Appl. **39**(3), 2947–2955
13. L. Aguilar, P. Melin, O. Castillo, Intelligent control of a stepping motor drive using a hybrid neuro-fuzzy ANFIS approach. Appl. Soft Comput. **3**(3), 209–219
14. P. Melin, O. Castillo, *Modelling, Simulation and Control of Non-Linear Dynamical Systems: An Intelligent Approach Using Soft Computing and Fractal Theory* (CRC Press, 2001)
15. P. Melin, O. Castillo, Intelligent control of complex electrochemical systems with a neuro-fuzzy-genetic approach. IEEE Trans. Industr. Electron. **48**(5), 951–955
16. Festo Didactic GmbH & Co. KG, Robotino Manual, 73770 Denkendorf, Germany, 2007
17. M. Njah, M. Jallouli, Wheelchair obstacle avoidance based on fuzzy controller and ultrasonic sensors, in *Proceedings of the International Conference on Computer Applications Technology (ICCAT'13)*, pp. 1–5, Sousse, Tunisia, January 2013

Part V
Fuzzy Logic Applications

Fuzzy Logic in Recommender Systems

Amita Jain and Charu Gupta

Abstract A recommender system studies the past behaviour of a user and recommends relevant and accurate items for the user from a large pool of information. For user '*u*' a recommender system filters relevant and accurate information by finding items which are similar to the target items (items searched by user) and by finding (other) similar users which co-relate to the user '*u*' interests and needs. For exhibiting this filtration, a recommender system uses the features of the items and maintains user profile which contains his past purchases, his buying pattern. These features and user profile imprecise, uncertainty and vague thus should be analysed carefully for optimal prediction. Fuzzy logic has been extensively used in the design of a recommender system to handle the uncertainty, impreciseness and vagueness in item features and user's behaviour. In this chapter, the use of fuzzy logic in recommender system as well as the analytical framework for analysis of the design of a recommendation system is discussed. It also presents the analysis of the growth of fuzzy logic in recommender system and its applications.

Keywords Collaborative filtering · Content based filtering · Dimensions of recommender system · Fuzzy logic · Recommender system

1 Introduction

Any business which involves electronic transaction is popularly known as E-commerce. As the World Wide Web (WWW) is growing day by day, the buying pattern of the people has drifted from brick and mortar stores to online buying [1]. So, data over the internet has increased manifolds.

A. Jain (✉)
Ambedkar Institute of Advanced Communication Technologies and Research, Delhi, India
e-mail: amita_jain_17@yahoo.com

C. Gupta
Bhagwan Parshuram Institute of Technology, Delhi, India
e-mail: charugupta0202@gmail.com

O. Castillo et al. (eds.), *Fuzzy Logic Augmentation of Neural and Optimization Algorithms: Theoretical Aspects and Real Applications*, Studies in Computational Intelligence 749, https://doi.org/10.1007/978-3-319-71008-2_20

Nowdays, online buying has encouraged companies to work out more and more options for the customers to select from. They are customizing the products and increasing the feature set to suit various needs of multiple customers. But all the features of the products may not be useful for every customer and the customer may get confused due to the information overloading of the various features. Subsequently, the customers are interested to analyse all the information of the related products before actually buying the most suitable product for them. *Recommender system (RS)* is a good solution to efficiently deal with this information overload [2].

A recommender system is an automated tool which suggests appropriate items to a user [3]. A recommender system consists of a set of users ($u_1 \ldots u_i$) and a set of items ($i_1 \ldots i_k$). According to the definition, a user in a recommender system is an entity which uses a recommender system [4]. An item can be a tangible or digital object which is suggested to a user through text message, e-mail [4]. All items have a set of features. For example, an item can refer to an electronic product, a car, a movie or a song which individually have some features of their own. For instance, a car has features like mileage, boot space, leg room, exteriors, interiors, turning radius and so on. These features define an item.

Further, e-commerce uses recommender system to suggest products/items to customers of their interest. The items which are suggested to the customer are based on various factors, like- previous buying behaviour of the customer (historical data), personal information of the customer, most rated items similar to customer's interest. Thus, a recommender system has the ability to give personalized recommendation [4] to the user. The advantage of such a system is that it saves valuable time of a customer and increases the loyalty of the company as well [5]. Further, this personalization in recommendation, broadly, comes from two categories - *explicit* or *implicit user behaviour*. For instance, explicit user preference is when a user u_i gives a rating r_j to an item; and implicit preference is when the system follows the clickstream information of the user. A reference model [6] to depict these preferences is given in Fig. 1.

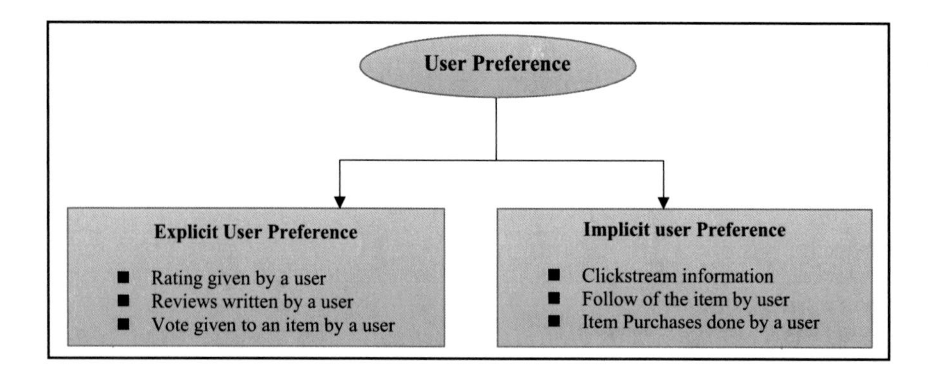

Fig. 1 Reference model of user preference

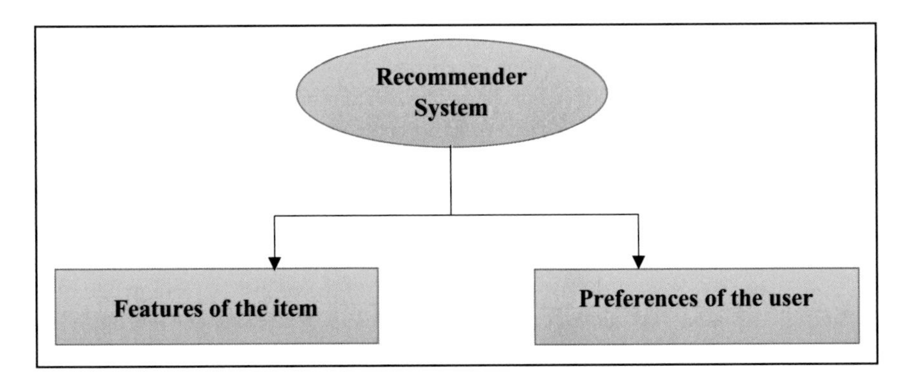

Fig. 2 Factors affecting a recommender system

A Recommender System works on the features of the item as well as on the preferences of the user, as given in Fig. 2. The performance/efficiency of a recommender system depends on these factors and how they are represented. Both these factors play a vital role in demonstrating the level of personalization to a user.

But these factors are subjective and imprecise. This poses a major challenge for recommender system to handle the uncertainty [7, 8]. It is also interesting to note that these uncertainties have an extent (low to high) to which the items have features. For example, a car can have a feature, such as *interiors* as very good, good, not so good, bad, and awful.

To handle this uncertainty, use of fuzzy logic in recommender system is highly desirable. This not only helps in modelling the vagueness in a logical manner but also gives a consistent justification and heuristic rules for designing a recommender system [9]. In literature, fuzzy logic has been used to design a recommender system [10, 11]. With the application of fuzzy logic [12, 13], the recommendation has seen the following major changes:

a. the fuzzy rules help to model the vagueness in user input in an efficient manner with the help of linguistic variables ("like"/"dislike"/"awful"/"good").
b. use of fuzzy logic with other techniques like clustering improves the determination of similar users, by finding the fuzzy similarity degree between two users,
c. the closeness between two fuzzy ratings (user input) over a particular product has been understood in an easy way with the help of fuzzy rules,
d. handling of user demographic data is better with fuzzy rules.

To understand the use of fuzzy logic in recommender system let us consider a recommender system which recommends an automotive to a user. To generate the recommendations which are most suitable to a user; the system inputs certain priorities on various parameters which the user can set. For instance, his budget, average kilometre coverage, fuel efficiency, styling, comfort, service, space, infotainment and so on. If a user is asked to set his parameter priorities on a scale of 1 (least important) to 5(most important), then instead of giving a crisp value to these parameters; he gives a degree of his likeness i.e., for styling he may set the value as 3–5. Human reasoning works better on range of values as it better captures the vagueness in deciding a particular parameter. Taking this as an advantage, the recommender system can use fuzzy rules to map user ratings, demographic data (such as age, income, and profession), and similarity of two users more efficiently.

This chapter aims at providing a support to the use and need of fuzzy logic in recommender system, and to motivate the readers to find new applications in various dimensions of the recommender system.

In this chapter, Sect. 2 discusses the idea behind the origin of recommendation. Section 3 explains the analytical framework of a recommender system. The understanding of the framework helps in analysing the future uses of recommender system with other techniques like fuzzy logic, genetic algorithms. In Sect. 4, a brief analysis of fuzzy logic in recommender system is presented. It gives an overview of the growth of fuzzy logic based recommender system in literature since 2003. A conclusive remark of this chapter is given in Sect. 5.

2 Idea of Recommendation—Social Information Reuse

Information reuse has been a key area of research. Social navigation is a branch of information reuse which tries to understand the interaction among people. Social navigation (now social network) refers to the idea of following the footsteps of others in order to find out what one wants [6].

The use of social navigation can be understood from a story of cavemen who used to hunt for food. Once one of them found a bright red apple and died after eating. The fellow cavemen *used this information* and did not eat that fruit. Humans have a tendency to learn from other's experience/actions [6]. With today's digital age, as the data over the internet rose to a level, it gave rise to two technologies—*information filtering and information retrieval.* The basic differences between these techniques [14] is summarized in Table 1. The difference between information

Table 1 Comparison of information filtering with information retrieval techniques based on different aspects

Aspect	Information filtering	Information retrieval
Users of the system	Use by a person/persons with long term interests	Use by a single person with one time goal
Information representation	User profiles with specifications	In the form of queries
Basic functioning	Distribution of information to groups or individuals	Collection and organization of information
Database type	Dynamic database	Static database
User related issue	Generally, for undefined group of people seeking advice with varied domains e.g. entertainment, product service etc.	For user's belonging to well defined groups with specific domains

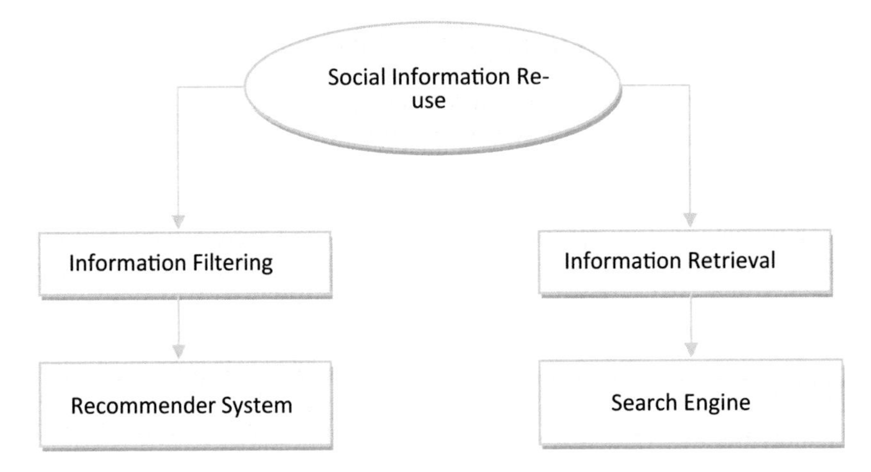

Fig. 3 Recommender system as an information filtering system

filtering and retrieval is important. It clearly defines the characteristics of a recommender system (*information filtering*) and a search engine (*information retrieval*). Although both information filtering and retrieval are seen as two sides of the same coin [14]. But it is always useful to understand the difference between the two.

From Fig. 3, it can be understood that a recommender system is an information filtering system which generates intelligent agents for searching remote, heterogeneous databases [15]. For instance, Netflix [16] recommends movies to the user based on his interest by going through a large set of databases in order to predict whether a particular item will be useful for a user or not based on user/item profile. Information retrieval systems are those which provide documents to users which satisfy his/her need for information [14]. System which are popularly used for information retrieval are search engines like Google.

A recommender system is considered as an information filtering system because it dynamically filters out vital information from large databases based on user's preference, behaviour or interest [17]. Sometimes there is a confusion regarding the differences between recommendation and prediction. In prediction, the focus is on estimating the degree to which an item will be liked by a user and it is often concerned with the browsing of specific products. A recommender system helps to suggest items/products most suitable to a user or items that he will like [18]. A recommender system has the ability to predict the items/products to a user based on his interest. The recommender system can be classified as—Content Based [19], Collaborative Filtering Based [20] or hybrid recommender system. A detailed classification of recommender system is given in the subsequent sections.

3 Analytical Framework of a Recommender System

A recommender system is an information filtering system which collects the opinion and experience data of users to find relevant data items *for a purpose* [21]. The role of a recommender system is to compute the recommendations and present them in a useful way to the user [19]. This section, explains the analytical framework consisting of various dimensions which are useful in understanding the basic design and functioning of a recommender system [6]. These dimensions are: *Domain of Recommendation; Purpose of Recommendation; Recommendation Context; Level of Personalization; Trustworthiness of Data; Interfaces and Recommendation Algorithms*. Figure 4 shows a diagrammatic view of these dimensions [6].

3.1 Domain of Recommendation

Domain of recommendation is concerned with the items to be recommended to the user. The item can be a single or a bundle or sequence of items? For example, the item can be a music piece/list of songs, news articles, or electronic product. It works on the principle of *interestingness of an item* with other items. It means, instead of recommending a particular item X, a bundle of items similar to X are recommended. Domain also deals with the ordering of the items in the recommendation

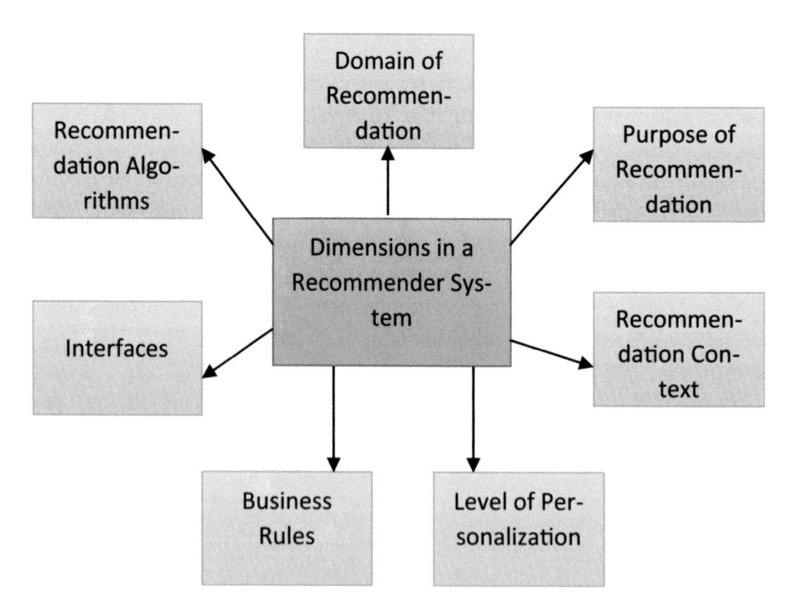

Fig. 4 Analytical framework of a recommender system

Fig. 5 A page from women fashion (western wear) from Amazon.in

list [23]. For instance, recommending a list of songs to a user, the ordering of the playlist plays an important role. It is usually populated with songs which the user has heard before. So the domain knowledge of what the user has liked before is important. Another example is when a user thinks of purchasing an item, the recommender system, by knowing the user's interest through selected item, can be given other options to choose from. For instance, Fig. 5 represents a page from *Amazon.in > Women Wear > Western Wear*. Here Amazon's recommender system outputs two ways of recommendation: first, "*frequently bought together*" and second, "*customers who bought this item also bought*". This gives the user a wider array of selected options from a large set of western wear. Also, the user gets the recommended list with *almost* the same price range (Fig. 5).

3.2 Purpose of Recommendation

In the entire process of recommendation, the aim is to output the most desirable recommendations to the user based on his interest. In purpose of recommendation, the idea is to make aware the community/educate for a particular facility. For example, TripAdvisor (a recommender system and a travel review site) educates the customer to book the travel to a place which is top ranked or give a recommendation score to all the hotels which are similar to the customer's query. It never compels a customer to book hotels/restaurants/flights. This system is efficient and powerful enough to educate the customer beyond recommendations [6].

3.3 Recommendation Context

The recommendation context is a spring board for a recommender system. It seeks two important aspects of a recommender system. First, the *level of attention* a user has while the system recommends items to him and second, *degree of interruption* a user is willing to take. Now let us understand these aspects with the help of examples. If a user is listening to a song, then, he will dislike any kind of interruption, be it a list of further recommended songs. As he just wants to enjoy his favourite track. Also, the context of recommendation has to understand the level of attention the user has. For example, for a movie freak, if the system recommends some good movies (the definition of "good" is handled by the recommendation algorithms; explained later) based on the situation that person is in, like, he is alone or with his friends or with family. The recommended movies will change in every context (if not all).

3.4 Level of Personalization

A recommender system personalizes the information to increase the interest of the user. This personalization can be based on the age or gender or profession. For example, an online clothing store might recommend different clothing options based on different age groups like 18–25, 25–35 years and 35 above. It is observed that the level of personalization affects the user's interest in a recommendation system [6].

3.5 Business Rules

Apart from understanding the context of recommendation, it is important to know whether a recommender system is trustworthy or not. The trust on a system is governed by rules which run the purchasing properties of a customer. *For instance*, a customer should be made assured that the product he is buying is the best and has the lowest price as compared to other recommendation systems. Now the company has to decide on two factors—first, *trustworthiness of data* and second, its *profit making strategy*. It is also vital to know about the biases, a company has [6]. The biases are often called as business rules. These business rules are used to increase the purchase of products by a user. This can be achieved by showing complementary products and the help in finding related products also [23]. *For instance*, it is very uncommon to see, that a shoe seller would admit that he does not sell the shoe that the user is looking for. Recommendations are governed by business rules, after all, these are generating revenue for the company. So, in short it can be said that a recommender system should be able to handle the issues of transparency and the trust worthiness of the users, who are giving ratings for a particular item.

3.6 *Interfaces*

In this dimension, the idea is to understand, what the final output of a recommendation system is. It can be in the form of *prediction, filtration.* Prediction gives specific scores for an item. Filtration gives the best result from a given search list. A recommender system can be an agent which gives helpful recommendation; it can be give a dialog interface where the user can show his like or dislike for a recommendation; or it can act as a search tool. It is also concerned with the type of input from the user—either implicit or explicit. This has been explained in Sect. 1. For a thorough discussion on interfaces readers can have a look at [19, 22].

3.7 *Recommendation Algorithms*

The recommendation algorithms can broadly be classified as: *Non-personalized summary statistics, content based filtering, collaborative filtering and hybrid.* These algorithmic models are explained in this section. Figure 6 depicts the broad classification of recommender system.

a. Content Based Filtering

Summary statistics are similar to calculating the average number of users who have rated a particular item. For example, how good a movie is? does not require any personalization. It can be computed without the use of a model. The advantage of using summary statistics is when many people rate a particular item/product the impact of each rating is low; this helps in dampening any individual outlier rating. But there are issues which need to be addressed while using *non-personalized*

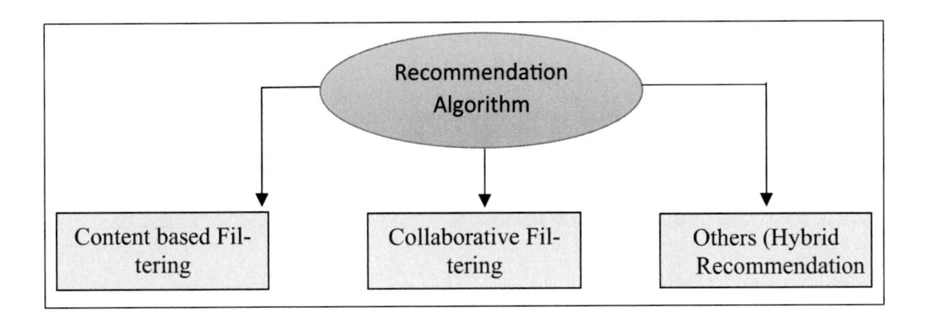

Fig. 6 Types of recommendation algorithms

summary statistics, such as, people with *dissimilar taste* when rate a particular item, or people with *strong opinions* moves the average rating to a higher value or sometimes the averages can be *biased* (when a user self-rate a movie/product/any other item). This method lacks the consideration of context, which is important for recommendation. For example, if someone at *Subway (a global restaurant chain)* wants a recommendation for a sauce, average statistics will say a tomato ketchup; but if he asks for the most likable sauce associated with a veg burger, then this models fails to find this association. Also, Non-personalized systems give similar recommendations to all users without considering individual tastes/preferences. This leads to designing recommender system where Product Association Recommendations are considered. The content based systems consider *Product Association Recommendations* (PAR). It helps to find other products that go well together with the target product. The co-purchasing of a product is a major consideration which is used to satisfy business rules. The rules have certain goals [6, 23]. The aim is to increase the number of products a user buys. It is also to show user complementary products to help him find the most related products. For example, Fig. 7 shows a webpage from *ebay.in* (online commerce company), where a user wishes to purchase I-phone (target product). The highlighted "similar items" in Fig. 5 show the content similar to the target product.

A major limitation of this method is the inability of the system to show the user "*the most popular*" products. It also gives the same recommendation to every user and does not look at individual differences. In scenarios where the user is visiting just once, it is very hard to build a profile. Similarly, if the user's taste change rapidly, this recommendation may not be very useful as the recommended list of items will be out of date.

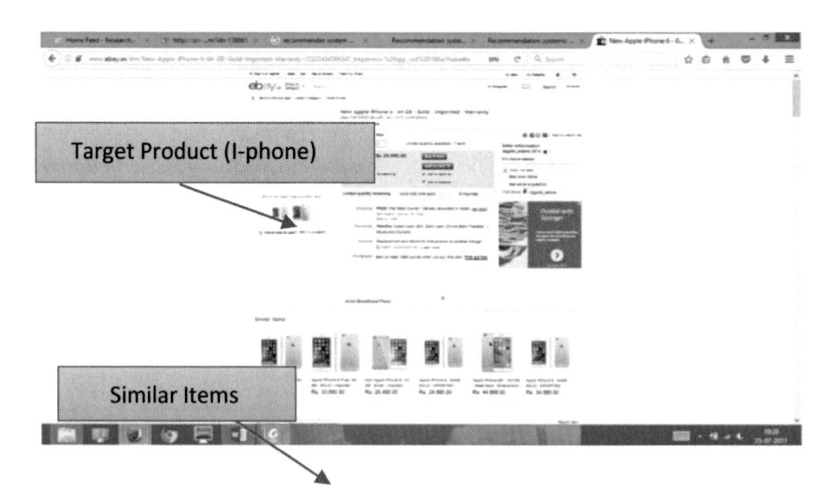

Fig. 7 A webpage from ebay.in for demonstrating content based recommendation

b. Collaborative Filtering

To overcome the limitation of content based filtering, a *user-user* and an *item-item* collaborative filtering method is used. Collaborative filtering (CF)-based recommendation techniques enable the users to choose on the basis of opinions of other people who have similar interests. The collaborative filtering is of two types—*user-user and item-item*. In user-user filtering, the recommendations are based on users who share similar interests. In item-item, the user is flooded with recommendations for the items that are loved by other users in the past who have shared similar interests earlier. Similarity measures are used to find the likeliness of two users. The similarity measures used are: cosine similarity, Pearson co-relation or adjusted cosine. These measures can be studied in [18, 20]. For using these measures, only those users who have rated both the items are considered. This poses a bottleneck for new user.

In collaborative filtering the prediction for a user '*i*' in personalized and non-personalized scenarios is given by Eq. 1.

$$P(a,i) = \sum_{u=1}^{n} \frac{r(u,i)}{n} \tag{1}$$

where, '*p*' is the prediction for user '*a*' for item '*i*' and the prediction is done for user '*a*' based on other users '*n*', *r(u, i)* is the rating by user '*u*' in item '*i*'.

Now there are two considerations, first, if item '*i*' is not liked by everyone and they have different opinions about item '*i*'; second, if the ratings used for item '*i*' are not on the same scale. These situations have to be handled differently.

In the first situation, for predicting item '*i*' to user '*a*', if the item is not drawn evenly from all users then, we use a personalized weight for every user and sum it over all. This is given by Eq. 2.

$$P(a,i) = \frac{\sum_{u=1}^{n} r(u,i) * w(a,u)}{\sum_{u=1}^{n} w(a,u)} \tag{2}$$

where, *r(u, i)* is the rating by user '*u*' for item '*i*', and *w(a, u)* is the weighted agreement between user '*a*' and user '*u*'. It is given by the similarity between user '*a*' and user '*u*'.

The second situation, where ratings are not on the same scale, the difference of ratings of user '*u*' for item '*i*' and the average rating for user '*u*' is taken as given by Eq. 3.

$$P(a,i) = \sum_{u=1}^{n} r(u,i) - r(u)/n + r(a) \tag{3}$$

where *r(u, i)* is the user item pair and *r(a)* gives the deviation which is added to the average rating of user '*a*'.

Collaborative filtering assumes that the past agreements of the users are helpful in predicting their future agreements. It also considers that the individual tastes of the users are stable and do not change over time and are in sync with one another. So, in collaborative filtering the system has to carefully work in the domain of agreement. Despite this assumption, there are situations when people agree on one domain and not on another. For instance, some people are happy having a discussion on politics but they might not necessarily agree on technical discussions. For such cases, the prediction for item '*i*' are given by Eqs. 2 and 3.

c. Hybrid Approaches

Both Content based and Collaborative systems have some limitations. Content based systems give recommendation on content profile of item. In some cases, it is difficult to generate content profile for every item. A content based system in unable to select items of interest to user, if the user has not given any evidence of it (*over specialization*). The content based systems are unable to distinguish between subjective information (*subjective domain*).

In collaborative filtering systems these limitations are overcome by adding user profile which can handle the *over specialization* problem and the *subjective domain* problem. But collaborative systems have certain problems of their own. It suffers from the early rater problem, gray sheep and data sparsity problem. A collaborative filtering system cannot make predictions for new items which have not been rated by any users. Similarly it is difficult to find accurate predictions for new users who have rated very few items. As discussed previously in collaborative filtering systems, the base assumption is that the tastes of the users do not change over time. But in reality, all users cannot have overlapping characteristics. Everyone cannot agree on every domain. So it is challenging to find accurate recommendations for gray sheep (user or group of users who do not have consistent agreement). Also, predicting those items which are not rated by enough number of users is also a difficult task because the data for that particular item is sparse [24]. Hybrid recommendation systems are the system which use the strengths of both the systems to efficiently handle recommendations [25–28].

4 Fuzzy Logic in Recommender System

Fuzzy logic [12, 13] gives a wide spectrum of methods to analyse the uncertainty in data. It is very helpful in handling data with imprecise information and gradualness of user preference [9]. Over the years, use of fuzzy logic to various areas has been studied. In this section, an analysis of the growth of fuzzy logic based recommender system is presented. It also discusses the applications where fuzzy logic based recommender system has been used.

4.1 *Fuzzy Logic Based Recommender System: An Analysis*

This sub-section presents the statistical analysis of fuzzy logic in recommender system. The data is collected and analysed from *web of science* (*WoS*) database to identify the growth of fuzzy logic based recommender systems. This analysis aims to provide the reader with the growth of fuzzy logic in recommender system, its major areas of applications from the time of its inception. It is interesting to note that fuzzy logic in recommender system has been extensively used in Computer Science as compared to other areas. A summary statistics has been shown in Table 2 (correlated from WoS). The research articles, since 2003, where fuzzy logic has been studied with recommender system is analysed.

Within Computer Science, the use of fuzzy logic has been studied in various areas (not limited to) like automatic group recommendations [29], knowledge based recommendations [30], software project management [31], E-Learning [32], multi-criteria collaborative filtering [33], trust model [34] and so on. The growth of these areas has seen various methodologies which are used for designing an application centric recommender system.

Recommender System based on fuzzy logic has been in use since, 2008. The year wise growth is depicted in Fig. 8. The growth of research in fuzzy logic based recommender system suggests a gradual increase from 2003 to 2012. It is the time when the potential of these systems was realized. Then a sudden increase in the research interest in 2013 is seen. It is during this time, that fuzzy logic based recommender system was applied to various applications like consensus ranking [35], item and trust-aware collaborative filtering [36], co-relation based similarity [37], competence recommendation [38], situation-aware collaborative filtering [39], tourism system [26], telecom products [27], personalized e-services [40, 41], stock market [42], movie recommendation [43].

Table 2 Use of fuzzy logic based recommender system in various research areas

S.No	Research area	Percentage
1	Computer science	92.31
2	Engineering	25.00
3	Operations research management science	15.39
4	Mathematics	6.73
5	Business economics	3.85
6	Telecommunications	3.85
7	Automation control systems	1.92
8	Education educational research	1.92
9	Information science library science	1.92

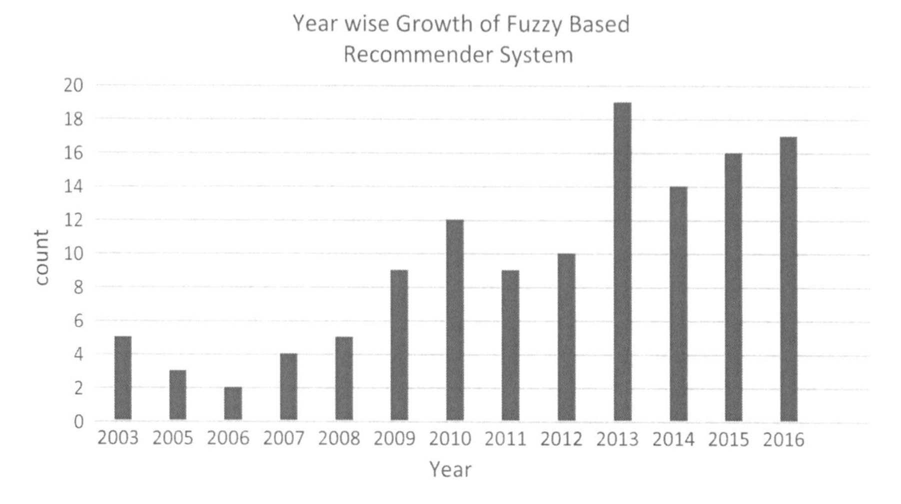

Fig. 8 Year wise growth of research in fuzzy logic based recommender system

4.2 *Applications of Fuzzy Logic in Recommender Systems*

As seen in Sect. 4.1, the growth of fuzzy logic based recommender system has been on an increase since its inception. In this section, the rise in the growth has been analysed with respect to the work done in fuzzy logic based recommendation since 2003. Yager in his seminal work [9], defined a fuzzy set A in X is defined by a membership function, which is defined as: μ_A (x): x € X → [0,1] where X is the domain space. So set A is characterized by A = {x, μ_A (x), x € X}.The membership function can have different interpretations based on the applications. For instance, it can be the degree of similarity or user preference [41, 42].

Since the idea of fuzzy logic based recommender system in 2003, this area of research has seen tremendous changes in the applications where the use of recommender system can be very well understood. The use of fuzzy sets to analyse recommender system design was a breakthrough to the world of recommendation [9]. The uncertainty in recommending accurate items/products to users has drawn attention. The uncertainty in representation of objects was studied. This representation implies that whether a system is content or collaborative, both need a vector of ratings for finding out the correct user interest. The initial methodology was developed for reclusive systems. In 2005, the use of fuzzy logic was studied with data mining techniques for recommender system. In [10] dynamic fuzzy cluster was proposed which used dynamic degree of membership for a user preference. In [11] fuzzy logic based recommender system monitored two considerations. These were—*consumers past purchases* and his *specific requirements*. This system aimed at guiding the consumers in electronic products. The initial work in 2006 used fuzzy sets and related reasoning techniques to model the user preferences and analysed case based recommendation system [44]. Following the same year, fuzzy association rules and multiple similarity measures (FARAMS) were used to design a recommender system using collaborative filtering [45]. The technique was simple and effective but suffered limitations with de-fuzzification process.

By this time, fuzzy logic was realised as a potential breakthrough in dealing with uncertainty of data in recommender system. Researchers focused on the granularity of data. The granularity in creating user profiles and for the items which remain unsold due to non-repetitiveness of the stock. In [46] a fuzzy hierarchical clustering algorithm was used to cluster similar users. The users within a cluster shared common characteristics. The algorithm worked on log data. This method was efficient than previous modified versions of fuzzy c-means. Another interesting property for collaborative filtering system was explored in [47] which considered irregularities in rating of item which could not be sold "repetitively". This approach used graded membership for domains of similar users to eliminate the possibility of *one-and-only items*.

After analysing fuzzy logic in recommender system, researchers realised that there are variable entities such as user demographics in recommender system which can be represented with a graph. Then, probabilistic graphical models were used with recommender system. In [48, 49], a combination of Bayesian networks and fuzzy set theory is used to recommend new items to the users based on the past voting patterns of users with similar taste. Along the statistical models, researchers began to use the modelling of user and rating with fuzzy linguistics. In [50], the information in technology transfer office [25] is used to design a recommender system which uses fuzzy linguistic modelling. It was studied extensively for managing subjectivity [51], for acquisition of user profiles [52], managing digital libraries [53].

Another important aspect of recommender system was dealt with fuzzy sets. Fuzzy sets were used to design a system which could adapt to user's changing behaviour. This changing behaviour (with time) brings dynamicity in the model. In [54] fuzzy set theoretic method were used to handle the vagueness and imprecision in data. It extends the fuzzy modelling proposed in [9]. It was designed for content based recommendation. Representation and aggregation methods were empirically tested on movie dataset. In [55] quality of recommendation was improved by maintaining recall with high precision. The models developed for managing the dynamics of recommender system proved to be efficient.

After 2011, Fuzzy logic based recommender systems found new applications in various domains. For instance, in [56], fuzzy logic based recommender system was used for tag-clouds. The fuzzy theory worked on the process of finding the strength of tag-item connections. It is observed that there is an increase in the growth of fuzzy logic based recommender system using collaborative filtering and hybrid systems [26, 34, 36, 41, 43]. Now multi-criteria collaborative filtering has also been used in tourism domain [57] to understand user preferences in multiple aspects. This also deals with the dynamic nature of recommender system. The future of fuzzy logic based recommender system is witnessing attention in both collaborative and hybrid systems with soft computing techniques as well. Collaborative filtering is being studied with fuzzy theory to understand novel video recommendation [58], with swarm intelligence techniques [59], for interior design drawing [60], social network [61].

5 Conclusion

The introduction of fuzzy logic based recommender system in 2003 by Yager, has attracted attention of many researchers to advance into this field to find new methods to deal with uncertainty in data, user preference and user demographics. This uncertainty is inherent in human behaviour and a careful analysis and understanding of the basic functioning of the recommender system definitely helps in predicting accurate recommendations for the user. This chapter, described the recommender system as an information filtering system with a discussion on the analytical framework of the recommender system. It enlightened the reader about the theory and applications of recommender system. It, also, qualitatively and quantitatively analysed the growth of fuzzy logic based recommender system and its applications from movies, tourism, stocks to social network. This chapter clearly explained fuzzy logic based recommender system (reclusive, collaborative filtering and hybrid systems) and its applications in various domains.

References

1. A. Soergel, http://www.usnews.com/news/articles/2016-12-20/with-online-sales-booming-is-brick-and-mortar-on-the-way-out. US News, (2016). Accessed 20 June 17
2. J. Lu, D. Wu, M. Mao, W. Wang, G. Zhang, Recommender system application developments: a survey. Decis. Support Syst. **74**, 12–32 (2015)
3. J. Bobadilla, F. Ortega, A. Hernando, A. Gutierrez, Recommender systems survey. Knowl. Based Syst. **46**, 109–132 (2013)
4. D. Kotkov, S. Wang, J. Veijalainen, A survey of serendipity in recommender systems. Knowl. Based Syst. **111**, 180–192 (2016)
5. F. Ricci, L. Rokach, B. Shapira, P.B. Kantor, *Recommender Systems Handbook*, 1 edn. (Springer, 2011)
6. http://www.coursera.org/specializations/recommender-systems. Accessed 12 Sept 2016
7. G.J. Klir, T.A. Folger, *Fuzzy Sets, Uncertainty, and Information* (Prentice-Hall, NJ, 1988)
8. A. Zenebe, A.F. Norcio, Representation, similarity measures and aggregation methods using fuzzy sets for content-based recommender systems. Fuzzy Sets Syst. **160**, 76–94 (2009)
9. R.R. Yager, Fuzzy logic methods in recommender systems. Fuzzy Sets Syst. **136**, 133–149 (2003)
10. S.H. Min, I. Han, Dynamic fuzzy clustering for recommender systems, in *Advances in Knowledge Discovery and Data Mining, Proceedings*. Book Series: Lecture Notes in Artificial Intelligence, vol. 3518. pp. 480–485 (2005)
11. Y.K. Cao, Y.F. Li, X.F. Liao, Applying fuzzy logic to recommend consumer electronics, in *Distributed Computing and Internet Technology, Proceedings*. Book Series: Lecture Notes in Computer Science, vol. 3816, pp. 278–289 (2005)
12. L.A. Zadeh, Fuzzy sets. Inf. Control **8**, 338–353 (1965)

13. L.A. Zadeh, Fuzzy sets as a basis for a theory of possibility. Fuzzy Sets Syst. **1**(1), 3–28 (1978)
14. N.J. Belkin, W.B. Croft, Information filtering and information retrieval: two sides of the same coin? Commun ACM. **35**(12), 29–38 (1992)
15. J.A. Konstan, J. Riedl, Recommender systems: from algorithms to user experience. User Model User-Adapt Inter. **22**, 101–123 (2012)
16. F.M. Harper, J.A. Konstan, The movielens datasets: history and context. ACM Trans. Interact. Intell. Syst. (2015)
17. C. Pan, W. Li, Research paper recommendation with topic analysis. IEEE Comput. Des. Appl **4**, V4–264 (2010)
18. J.H. Herlocker, J.A. Konstan, L.G. Terveen, J.T. Riedl, Evaluating collaborative filtering recommender systems. ACM Trans Inf Syst. **22**(1), 5–53 (2004)
19. C. Basu, H. Hirsh, W. Cohen, Recommendation as classification: using social and content-based information in recommendation, in *Recommender System Workshop'98*, pp. 11–15 (1998)
20. N. Good, B. Schafer, J. Konstan, A. Borchers, B. Sarwar, J. Herlocker, J. Riedl, Combining collaborative filtering with personal agents for better recommendations, in *Proceedings of the AAAI'99 Conference*, pp. 439–446 (1999)
21. P. Resnick, H.R. Varian, Recommender systems. Special Issue Commun. ACM. **40**(3) (1997)
22. J.B. Schafer, J. Konstan, J. Riedl, Recommender systems in e-commerce, in *Proceedings of ACM E-Commerce 1999 Conference* (1999)
23. J. Rosati, P. Ristoski, T. Di Noia, R. De Leone, H. Paulheim, RDF graph embeddings for content-based recommender systems, in *10th ACM Conference on Recommender Systems*, Boston, MA, USA. pp. 23–30 (2016)
24. A.M. Rashid, I. Albert, D. Cosley, S.K. Lam, S.M. McNee, J.A. Konstan, J. Riedl, Getting to know you: learning new user preferences in recommender systems, in *Proceedings of the International Conference on Intelligent User Interface (IUI2002)* (2002)
25. C. Porcel, L.A. Tejeda, M.A. Martinez, A hybrid recommender system for the selective dissemination of research resources in a technology transfer office. Inf. Sci. **184**(1), 1–19 (2012)
26. J.P. Lucas, N. Luz, M.N. Moreno, A hybrid recommendation approach for a tourism system. Expert Syst. Appl. **40**(9), 3532–3550 (2013)
27. Z. Zhang, H. Lin, K. Liu, A hybrid fuzzy-based personalized recommender system for telecom products/services. Inf. Sci. **235**, 117–129 (2013)
28. P. Vashisth, P. Khurana, P. Bedi, A fuzzy hybrid recommender system. J. Intell. Fuzzy Syst. **32**(6), 3945–3960 (2017)
29. S.R.D. Queiroz, F.D.T. de Carvalho, G.L. Ramalho, Making recommendations for groups using collaborative filtering and fuzzy majority, in *Advances in Artificial Intelligence*, Proceedings, Book Series: Lecture Notes in Artificial Intelligence, vol. 2507. pp. 248–258 (2002)
30. L. Martinez, M.J. Barranco, L.G. Perez, A knowledge based recommender system with multigranular linguistic information. Int. J. Comput Intell. Syst. **1**(3), 225–236 (2008)
31. R. Colomo-Palacios, I. Gonzalez-Carrasco, Lopez-Cuadrado J. Luis, RESYSTER: a hybrid recommender system for scrum team roles based on fuzzy and rough sets. Int. J. Appl. Math. Comput. Sci. **22**(4), 801–816 (2012)
32. M. Ferreira-Satler, F.P. Romero, V.H. Menendez-Dominguez, Fuzzy ontologies-based user profiles applied to enhance e-learning activities. Soft. Comput. **16**(7), 1129–1141 (2012)
33. M. Nilashi, O. bin Ibrahim, N. Ithnin, Hybrid recommendation approaches for multi-criteria collaborative filtering. Expert Syst. Appl. **41**(8), 3879–3900 (2014)
34. E. Majd, V. Balakrishnan, A trust model for recommender agent systems. Soft. Comput. **21** (2), 417–433 (2017)
35. T.C.K. Huang, Recommendations of closed consensus temporal patterns by group decision making. Knowl. Based Syst. **54**, 318–328 (2013)

36. C. Birtolo, D. Ronca, Advances in clustering collaborative filtering by means of fuzzy C-means and trust. Expert Syst. Appl. **40**(17), 6997–7009 (2013)
37. Vibhor Kant, Kamal K. Bharadwaj, Integrating collaborative and reclusive methods for effective recommendations: a fuzzy Bayesian approach. Int. J. Intell. Syst. **28**(11), 1099–1123 (2013)
38. Serrano-Guerrero, F.P. Romero, J.A. Olivas, Hiperion: a fuzzy approach for recommending educational activities based on the acquisition of competences. Inf. Sci. **248**, 114–129 (2013)
39. G. Castellano, M.G.C.A. Cimino, A.M. Fanelli, A collaborative situation-aware scheme based on an emergent paradigm for mobile resource recommenders. J. Ambient Intell. Humanized Comput. **4**(4), 421–437 (2013)
40. J. Lu, Q. Shambour, Y. Xu, A web-based personalized business partner recommendation system using fuzzy semantic techniques. Comput. Intell. **29**(1), 37–69 (2013)
41. Y.C. Hu, A novel non additive collaborative-filtering approach using multi-criteria ratings. Math. Problems Eng. (2013)
42. P. Paranjape-Voditel, U. Deshpande, A stock market portfolio recommender system based on association rule mining. Appl. Soft Comput. **13**(2), 1055–1063 (2013)
43. I.C. Wu, W.H. Hwang, A genre-based fuzzy inference approach for effective filtering of movies. Intell. Data Anal. **17**(6), 1093–1113 (2013)
44. D. Dubois, E. Huellermeier, H. Prade, Fuzzy methods for case-based recommendation and decision support. J. Intell. Inf. Syst. **27**(2), 95–115 (2006)
45. C.W.K. Leung, S.C.F. Chan, F.L Chung, A collaborative filtering framework based on fuzzy association rules and multiple-level similarity. Knowl. Inf. Systs. **10**(3), 357–381 (2006)
46. B. Lazzerini, F. Marcelloni, A hierarchical fuzzy clustering-based system to create user profile. Soft. Comput. **11**(2), 157–168 (2007)
47. C. Cornelis, J. Lu, X. Guo, One-and-only item recommendation with fuzzy logic techniques. Inf. Sci. **177**(22), 4906–4921(2007)
48. L.M. de Campos, J.M. Fernandez-Luna, J.F. Huete, A collaborative recommender system based on probabilistic inference from fuzzy observations. Fuzzy Sets Syst. **159**(12), 1554–1576 (2008)
49. V. Kant, K.K. Bharadwaj, Integrating collaborative and reclusive methods for effective recommendations: a fuzzy Bayesian approach. Int. J. Intell. Syst. **28**(11), 1099–1123 (2013)
50. C. Porcel, A.G. Lopez-Herrera, E. Herrera-Viedma, A recommender system for research resources based on fuzzy linguistic modeling. Expert Syst. Appl. **36**(3), 5173–5183 (2009)
51. E. Herrera-Viedma, A.G. Lopez-Herrera, A review on information accessing systems based on fuzzy linguistic modelling. Int. J. Comput. Intell. Syst. **3**(4), 420–437 (2010)
52. C. Porcel, E. Herrera-Viedma, Dealing with incomplete information in a fuzzy linguistic recommender system to disseminate information in university digital libraries. Knowl. Based Syst. **23**(1), 32–39 (2010)
53. J. Serrano-Guerrero, E. Herrera-Viedma, J.A. Olivas, A google wave-based fuzzy recommender system to disseminate information in university digital libraries 2.0. Inf. Sci. **181**(9), 1503–1516 (2011)
54. A. Zenebe, A.F. Norcio, Representation, similarity measures and aggregation methods using fuzzy sets for content-based recommender systems. Fuzzy Sets Syst. **160**(1), 76–94 (2009)
55. M. Kiewra, N.T. Nguyen, A fuzzy-based method for improving recall values in recommender systems. J. Intell. Fuzzy Syst. **20**(1–2), 89–104 (2009)
56. M.Z. Reformat, R.R. Yager, Tag-based fuzzy sets for criteria evaluation in on-line selection processes. J. Ambient Intell. Humanized Comput. **2**(1), 35–51 (2011)
57. M. Nilashi, O. bin Ibrahim, N. Ithnin, A multi-criteria collaborative filtering recommender system for the tourism domain using expectation maximization (EM) and PCA-ANFIS. Electron. Commer. Res. Appl. **14**(6), 542–562 (2015)
58. M. Ramezani, F. Yaghmaee, A novel video recommendation system based on efficient retrieval of human actions. Phys. A-Stat. Mech. Appl. **457**, 607–623 (2016)
59. R. Katarya, O.P. Verma, A collaborative recommender system enhanced with particle swarm optimization technique. Multimedia Tools Appl. **75**(15), 9225–9239 (2016)

60. Kuo-Sui Lin, Fuzzy similarity matching method for interior design drawing recommendation. Rev. Socionetwork Strat. **10**(1), 17–32 (2016)
61. G. Posfai, G. Magyar, L.T. Koczy, A fuzzy information propagation algorithm for social network based recommender systems, in *Advances in Intelligent Systems and Computing (AISC)*, Vol. 462 (Springer, 2017)

Money Management for a Foreign Exchange Trading Strategy Using a Fuzzy Inference System

Amaury Hernandez-Aguila, Mario Garcia-Valdez and Oscar Castillo

Abstract Trading a financial market involves the use of several tools that serve different purposes necessary to understand the behaviour of that particular market. One of the tools a trader must use is a method that determines the lot size of a trade, i.e., how many units are going to be bought or sold for a trade. Frequently, traders will use their experience and subjective deduction capabilities to determine the lot size, or simple mathematical formulas based on how much profit or loss they have realized during a particular period of time. This work proposes using a fuzzy inference system to determine the lot size, which uses input variables that any trading strategy should have access to (which means that any existing strategy can implement the proposed method). The experiments in this work compare basic trading strategies based on simple moving averages, one with a fixed lot size for every trade performed, and another one which uses the fuzzy inference system to establish a dynamic lot size. The results show that the dynamic lot size using the fuzzy inference system can help a trading strategy perform better.

Keywords Fuzzy inference systems · Trading strategy · Money management

1 Introduction

A financial market (for example, a stock market or a foreign exchange market) depicts the price movements for a particular item being traded. This means that if the Google stock market shows that a share is being sold for $50 USD, one has to pay said quantity in order to trade that market. The price for this market will change

A. Hernandez-Aguila (✉) · M. Garcia-Valdez · O. Castillo
Tijuana Institute of Technology, Tijuana, BC, Mexico
e-mail: amherag@tectijuana.mx

M. Garcia-Valdez
e-mail: mario@tectijuana.mx

O. Castillo
e-mail: ocastillo@tectijuana.mx

O. Castillo et al. (eds.), *Fuzzy Logic Augmentation of Neural and Optimization Algorithms: Theoretical Aspects and Real Applications*, Studies in Computational Intelligence 749, https://doi.org/10.1007/978-3-319-71008-2_21

over time, and the price could increase or decrease. If the price increases, this means that a trader will have a profit equal to the current price minus $50 USD (the price at which the share was bought). An analogy to this process is easy to create: if one buys a car for $2000 USD, and sells it for $2500 USD later on, a profit of $500 USD was made by this transaction. But as the market price can increase, it can also decrease. To understand how the selling process occurs, one can imagine that a lot of cars has already been acquired. If a car is sold for $2500 USD and one buys another car for $2000 USD later on, a profit of $500 USD was still made, even though the initial operation was selling a car, instead of buying one.

As one can make a profit by trading a financial market, one can also end in a loss. To explain this outcome, one can continue with the previous analogy: if someone buys a car for $2000 USD, and after some time this person notices that this particular car model keeps getting cheaper, the trader can decide to sell the car for $1500 USD to cut the losses (the person decides to sell the car before the market price decreases even more). This transaction would result in a loss of $500 USD. In the case of a loss in a selling transaction, one can imagine that someone has already a lot of cars and decides to sell one for $2000 USD. If the price has been getting higher over time, and this person decides that it is time to restock the car lot with the car model that was sold before, but the current market price is $2500 USD, restocking that car will result in a loss of $500 USD.

Another variable that can be introduced to this buy/sell process is the lot size, e.g. how many cars are going to be bought/sold per transaction (following the previously mentioned analogy). Taking this variable into consideration, a trader can decide to buy a small lot of cars if there is a strong possibility of being wrong with the decision of buying. Similarly, if the trader considers that there is a strong possibility of being right with the decision of buying in the market, or if there is suspicion that the price will greatly rise, then the trader can decide to buy a bigger lot of cars. But how can a trader determine the lot size to buy or sell? Surprisingly, there are still many sources that point out that in order to determine this lot size, one can use intuition or simple mathematical formulas. One of the most popular methods to determine this lot size is to use a fixed percentage of the trading capital per trade [1]. This way, a trader will risk smaller and smaller amounts if the trading account has suffered losses in the past trades, and will risk bigger amounts if the trading account has made profits. In other words, the trader should risk more if profits have been made, and risk less otherwise. Another way of managing this risk would be to diversify the trades over several markets [2]. The concept that describes the aforementioned process is called money management.

"Put two rookie traders in front of the screen, provide them with your best high-probability set-up, and for good measure, have each one take the opposite side of the trade. More than likely, both will wind up losing money. However, if you take two pros and have them trade in the opposite direction of each other, quite frequently both traders will wind up making money—despite the seeming contradiction of the premise. What's the difference? What is the most important factor separating the seasoned traders from the amateurs? The answer is money management." [3] The reader can find many literature works which state similar ideas as the one expressed before, such as [1, 4, 5]. The basic idea is that if the trader is not sure about the trend that the market will take, a smaller lot size should be traded, and if there is a strong possibility of a trend to follow, the lot size should be bigger. If this principle is followed, a trader could end up making profits even if most of the trades were lost. For example, a trader could lose 9 trades where 10 units were lost in each trade, but the trader determined that for a particular trade, a bigger lot had to be traded. A profit of 100 units for a single good trade could result in a profit for the total 10 trades.

As can be noted in the previous paragraph, the determination of the lot size can be expressed in a fuzzy manner: "if the possibility of a trend is *high*," "the lot size will be *small*," "the trading account has made a *big* profit," etc. This is why the present work proposes the use of a fuzzy inference system as the foundation for a money management strategy. This system will receive input variables that can be obtained from any trading strategy, and will output the lot size that should be traded under those circumstances. Section 4 presents the details underlying this method. In order to test the proposed method, an experiment was carried out and its design is presented in Sect. 5, while the results of this experiment are presented in Sect. 6. If the reader is unfamiliar with the concepts involved in the trading strategy used for the experiments, a series of preliminary concepts are presented in Sect. 2, and some related works are presented in Sect. 3. Finally, Sects. 7 and 8 present some conclusions and ideas for future works, respectively.

2 Preliminaries

In the following Subsections, some basic concepts are discussed. These concepts are related to the proposed method presented in Sect. 4.

2.1 Trading Strategy

A trading strategy is a set of tools used by a trader in order to determine when a particular type of trade should be executed. The type of a trade can be a buy or sell now order (which is executed immediately), or an entry order (which will be executed after the market price reaches a determined threshold). These orders can include other operations, like a stop loss or take profit orders (the trade will stop after a specific amount of units has been gained or lost), and trailing stops (a stop loss which gets constantly updated depending on the price movements). A trading strategy also incorporates a money management strategy, which determines the size of the lot to be traded.

The presented method in this work focuses on the money management area, i.e. determining how many units are going to be traded for each executed order.

2.2 Simple Moving Average

A moving average (MA) is a common operation applied to time series. In general, an MA uses a certain number of past data points in a time series to obtain an average for the current data point. For example, for a time series TS = [1, 4, 6–9], an MA which takes the past 3 data points as an average would calculate the following points: 2, 3, 4, 5. The first average, 2, would represent the average of 1, 2, and 3. The second average, 3, would represent the average of 2, 3, and 4, and so on. If these averages are unweighted, the operation is known as a simple moving average (SMA), and is described by (1)

$$\begin{aligned} SMA &= \frac{p_M + p_{M-1} + \cdots + p_{M-(n-1)}}{n} \\ &= \frac{1}{n}\sum_{i=0}^{n-1} p_{M-i} \end{aligned} \tag{1}$$

The use of MAs in trading strategies is a common practice. The purpose of applying them to a financial market time series is to obtain a smoother curve which represents the general direction of the market (uptrend, downtrend or stagnation) [4]. The most common way of using MAs to determine a trade opportunity is to use two MAs, one using a small number of data points for the average (often called a short MA), and another one using a bigger number (often called a long MA) [4].

2.3 Volatility

The volatility of a market is obtained by calculating the average number of units that a period is presenting. For example, if one is observing the chart of a financial market with a 15-min timeframe (which means that each data point represents the resulting movements in the last 15 min), and every data point presents movements of 50 units, then the market has a volatility of 50 units. Similar to MAs, the volatility can be obtained taking into consideration only the past n periods or data points.

3 Related Work

According to some sources, most traders end losing money by trading on financial markets. A study that discusses more about this subject can be found in the work by Odean and Barber [10]. Nevertheless, this study also states that the bigger companies do make money on the long run by trading on financial markets. A hypothesis which explains this phenomenon is that there are many inexperienced individuals trading on the market, traders that are not using the correct tools and the correct trading process in order to make a profit. Furthermore, the psychology in the trading process can also influence the trader's decisions, and often makes the trader lose money [4]. Automatized trading strategies can help with the psychological factor, such as the trading strategy proposed by Harris [11]. On the other hand, in order to have a correct trading strategy, one has to have several tools to understand different aspects of the trading process. For example, a one has to use an effective money management strategy [12], and understand the profitability and risk one is taking while trading [11].

The present work uses a trading strategy based on moving averages (simple moving averages, to be specific). Several works which use MAs exist, and they have been shown to work on specific markets before. As examples of such works, one can read the work by Pätäri and Vilska [13], who applied a trading strategy to predict the Finnish market. Similarly, Nguyen et al. [14], and Hung and Zhaojun, have applied similar techniques to forecast the French and the Vietnamesse stock markets, respectively. More complex methods exist involving SMAs, for example, the work by Wang et al. [15] use genetic algorithms to optimize the trading rules based on SMAs.

To mention a few other techniques to forecast the market, one can use neural networks, which has grown to be a popular technique. As an example, one can read the work by Chan and Teong [8]. Fuzzy systems have been used to forecast financial markets too, and an example is the work by Leu and Chiu [16]. Hybrid systems usually perform better than applying a single technique, as in the works by Yu and Wang [17], and Wei et al. [3].

Regarding techniques to determine the lot size to be traded, i.e. money management, the authors of the present work could not find methods involving computational intelligence techniques. Most of the techniques involve using the intuition of the trader, or simple mathematical formulas, as is explained in the work by Hakim [18].

4 Proposed Method

This work proposes a method to determine the amount of units, or lot size, that a trader or trading strategy should use for a particular trade, given certain input variables. In order to calculate an inference involving these input variables, a fuzzy inference system is proposed. The input variables should be obtainable in any trading strategy. The proposed method was purposely designed like this in order to be portable to any already existing trading strategy.

This Section will explain first the input variables used, and the fuzzy inference system will follow.

4.1 Percentage of Change

The percentage of change is a simple calculation which involves the average profit a trading strategy has made in the past n trades, relative to the last m averages. For example, if the last 4 average profits are 45, 50, 60 and 50, this last average profit (50) represents $\sim$83% of the maximum obtained in the last 3 average profits (50, 60, and 50). In this case, $n = 4$ and $m = 3$. With this method, a good representation of how well a method is performing during a period of time can be obtained. Furthermore, the method obtains an output which ranges from -1 to 1, which is useful to quickly understand how well the trading strategy is performing.

4.2 Probability of Trend

Any trading strategy should give as an output whether a trade should be on a long position (buying) or on a short position (selling). But there should also exist the possibility to determine how strong that signal is. This strength will be called "probability of trend" in the present work, and should range from 0 to 1. For example, in the case of using MAs to determine the direction of the market, one can use the ratio between the strength of the long-period MA and the short-period MA to obtain a value between 0 and 1.

4.3 Volatility

As explained in Sect. 2.3, the volatility of a market represents the average amount of units of the movements in the past n periods. Similar to how the proposed method calculates the percentage of change (Sect. 4.1), the volatility is calculated as a quantity relative to the last m averages. For example, if the last 5 volatilities in the market, taking into consideration an arbitrary value for n, are 30, 40, 60 and 45, the last value (45) will be divided by the maximum value in the last m averages. For example, if $m = 3$, then the volatility for the last period will be equal to 45/60 = 0.75.

4.4 Fuzzy Inference System

A Mamdani fuzzy inference system was used in order to calculate a lot size. Three input variables were used: percentage of change, probability of trend, and volatility. The domains of these input variables and the membership functions are described below.

For the percentage of change, the domain of the input value can range from -1 to 1. This linguistic variable is represented by four membership functions: *high loss*, *low loss*, *low profit*, and *high profit*. All the membership functions are Gaussian membership functions. The means for each of the membership functions are as follow: −1.0, −0.5, 0.5, and 1.0, respectively. *High loss* and *high profit*, both have a standard deviation of 0.33. In the case of low loss and low profit, both have a standard deviation of 0.15. A graphical representation of this input variable is shown in Fig. 1.

For the next input variables, and the single output variable, each of them have the same graphical representation as is shown in Fig. 2.

The probability of change has the following three adjectives: *low probability*, *medium probability*, and *high probability*. In the case of volatility, the three following adjectives: *low volatility*, *medium volatility*, and *high volatility*. Finally, for the single output variable, the adjectives are: *low lot size*, *medium lot size*, and *high lot size*. For each of these variables, the means of the adjectives, in order, are 0.0, 0.5, and 1.0, while the standard deviations are all equal to 0.15. The defuzzification method to obtain the output is center of area.

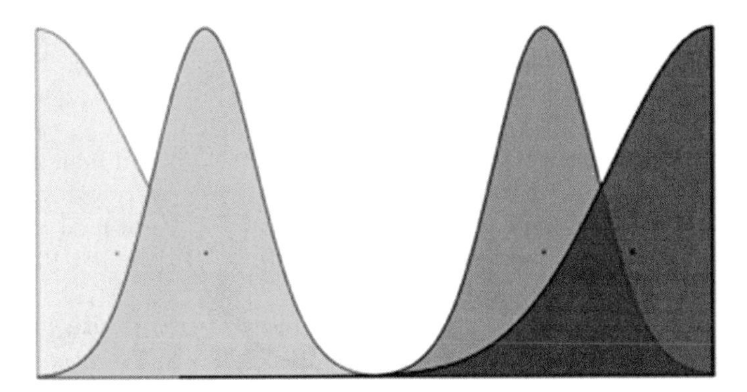

Fig. 1 Graphical representation of the membership functions for the percentage of change

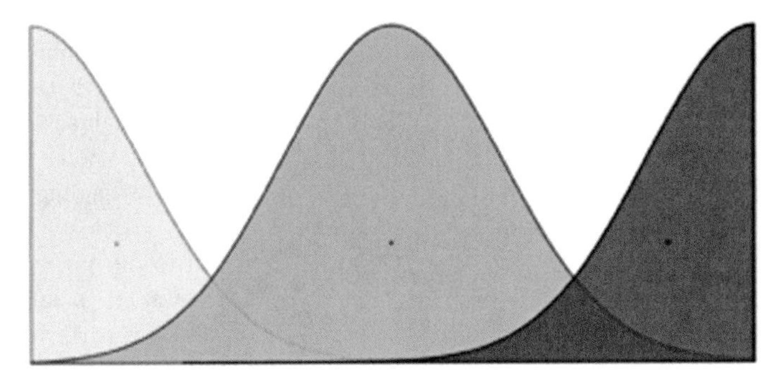

Fig. 2 Graphical representation of the membership functions for the probability of trend, volatility, and lot size

The fuzzy rules for the fuzzy inference system are as follows:

- if percentage-of-change is high-loss, then lot-size is low-lot-size
- if percentage-of-change is low-loss, then lot-size is medium-lot-size
- if percentage-of-change is low-profit, then lot-size is medium-lot-size
- if percentage-of-change is high-profit, then lot-size is high-lot-size
- if probability-of-trend is low-probability, then lot-size is low-lot-size
- if probability-of-trend is medium-probability, then lot-size is medium-lot-size
- if probability-of-trend is high-probability, then lot-size is high-lot-size
- if volatility is low-volatility, then lot-size is low-lot-size
- if volatility is medium-volatility, then lot-size is medium-lot-size
- if volatility is high-volatility, then lot-size is high-lot-size

5 Experiments

An implementation of the proposed method was done in the programming language Clojure. Data was gathered from a ForEx broker called Oanda, which provides a free-to-access API. The financial market chosen for the experiment was the EUR vs USD, with a timeframe of 15 min per session. The data was obtained from January 1st 2016 to January 1st 2017.

In order to obtain the percentage of change, the average profit from the last 10 trades was calculated. To obtain a relative value, the last 10 averages were used.

In the case of the volatility, the average of the last 10 prices was used to calculate the volatility at a certain session. To obtain a relative value, the last 10 averages were used.

Before explaining how the probability of trend was determined, the trading strategy needs to be explained. Two SMAs were used to determine when to place a trade in the following manner: if the short SMA crosses above the long SMA, a buying position is executed; if the short SMA crosses below the long SMA, a selling position is executed. The price value of the short SMA is subtracted to the previous price value of the short SMA, and is divided by the price value of the long SMA minus the previous price value of the long SMA, and the absolute value is obtained at the end. This will result in a value that ranges from 0 to 1. The short SMA takes the last 10 price values to calculate the average for each session, while the long SMA takes the last 200 price values. For example, Fig. 3 shows a situation where the probability of trend is very high, while Fig. 4 shows another situation where the probability of trend is very low.

The three variables explained above are given as input to the fuzzy inference system (which was constructed by using Wagner's Juzzy toolkit [19]), and a lot size is calculated. This method was then compared to the same trades performed with a fixed lot size. A total of 799 trades were performed.

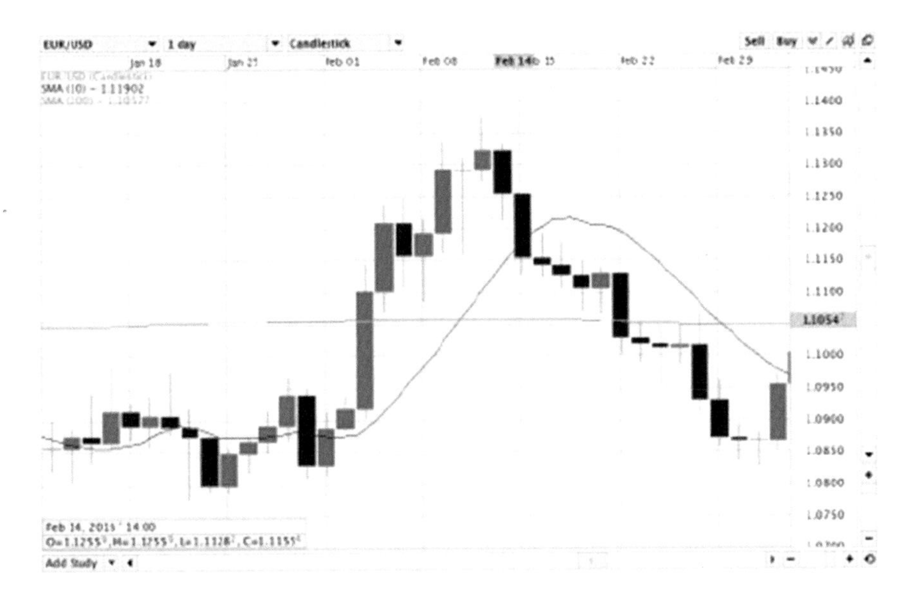

Fig. 3 Situation presenting a high probability of trend

Fig. 4 Situation presenting a low probability of trend

6 Results

The profits generated by using a fixed or constant lot size are presented in Fig. 5, while the profits generated by using a dynamic lot size, determined by the percentage of change, the volatility, and the probability of trend, are shown in Fig. 6. The profit is represented by PIPs, where a PIP is the minimum amount the market can move in certain direction. Figure 7 presents a comparison between the two methods.

Fig. 5 Profit from the trades performed by using a constant lot size

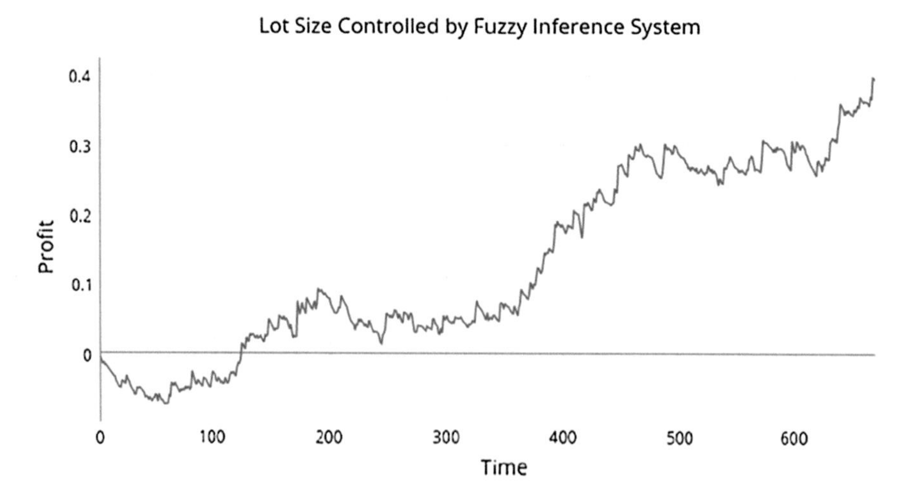

Fig. 6 Profit from the trades performed by using a lot size determined by the proposed method

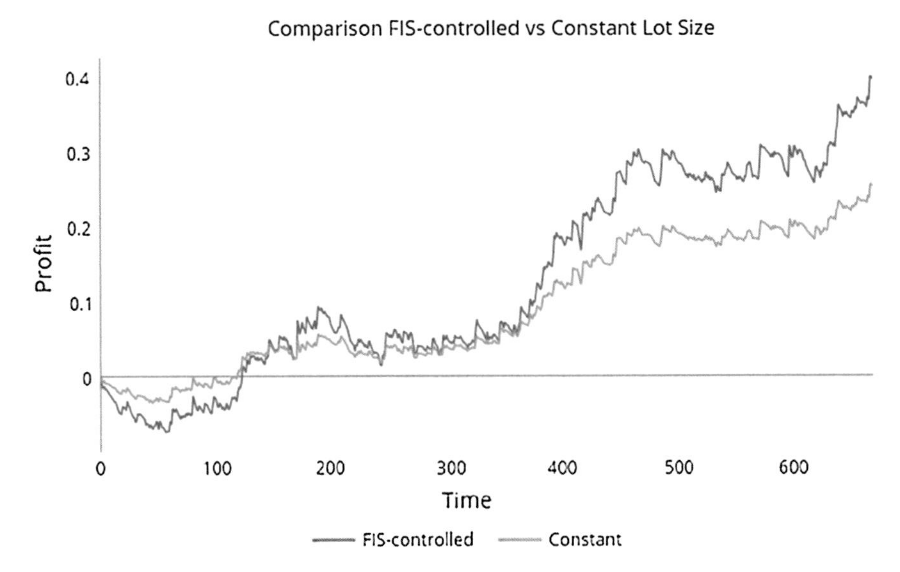

Fig. 7 Comparison between the profits generated by the trading strategy using a constant lot size versus a dynamic lot size

7 Conclusions

It is interesting to note that the proposed method presented bigger losses in the first 100–150 trades, as is shown in Fig. 7. Nevertheless, on the long run, the proposed method achieved greater profits than the method which used a constant lot size.

Although it would be very naïve to state that the presented experiment is a definitive proof that the proposed method will always work, it can be concluded that the proposed method is a good candidate to be tested as a money management strategy.

8 Future Work

The proposed method should be tested with a far greater amount of data points, and in other financial markets. For example, other currency pairs could be used, as well as different stock markets.

References

1. N.J. Balsara, *Money Management Strategies for Futures Traders*, vol. 4 (Wiley, 1992)
2. B.A. McDowell, A *Trader's Money Management System: How to Ensure Profit and Avoid the Risk of Ruin*, vol. 406 (Wiley, 2010)
3. L.-Y. Wei, C.-H. Cheng, H.-H. Wu, A hybrid ANFIS based on n-period moving average model to forecast TAIEX stock. Appl. Soft Comput. **19** (2014)
4. A. Elder, *Trading for a Living: Psychology, Trading Tactics, Money Management*, vol. 31 (Wiley, 1993)
5. H. Till, *Case Studies and Risk Management in Commodity Derivatives Trading* (EDHEC-Risk Institute, 2011)
6. R.P. Barbosa, O. Belo, Autonomous forex trading agents, in *Industrial Conference on Data Mining* (Springer, Berlin, 2008)
7. S. Basak, D. Makarov, Strategic asset allocation in money management. J. Finance **69**(1) (2014)
8. K.C.C. Chan, F.K. Teong, Enhancing technical analysis in the Forex market using neural networks, in *Proceedings of IEEE International Conference on Neural Networks*, vol. 2 (IEEE, 1995)
9. C.W.J. Granger, P. Newbold, *Forecasting Economic Time Series* (Academic Press, 2014)
10. T. Odean, B.M. Barber, Just How Much Do Individual Investors Lose by Trading? Law and Economics Workshop (2007)
11. M. Harris, *Profitability and Systematic Trading: A Quantitative Approach to Profitability, Risk, and Money Management*, vol. 342 (Wiley, 2008)
12. R. Vince, *The Mathematics of Money Management: Risk Analysis Techniques for Traders*, vol. 18 (Wiley, 1992)
13. H.T. Nguyen, H.V.D. Pham, H. Nguyen, The profitability of the moving average strategy in the french stock market. J. Econ. Dev. **16**(2) (2014)
14. N.H. Hung, Y. Zhaojun, Profitability of applying simple moving average trading rules for the Vietnamese stock market. J. Bus. Manag. **2**(3) (2013)
15. L. Wang et al., Generating moving average trading rules on the oil futures market with genetic algorithms. Math. Probl. Eng. (2014)
16. Y. Leu, T.-I. Chiu, An effective stock portfolio trading strategy using genetic algorithms and weighted fuzzy time series, in *2011 15th North-East Asia Symposium on Nano, Information Technology and Reliability (NASNIT)*, IEEE (2011)
17. L. Yu, Lean, K.K. Lai, S. Wang, Designing a hybrid AI system as a forex trading decision support tool, in *17th IEEE International Conference on Tools with Artificial Intelligence (ICTAI'05)*, IEEE (2005)
18. H. Hakim, Forex Trading and Investment. Diss., Worcester Polytechnic Institute, 2012
19. C. Wagner, Juzzy-a java based toolkit for type-2 fuzzy logic, in *2013 IEEE Symposium on Advances in Type-2 Fuzzy Logic Systems (T2FUZZ)*, IEEE (2013)

Generation and Reduction of Fuzzy Sets with PG-Means and Fuzzy Similarity Measures

Arturo Garcia-Garcia, Andres Mendez-Vazquez and Marek Z. Reformat

Abstract The probabilistic clustering techniques can be applied to generate fuzzy sets in situations where there is little or no information about data. Quite often, they generate a huge number of clusters. These clusters can be interpreted as fuzzy sets in a process of building a fuzzy system. A large number of fuzzy sets introduce noise to the fuzzy system, hence the need to reduce their number. Fuzzy Similarity Measures (FSMs) are widely used for comparison of fuzzy sets. Multiple FSMs have been proposed so far, but identifying a single FSM that is the most suitable for a given task is not always a straightforward process. On many occasions, FSMs are used to reduce a number of fuzzy sets. In this paper, we present the results of analyzing suitability of FSMs to reduce number of fuzzy sets and fuzzy if-then rules for an image segmentation problem. We use a PG-Means algorithm to generate fuzzy sets on both input and output variables. We propose and apply algorithms utilizing FSMs to reduce the number of fuzzy sets and rules. The paper includes a case study investigating the application of the proposed method on two images.

Keywords PG-means · Fuzzy sets · Fuzzy similarity measures
Fuzzy sets reduction

A. Garcia-Garcia (✉) · A. Mendez-Vazquez
CINVESTAV, Computer Science, Zapopan, Jalisco, Mexico
e-mail: aggarcia@gdl.cinvestav.mx

A. Mendez-Vazquez
e-mail: amendez@gdl.cinvestav.mx

M. Z. Reformat
University of Alberta, Edmonton, Canada
e-mail: marek.reformat@ualberta.ca

O. Castillo et al. (eds.), *Fuzzy Logic Augmentation of Neural and Optimization Algorithms: Theoretical Aspects and Real Applications*, Studies in Computational Intelligence 749, https://doi.org/10.1007/978-3-319-71008-2_22

1 Introduction

Generation and reduction of fuzzy sets is one of the main problems associated with a construction of rule-based fuzzy systems. Application of Mixture Models (MM) to generate fuzzy sets allows to eliminate the need of domain experts providing input regarding construction of fuzzy sets [3]. However, a close 'relation' of an MM algorithm to the processed data can result in a model with highly overlap mixtures. Thus, the transformation from distribution mixtures to fuzzy sets into a highly-partitioned space with a high degree of overlapping between fuzzy sets. Then, if the degree of overlapping is high, we could end up with a system with a high level of noise. Therefore, the application of Fuzzy Similarity Measures (FSMs) and Inclusion Fuzzy Degrees (IFDs) could help to generate and trim fuzzy sets to overcome this problem.

Chia-Feng et al. [1] describe a method that builds an auto-generated fuzzy system with learning abilities. To build such a system, the authors used an On-Line Self-Aligning Clustering (OSAC) for constructing fuzzy sets, and an Ant and Particle Swarm Cooperative Optimization (APSCO) for building fuzzy rules. Bellaaj et al. [2] proposed a method for reducing a number of fuzzy rules using similarity concepts and interpolation techniques. The method begins with computing similarity degrees between rules from an initial rule base. Next, it removes rules with a high similarity values. When such a rule-based system encounters new data, it is capable of producing new rules using a simple method: when a minimum number of rules are fired, a new rule using interpolation is added. A drawback of the method is its sensitivity to outliers. Garcia and Mendez [3] present a method based on probabilistic clustering to obtain fuzzy sets defined on all input and output variables.

Fuzzy Similarity Measures (FSMs) have already been proposed for comparing rules, and there is a large number of works dedicated to this topic. For example, one FSM uses the similarity operator proposed by Dubois et al. [4]. Another FSM is proposed by Pappiset et al. as a fuzzification of the Jaccard index. In another work, Dubois et al. [4] offers two FSM that also use the Jaccard index. Wang [5] proposed to apply a min-max split for each element in X. This can be interpreted as the utilization of local similarities into a weighted sum for the determination of similarity values. Lee-Kwang et al. [6] used a max-min function to first calculate the min values between two fuzzy membership values at the same point x, then calculate the maximum value from them. Some examples are given by Wang et al. [7] which use implicators over fuzzy membership values with infimum and minimum operators. Turksen et al. [8] proposed a FSM based on a distance measure, while Santini et al. [9] introduced a FSM based on a normalized measure.

Additionally, such concepts as Similarity of Rule Premises (SRP), Similarity of Rule Consequents (SRC), and Similarity between Rules (SR) have been applied for comparing fuzzy if-then rules. For example, Jin et al. [10] used a specific FSM to preserve the completeness of the fuzzy partitioning of input variables, as well as to preserve the diversity of fuzzy subsets. Consequently, the authors calculated the similarity of rule premises using an FSM together with a Genetic Algorithm and a gradient method. Chen et al. [11] used an FSM for rule-base self-extraction and simplification using a Gradient descent algorithm. Following these ideas, Bellaj et al. [2] proposed a way to measure similarity between two fuzzy triangular sets. The approach of Jin [12] and Chen and Linkens [11] involves a reduction of rules using fuzzy similarity along with an interpolation of Takagi-Sugeno-Kang (TSK) fuzzy systems [13].

The inclusion for fuzzy sets was originally introduced by Zadeh [14] in 1965, and there is a large corpus of work on the literature that use fuzzy inclusion. Cornelis and Kerre in [15] presented a notion of a graded inclusion indicator for intuitive fuzzy sets. Bordogna et al. in [16] applied some properties and semantic aspects of fuzzy inclusion to interpret queries in Data Base Management Systems and IRS. Another work regarding the use of fuzzy inclusion and database queries is proposed by Boscand Pivert in [17] where they introduced a fuzzy inclusion indicator which models a more drastic behavior than those based on R-implications or S-implications.

More specific works regarding the definition and use of fuzzy inclusion are given by Sinha and Dougherty [18]. They generalized the notion of set inclusion for fuzzy sets and defined certain properties that an indicator for fuzzified set inclusion should accomplish. Cornelis et al. in [19] revisited the indicators proposed by Sinha and Dougherty and proposed a framework for the axiomatic characterization of inclusion grades for fuzzy sets. Beg and Ashraf, in [20], discussed some of the fuzzy inclusion degrees and verified the accomplishment of some of the properties for a fuzzy inclusion degree function. Wygralak in [21] proposed a definition of fuzzy inclusion and demonstrated some properties of fuzzy inclusion.

The work presented here focuses on the application of FSMs to reduce a number of fuzzy sets in input and output domains generated using PG-Means. Two algorithms (Sect. 3.2) for reduction a number of fuzzy sets based on similarity measures between fuzzy sets are described. A fuzzy similarity based hierarchical clustering of rules is used in a process of reducing fuzzy if-then rule sets (Sect. 3.3). Once clusters of rules are determined, ‘representative’ rules of cluster are determined. Those rules constitute a reduced set of rules. This process is applied to a task of image segmentation (Sect. 5).

2 Basic Theory

2.1 Fuzzy Similarity Measures

A similarity measure is considered as a comprehensive measure of a fuzzy system [2] where such a measure is defined between fuzzy sets, its respective backgrounds and rules. A Fuzzy Similarity Measure (FSM) is a real-valued function that quantifies the resemblance between fuzzy sets. The following definition is used here.

Definition 1 A similarity measure is a function defined between the collection of fuzzy sets and the interval $[\mathbf{0}, \mathbf{1}]$:

$$S : F(X) \times F(X) \rightarrow [0, 1] \tag{1}$$

where $\boldsymbol{F}(\boldsymbol{X})$ is the set of all fuzzy sets defined on the universe of discourse X.

Further, we can provide a few basic properties of such a function.

Definition 2 Given $\boldsymbol{A}, \boldsymbol{B} \in \boldsymbol{F}(\boldsymbol{X})$, a fuzzy similarity has the following properties [22]:

1. $S(A, B) = S(B, A), \forall A, B \in F(X)$.
2. $S(A^c, B^c) = S(A, B), \forall A, B \in F(X)$.
3. $0 \leq S(A, B) \leq 1, \forall A, B \in F(X)$.
4. $A = B$ if and only if $S(A, B) = 1$.
5. If $S(A, B) = 0$, then either $A \cap B = \emptyset$ or $A^c \cap B^c = \emptyset$, or $B = 1 - A$.
6. For $0 \leq \epsilon \leq 1$, if $S(A, B) \geq \epsilon$, we say that the two fuzzy sets A and B are $\varepsilon -$ *similar*.

The topic of FSM has been addressed in multiple publications [2, 5, 6, 11, 12, 22–28]. Among many different FSMs, we have selected four different measures[1] for the experiments presented in the paper:

- SIM_1:

$$1 - \max_{x \in X}(|\mu_A(x) - \mu_B(x)|) \tag{2}$$

- SIM_2:

$$1 - \frac{\sum_{x \in X}(|\mu_A(x) - \mu_B(x)|)}{\sum_{x \in X}(\mu_A(x) + \mu_B(x))} \tag{3}$$

- SIM_3:

$$\frac{\sum_{x \in X} min(\mu_A(x), \mu_B(x))}{\sum_{x \in X} max(\mu_A(x), \mu_B(x))} \tag{4}$$

- SIM_4:

$$\max_{x \in X}\{\min(|\mu_A(x), \mu_B(x)|)\} \tag{5}$$

[1]A comprehensive analysis of FSM is presented in the paper 'Experimental Evaluation and Comparison of Fuzzy Similarity Measures' to be submitted to IEEE Transactions of Fuzzy Systems.

This selection was dictated by the different behaviors that these similarities have. Unfortunately, limited space prevents us from providing details. In the paper, we also use such measures as Similarity of Rule Premise (SRP), Similarity of Rule Consequent (SRC) and Similarity Rules (SR) that have been proposed in Garcia-Garcia et al. [37]. The definitions of SRP, SRC and SR are listed below:

- SRP:

$$SRP(i,j) = \frac{\sum_{k=1}^{\#Inputs} S\left(A_{i,k}, A_{j,k}\right)}{\#Inputs} \tag{6}$$

- SRC:

$$SRC(i,j) = S\left(B_i, B_j\right) \tag{7}$$

- SR:

$$SR(i,j) = \min(SRP(i,j), SRC(i,j)) \tag{8}$$

2.2 *Fuzzy Inclusion Degree Measure*

Fuzzy inclusion means that a given fuzzy set contains another fuzzy set to a degree between 0 and 1. Thus, a measure of fuzzy inclusion is a fuzzy set in $F(X) \times F(X)$ [29]. This measure is called Fuzzy Inclusion Degree (FID). The following definition is used here:

Definition 3 An inclusion fuzzy degree is a function defined between the collection of fuzzy sets and the interval $[\mathbf{0}, \mathbf{1}]$:

$$I : F(X) \times F(X) \to [0, 1] \tag{9}$$

Where $\boldsymbol{F}(\boldsymbol{X})$ is the set of all fuzzy sets defined on the universe of discourse X. Further, we can provide a few basic properties that such a function has to satisfy.

Definition 4 Given $\boldsymbol{A}, \boldsymbol{B} \in \boldsymbol{F}(\boldsymbol{X})$, a fuzzy inclusion degree has the following properties [18, 24]:

1. $I(X, \varnothing) = 0$.
2. $I(A, B) = 1$, iff $A \subseteq B$ in Zadeh's sense.
3. $I(A, B) = 1$, iff $\mathcal{O}(A) \cap \mathcal{Z}(B) \neq \varnothing$.
4. If $B \subseteq C$, then $I(A, B) \leq I(A, C)$. In other words, the indicator is a non-decreasing function in the 2nd variable.

5. If $B \subseteq C$, then $I(C,A) \leq I(B,A)$. In other words, the indicator is a non-increasing function in the 1st variable.
6. For all $A,B,C \in F(X)$ if $A \subseteq B \subseteq C$, then $I(C,A) \leq I(B,A), I(C,A) \leq I(C,B)$.
7. $I(A,B) = I(B^c, A^c)$.

There are two unary operations *one* $\mathcal{O}$ and *zero* $\mathcal{Z}$ as follows: $\mathcal{O}, \mathcal{Z} : [0,1]^{\mathcal{U}} \rightarrow [0,1]^{\mathcal{U}}$ such that

$$\mathcal{O}(A) = \{x|\mu_A = 1\} \text{ and } \mathcal{Z}(A) = \{x|\mu_A = 0\} \tag{10}$$

The topic of FID has been addressed in multiple research publications [14, 18, 24, 28–36]. Among many different IFDs, we have selected the simplest inclusion degree for the experiments presented in the paper:

- INC_1:

$$\frac{|A \cap B|}{|A|} \tag{11}$$

2.3 *PG-Means*

The PG-Means [38] is a method used to find the mixture models that fit a given data without any previous knowledge regarding a number of mixtures. This method uses the Gaussian Mixture model with Expectation-Minimization training. It starts with a simple model (single mixture) and increases a number of models—k—by one at each iteration until it finds a mixture model that fits the data very well. At the end of each iteration, the method verifies if the data fits the model or if there is a need for adding one more model. The verification process means a projection of both data and Mixture Gaussian model on each dimension, and application of the Kolmogorov-Smirnov test. The iterative algorithm is as follows:

Algorithm I: PG-Means Algorithm [38]

1: Let $k \leftarrow 1$. Initialize the cluster with the mean and covariance of X.
2: **for** $i = 1 \ldots p$ **do**
3: Project X and the model to one dimension with the same projection.
4: Use the KS test at significance level α to test if the projected model fits the projected dataset.
5: If the test rejects the null hypothesis, then break out of the loop.
6: **end for**
7: **if** any test rejected the null hypothesis **then**
8: **for** $i = 1 \ldots 10$ **do**
9: Initialize $k+1$ clusters as the k previously learned plus one new cluster.
10: Run EM on the $k+1$ clusters.
11: **end for**
12: Retain the model of $k+1$ clusters with the best likelihood.
13: Let $k \leftarrow k+1$, and go to step 2.
14: **end if**
15: Every test accepts the null hypothesis; stop and return the model.

3 Fuzzy Sets Generation and Fuzzy Sets Reduction

In the literature, one can find multiple approaches proposed for generating fuzzy sets without prior information or knowledge about them. Some of them use a randomized algorithm or special heuristics. Others create fuzzy sets using a top-down probabilistic clustering algorithm, like Dirichlet Mixture Estimation (DME). Here, we describe our two algorithms for creating and minimizing a number of fuzzy sets.

3.1 Fuzzy Sets Generation

The generation of fuzzy sets for linguistic variables is done using the PG-Means algorithm. To illustrate how it works, we apply PG-Means to generate fuzzy sets on input and output spaces of an image segmentation problem. We want to build fuzzy sets for three input components representing an RGB image, as well as fuzzy sets for the output component that, in this case, represents color-based segmentation of image.

First, an image in RGB is mapped into the HSV color space. Following this, the image is split into its color components. For each color component, the PG-Means is applied to obtain a set of Multivariate Gaussians that should fit the Gaussians Mixture Model (GMM) for each component. The PG-Means algorithm starts with two Gaussian functions and uses the EM algorithm to determine an initial GMM. The model together with the data are projected into a single dimension. The Kolmogorov-Smirnov (KS) test is used to determine whether the projected model fits the projected data. This process is repeated several times for a single model. If any test is rejected the algorithm adds one more cluster and starts again with the EM learning. It is a bottom-up algorithm.

The Gaussians model that is learned by this algorithm overestimates the number of Gaussians. Thus, some Gaussians are overlapped. In this work, FSMs and Fuzzy Inclusion Degrees are used to reduce the number of fuzzy sets.

3.2 Fuzzy Sets Reduction

In order to reduce a number of the fuzzy sets, we need to delete sets that are similar to other fuzzy sets. The following algorithm is proposed to choose which fuzzy sets should be deleted and which ones should be kept. This algorithm is for the input variable:

Algorithm II: Reduction of Fuzzy Sets of the Input Variable

```
1: For each input variable
2: for c in {c_h, c_s, c_v} do
3:    Obtain the Similarity Matrix M_c^S and the Inclusion Degree Matrix M_c^I
4:    M_c^S ← CalculateSM(F(X_c))
5:    M_c^I ← CalculateIDM(F(X_c))
6:    where F(X_c) is the set of all fuzzy sets of X_c
7:    for A, B in F(X_c) and A! = B do
8:       simAB ← M_c^S(A, B)
9:       incDegAB ← M_c^I(A, B)
10:      incDegBA ← M_c^I(B, A)
11:      if incDegAB > thresholdIncDeg||incDegBA > thresholdIncDeg then
12:         if incDegAB > thresholdIncDeg then
13:            Delete Fuzzy Set A from F(X_c)
14:         else
15:            Delete Fuzzy Set B from F(X_c)
16:         end if
17:      else
18:         if simAB > thresholdSim then
19:            if incDegAB < incDegBA then
20:               Delete Fuzzy Set B from F(X_c)
21:            else
22:               Delete Fuzzy Set A from F(X_c)
23:            end if
24:         end if
25:      end if
26:   end for
27: end for
```

Algorithm II represents our approach for deleting fuzzy sets of input variables. First, two fuzzy sets are selected. The inclusion degrees calculated for these two sets, i.e., $I(A,B)$ and $I(B,A)$, are used to determine which fuzzy set should be deleted. We consider two cases: (1) when one of the inclusion degrees is greater than a threshold, and (2) when similarity between both sets is greater than a (different) threshold. In the first case, if $I(A,B) > threshold$ then the fuzzy set A is deleted, and if $I(B,A) > threshold$ then the fuzzy set B is deleted. The second case occurs when $I(A,B) < threshold$ and $I(B,A) < threshold$. Now, a similarity measure is used to determine which fuzzy set should be deleted. If the value of similarity measure between those fuzzy sets is greater than a threshold, one of the fuzzy sets is deleted. Again, we use the inclusion degree to determine which of the fuzzy sets is deleted: if $I(A,B) < I(B,A)$ then fuzzy set B is deleted, otherwise the fuzzy set A is deleted. In the scenario that both cases are not satisfied, none of the fuzzy sets is deleted.

We also encounter a similar problem in the output variable space. We have too many clusters just like in the case of input variables. To reduce a number of such clusters and to obtain a smaller number of fuzzy sets for the output variable, we propose the following algorithm:

Algorithm III: Reduction of Fuzzy Sets of the Output Variable

1: $\mathbf{X} = \{\mathbf{X}_1, \cdots, \mathbf{X}_d\}$
2: $\boldsymbol{\mu} = \{\mu_1, \cdots, \mu_N\}$ and $\boldsymbol{\Sigma} = \{\Sigma_1, \cdots, \Sigma_N\}$ with dimension d.
3: For each component obtain FSM matrices ($\mathcal{M}_i^S$) and IFD matrices ($\mathcal{M}_i^I$)
4: **for** i in $\{1, \cdots, d\}$ **do**
5: $\quad \mathcal{M}_{\mathcal{F}S}(X_i)$ Obtaining marginals and keep them as Gaussian Fuzzy Sets
6: $\quad$ Obtain the Similarity Matrix M_c^S and the Inclusion Degree Matrix M_c^I
7: $\quad \mathcal{M}_i^S \leftarrow CalculateSM(\mathcal{M}_{\mathcal{F}S}(X_i))$
8: $\quad \mathcal{M}_i^I \leftarrow CalculateIDM(\mathcal{M}_{\mathcal{F}S}(X_i))$
9: **end for**
10: Obtain a FSM and IFD matrices of the clusters
11: **for** i, j in $\{1, \cdots, N\}$ **do**
12: $\quad \mathcal{M}^S(i,j) \leftarrow \frac{1}{d} \sum_{k=1}^{d} \mathcal{M}_k^S(i,j)$
13: $\quad \mathcal{M}_c^I \leftarrow \frac{1}{d} \sum_{k=1}^{d} \mathcal{M}_k^I(i,j)$
14: $\quad$ Obtain the distance matrix between means D.
15: $\quad D(i,j) \leftarrow dist(\mu_i, \mu_j)$
16: **end for**
17: Reduction criteria
18: **for** i, j in $\{1, \cdots, N\}$ **do**
19: $\quad$ **if** $D(i,j) < \epsilon_{dist}$ **then**
20: $\quad\quad sim_{i,j} \leftarrow M^S(i,j)$
21: $\quad\quad incDeg_{i,j} \leftarrow M^I(i,j)$
22: $\quad\quad incDeg_{j,i} \leftarrow M^I(j,i)$
23: $\quad\quad$ **if** $incDeg_{i,j} > thresholdIncDeg || incDeg_{j,i} > thresholdIncDeg$ **then**
24: $\quad\quad\quad$ **if** $incDeg_{i,j} > thresholdIncDegND$ **then**
25: $\quad\quad\quad\quad$ Delete Cluster i
26: $\quad\quad\quad$ **else**
27: $\quad\quad\quad\quad$ Delete Cluster j
28: $\quad\quad\quad$ **end if**
29: $\quad\quad$ **else**
30: $\quad\quad\quad$ **if** $sim_{i,j} > thresholdSimND$ **then**
31: $\quad\quad\quad\quad$ **if** $incDeg_{i,j} < incDeg_{j,i}$ **then**
32: $\quad\quad\quad\quad\quad$ Delete Cluster j
33: $\quad\quad\quad\quad$ **else**
34: $\quad\quad\quad\quad\quad$ Delete Cluster i
35: $\quad\quad\quad\quad$ **end if**
36: $\quad\quad\quad$ **end if**
37: $\quad\quad$ **end if**
38: $\quad$ **end if**
39: **end for**
40: run EM with the new number of clusters to get the final model.

To determine the output fuzzy variables, we use the marginals as in [3]. A i-th fuzzy set is obtained by converting the Multivariate Gaussian of the i-th cluster to a one dimensional Gaussian membership function. In order to do so, we propose two approaches. In the first one, the mean of a Gaussian function is the cluster's index (i.e. $\mu_i = i$), while its variance is a constant $a(\sigma_i = a)$. In the second approach, the means of the function is the cluster's index (i.e. $\mu_i = i$), and the variance is given by:

$$\sigma_i = \frac{\sum_{j=1}^{N_m} \sigma_{i,j}}{N_m} \tag{12}$$

where $\sigma_{i,j}$ is the means of the j-th marginal of the i-th cluster, and N_m is the number of marginals.

3.3 Fuzzy Rule Generation and Reduction

In [37], Garcia et al. has proposed the use of fuzzy similarity between fuzzy sets, premises and consequences of rules, as well as hierarchical clustering for determining an appropriate set of fuzzy rules. The first step is to generate an initial set of rules. The previously generated fuzzy sets, both for the input variables and output variables, are the input to the process of building an initial rule base. This process consists of two steps:

- First, based on the results of the fuzzification of input variables, we determine the antecedent fuzzy sets by identifying which sets are activated based on a specific threshold value [37].
- Second, we determine the consequent fuzzy sets via selecting a component with the highest probability assigned to it by PG-Means after application of Algorithms II and III.

Following this, we reduce the number of rules using FSM, SRP, SRC and SR together with hierarchical clustering. We calculate the similarity between the fuzzy sets (Sect. 2.1) of each rule. As the result, we build a similarity matrix M of SR, where each element $m_{i,j}$ is an SR value representing similarity between the i-th and j-th rules. In order to use the similarity matrix to reduce the rule base, we construct another matrix M' by $m'_{i,j} = 1 - m_{i,j}$. This is a dissimilarity matrix.

The dissimilarity matrix M' is the input to a hierarchical clustering process. The obtained hierarchy, called dendrogram, is further used to reduce the set of rules. We apply different cut-off values to 'cut' the dendrogram at different levels, and based on that identify subsets, i.e., clusters of fuzzy rules. Each such cluster k is associated with a sub-matrix M_k extracted from the rule similarity matrix M. M_k contains similarities between rules that belong to this cluster. For each row i of M_k, we calculate the means and the variance of similarity values SR. These values indicate how similar the i-th rule is when compared with other rules of the sub-matrix. The highest value of the means denotes the rule that is the most similar to other rules. On the other hand, the variance value is used to break a tie when more rules have the maximum means value. Such a process is repeated for each cluster. Once each cluster is 'replaced' with a single rule, a reduced set of fuzzy if-then rules is determined.

4 Experimental Set-Up

4.1 Data Description

The use of multiple FSMs for fuzzy rule reduction is illustrated with a process of color image segmentation. Two images are used here: *STONE*, Fig. 1a; and *SCISSORS*, Fig. 1b.

As the first step, we transform each image into a three-dimensional data set with dimensions: H—hue; S—saturation; and V—value. In the second step, we use PG-Means to generate fuzzy sets for each of three dimensions (input variables). In order to determine fuzzy sets for the output variable the image segmentation is performed. We use Algorithm II to reduce the fuzzy sets for the three dimensions H, S and V; and Algorithm III to reduce the fuzzy sets for the output variable. As the final step, we use a hierarchical clustering process together with the algorithm proposed in [37] to reduce a number of fuzzy rules.

We use four different similarities (Sect. 2.1) when we apply Algorithms II and III. An example of initial and reduced fuzzy sets is shown in Fig. 2. The sets are obtained when algorithms with the similarity $\mathbf{SIM_4}$ arreused for STONE, Fig. 2a, and for SCISSORS, Fig. 2b.

4.2 Performance Measure

In order to evaluate performance of an image segmentation using fuzzy if-then rules we apply the Combine Error Rate (CER) [39]. This index is an average of the total false positive and false negative rates taken across each class:

$$CER = \frac{\sum_i \frac{n_i}{n} FP_i + \sum_i \frac{n_i}{n} FN_i}{2} \tag{13}$$

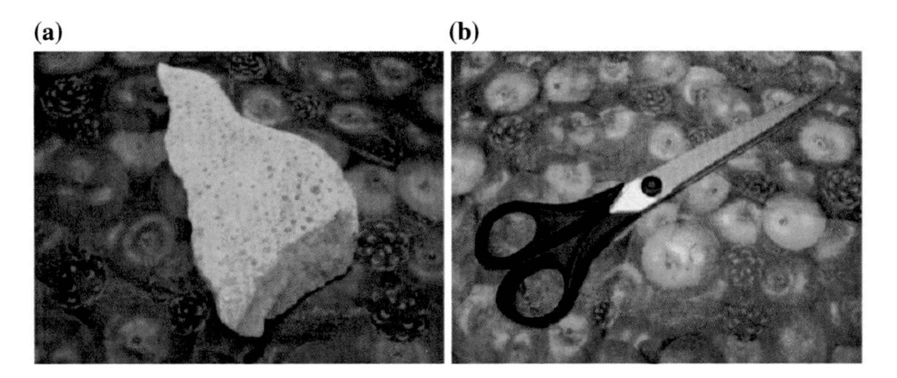

Fig. 1 Original images of *STONE* (**a**) and *SCISSORS* (**b**)

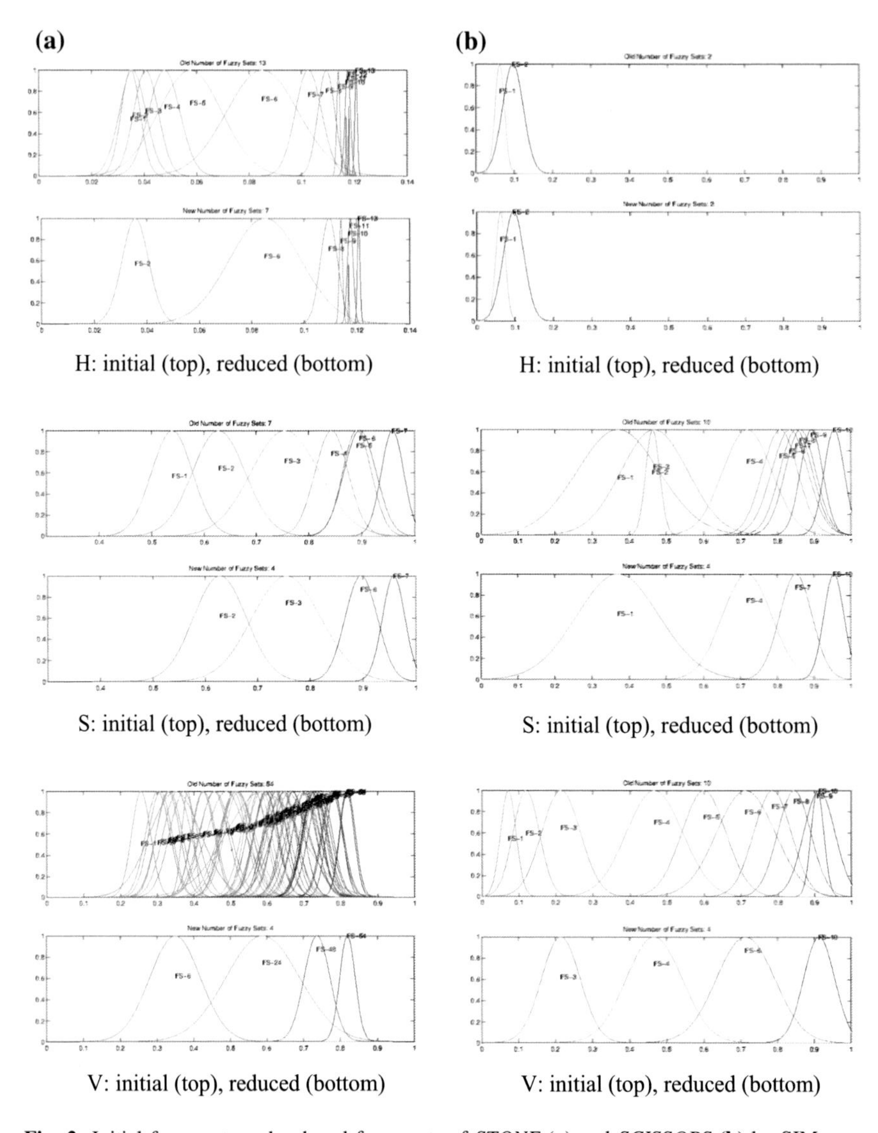

Fig. 2 Initial fuzzy sets and reduced fuzzy sets of *STONE* (**a**) and *SCISSORS* (**b**) by SIM_4

where i represents a number of segments, n—a total number of data points, n_i—a number of data points of segment i, FP_i—a number of false positives for the segment i, and FN_i—a number of false negatives for the segment i.

5 Experiments and Results

5.1 Input Fuzzy Sets

In order to generate fuzzy sets, we transform RGB images into the HSV space. Then, we run the PG-Means. Figure 3 shows the histogram and the Mixture Gaussian Model obtained after applying the PG-Means algorithm for the image *STONE*. Figure 3 shows histograms for each component (H, S, and V) in black, as well as the Gaussian Mixtures shown in red. As it is known, the Gaussian Mixture is obtained by the sum of distributions of each component $(p(x|i))$ multiplied by its mixture proportion $(p(i)), p(x) = \sum_{i=1}^{k} p(x|i) * p(i)$. As we can see, the mixture models fit data very well.

The Multivariate Gaussians 'associated' with the results of PG-means (Fig. 3) are shown in the top plots for H, S, and V in Fig. 2. Their reduced sets obtained using Algorithm II are illustrated in the bottom plots, Fig. 2.

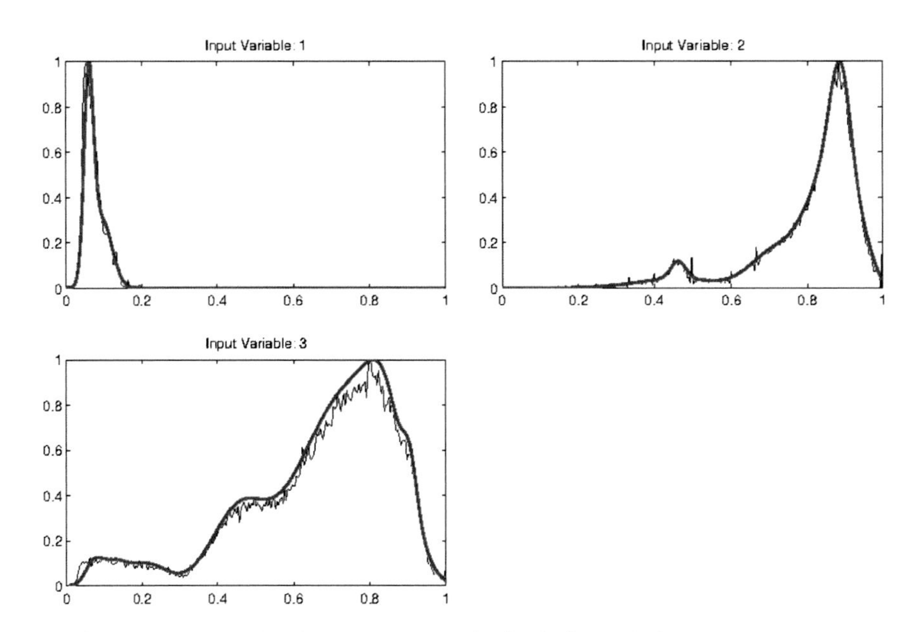

Fig. 3 Histograms and the mixture gaussian model for the image *STONE*

Table 1 Number of input fuzzy sets for both images: before and after reduction

	STONE image				*SCISSORS* image			
	$\mathbf{SIM_1}$	$\mathbf{SIM_2}$	$\mathbf{SIM_3}$	$\mathbf{SIM_4}$	$\mathbf{SIM_1}$	$\mathbf{SIM_2}$	$\mathbf{SIM_3}$	$\mathbf{SIM_4}$
H: original no of sets	13	13	13	13	2	2	2	2
Reduced no of sets	11	10	11	7	2	2	2	2
S: original no of sets	7	7	7	7	10	10	10	10
Reduced no of sets	6	6	6	4	7	6	6	4
V: original no of sets	54	54	54	54	10	10	10	10
Reduced no of sets	13	8	11	4	9	8	9	4

Table 1 contains the number of input fuzzy sets—initial and reduced—for each dimension of HSV space for both images. The reduction has been obtained after application of Algorithm II with the following values of parameters: the inclusion degree—*cutIncDeg* = 0.9, and the similarity threshold—*cutSim* = 0.55. As it can be seen, a meaningful reduction of fuzzy sets occurs in almost all cases. The case with the similarity SIM_4 leads to the most substantial reduction of input fuzzy sets. Among other results, the results obtained with $\mathbf{SIM_2}$ are also encouraging.

5.2 *Output Fuzzy Sets*

The HSV representation of images together with the PG-Means is used in an image segmentation process. The obtained numbers of clusters are considerably high: especially for *STONE*—26 for *STONE*, and relatively high for *SCISSORS*—7. Therefore, we need to decrease those numbers by applying Algorithm III. Additionally, we use (Eq. 12) to determine a final set of output fuzzy sets.

As in the case of the input variables, the cut-off values used for the reduction of the fuzzy sets have been determine experimentally: for the distance epsilonDistance = 0.25, for the inclusion degree *cutIncDegND* = 0.45, and for the similarity cutSimND = 0.35. These cut-off values have been used together with FSMs: $\mathbf{SIM_2}$, $\mathbf{SIM_3}$ and $\mathbf{SIM_4}$. For $\mathbf{SIM_1}$, the cut-off values are: for the distance *epsilonDistance* = 0.25, for the inclusion degree *cutIncDegND* = 0.275, and for the similarity *cutSimND* = 0.15.

Figure 4 shows belonging of pixels to clusters in a 3D cloud before and after reduction of number of clusters.

Table 2 illustrates the reduction in the number of fuzzy sets of the output variable. Once again, the most substantial reduction of fuzzy sets is obtained when $\mathbf{SIM_4}$ is used. Also, the application of $\mathbf{SIM_2}$ gives encouraging results.

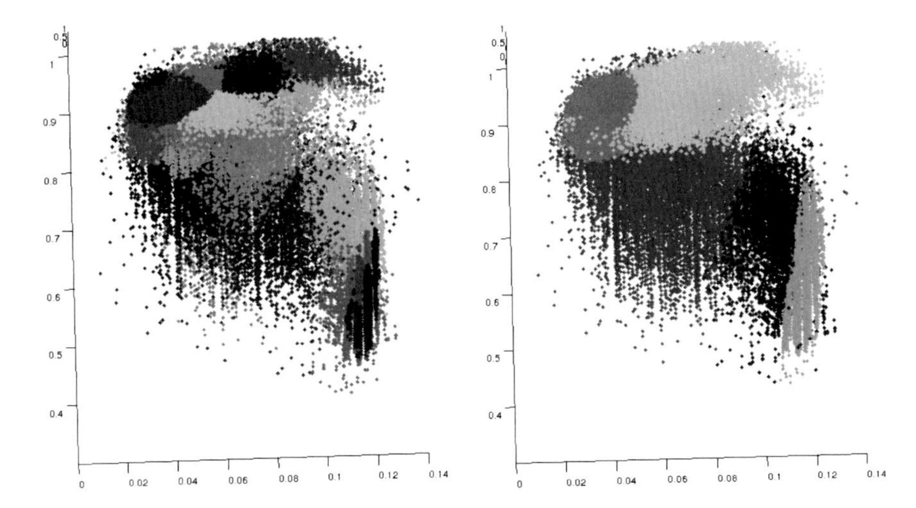

Fig. 4 Reduction of fuzzy sets for the output variable; left image shows the initial clustering while the right image shows the final clustering

Table 2 Number of output fuzzy sets for both images

	STONE image				*SCISSORS* image			
	SIM$_1$	**SIM$_2$**	**SIM$_3$**	**SIM$_4$**	**SIM$_1$**	**SIM$_2$**	**SIM$_3$**	**SIM$_4$**
Original no of sets	26	26	26	26	7	7	7	7
Reduced no of sets	7	5	9	3	3	3	5	3

Table 3 Number of fuzzy rules and obtained values of CER for both images

	STONE image				*SCISSORS* image			
	SIM$_1$	**SIM$_2$**	**SIM$_3$**	**SIM$_4$**	**SIM$_1$**	**SIM$_2$**	**SIM$_3$**	**SIM$_4$**
No. rules	2384	1084	2368	187	234	180	305	69
CER value	0.5055	0.500	0.496	0.572	0.500	0.471	0.510	0.407

5.3 *Original Fuzzy Rules*

The reduced number of input and output fuzzy sets allows us to generate, automatically, fuzzy if-then rules. The information about those rules for both images, as well as the values of the CER index obtained using these rules in the segmentation process are shown in Table 3.

5.4 Reduction of Fuzzy Rules

Hierarchical Clustering. The first stage of the rule reduction process is to identify clusters of similar rules. In order to achieve this, we construct matrices of similarity between fuzzy if-then rules for both images. We build four similarity matrices using the four similarity measures, Eqs. 6, 7, and 8, presented in Sect. 2.1. First, we take two rules and calculate similarity values between fuzzy sets of the same linguistic variable using one of the equations Eqs. 2, 3, 4, or 5. Second, we calculate a similarity value of the premise (antecedent) of rules, Eq. 6, as well as a similarity of the consequence, Eq. 7. Again, we use one of the equations Eqs. 2, 3, 4, or 5 for similarity calculations. Finally, we calculate a similarity value between the rules using Eq. 8. We repeat this process for all rules of the rule-based system. As the result, we obtain a similarity matrix $\boldsymbol{M}$, and the equivalent dissimilarity matrix $\boldsymbol{M}'$. The dissimilarity matrix is used to perform hierarchical clustering.

The result of a clustering process, i.e., a dendrogram, that represents a hierarchy of clusters for the image *SCISSORS*, is shown in Fig. 5. The presented hierarchy is obtained for the similarity $\mathbf{SIM_4}$.

Image Segmentation. The obtained hierarchies of clusters provide us with a means to reduce number of rules. The clusters—identified for a given cut-off point—represent groups of rules that are similar to each other. In the following steps, each such a group is replaced by a single rule. In the work presented here, we use two sets of pre-selected cut-off values depending on the image, i.e., one set for *STONE*, and one for *SCISSORS*. For the image: *SCISSORS*, we use five pre-selected cut-off points that allows us to determine a specific number of clusters. Figure 5 shows five lines representing cut-off values resulting in 15, 25, 35, 45 and 50 clusters. For the image *STONE*, on the other hand, we use five pre-selected cut-off points that allows us to end-up with 25, 50, 65, 75 and 90 clusters. Once we have identified clusters of rules,

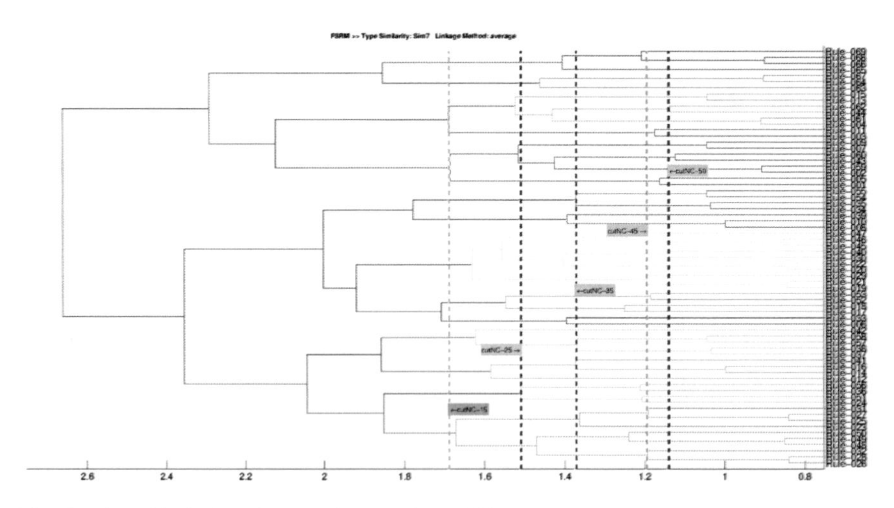

Fig. 5 Hierarchical clustering: dendrogram for *SCISSORS* (b)

Table 4 Results of fuzzy if-then rules reduction process for both images

	STONE image				*SCISSORS* image			
	SIM$_1$	**SIM$_2$**	**SIM$_3$**	**SIM$_4$**	**SIM$_1$**	**SIM$_2$**	**SIM$_3$**	**SIM$_4$**
No of rules								
Initial	2384	1084	2368	187	234	180	305	69
After reduction	90	50	90	25	45	15	15	15
% of initial value	**3.78**	**4.61**	**3.80**	**13.37**	**19.23**	**8.33**	**4.92**	**21.74**
CER values								
	SIM$_4$	**SIM$_2$**	**SIM$_2$**	**SIM$_4$**	**SIM$_2$**	**SIM$_4$**	**SIM$_4$**	**SIM$_4$**
Initial	0.505	0.500	0.496	0.572	0.500	0.470	0.510	0.407
After reduction	0.427	0.473	0.491	0.433	0.491	0.322	0.427	0.449
Difference	**0.078**	**0.027**	**0.005**	**0.139**	**0.009**	**0.148**	**0.083**	−0.042

Bold values represent the percentage of surviving rules after the algorithm has been applied

we find a representative rule for each cluster. The method presented in Sect. 3 together with the similarity matrix $\boldsymbol{M}$ are used to choose a single rule among all rules of a given cluster. The applied selection process leads to identifying a rule that is the most similar to other rules in the cluster.

We have performed a series of rule reduction experiments involving all combinations of number of rules and different similarities. We present, Table 4, only the best results for each similarity obtained after the reduction process. In this case, the best results mean the biggest reduction in a number of rules without decreasing the values of CER performance index. The row 'SIM for reduction' indicates the kind of similarity (Sect. 2.1) used to reduce a number of rules. As before, it can be seen that the best results are obtained with similarities **SIM$_4$** and **SIM$_2$**.

Figures 6 and 7 show the results of segmentation obtained using different sets of if-then rules. In the first row, images: (a–d), we include the segmentation obtained using the full sets of rules generated based on the reduced, by Algorithm III, number of fuzzy sets on the output variable. The second row, images: (e–h), shows the segmentation given by the full sets of rules generated based on the reduced number of fuzzy sets on both input and output. Finally, the third row, images: (i–l), shows the segmentation obtained using the reduced number of if-then rules generated with the reduced number of input and output fuzzy sets.

Let us take a look at the results of the application of proposed methodology for reduction of number of rules for an image segmentation task, Table 4. Firstly, we would like to focus on the main goal of this work—reduction of rules (part of the table: *No of rules*). As it can be seen in each case, a substantial reduction is obtained. The largest reduction for STONE image is seen for the similarities **SIM$_1$** and **SIM$_3$** where the number of reduced rules is about 3.8% of the original rule base. At the same time, we do not experience any degradation of performance—the values of CER for both mentioned cases are slightly lower than the ones obtained with the original rule base (part of the table: *CER values*). For SCISSORS, the best

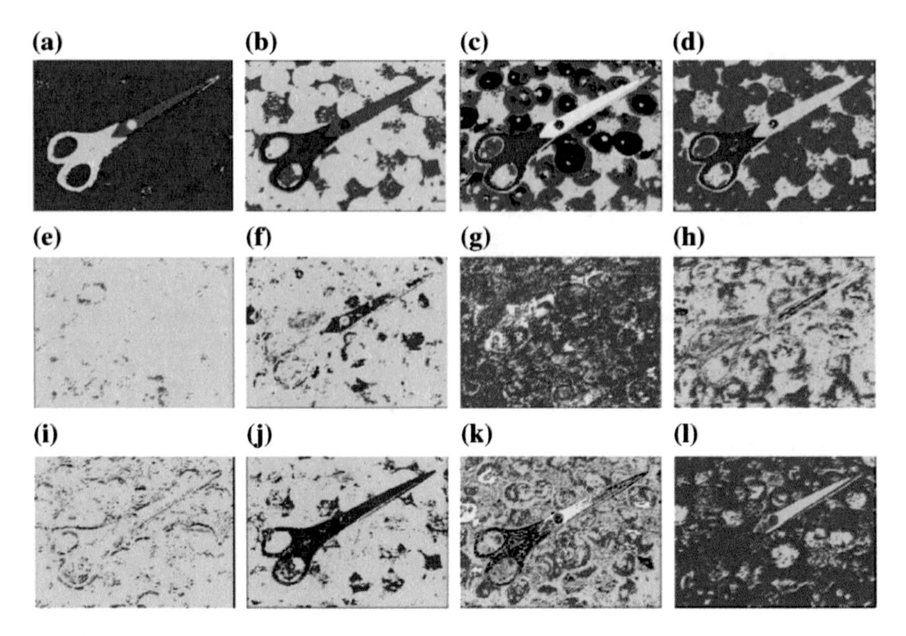

Fig. 6 Segmentation of the image *SCISSORS*: images (**a–d**) after reduction of output fuzzy sets with the full set of if-then rules; images (**e–h**) after reduction of input and output fuzzy set and the full set of rules; and images (**i–l**) after reduction of input and output fuzzy systems, as well as reduction of the fuzzy if-then of rules

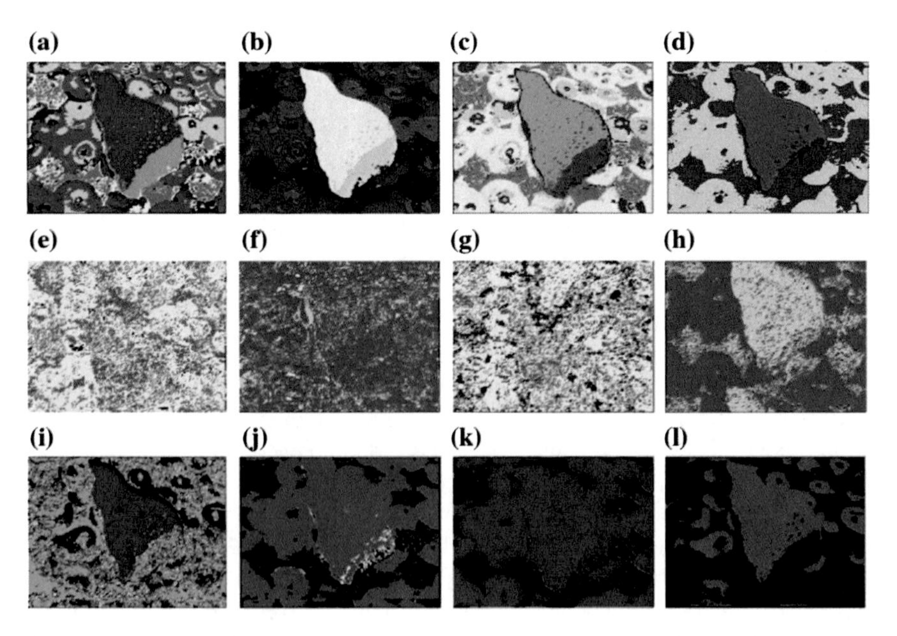

Fig. 7 Segmentation of the image *STONE*: images (**a–d**) after reduction of output fuzzy sets with the full set of if-then rules; images (**e–h**) after reduction of input and output fuzzy set and the full set of rules; and images (**i–l**) after reduction of input and output fuzzy systems, as well as reduction of the fuzzy if-then of rules

results are obtained when similarities $\mathbf{SIM_2}$ and $\mathbf{SIM_3}$ have been applied—the reduced numbers of rules represent 8.33% and 4.92% of the original number of rules, respectively. Also, the values of CER are improved. The only exception is the system with $\mathbf{SIM_4}$. In this case we observe a slight decrease in the value of CER.

6 Conclusion

An automatic generation of fuzzy sets with multi-dimensional data is an attractive part of a process of constructing fuzzy if-then rule-based systems. However, a disadvantage of the automatic approach is the generation of a large number of fuzzy sets in each dimension what leads to building rule-based systems with many rules. The paper proposes a methodology for effective reduction of generated rules. The method is based on Fuzzy Similarity Measures (FSMs) that are: (1) used to determine overlapping fuzzy sets in input and output domains of constructed systems, and (2) applied for building similarity matrix to perform hierarchical clustering of rules. The obtained clusters are replaced by single rules. Those rules form reduced rule-based systems.

The approach has been applied to an image segmentation task. Four different FSMs have been used to construct and reduce rule-based systems for segmentation of two different images. The experimental results—the reduced number of rules represents 4–20% of initial number of rules—show that simplified sets of rules do not decrease the performance of automatically generated fuzzy rule-based systems.

Acknowledgements The authors would like to thank prof. Greg Hamerly, from the Baylor University, for letting them use the PG-Means program he developed with Yu Feng [38].

References

1. C.-F. Juang, C.-Y. Wang, A self-generating fuzzy system with ant and particle swarm cooperative optimization. Expert Syst. Appl. **36**, 5362–5370 (2009)
2. H. Bellaaj, R. Ketata, M. Chtourou, A new method for fuzzy rule base reduction. J. Intell. Fuzzy Syst. **25**(3), 605–613 (2013)
3. A. Garcia-Garcia, A. Mendez-Vazquez, Learning fuzzy rules through ant optimization, lasso and dirichlet mixtures, in *2014 IEEE International Conference on Fuzzy Systems (FUZZ-IEEE)*, July 2014, pp. 2558–2565
4. D. Dubois, H. Prade, A unifying view of comparison indices in a fuzzy set-theoretic framework. Fuzzy Sets Possibility Theory: Recent Dev.(Pergamon, New York, 1982)
5. W.-J. Wang, New similarity measures on fuzzy sets and on elements. Fuzzy Sets Syst. **85**(3), 305–309 (1997)
6. H. Lee-Kwang, Y.-S. Song, K.-M. Lee, Similarity measure between fuzzy sets and between elements. Fuzzy Sets Syst. **62**(3), 291–293 (1994)
7. X. Wang, B.D. Baets, E. Kerre, A comparative study of similarity measures. Fuzzy Sets Syst. **73**(2), 259–268 (1995)

8. I. Turksen, Z. Zhong, An approximate analogical reasoning schema based on similarity measures and interval-valued fuzzy sets. Fuzzy Sets Syst. **34**(3), 323–346 (1990)
9. S. Santini, R. Jain, Similarity measures. Pattern Anal. Mach. Intell. IEEE Trans. **21**(9), 871–883 (1999)
10. Y. Jin, W. von Seelen, B. Sendhoff, On generating fc3 fuzzy rule systems from data using evolution strategies. Syst. Man Cybern. Part B: Cybern. IEEE Trans. **29**(6), 829–845 (1999)
11. M.-Y. Chen, D. Linkens, Rule-base self-generation and simplification for data-driven fuzzy models. Fuzzy Sets Syst. **142**(2), 243–265 (2004)
12. Y. Jin, Fuzzy modeling of high-dimensional systems: complexity reduction and interpretability improvement. Fuzzy Syst. IEEE Trans. **8**(2), 212–221 (2000)
13. T. Takagi, M. Sugeno, Fuzzy identification of systems and its applications to modeling and control. Syst. Man Cybern. IEEE Trans. SMC-15(1), 116–132 (1985)
14. L.A. Zadeh, Fuzzy sets. Inf. Control **8**(3), 338–353 (1965)
15. C. Cornelis, E. Kerre, Inclusion Measures in Intuitionistic Fuzzy Set Theory. (Berlin, Heidelberg: Springer Berlin Heidelberg, 2003), pp. 345–356. [Online]. Available: http://dx.doi.org/10.1007/978-3-540-45062-728
16. G. Bordogna, P. Bosc, G. Pasi, Fuzzy inclusion in database and information retrieval query interpretation, in *Proceedings of the 1996 ACM Symposium on Applied Computing, ser. SAC'96*. New York, NY, USA: ACM, 1996, pp. 547–551. [Online]. Available: http://doi.acm.org/10.1145/331119.331451
17. P. Bosc, O. Pivert, On a reinforced fuzzy inclusion and its application to database querying. (Berlin, Heidelberg: Springer Berlin Heidelberg, 2012), pp. 351–360. [Online]. Available: http://dx.doi.org/10.1007/978-3-642-31709-536
18. D. Sinha, E.R. Dougherty, Fuzzification of set inclusion: theory and applications. Fuzzy Sets Syst. **55**(1), 15–42 (1993). [Online]. Available: http://www.sciencedirect.com/science/article/pii/016501149390299W
19. C. Cornelis, C. V. der Donck, E. Kerre, Sinhadougherty approach to the fuzzification of set inclusion revisited. Fuzzy Sets Syst, **134**(2), 283–295 (2003). [Online]. Available: http://www.sciencedirect.com/science/article/pii/S0165011402002257
20. I. Beg, S. Ashraf, Fuzzy inclusion and design of measure of fuzzy inclusion. RIMAI J., **8** (2012)
21. M. Wygralak, Fuzzy inclusion and fuzzy equality of two fuzzy subsets, fuzzy operations for fuzzy subsets, Fuzzy Sets Syst. **10**(13), 157–168 (1983). [Online]. Available:www.sciencedirect.com/science/article/pii/S0165011483801122
22. A. Pal, B. Mondal, N. Bhattacharyya, S. Raha, Similarity in fuzzy systems. J. Uncertainty Anal. Appl. **2**(1) (2014)
23. C.P. Pappis, N.I. Karacapilidis, A comparative assessment of measures of similarity of fuzzy values. Fuzzy Sets Syst. **56**(2), 171–174 (1993)
24. W. Zeng, H. Li, Inclusion measures, similarity measures, and the fuzziness of fuzzy sets and their relations. Int. J. Intell. Syst. **21**(6), 639–653 (2006)
25. T. Liao, Z. Zhang, A review of similarity measures for fuzzy systems, in Fuzzy *Systems, 1996. Proceedings of the Fifth IEEE International Conference on*, vol. 2, Sep 1996, pp. 930–935
26. M. Alamuri, B. Surampudi, A. Negi, A survey of distance/similarity measures for categorical data, in *Neural Networks (IJCNN), 2014 International Joint Conference on, July 2014*, pp. 1907–1914
27. S. Raha, N. Pal, K. Ray, Similarity-based approximate reasoning: methodology and application. Syst. Man Cybern. Part A: Syst. Humans IEEE Trans. **32**(4), 541–547 (2002)
28. H. L. Capitaine, C. Fr′elicot, Towards a unified logical framework of fuzzy implications to compare fuzzy sets, in *Proceedings of the Joint 2009 International Fuzzy Systems Association World Congress and 2009 European Society of Fuzzy Logic and Technology Conference*, Lisbon, Portugal, July 20–24, 2009, pp. 1200–1205
29. V.R. Young, Fuzzy subsethood. Fuzzy Sets Syst. **77**(3), 371–384 (1996)

30. J. Goguen, The logic of inexact concepts. Synthese **19**(3–4), 325–373 (1969). [Online]. Available: http://dx.doi.org/10.1007/BF00485654
31. D. Dubois, H. Prade, *Fuzzy sets and systems—theory and applications* (Academic press, New York, 1980)
32. M. Wygralak, Fuzzy inclusion and fuzzy equality of two fuzzy subsets, fuzzy operations for fuzzy subsets. Fuzzy Sets Syst. **10**(1), 157–168 (1983)
33. M. Nachtegael, H. Heijmans, D. Van der Weken, E. Kerre, Fuzzy adjunctions in mathematical morphology, in *Proceedings of JCIS*, 2003, pp. 202–205
34. P. Bosc, A. Hadjali, O. Pivert, Graded tolerant inclusion and its axiomatization, in *Proceedings of the 12th International Conference on Information Processing and Management of Uncertainty in Knowledge-Based Systems (IPMU)*, 2008
35. I. Beg, S. Ashraf, Fuzzy relational calculus. Bull. Malays. Math. Sci. Soc. 2 (2013)
36. I. Beg, S. Ashraf, Kleene's fuzzy similarity and measure of similarity. Ann. Fuzzy Math. Inform. **6**(2), 251–261 (2013)
37. A. Garcia-Garcia, M. Reformat, A. Mendez-Vazquez, Similarity-based method for reduction of fuzzy rules, in *2014 North American Fuzzy Information Processing Society (NAFIPS)*, October 2016
38. Y.F.G. Hamerly, PG-means: learning the number of clusters in data. Adv. Neural. Inf. Process. Syst. **19**, 393–400 (2007)
39. A. Rosenberg, Automatic detection and classification of prosodic events, Ph.D. Dissertation, Columbia University, 2009

Fuzzy Optimized Classifier for the Diagnosis of Blood Pressure Using Genetic Algorithm

Juan Carlos Guzmán, Patricia Melin and German Prado-Arechiga

Abstract We propose to optimize the fuzzy rules, which are based on an expert, the objective is to classify the blood pressure level in a correct way with the necessary number of rules and not to have some type of mistake at the moment of giving the diagnosis, since the use of unnecessary rules could cause a confusion in the fuzzy classifier. The fuzzy classifier is only part of the neuro fuzzy hybrid model, which uses techniques such as: neural networks, fuzzy logic and evolutionary computation, in this latter technique, genetic algorithms are used, which use individuals as possible solutions and thus obtain the best solution, in this case find the appropriate number of fuzzy rules for fuzzy system. This study aims to model blood pressure for 24 h and obtain the trend per patient, once this trend is obtained, this information enters a fuzzy system based on rules given by an expert, who will be classified into some of the blood pressure levels based on level European guide and finally give us a diagnosis per patient.

Keywords Fuzzy system · Blood pressure · Diagnosis

1 Introduction

At present, using intelligent systems in the medical area has increased considerably, since the use of these systems, helps to perform quick and prescriptive diagnostics and this helps to avoid diseases and in the worst case helps to detect the disease in an early stage and be able to take the correct treatment and thus avoid the death of patients [10, 12].

This is why in this work is focused on the diagnosis of blood pressure, which will allow us to know what type of blood pressure level the patient has, based on the

J. C. Guzmán · P. Melin (✉)
Tijuana Institute of Technology, Tijuana, BC, Mexico
e-mail: pmelin@tectijuana.mx

G. Prado-Arechiga
Excel Medical Center, Philadelphia, USA

O. Castillo et al. (eds.), *Fuzzy Logic Augmentation of Neural and Optimization Algorithms: Theoretical Aspects and Real Applications*, Studies in Computational Intelligence 749, https://doi.org/10.1007/978-3-319-71008-2_23

information of a 24-h monitoring with 45 samples that was taken to patient and with the help of techniques of artificial intelligence [3, 6, 11], in this case the neural networks, the tendency of this monitoring was obtained [14], which will be classified by the fuzzy system, which has fuzzy rules, which are based on an expert and optimized with a technique of evolutionary computation, in this case we used genetic algorithms which helps us to find the correct number of fuzzy rules and avoid confusion to the system and thus obtain the precise diagnosis.

Constant blood pressure monitoring can help considerably, since this information is important to take control of the patient and help doctors to make a decision using smart systems, in this case the classifier is focused on helping in a fast way to give an accurate diagnosis, based on information provided by the patient, during continuous monitoring.

It is for this reason, that with the help of a cardiologist and the student's collaboration, a database with 24-h monitoring of 30 patients was obtained, it is worth mentioning that this database is growing day by day since only it is possible to obtain a monitoring by day [5].

The idea of optimizing the rules of the fuzzy classifier is to obtain better results, using less fuzzy rules, since the use of unnecessary rules, can confuse the classifier when classifying the blood pressure level, for this we chose to use genetic algorithms, which based on individuals helps us to find the best solution, in this case, the best rules to use in the fuzzy system [4, 5].

The paper is organized as follows: in Sect. 2 a methodology is presented, in Sect. 3 Proposal method and the classifier optimization using genetic algorithm is presented, in Sect. 4 Simulation results of the optimized classifier is presented in Sect. 5 the conclusion obtained after performing the experiments with the optimized classifier is presented.

2 Methodology

2.1 *Blood Pressure*

Blood pressure is the force exerted by the blood against the walls of the arteries. Each time the heart beats, it pumps blood into the arteries, which is when its pressure is higher. This is called systolic pressure. When your heart is at rest between one beat and another, your blood pressure decreases. This is called diastolic pressure.

Both numbers, systolic and diastolic pressure, are used to read blood pressure. In general, systolic pressure is mentioned first or above the diastolic. A reading with values of:

119/79 or less is considered normal blood pressure
140/90 or higher is considered high blood pressure.

Between 120 and 139 for the highest number, or between 80 and 89 for the lowest number is prehypertension. Prehypertension means that you can develop high blood pressure unless you take action.

High blood pressure does not usually have symptoms, but can cause serious problems such as strokes, heart failure, infarction and kidney failure.

You can control your blood pressure by using healthy lifestyle habits such as exercise and DASH diet and, if necessary, medications.

2.2 *Definitions and Classification of Office Blood Pressure Levels*

In Table 1 the categories of blood pressure are shown in order of lowest level to the highest level and finally isolated systolic pressure is shown and should be represented as grade 1, grade 2 and grade 3, this based on the ranges specified in the systolic pressure [9, 13].

2.3 *Genetic Algorithms*

Genetic Algorithms (GA) are adaptive methods that can be used to solve search and optimization problems. They are based on the genetic process of living organisms.

A genetic algorithm consists of a mathematical function that takes as input the copies and returns as outputs which of them must generate offspring for the new generation. Genetic operators: Selection or reproduction, Crossing, Mutation.

3 Proposal Method and the Classifier Optimization Using Genetic Algorithm

Figure 1 shows the General neuro fuzzy hybrid model, which consists of the following modules: first the data module, which represents the information obtained in the 24-h monitoring of the patient, this values enter to the second module called

Table 1 Classification of office blood pressure levels

Category	Systolic		Diastolic
Hypotension	<90	And/or	<60
Optimal	<120	And	<80
Normal	120–129	And/or	80–84
High normal	130–139	And/or	85–89
Grade 1 hypertension	140–159	And/or	90–99
Grade 2 hypertension	160–179	And/or	100–109
Grade 3 hypertension	≥ 180	And/or	≥ 110
Isolated systolic hypertension	≥ 140	And	<90

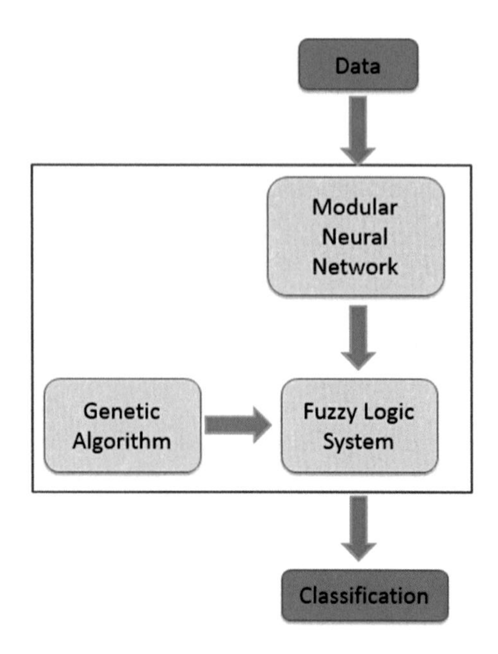

Fig. 1 General neuro fuzzy hybrid model

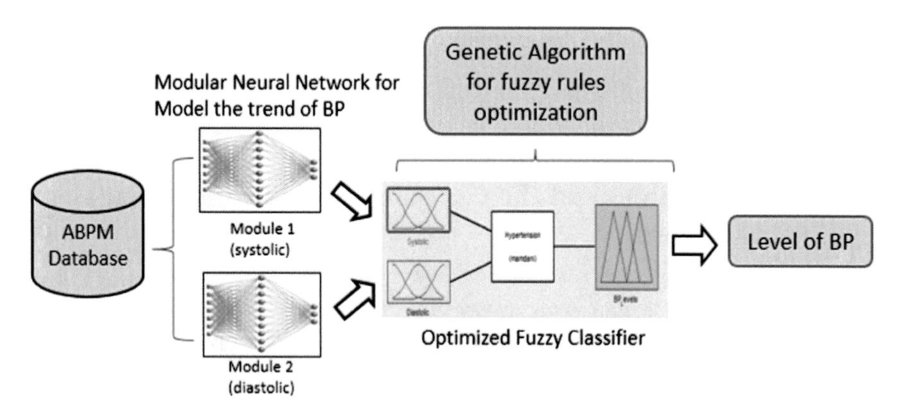

Fig. 2 Specific neuro fuzzy hybrid model

modular neural network, which will process the information and finally give the trend of monitoring, once the trend is obtained, this will be sent to the fuzzy system which is optimized by genetic algorithms module, which optimizes the rules and membership functions and this helps to have a better classification when giving the diagnosis of which level of blood pressure corresponds to the input data [1, 2, 7, 8].

In the specific neuro fuzzy hybrid model illustrated in Fig. 2 we can observe in a more specific way how the model is divided, which starts with the database module

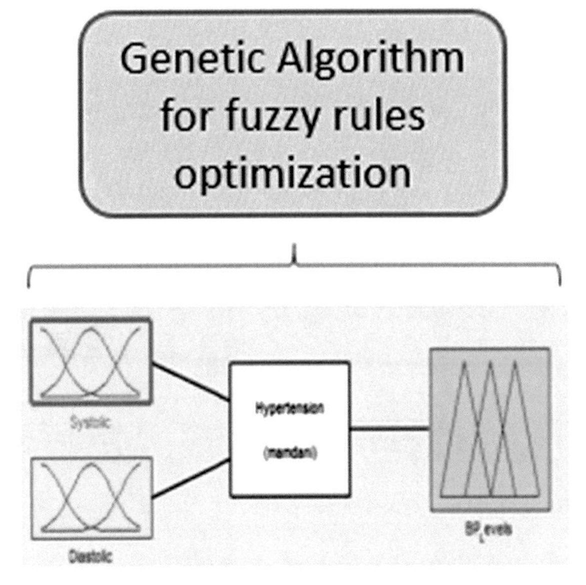

Fig. 3 Fuzzy classifier optimized by genetic algorithm

of the 24 h monitoring, after which we have the modular neural network, which Consists of two modules, module 1 is the systolic pressure and module 2 diastolic pressure, then the trend of this database enters to the optimized fuzzy system and finally gives a diagnosis of blood pressure level.

The strong part of this work is done in the fuzzy classifier, which is optimized by genetic algorithms, this is shown in Fig. 3:

3.1 Design of the Optimized Classifier Using GA

After some experiments we opted to optimize the fuzzy classifier, we made the decision to use genetic algorithms, for this we must have a structure of the chromosome. The chromosome must have genes, in this case it consists of 122 genes, which determine the points of membership functions and the number of rules.

First genes 1–72 (real numbers) allow moving the points of the membership functions of the inputs and output, genes 73–121 are the total of rules, and finally we have the 122 gene, it will help to reduce the Number of rules activating and deactivating them. Then we can see Fig. 4 which shows the structure of the chromosome.

The optimized classifier as shown in Fig. 5 consists of two inputs and one output, seven membership functions for each of the inputs as shown in Figs. 6 and

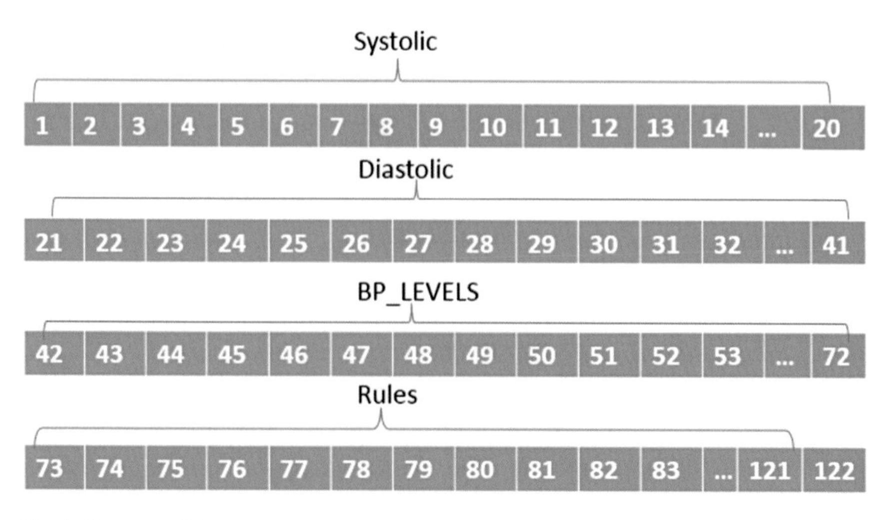

Fig. 4 Structure of the chromosome

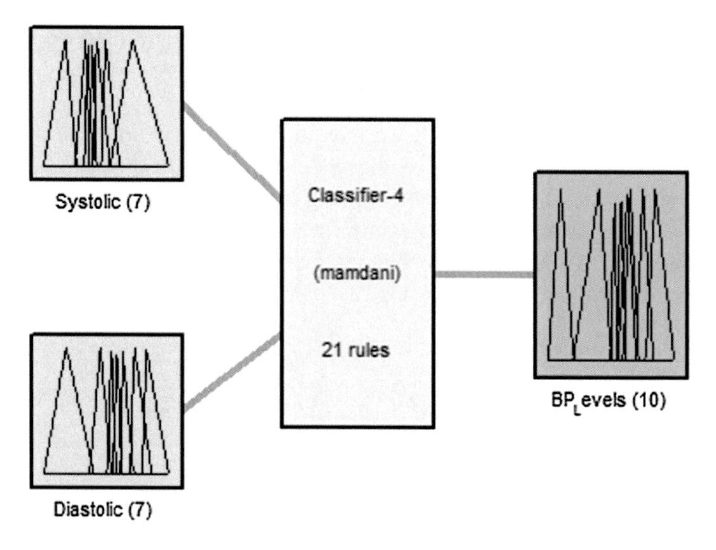

Fig. 5 Structure of the optimized classifier

7 respectively and 10 membership functions for the output as shown in Fig. 8, consists of 21 rules as illustrate in Fig. 9 and is Mandani type.

This fuzzy classifier is based on the table of European Union blood pressure levels and by an expert, this analysis results in the following blood pressure levels: hypertension, optima, normal, high normal, grade 1, 2 and 3 of hypertension and finally isolated systolic hypertension divided into grade 1, 2 and 3.

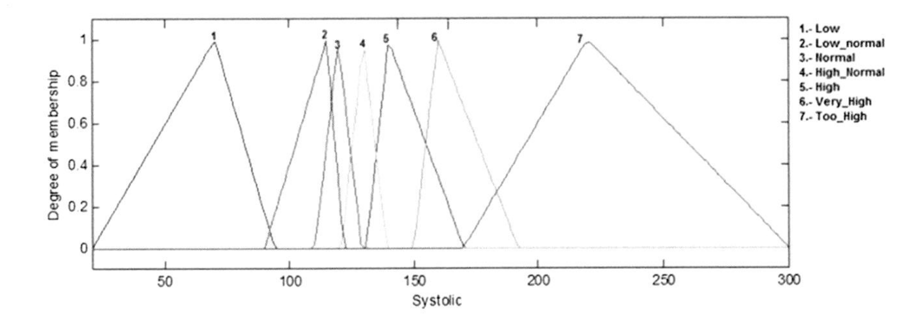

Fig. 6 Systolic input for the optimized classifier

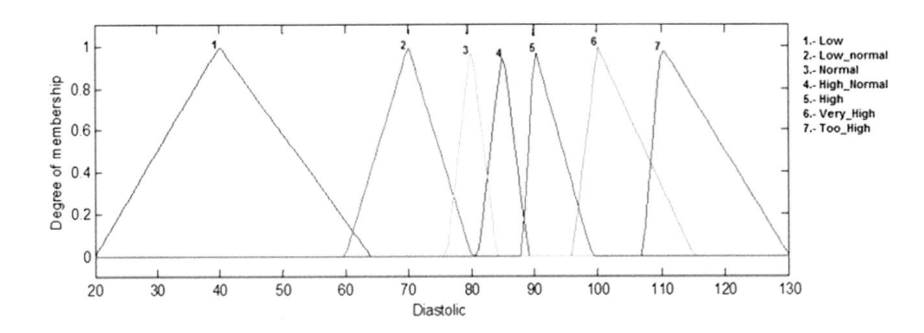

Fig. 7 Diastolic input for the optimized classifier

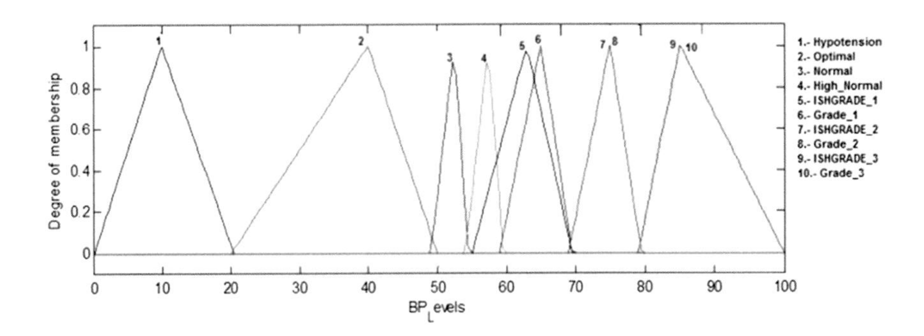

Fig. 8 BP_Levelsis the output of the optimized classifier

1. If (Systolic is Low) and (Diastolic is Low) then (BP_Levels is Hypotension)
2. If (Systolic is Low_Normal) and (Diastolic is Low_Normal) then (BP_Levels is Optimal)
3. If (Systolic is Normal) and (Diastolic is Normal) then (BP_Levels is Normal)
4. If (Systolic is High_Normal) and (Diastolic is High_Normal) then (BP_Levels is High_Normal)
5. If (Systolic is High) and (Diastolic is High) then (BP_Levels is Grade_1)
6. If (Systolic is Very_high) and (Diastolic is Very_High) then (BP_Levels is Grade_2)
7. If (Systolic is too_high) and (Diastolic is Too_High) then (BP_Levels is Grade_3)
8. If (Systolic is Very_high) and (Diastolic is High) then (BP_Levels is Grade_2)
9. If (Systolic is too_high) and (Diastolic is Very_High) then (BP_Levels is Grade_3)
10. If (Systolic is too_high) and (Diastolic is High) then (BP_Levels is Grade_3)
11. If (Systolic is High) and (Diastolic is Very_High) then (BP_Levels is Grade_2)
12. If (Systolic is High) and (Diastolic is Too_High) then (BP_Levels is Grade_3)
13. If (Systolic is Very_high) and (Diastolic is Too_High) then (BP_Levels is Grade_3)
14. If (Systolic is High) and (Diastolic is Normal) then (BP_Levels is ISHGRADE_1)
15. If (Systolic is High) and (Diastolic is High_Normal) then (BP_Levels is ISHGRADE_1)
16. If (Systolic is Very_high) and (Diastolic is Normal) then (BP_Levels is ISHGRADE_2)
17. If (Systolic is Very_high) and (Diastolic is High_Normal) then (BP_Levels is ISHGRADE_2)
18. If (Systolic is too_high) and (Diastolic is Normal) then (BP_Levels is ISHGRADE_3)
19. If (Systolic is too_high) and (Diastolic is High_Normal) then (BP_Levels is ISHGRADE_3)
20. If (Systolic is Normal) or (Diastolic is Normal) then (BP_Levels is Normal)
21. If (Systolic is High_Normal) or (Diastolic is High_Normal) then (BP_Levels is High_Normal)

Fig. 9 Fuzzy rules for the optimized classifier

4 Simulation Results of the Optimized Classifier

The results were performed using the 24-h monitoring database of 30 patients and through this information we obtain the classification accuracy rate: 100% and classification error rate: 0% as show in the Table 2. The classification accuracy rate and error rate was obtained by the next equations:

Table 2 Shows the results of the 30 patients who were monitored and classified in the optimized classifier

Patient	Systolic	Diastolic	Optimized classifier	Fuzzy percentage	ESH BP_Leves table
1	139	84	High normal	61.3	High normal
2	135	90	Grade 1	62.5	Grade 1
3	160	98	Grade 2	74	Grade 2
4	177	110	Grade 3	84.3	Grade 3
5	142	85	ish_grade 1	61.3	Ish grade 1
6	160	89	ish_grade 2	71.8	Ish_grade 2
7	182	89	ish_grade 3	83.2	Ish_grade 3
8	85	50	Hypotension	10.2	Hypotension
9	110	70	Optima	36.6	Optima
10	125	82	Normal	55.2	Normal
11	135	85	High normal	60.8	High normal

(continued)

Table 2 (continued)

Patient	Systolic	Diastolic	Optimized classifier	Fuzzy percentage	ESH BP_Leves table
12	159	94	Grade 1	71.8	Grade 1
13	175	105	Grade 2	79.3	Grade 2
14	180	110	Grade 3	84	Grade 3
15	110	80	Normal	52	Normal
16	128	89	High normal	56.9	High normal
17	158	77	ish_grade1	66.4	Ish grade 1
18	150	108	Grade 2	82.9	Grade 2
19	199	95	Grade 3	87.8	Grade 3
20	179	99	Grade 2	81.6	Grade 2
21	181	100	Grade 3	82.6	Grade 3
22	210	90	Grade 3	87.4	Grade 3
23	140	100	Grade 2	75.8	Grade 2
24	159	120	Grade 3	87.7	Grade 3
25	178	115	Grade 3	87.8	Grade 3
26	140	80	Ish grade 1	59.7	Ish grade 1
27	150	89	Ish grade 1	65.2	Ish grade 1
28	179	80	Ish grade 2	73.8	Ish grade 2
29	179	89	Ish grade 2	81.6	Ish grade 2
30	199	82	Ish grade 3	77.8	Ish grade 3

$$\text{Classification Accuracy Rate} = \frac{\text{Number of Training Instances Correctly Classified}}{\text{Number of Training instances}}$$

$$\text{Classification Error Rate} = \frac{\text{Number of Training Instances Incorrectly Classified}}{\text{Number of Training Instances}}$$

5 Conclusions

The use of fuzzy logic, neural networks and genetic algorithms, is nowadays very useful for performing diagnostic systems in medicine, in this case, fuzzy classifier is focused on diagnosing of blood pressure levels, in a fast way and more accurately, using a 24-h monitoring database of a patient gives us information which can be processed by neural networks and help to obtain an adequate trend that at the time of being classified is as reliable as possible, It is for this reason that great results have been obtained so far with observed patients. We are collecting 24-h monitoring day by day to have more experiments and thus to achieve greater reliability. Future work could include optimization with hybrid approaches, like in [15].

Acknowledgements We would like to express our gratitude to the CONACYT and Tijuana Institute of Technology for the facilities and resources granted for the development of this research.

References

1. A.M. Abdelbar, S. Abdelshahid, D.C. Wunsch, Fuzzy PSO: A generalization of particle swarm optimization, in *Proceedings of International Joint Conference on Neural Networks*, vol. 2, pp. 1086–1091 (2005)
2. A.A. Abdullah, Z. Zakaria, N.F. Mohammad, Design and development of fuzzy expert system for diagnosis of hypertension, in *Proceedings of 2011 International Conference on Intelligent Systems, Modelling and Simulation, ISMS 2011*, pp. 113–117 (2011)
3. Z. Abrishami, I. Azad, Design of a fuzzy expert system and a multi-layer neural network system for diagnosis of hypertension, vol. 4, pp. 138–145, October 2015
4. F. Başçiftçi, A. Eldem, Using reduced rule base with expert system for the diagnosis of disease in hypertension. Med. Biol. Eng. Comput. **51**(12), 1287–1293 (2013)
5. S. Das, P.K. Ghosh, Hypertension diagnosis : a comparative study using fuzzy expert system and neuro fuzzy system, in *Proceeding of IEEE International Conference on Fuzzy Systems (FUZZ-IEEE), Hyderabad*, July 7–10, 2013, pp. 1–7
6. X.Y. Djam, Y.H. Kimbi, Fuzzy expert system for the management of hypertension. Pac. J. Sci. Technol. **12**(1), pp. 390–402 (2011)
7. A. Kaur, A. Bhardwaj, Genetic neuro fuzzy system for hypertension. Int. J. Comput. Sci. Inf. Technol. **5**(4), 4986–4989 (2014)
8. R. Kaur, A. Kaur, Hypertension diagnosis using fuzzy expert system, in *International Journal Engineering Research and Applications*, pp. 14–18 (2014)
9. G. Mancia et al., 2013 ESH/ESC guidelines for the management of arterial hypertension. Blood Press. **22**(4), 193–278 (2013)
10. P. Melin, J.C. Guzman, G. Prado-Arechiga, [PP.08.10] Artificial intelligence utilizing neuro-fuzzy hybrid model for the classification of blood pressure. J. Hypertens. **34** (2016)
11. P. Melin, G. Prado-Arechiga, J.C. Guzman, PS 05-07 Classification of blood pressure based on a neuro-fuzzy hybrid computational model. J. Hypertens. **34** (2016)
12. P. Melin, G. Prado-Arechiga, M. Pulido, I. Miramontes, OS 26-01 Classification of arterial hypertension using a computational model based on artificial modular neural networks. J. Hypertens. **34** (2016)
13. P. Srivastava, A. Srivastava, A. Burande, A. Khandelwal, A note on hypertension classification scheme and soft computing decision making system. ISRN Biomath. **2013**, 11 (2013)
14. B.B. Sumathi, Pre-diagnosis of hypertension using artificial neural network. **11**(2) (2011)
15. F. Valdez, P. Melin, O. Castillo, Evolutionary method combining particle swarm optimization and genetic algorithms using fuzzy logic for decision making, in *IEEE International Conference on Fuzzy Systems*, pp 2114–2119 (2009)

A New Model Based on a Fuzzy System for Arterial Hypertension Classification

Martha Pulido, Patricia Melin and German Prado-Arechiga

Abstract In this paper, a method is proposed for classification of the blood pressure of patient (systolic pressure and diastolic pressure). This technique consists on a creating fuzzy system for the classification of the arterial hypertension. The fundamental idea of this paper on achieving Classification of the arterial hypertension of a patient so that the doctor can provide a more accurate Diagnosis, Prevent and control of risk factors that may effect of the patient.

Keywords Systolic · Diastolic · Classification · Fuzzy system
Patient · Arterial hypertension

1 Introduction

The prevalence of hypertension in a population or, what is the same, the proportion of individuals that are hypertensive at a given time, varies widely depending on certain factors, which basically include the population age group considered, the methodology Used in the measurement of blood pressure, the number of readings performed and, above all, the limit chosen to differentiate normal tension from hypertension.

This last factor is the most determinant when estimating the prevalence. In fact, choosing the classic limit of 160/95 mmHg or the current limit of 140/90 mmHg definitively determines the prevalence of hypertension in the population. Internationally, using the limits of 160/95 mmHg, the prevalence recorded range from 10 to 20% of the adult population (18 years and over). If the limits used are 140/90 mmHg, the prevalence may increase up to 30%.

In Spain, there is a wealth of information from the pioneering studies of the 1980s. Almost all of those studies, and those published subsequently, show prevalence rates of around 20% using the 160/95 criterion MmHg, which rises to about 35% in some cases when using the 140/90 mmHg. In the only national

M. Pulido · P. Melin (✉) · G. Prado-Arechiga
Tijuana Institute of Technology, Tijuana, Mexico
e-mail: pmelin@tectijuana.mx

O. Castillo et al. (eds.), *Fuzzy Logic Augmentation of Neural and Optimization Algorithms: Theoretical Aspects and Real Applications*, Studies in Computational Intelligence 749, https://doi.org/10.1007/978-3-319-71008-2_24

cross-sectional study conducted in 1990, which included about 2000 randomly selected individuals, the prevalence rate was 45%. This so unusually high rate, even applying the criterion of 140/90 mmHg and has its explanation in the fringe. [1].

The adult population considered in the study, which was from 35 to 64 years, while the studies mentioned above covered ages from 18 years, including or not, depending on the case, individuals over 65 years. Thus, as a whole, we can estimate that the prevalence in our country oscillates around 20% in the adult population (18–65 years) if the criterion of 160/95 mmHg is used and that it rises to more than 30% when using the 140/90 mmHg [2].

In the area of intelligent computing there have been methods applied to medicine, such as neural networks, and Fuzzy Systems [3] and that have been applied in a wide variety of areas, such as automatic control, digital signal processing, communications, expert systems, medicine, etc. However, the most significant applications of fuzzy systems have focused specifically on the area of automatic control. Essentially a **fuzzy** system, is a knowledge-based structure defined through a set of fuzzy "if-then" rules, which contain a Fuzzy logical quantification of the expert's linguistic description of how to perform adequate control [4–13].

In this study, we present fuzzy logic for the classification of arterial hypertension, since it is important to identify patients with high normal blood pressure or who have an undiagnosed disease, and to take into account the risk approach and the vision with which we face the different diseases, in this case, to design strategies to predict, protect the individual, the family and community.

2 Arterial Hypertension

Hypertension Arterial (AH) is a syndrome characterized by elevated Blood Pressure (BP) and its consequences. In only 5% of the cases is a cause (secondary AH); In the rest, an etiology cannot be demonstrated (primary AH); But it is believed, each day more, that there are several processes still not identified, and with genetic, those that result in elevation of BP. Hypertension is a very important risk factor for the future development of vascular disease (cerebrovascular disease, coronary heart disease, heart or kidney failure) [14–16].

The relationship between BP numbers and cardiovascular risk is continuous (at a higher level, higher morbidity and mortality), and there is no dividing line between normal and pathological blood pressure. The definition of arterial hypertension is arbitrary. The threshold chosen is that from which the benefits obtained with the intervention surpass those of the non-performance. Over the years, cut-off values have been decreasing as more data have been obtained regarding the prognostic value of hypertension and the beneficial effects of its treatment.

In most patients with high blood pressure, no cause can be identified, this is called primary hypertension. It is estimated that approximately 95% of patients with hypertension have primary hypertension. The alternative term, essential hypertension, is less appropriate from a linguistic point of view, since the essential word usually denotes something that is beneficial to the individual. The term "benign hypertension" should also be avoided because hypertension always carries the risk of premature cardiovascular disease [17].

The term secondary hypertension is used when hypertension is produced by an underlying, detectable mechanism. There are numerous physiological conditions such as renal artery stenosis, pheochromocytoma and aortic coarctation, which can produce arterial hypertension. In some of these cases, elevated blood pressure is reversible when the underlying disease is treated successfully. It is important to identify the small number of patients with a secondary form of hypertension, because in some cases there is a clear possibility of hypertensive disease [18–22].

3 Methodology

3.1 Data Collection

We collected patient data with the Ambulatory blood pressure monitoring (ABPM) in accordance with the FDA (Food and Drug Administration) regulation and the BHS (British Hypertension Society), which is a tool that can help diagnose hypertension. The main goal is to model the 24-h ABPM patterns in patients with

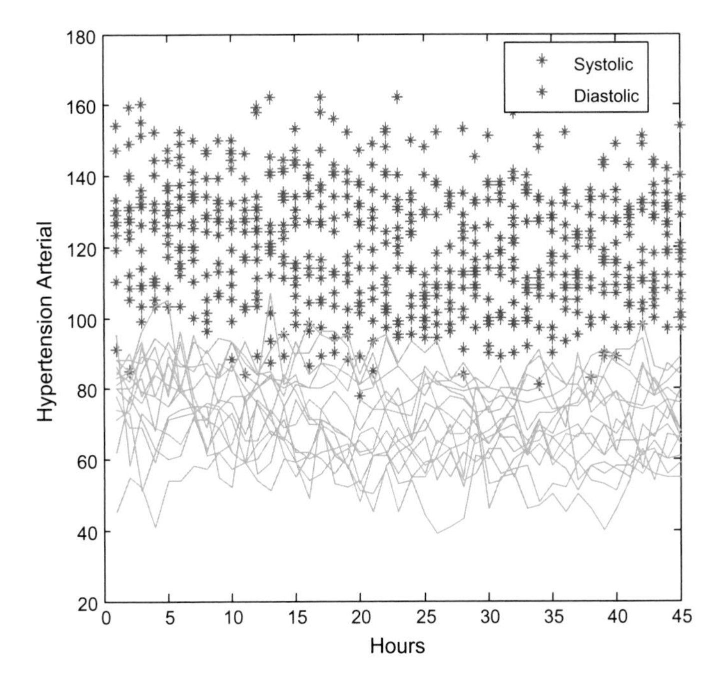

Fig. 1 Data collection of 40 patients

the FLs and classify the BP of the patient using, The fuzzy classification using data testing with 45 records of 40 patients in the database, in others words, the input with the records of systolic pressure and diastolic pressure, as shown in Fig. 1.

3.2 *The Proposed Model*

Figure 2 illustrates the Model of the Fuzzy Classifier, where we have two inputs the first is the Systolic Pressure and the second input second is the Diastolic Pressure. To perform the test we used 45 records of 40 patients, for each of the fuzzy classifier and output of the classification of the Arterial Hypertension.

Figure 3 shows the first input of the fuzzy system which is the Systolic Pressure granulated in three Gaussian membership functions, (optimal, Normal High), is of Mamdani type, the logic of the fuzzy classifier is of the type-1 and the range (0–300).

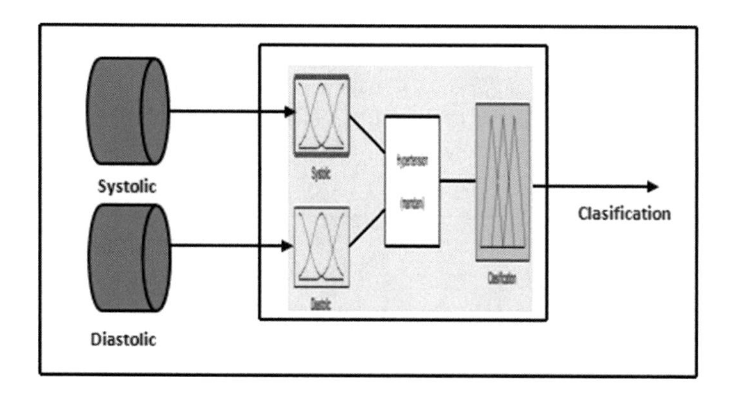

Fig. 2 Model of the Fuzzy system classification

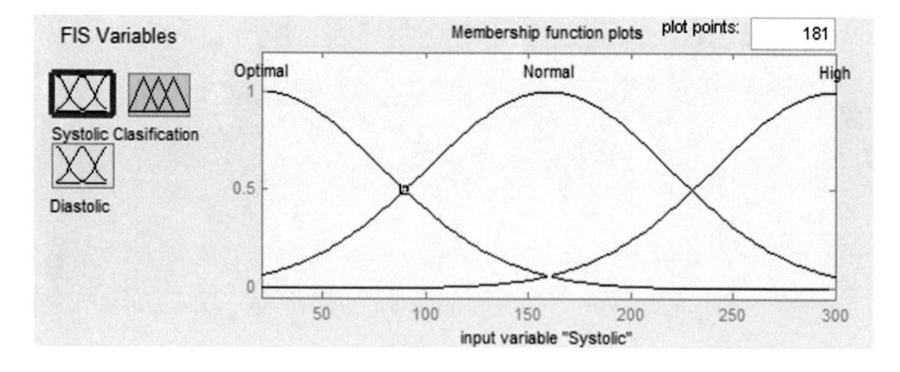

Fig. 3 The first input is the Systolic Pressure

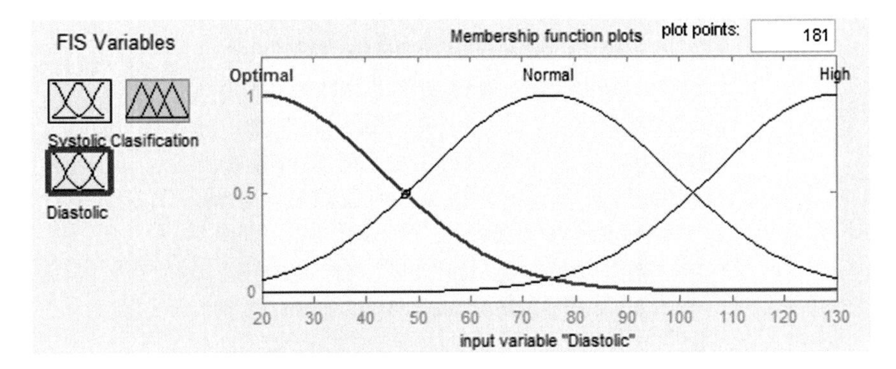

Fig. 4 The second input Diastolic

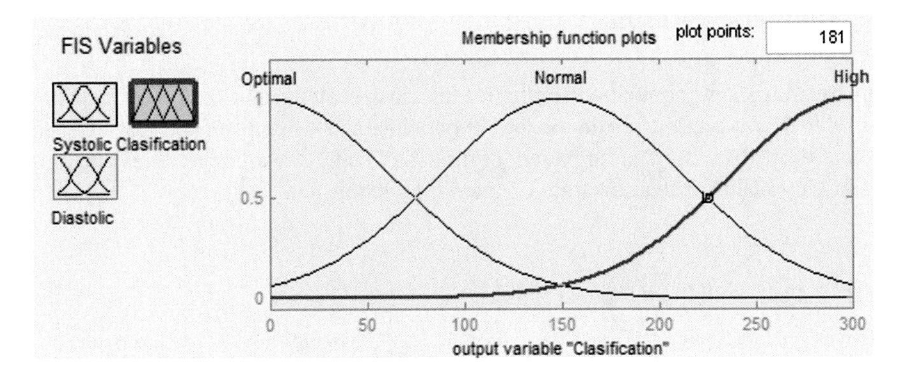

Fig. 5 The "classification" output variable

Figure 4 illustrates the second input which is the Diastolic Pressure granulated in three Gaussian membership functions (optimal, Normal High), The fuzzy system is of Mamdani type, the logic of the fuzzy classifier is type-1 and range (0–130).

Figure 5 shows the classification output classification granulated into three Gaussian membership functions, (optimal, Normal High). The fuzzy system is of Mamdani type, the logic of the fuzzy classifier is type-1 and the range (0–300).

Figure 6 represents the 9 possible rules of the fuzzy classifier we have two inputs in the fuzzy system with 3 membership functions, and the output with 2 membership functions.

1. If (Systolic is Low) and (Diastolic is Low) then (Clasification is Low) (1)
2. If (Systolic is Low) and (Diastolic is Normal) then (Clasification is Low) (1)
3. If (Systolic is Low) and (Diastolic is High) then (Clasification is Normal) (1)
4. If (Systolic is Normal) and (Diastolic is Low) then (Clasification is Normal) (1)
5. If (Systolic is Normal) and (Diastolic is Normal) then (Clasification is Normal) (1)
6. If (Systolic is Normal) and (Diastolic is High) then (Clasification is High) (1)
7. If (Systolic is High) and (Diastolic is Low) then (Clasification is High) (1)
8. If (Systolic is High) and (Diastolic is Normal) then (Clasification is High) (1)
9. If (Systolic is High) and (Diastolic is High) then (Clasification is High) (1)

Fig. 6 Possible Fuzzy rules

4 Simulation Results

This section shows simulation results for the classification of arterial hypertension, In Table 1, we present results of the 40 patients, and to perform the test of each patient we use 45 samples for patient of the systolic and diastolic pressures and as a result we obtain the classification of arterial hypertension.

Table 1 Classification of the arterial hypertension

Persons	Systolic	Diastolic	Systolic classification	Diastolic classification	Classification
Person 1	117	76	117	76	Optimal
Person 2	118	77	115	75	Optimal
Person 3	107	74	107	74	Optimal
Person 4	122	75	122	75	Normal
Person 5	114	66	114	61	Optimal
Person 6	141	81	142	81	High
Person 7	106	62	112	64	Optimal
Person 8	120	81	120	81	Normal
Person 9	107	61	112	58	Optimal
Person 10	130	74	129	73	Normal
Person 11	116	73	116	73	Optimal
Person 12	134	62	130	77	Normal
Person 13	135	82	136	81	Normal
Person 14	121	77	120	72	Optimal
Person 15	109	63	108	63	Optimal
Person 16	123	71	124	70	Normal

(continued)

Table 1 (continued)

Persons	Systolic	Diastolic	Systolic classification	Diastolic classification	Classification
Person 17	125	77	126	76	Normal
Person 18	106	65	106	65	Optimal
Person 19	110	68	112	68	Optimal
Person 20	123	76	123	76	Normal
Person 21	115	72	114	72	Optimal
Person 22	112	71	112	71	Optimal
Person 23	122	76	122	76	Normal
Person 24	117	68	116	68	Optimal
Person 25	121	74	121	74	Optimal
Person 26	129	82	130	82	Normal
Person 27	121	63	121	63	Optimal
Person 28	112	72	112	72	Optimal
Person 29	123	82	122	82	Normal
Person 30	95	61	95	61	Optimal
Person 31	106	65	109	77	Optimal
Person 32	110	69	115	75	Optimal
Person 33	116	67	127	69	Normal
Person 34	130	86	125	80	Normal
Person 35	117	73	119	78	Optimal
Person 36	117	54	117	61	Optimal
Person 37	113	72	108	70	Optimal
Person 38	132	86	123	80	Normal
Person 39	128	81	127	81	Normal
Person 40	131	85	133	84	Normal

4.1 Statistical T Student Test for the Classification

In Table 2 the results with the t student test for the Systolic and Diastolic Pressure are presented. The number of samples used for the Real Data is also of 40 and for the Classification method are 40 samples with a confidence interval of 99%, and assume, and that the test is good since the result of the classification is similar to the real data.

Table 2 Statistics Tests

Classification	Samples data	Samples classification	T value	P value
Systolic	40	40	−0.10	0.925
Diastolic	40	40	−0.23	0.820

5 Conclusions

This paper used a fuzzy classifier of Mamdani type and tests were performed with Gaussian membership functions to classify blood pressure levels (systolic and diastolic) of 40 patients. It is very important to apply a method to classify a patient's arterial hypertension as it helps cardiologists diagnose and prevent diseases. As future work, we could consider optimization with other methods, like in [23, 24], or different hybrid approaches, such as in [25–28], and also type-2 fuzzy logic, like in [29–31].

Acknowledgements We would like to express our gratitude to the CONACYT, Tijuana Institute of Technology for the facilities and resources granted for the development of this research.

References

1. U. Keil, K. Kuulasmaa, The project: geographical variation in the major risk factors of coronary heart disease in men and women aged 35–64 years. Rapp Trimest Statis Sanit Mond 41, 115–139 (1988)
2. S. Mendis, L.H. Lindholm, G. Mancia, J. Whiwort, M. Alderman et al., World Health Organization (WHO) and International Society of Hypertension (ISH) risk prediction charts: assessment of cardiovascular risk for prevention and control of cardiovascular disease in low an middle income countries. J Hypertens. **25**(8), 1578–1582 (2007)
3. L.A. Zadeh, Knowledge representation in Fuzzy Logic. IEEE Trans. Knowl. Data Eng. **1**, 89 (1989)
4. R.L. Sacco, The 2006 William Feinberg lecture: shifting the paradigm from stroke to global vascular risk estimation. Stroke **38**, 1980–1987
5. G. Beevers, G.Y.H. Lip, E. O'Brien, Blood pressure measurement part I sphygmomanometry: factors common to all techniques. Br. Med. J. **322**, 981 (2001)
6. P.M. Kearney, M. Whelton, K. Reynaldos, P.K. Whelton, Wordwide prevalence of hypertension: a systematic rewiew. J. Hypertens. **22**(1), 21–24 (2004)
7. R.B. Agustino, M.W. Russel, D.M. Huse, C. Ellison, H. Silbershatz et al, Primary and subsequent coronary risk appraisal: new results the Framingham study. Am. Heart J. **139**, 272–281 (2000)
8. R. Samant, R. Srikantha, Evaluation of artificial neural networks in prediction of essential hypertension. Int. J. Comput. Appl. **14**, 11–21 (2013)
9. M. Pulido, P. Melin, G. Prado-Arechiga, A new method based on modular neural network for arterial hypertension diagnosis, in *Nature-Inspired Design of Hybrid Intelligent Systems*, Springer, Ed., Springer, Germany, 2017) pp. 195–205
10. P. Melin, G. Prado-Arechiga, M. Pulido, I. Miramontes, Classification using a computational model based on artificial modular neural networks. J. Hypertens. (2017)
11. P. Melin, G. Prado-Arechiga, M. Pulido, I. Miramontes, Classification using a computational model base on artificial modular neural networks. J. Hypertens. (2016)
12. P. Melin, M. Pulido, I. Miramontes, G. Prado-Arechiga, A new method based on artificial modular neural networks for classification of arterial. J. Hypertens. (2016)
13. J.C. Guzman, P. Melin, G. Prado-Arechiga, Neuro-fuzzy hybrid model for the diagnosis of blood pressure, in *Nature-Inspired Design of Hybrid Intelligent Systems*, Springer, Ed. (Springer, Germany, 2017), pp. 573–582

14. N. Shehu, S.U. Gulumbe, H.M. Liman, Comparative study between conventional statistical methods and neural networks in predicting hypertension status. Advances in Agriculture, Sciences and Engineering Research (2013)
15. D.L. Simel, Approach to the patient: history and physical examination, in *Goldman's Cecil Medicine* L. Goldman, A.I. Schafer, eds.
16. X.Y. Djam, Y.H. Kimbi, Fuzzy expert system for the management of hypertension. Pac. J. Sci. Technol. **11**, 1 (2011)
17. *Harrison Principles of Internal Medicine, Tachyarrhythmias* 6th edn., Chapter 214. (McGraw-Hill, 2016)
18. A.A. Abdullah, Z. Zakaria, N.F. Mohammad, Design and development of Fuzzy Expert System for diagnosis of hypertension, in *International Conference on Intelligent Systems, Modelling and Simulation*, vol. 56, no. 5–6 (Univeristy Malaysia Perlis, Jejawi, Malaysia, IEEE, 2011), pp. 26–30
19. A.A. Abdullah, Z. Zakaria, N.F. Mohammad, Design and development of Fuzzy Expert System for diagnosis of hypertension, in *International Conference on Intelligent Systems, Modelling and Simulation*, vol. 10 (IEEE, 2011), pp. 131–141
20. Accord Study Group, Effects of intensive blood-pressure control in type 2 diabetes. N. Engl. J. Med. **2010**(362), 1575–1585 (2010)
21. E.J. Battegay, G.Y. Lip, G.L. Bakris, *Hypertension principles and practice* (Taylor & Francis, Boca Raton, FL, 2005)
22. T. Pickering, Shimbo P. Daichi, D. Haas, Ambulatory blood-pressure monitoring. N. Engl. J. Med. **354**(22), 2368–2374 (2006)
23. A. Sombra, F. Valdez, P. Melin, O. Castillo, A new gravitational search algorithm using fuzzy logic to parameter adaptation, in *IEEE Congress on Evolutionary Computation* (Cancun, México, 2013), pp. 1068–1074
24. F. Valdez, P. Melin, O. Castillo, Evolutionary method combining particle swarm optimization and genetic algorithms using fuzzy logic for decision making, in *IEEE International Conference on Fuzzy Systems* (2009), pp. 2114–2119
25. O. Castillo, P. Melin, E. Ramírez, J. Soria, Hybrid intelligent system for cardiac arrhythmia classification with Fuzzy K-Nearest Neighbors and neural networks combined with a fuzzy system. Expert Syst. Appl. **39**(3), 2947–2955
26. L. Aguilar, P. Melin, O. Castillo, Intelligent control of a stepping motor drive using a hybrid neuro-fuzzy ANFIS approach. Appl. Soft Comput. **3**(3), 209–219
27. P. Melin, O. Castillo, in *Modelling, Simulation and Control of Non-Linear Dynamical Systems: An Intelligent Approach Using Soft Computing And Fractal Theory* (CRC Press, 2001)
28. P. Melin, O. Castillo, Intelligent control of complex electrochemical systems with a neuro-fuzzy-genetic approach. IEEE Trans. Ind. Electron. **48**(5), 951–955
29. G.M. Mendez, O. Castillo, Interval type-2 TSK fuzzy logic systems using hybrid learning algorithm, in *The 14th IEEE International Conference on Fuzzy Systems, 2005. FUZZ'05*, pp. 230–235
30. O. Castillo, P. Melin, Design of intelligent systems with interval type-2 fuzzy logic, in *Type-2 Fuzzy Logic: Theory and Applications*, pp. 53–76
31. P. Melin, C.I. Gonzalez, J.R. Castro, O. Mendoza, O. Castillo, Edge-detection method for image processing based on generalized type-2 fuzzy logic. IEEE Trans. Fuzzy Syst. **22**(6), 1515–1525

A Takagi–Sugeno-Kang Fuzzy Model Formalization of Eelgrass Leaf Biomass Allometry with Application to the Estimation of Average Biomass of Leaves in Shoots: Comparing the Reproducibility Strength of the Present Fuzzy and Related Crisp Proxies

Hector Echavarria-Heras, Cecilia Leal-Ramirez, Juan Ramón Castro-Rodríguez, Enrique Villa Diharce and Oscar Castillo

Abstract The identification of the functional relationship that regulates the variation of individual leaf biomass in terms of related area in eelgrass, allows the derivation of convenient proxies for a nondestructive estimation of the average biomass of the leaves in shoots. The concourse of these assessment methods is fundamental for assessing the performance of restoration efforts for this species that are based on transplanting techniques. Prior developments proposed proxies for a nondestructive estimation of aforementioned average biomass of leaves in shoots derived from allometric models for the dependence of leaf biomass in terms of linked area. The reproducibility power of these methods is highly dependent on analysis method and data quality. Indeed, previous results show that allometric proxies for average biomass of leaves in shoots produced by parameter estimates fitted from quality controlled data via nonlinear regression yield the highest

H. Echavarria-Heras (✉) · C. Leal-Ramirez
Centro de Investigación Científica y de Educación Superior de Ensenada, Carretera Ensenada-Tijuana No 3918, Zona Playitas, Código Postal 22860 Ensenada, BC, Mexico
e-mail: hetxavar@cicese.mx

J. R. Castro-Rodríguez
Facultad de Ciencias Químicas e Ingeniería, UABC, Calzada Universidad 14418, Parque Industrial Internacional, 22390 Tijuana, Baja California, Mexico

E. V. Diharce
Centro de Investigación en Matemáticas, A.C. Jalisco s/n, Mineral Valenciana, Código Postal 36240 Guanajuato, GTO, Mexico

O. Castillo
Instituto Tecnológico de Tijuana, Calz del Tecnológico s/n, Tomas Aquino, 22414 Tijuana, Baja California, Mexico
e-mail: ocastillo@tectijuana.mx

O. Castillo et al. (eds.), *Fuzzy Logic Augmentation of Neural and Optimization Algorithms: Theoretical Aspects and Real Applications*, Studies in Computational Intelligence 749, https://doi.org/10.1007/978-3-319-71008-2_25

reproducibility strength. Nevertheless, the use of data processing entails subtleties mainly related to the subjectivity of the criteria for the rejection of inconsistent replicates in raw data. Here we introduce efficient- data quality control- free surrogates derived from a first order Takagi-Sugeno-Kang fuzzy model aimed to approximate the mean response of eelgrass leaf biomass depending on associated area. A comparison of the performances of the allometric and the fuzzy model constructs identified using available raw data shows that the Takagi-Sugeno-Kang paradigm for individual leaf biomass in terms of related area produced the most precise proxies for observed average biomass of leaves in shoots. The present results show how gains derived from the outstanding approximation capabilities of the first order Takagi-Sugeno-Kang fuzzy model for the nonlinear dynamics can be extended to the realm of eelgrass allometry.

Keywords Eelgrass conservation · Nondestructive assessments
Allometric models · Takagi-Sugeno-Kang fuzzy model

1 Introduction

Globally, physical disturbances induced by anthropogenic influences are behind the rapid rates of disappearance of seagrass meadows [1]. The threatened species *Zostera marina*, commonly known as eelgrass, distributes in estuarine and near-shore environments worldwide, forming highly productive meadows that deliver a wide range of ecological services thereby substantially contributing to human welfare [2]. The ecological relevance of eelgrass has prompted conservation efforts that often rely on transplanting as an effective remediation procedure [3, 4]. Nevertheless, as it is generally observed in transplanted seagrass plots a substantial extent of effort is still required to advance cost-effective means for hindering grazing, as well as, physical disruption as a whole [5]. Another concern related to the practice of seagrass transplanting pertains to the suitability of monitoring methods aimed to assess the reinstatement of the functions and values observed in natural populations. These assessments require the measurement of standing stock and productivity over the whole year cycle. But, conventional techniques for the measurement of these variables rely on destructive sampling, and for transplanted plots these procedures can have a significant effect in shoot density. Therefore, an appraisal of the effectiveness of eelgrass transplanting methods should be preferably done by means of nondestructive procedures.

In this work, we focus on approaches aimed to a non-destructive estimation of the average biomass in eelgrass shoots observed at a time t. This variable, that is considered as an important component of the related standing stock, will be denoted here by means of the symbol $w_m(t)$. In order to describe the paradigms of indirect assessment for $w_m(t)$ that we account for here requires to start by providing a mathematical representation of this aggregate. For that aim, let $w(t)$ denote the biomass of an individual eelgrass leaf at time t, and a subscript s to label a generic eelgrass shoot assumed to hold a number $nl(s)$ of leaves. Then, if the observed leaf biomass in a shoot s is represented by means of $w_s(t)$, we will have,

$$w_s(t) = \sum_{nl(s)} w(t), \tag{1}$$

where summation of the leaves that the shoot s holds is indicated by means of $\sum_{nl(s)}$. Therefore, the average for the leaf biomass of a number $ns(t)$ of shoots collected at a time t is calculated through

$$w_m(t) = \frac{\sum_{ns(t)} w_s(t)}{ns(t)}, \tag{2}$$

with $\sum_{ns(t)}$ standing for summation over the number $ns(t)$ of collected shoots.

So far, methods for an indirect estimation of the average biomass in eelgrass shoots have hinged on proxies derived from an allometric model for the biomass of an individual leaf in terms of the related area [6]. Namely

$$w(t) = \beta a(t)^{\alpha} \tag{3}$$

with $a(t)$ standing for individual eelgrass leaf area measured at time t, and being α and β parameters. Using Eq. (3) we can readily derive an allometric substitute for the observed value of the average leaf biomass in shoots $w_m(t)$ denoted through $w_m(\alpha, \beta, t)$. Indeed, this proxy becomes

$$w_m(\alpha, \beta, t) = \frac{\sum_{ns(t)} w_s(\alpha, \beta, t)}{ns(t)}, \tag{4}$$

where

$$w_s(\alpha, \beta, t) = \sum_{nl(s)} \beta a(t)^{\alpha} \tag{5}$$

stands for the corresponding allometric surrogate for $w_s(t)$, the biomass of the leaves in shoot s, given by Eq. (1).

But, instead of staying with the allometric model of Eq. (3), it may be alternately assumed that there exists an unknown generalized nonlinear functional relationship describing the variability of leaf biomass $w(t)$ in terms of area $a(t)$, that is, it can be formally assume that

$$w(t) = f(a(t)), \tag{6}$$

where the function $f : R \rightarrow R$ is continuous and taking positive values. Now, available data consist of a suitable number of replicate pairs $(w(t), a(t))$ of leaf biomass and area values measured at different sampling times t. So, we may take advantage of the outstanding estimation capabilities of the first order Takagi-Sugeno-Kang (TSK) fuzzy model to conceive an approximation for the mean response $E(f(a(t)))$ of $w(t)$, assuming that this variable is expressed by means of Eq. (6). Indeed, as it is elaborated in the methods section, it is possible to adapt a TSK paradigm whose general output that approximates $w(t)$ is denoted here by means of the symbol $w_{TSK}(t)$. Formally

$$w_{TSK}(a(t)) = \sum_{i} \varphi^{i}(a(t))f^{i}(a(t)), \tag{7}$$

where $\sum_i$ indicates summation over an index $i = 1, 2, \ldots, m$, and $\varphi^{i}(a(t))$ and $f^{i}(a(t))$ stand one to one for the *ith* normalized firing strength and consequent term of a first order TSK fuzzy model. By mimicking the steps establishing Eqs. (1, 2) the use of the $w_{TSK}(t)$ proxies allow to obtain a TSK fuzzy model approximation for the average of leaf biomass in shoots $w_m(t)$. This proxy is denoted here through the symbol $w_{mTSK}(t)$ and takes the form

$$w_{mTSK}(t) = \frac{\sum_{ns(t)} w_{sTSK}(t)}{ns(t)}, \tag{8}$$

where

$$w_{sTSK}(t) = \sum_{nl(s)} w_{TSK}(t) \tag{9}$$

symbolizes the TSK approximation to $w_s(t)$.

Previous accounts sustain that factors like analysis method or data quality may influence the reproducibility features of allometric surrogates for the observed values of eelgrass aggregates [7]. With the aim of comparing the performances of the $w_m(\alpha, \beta, t)$ and $w_{mTSK}(t)$ devices while mimicking the observed values of $w_m(t)$, we carried on an assessment of the scope of the influences of analysis method and data quality on the reproducibility capabilities of the considered proxies. Data quality control procedures involved a Mean Absolute Deviation (MAD) criteria for the removal of inconsistent replicates of observed leaf biomass measurements in the crude data [7]. Parameter estimates for the model of Eq. (3) sustaining the $w_m(\alpha, \beta, t)$ allometric proxies were obtained by using the traditional analysis method of allometry and also by means of nonlinear regression. In turn, for the $w_{mTSK}(t)$ construct, the identification of the parameters of the membership functions of the antecedents of the rules was carried away by means Subtractive Clustering (SC) [8, 9]. Then, a Recursive Least Squares (RLS) routine [10, 11] produced estimates for parameters involved in the consequents of the rules. The related identification tasks were performed on both the crude and the quality controlled data sets. Assessments of performances of the $w_m(\alpha, \beta, t)$ and $w_{mTSK}(t)$ devices involved the values of Lin's concordance correlation coefficient $\hat{\rho}$ [12], explained in Appendix 1 and which provides a measure of the competences of these proxies to reproduce observed $w_m(t)$ values. The present results show that in order to grant suitable reproducibility levels of the $w_m(\alpha, \beta, t)$ paradigm, besides assuring quality controlled data, we must also consider a suitable analysis method for the identification of the parameters in Eq. (3). Nevertheless, the involvement of data quality control is questionable because the elimination of inconsistencies in replicates is based on subjective judgments that can lead to forged reduction of variability and concomitant biased results. Our results show that while using the original raw data for the identification of the structure of the models sustaining the $w_m(\alpha, \beta, t)$ or the $w_{mTSK}(t)$ approximations, the proposed TSK paradigm yield the highest reproducibility of observed $w_m(t)$ values.

2 Methods

2.1 Data

The present data was collected from a coastal lagoon located in San Quintin Bay, Mexico. Raw data includes a total of n_{rw} measurements of leaf biomasses $w(t)$ [g], and accompanying widths $h(t)$ [mm] and lengths $l(t)$ [mm]. The product of length times width $h(t) \cdot l(t)$ provided concomitant estimations of leaf area $a(t)$ [mm^2] [13].Then, as elucidated in Appendix 1, following Echavarria-Heras et al. [6], groups $G(a(t))$ containing a number $n(g)$ leaf biomass replicates were linked to a given leaf area measurement $a(t)$. Then, for groups $G(a(t))$ complying with the condition $n(g) \geq 10$, the Median Absolute Deviation (MAD) procedure [7, 14, 15], which sustains the criteria of inequality (29) for the removal of inconsistent replicates was adapted. Meanwhile, for groups $G(a(t))$ satisfying $n(g) < 10$, a straight data cleaning method removed replicates that were severely deviated from the power function-like trend akin to Eq. (3). This yield a processed data set with a number n_{qw} of remaining leaves after taking away inconsistent replicates (Appendix 2).

2.2 Model Identification Approaches

Model identification tasks performed here hinged initially on obtaining estimates of the parameters α and β in Eq. (3). These were obtained by means the traditional analysis method of allometry, which uses a log-transformation to map the crude data $(w(t), a(t))$ in arithmetic scale to a geometrical scale $(\log(w(t)), \log(a(t)))$. Alternatively, we produced estimates for α and β by fitting the model of Eq. (3) using direct nonlinear regression procedures on the original arithmetic scale of crude data [6].

In order to explain the identification procedures for the general output of the TSK fuzzy model adapted for the present analysis, we firstly need to elaborate on the structure of this construct. The formalities, as well as, notation conventions of the involved fuzzy inference system appear in Appendix 3. The related identification tasks abided by the general guidelines explained in [9] and [16, 17], that is, the structure of the connected TSK fuzzy model was acquired by achieving the following steps: (1) defining the input space X, by selecting out among the set of possible input variables those relevant to the specified allometric scaling problem, (2) identifying suitable linguistic terms describing the fuzzy partition of the input space, (3) specifying the forms of the membership functions, (4) setting the number of inference rules, (5) obtaining estimates of the parameters characterizing the membership functions of the antecedents, and (6) producing estimates for the parameters in the consequents of the rules. We completed these endeavors by firstly taking into account an input-output structure suitable for the description of the functional relationship that describes the variability of $w(t)$ in terms of $a(t)$.

The steps that allow the identification of a suitable structure for the TSK fuzzy model considered here are summarized by Eqs. (35)–(64). Moreover, in carrying out step (1) of conceiving a suitable input domain (cf. Eq. (35)), we conceived a domain X containing only one input descriptor variable x_1 and made the association $x_1 \rightarrow a(t)$, that is, the input variable coincides with observed leaf area $a(t)$. Concomitantly, the output domain Y is described by a response variable y_1, taken as leaf biomass, that is, we set the correspondence $y_1 \rightarrow w(t)$ (cf. Eq. (36)). Therefore, the case $n = 1$ and $p = 1$ holds, and according to Eq. (37), the input value $a(t)$ can be fuzzyfied by a set L_a containing linguistic terms $A_k(a(t))$ for $1 \leq k \leq q(1)$. This also describes the set A_X in Eqs. (38) or (39). Moreover, a particular linguistic term $A_k(a(t))$ is associated to a membership function $\mu A_k(a(t))$ in the set μ_X defined by Eq. (40), and this conforms the fuzzy partition (A_X, μ_X) of the input domain X. Furthermore, the form of the incumbent Cartesian product L_X is given by Eq. (45) while Eqs. (46) and (47) set the value m (the cardinality of L_X), which yields the number of inference rules to be given by

$$m = q(1). \tag{10}$$

And since we are dealing with the case $n = 1$, in order to acquire the ordering of the elements of the Cartesian product L_X established by Eq. (48) and required for the identification of the antecedent conjunctions of Eq. (49), it will suffice to think of a correspondence $i \rightarrow A_i(a(t))$. Hence, the antecedent conjunctions become

$$Q^i(a(t)) : [a(t) \; is \, A_i(a(t))] \tag{11}$$

for $i = 1, 2, \ldots, m$. Moreover, as stated by Eq. (51), the $Q^i(a(t))$ antecedent can be associated to a set of membership functions $\mu Q^i(a(t))$ having the form

$$\mu Q^i(a(t)) = \{\mu A_i(a(t))\} \tag{12}$$

for $i = 1, 2, \ldots, m$. Then, in order to conceive a TSK fuzzy inference system the $Q^i(a(t))$ antecedent of Eq. (11) must yield an inferential rule R^i expressed by

$$R^i : \left\{ \begin{array}{lll} \textit{if} & : & a(t) \; \textit{is} \, A_i\,(a(t)) \\ \textit{then} & : & w(t) = f^i(a(t)) \end{array} \right\} \tag{13}$$

for $i = 1, 2, \ldots, m$. Moreover, for $i = 1, 2, \ldots, m$, Eq. (61) sets linear consequent functions given by

$$f^i(a(t)) = p_1^i a(t) + p_2^i \tag{14}$$

And by virtue of Eq. (62) the firing strength $\vartheta^i(a(t))$ of the antecedent $Q^i(a(t))$ of rule R^i reduces to:

$$\vartheta^i(a(t)) = \mu A_i(a(t)), \tag{15}$$

for $i = 1, 2, \ldots, m$. Therefore, the normalized firing strength $\varphi^i(a(t))$ of Eq. (63) takes a form

$$\varphi^i(a(t)) = \frac{\mu A_i(a(t))}{\sum_i \mu A_i(a(t))}. \tag{16}$$

Hence, Eq. (64) explains the form of the overall TSK output $w_{TSK}(a(t))$ expressed by Eq. (7) as a weighted average of all rule outputs $\varphi^i(a(t))f^i(a(t))$.

Generally, the parameter estimation steps involved in identifying a fuzzy model are usually achieved by means of optimization methods [18, 19], taking into account both linguistic information obtained from expert knowledge and available data [20]. Nevertheless, for the present TSK fuzzy model estimation of the parameters of the membership functions of the antecedents of the rules are carried away by using Subtractive Clustering (SC). We also adapted a Recursive Least Squares (RLS) routine aimed to produce estimates for parameters in the consequents of the rules. See Appendix 4.

2.3 Assessment of Suitability of the $w_m(\alpha, \beta, t)$ and $w_{mTSK}(t)$ Proxies for $w_m(t)$

Firstly, the reproducibility capabilities of the $w_m(\alpha, \beta, t)$ device was explored. These studies were done by considering both, the set of unprocessed data ($n = n_{rw}$) and that remaining after quality control procedures ($n = n_{qw}$). Using Eq. (2) produced the values $w_m(t)$ for the average biomass in shoots at a sampling time t, for both, the set of raw and that of processed data. Then, estimates $\hat{\alpha}$ and $\hat{\beta}$ for the parameters α and β, in Eq. (3) were acquired at first by using the traditional approach of linearizing Eq. (3) by log-transforming the original data ($Log\,LR$), and then by fitting the model of Eq. (3) thought a direct iterative nonlinear least-squares method (NLR) [21]. Afterwards, for all cases we used Eq. (3) to produce the

$w_m(\alpha, \beta, t)$ proxy values for observed averages $w_m(t)$. The suitability of these surrogates to reproduce $w_m(t)$ values (for both the raw and the processed data sets) was assessed thought the related concordance correlation coefficient values, $\hat{\rho}_{Log\,LR}$ for the linear regression or $\hat{\rho}_{NLR}$ for the nonlinear regression method (see Appendix 2). Similarly, the values of $\hat{\rho}_{TSK}$, the concordance correlation coefficient for $w_m(t)$ and $w_{mTSK}(t)$ provided criteria for the assessment of the reproducibility power of the adapted TSK fuzzy model approximation device.

3 Results

3.1 Data

Figure 1 shows the dispersion pattern of observed leaf biomasses $w(t)$ and corresponding leaf areas $a(t)$. The number of available pairs of leaf biomass and related areas was $n_{rw} = 10412$. We can observe that the typical power function-like trend that associates to the scaling relationship of Eq. (3) is masked by a great variability of replicates.

We can spot a relatively greater proliferation of inconsistent replicates for leaves with areas under 1000 mm^{2}and also for those bigger than 5000 mm^2. This feature in the data could be explained by: (1) a lack of standardization of measurement procedures for both leaf biomass and leaf width of the smaller leaves. These being associated to tiny width values that bring relatively insignificant biomasses that

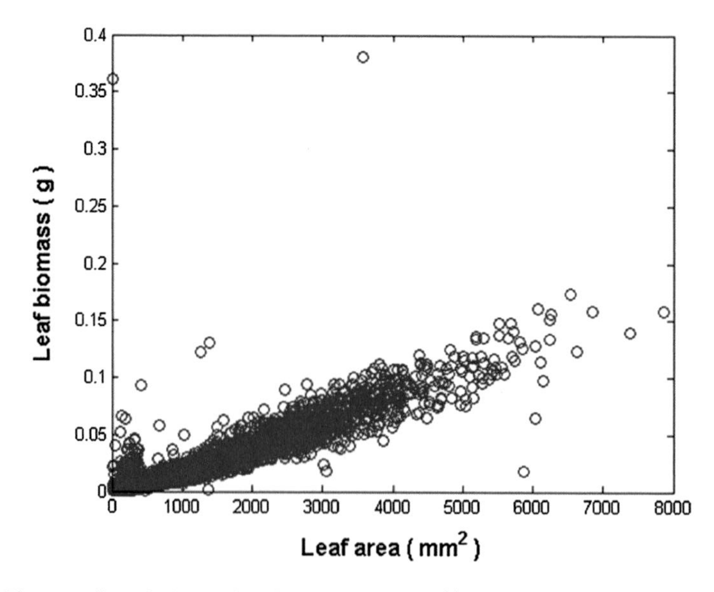

Fig. 1 The spreading of observed leaf biomass values $w(t)$ measured for a given area $a(t)$

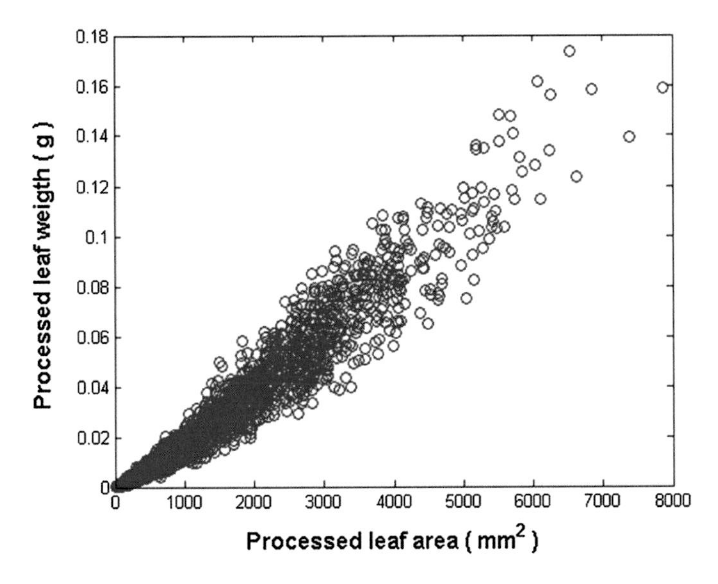

Fig. 2 The spreading of leaf biomass values $w(t)$ corresponding to a given leaf area value $a(t)$ that remained after data quality control procedures

could convey measurement bias imputable to precision of the analytical scale involved in dry weight estimation; and (2) longer leaves are older thus being extensively exposed to herbivory and drag forces, thereby explaining the presence of incomplete or damaged leaves in data, which certainly alter the true weight-to-area relationship. Figure 2 portraits the spreading of the $n_{qw} = 6094$ pairs of leaf biomass and area measurements remaining after removing inconsistent replicates by applying the data quality control procedures of Appendix 1. In Fig. 2 we can notice that the power function-like trend tied to the model of Eq. (3) is more clearly depicted.

3.2 *Model Identification Tasks*

Figure 3 provides a comparison of observed leaf biomass $w(t)$ values and counterparts projected by means of the allometric approximation defined by Eq. (3). In obtaining estimates for the allometric parameters α and β for this trial, we used linear regression on log-transformed data (*Log LR*). This produced values of $SEE_{Log\,LR} = 8.26 \times 10^{-3}$ and $\hat{\rho}_{Log\,LR} = 0.8910$ for the standard error of estimate and the tied concordance correlation coefficient respectively (Table 1).

Figure 4 displays processed values for $w(t)$ now compared with lines corresponding to their allometric approximations $w(\alpha, \beta, t)$ calculated using parameters fitted by linear regression on log-transformed processed data (*LogLR*).

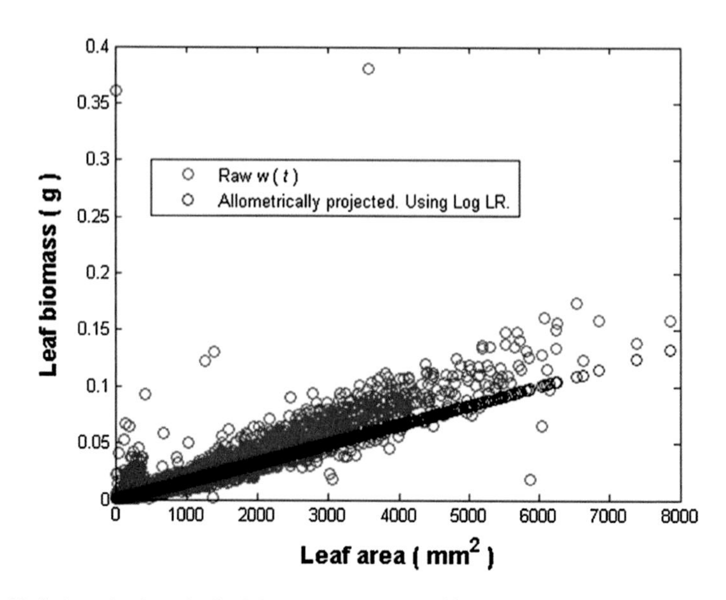

Fig. 3 Variation of values for individual leaf biomass $w(t)$ in the raw data and proxies projected by means of Eq. (3). The shown mean response curve associates to estimates for α and β fitted by linear regression on log-transformed data (*LogLR*)

Table 1 Results of identification procedures of the model of Eq. (3) and the TSK fuzzy approximation of Eq. (21)

Data set	Analysis method	Standard error of estimate (SEE)	$\hat{\rho}$	95% confidence limits $(\hat{\rho}_{min}, \hat{\rho}_{max})$
Raw	Log LR	8.26×10^{-3}	0.8910	0.8878–0.8941
Processed	Log LR	6.33×10^{-3}	0.9456	0.9437–0.9473
Raw	NLR	7.18×10^{-3}	0.9307	0.9282–0.9332
Processed	NLR	4.44×10^{-3}	0.9777	0.9766–0.9788
Raw	TSK FM	7.22×10^{-3}	0.9300	0.9274–0.9325
Processed	TSK FM	4.56×10^{-3}	0.9762	0.9750–0.9774

Included are standard error of estimate (SEE), concordance correlation coefficient $\hat{\rho}$ and associated 95% confidence limits

This assessment produced $SEE_{Log\,LR} = 6.33 \times 10^{-3}$ and $\hat{\rho}_{Log\,LR} = 0.9456$ (Table 1). We can assess that on spite of an improvement in goodness of fit derived from data processing log-linear regression as an analysis methods failed to produce consistent reproducibility results [6].

Figure 5 provides a comparison of observed $w(t)$ values and counterparts projected by means of Eq. (3). Estimates for the allometric parameters α and β for this trial, were fitted from raw data by means of nonlinear regression as an analysis method. This produced values of $SEE_{NLR} = 7.18 \times 10^{-3}$ and $\hat{\rho}_{NLR} = 0.9307$ for the

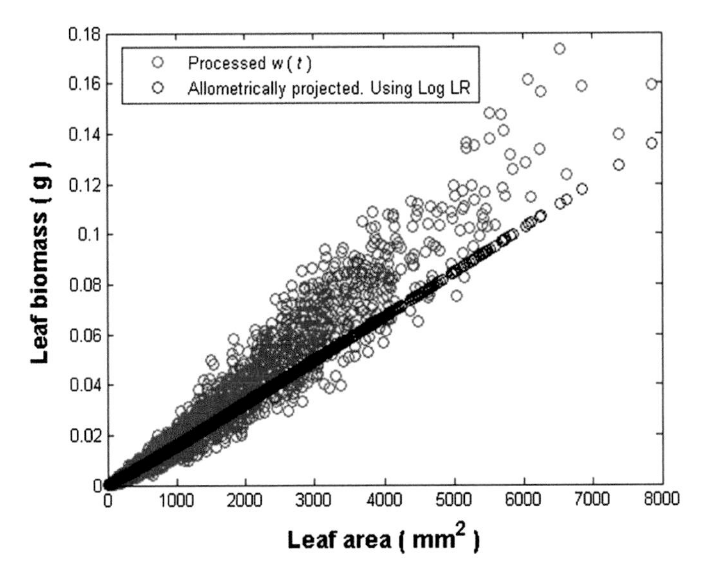

Fig. 4 Variation of individual leaf biomass $w(t)$ values in the processed data and proxies projected by means of the allometric approximation given by Eq. (4). In order to get estimates of the allometric parameters α and β we fitted the associated allometric model to the processed data using ($LogLR$) as an analysis method

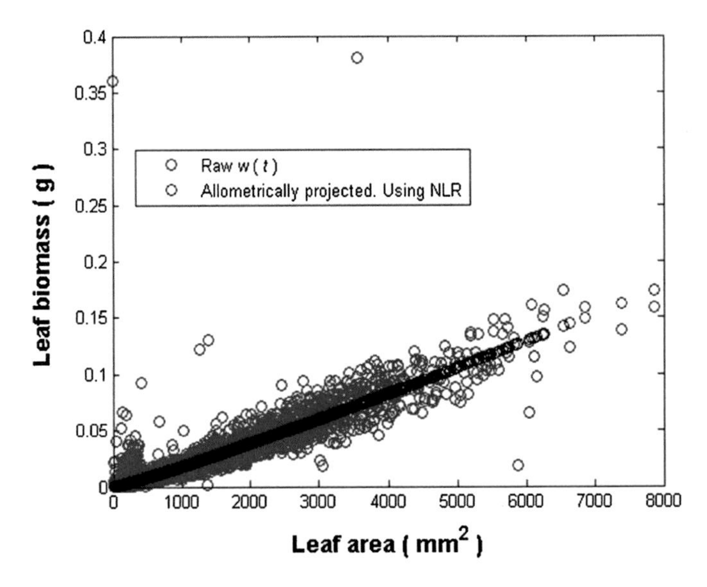

Fig. 5 Variation of individual leaf biomass $w(t)$ values in the observed data and proxies projected by means of Eq. (3). Producing the shown mean response curve required estimates for α and β fitted from the raw data set by means of nonlinear regression

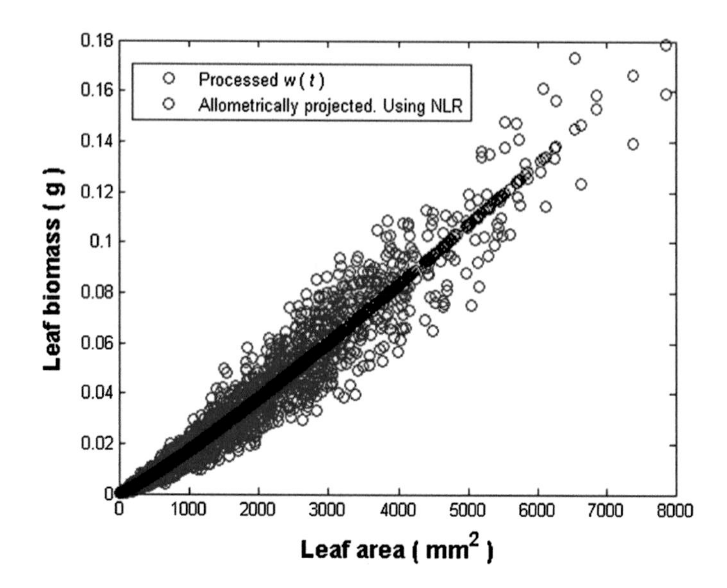

Fig. 6 Variation of individual leaf biomass $w(t)$ values in the processed data compared to substitutes projected by means of the allometric model of Eq. (3). In order to get estimates of the allometric parameters α and β we used (*NLR*) as an analysis method

tied standard error of estimate and the concordance correlation coefficient respectively (Table 1).

Figure 6 displays processed values for $w(t)$ now compared with values corresponding to proxies calculated using allometric parameters in Eq. (3) acquired by nonlinear regression and processed data. This trial produced $SEE_{NLR} = 4.44 \times 10^{-3}$ and $\hat{\rho}_{NLR} = 0.9777$ (Table 1). We can assess that as opposed to log-linear regression the nonlinear counterpart produced consistent reproducibility results [6].

Equations (10)–(16) provide the formal set up for the adaptation of the TSK proxy $w_{TSK}(a(t))$ defined by Eq. (7) and aimed to approximate observed eelgrass leaf biomass values $w(t)$. In order to achieve the step of identifying suitable linguistic terms defining the fuzzy partition of the input space X (in present settings characterized by leaf area values $a(t)$), it is necessary to fix the value of $q(1)$ in Eq. (10), which determines the cardinality m and the number of rules R^i. Application of the SC method to the present data identified 2 groups, therefore, $q(1) = m = 2$, that is, $L_X = \{A_k(a(t))|k = 1, 2\}$ with $A_1(a(t))$ and $A_2(a(t))$ taken as the fuzzy sets labeled as "small leaf area" and "large leaf area" respectively. So according to Eq. (40) we have $\mu_X = \{\mu A_1(a(t)), \mu A_2(a(t))\}$. Besides, the membership functions are assumed to have a Gaussian form, that is

$$\mu A_1(a(t)) = exp\left\{-\frac{1}{2}\left[\left(\frac{a(t)-\theta_{11}}{\sigma_{11}}\right)^2\right]\right\} \tag{17}$$

and

$$\mu A_2(a(t)) = exp\left\{-\frac{1}{2}\left[\left(\frac{a(t)-\theta_{12}}{\sigma_{12}}\right)^2\right]\right\}, \tag{18}$$

where $\theta_{11}, \theta_{12}, \sigma_{11}$, and σ_{12} are parameters to be identified from the data. This arrangement will also bring about antecedent conjunctions $Q^i(a(t)) : [a(t)\ is\, A_i(a(t))]$ and associated inference rules:

$$R^i : \left\{ \begin{array}{lll} if & : & a(t)\ is\, A_i\,(a(t)) \\ then & : & w(t) = f^i(a(t)) \end{array} \right\}, \tag{19}$$

with

$$f^i(a(t)) = p_1^i a(t) + p_2^i. \tag{20}$$

The total output $w_{TSK}(a(t))$ defined by Eq. (7) is explicitly given by

$$w_{TSK}(a(t)) = \varphi^1(a(t))f^1(a(t)) + \varphi^2(a(t))f^2(a(t)), \tag{21}$$

where

$$\varphi^i(a(t)) = \frac{\mu A_i(a(t))}{\sum_j \mu A_j(a(t))}, \tag{22}$$

for $i = 1, 2$. Estimates of the parameters of the membership functions obtained using the SC method working on the raw data set are

$$\begin{bmatrix} \theta_{11} \\ \theta_{12} \end{bmatrix}_{raw} = \begin{bmatrix} 351.99 \\ 930.10 \end{bmatrix} \text{ and } \begin{bmatrix} \sigma_{11} \\ \sigma_{12} \end{bmatrix}_{raw} = \begin{bmatrix} 358.91 \\ 2631.86 \end{bmatrix}, \tag{23}$$

while the values of estimates of parameters of the consequents $f^1(a(t))$ and $f^2(a(t))$ of the rules fitted from the raw data set using the RLS method of Appendix 4 are

$$\begin{bmatrix} p_1^1 & p_2^1 \\ p_1^2 & p_2^2 \end{bmatrix}_{raw} = \begin{bmatrix} 1.57e-5 & 3.55e-8 \\ 2.02e-5 & 7.25e-9 \end{bmatrix}. \tag{24}$$

Meanwhile, estimates of the parameters of the membership functions obtained from the processed data set using the SC method are

$$\begin{bmatrix} \theta_{11} \\ \theta_{12} \end{bmatrix}_{processed} = \begin{bmatrix} 939.20 \\ 382.46 \end{bmatrix} \text{and} \begin{bmatrix} \sigma_{11} \\ \sigma_{12} \end{bmatrix}_{processed} = \begin{bmatrix} 2731.63 \\ 386.15 \end{bmatrix}, \tag{25}$$

and the values of estimates of parameters of the consequents $f^1(a(t))$ and $f^2(a(t))$ of the rules fitted from the processed data using the RLS method of Appendix 4 are

$$\begin{bmatrix} p_1^1 & p_2^1 \\ p_1^2 & p_2^2 \end{bmatrix}_{processed} = \begin{bmatrix} 2.02e-5 & 5.35e-9 \\ 1.50e-5 & 2.68e-8 \end{bmatrix} \tag{26}$$

Figure 7 provides plots for the consequents $f^1(a(t))$ and $f^2(a(t))$ given by Eq. (20). The normalized firing strength factors $\varphi^1(a(t))$ and $\varphi^2(a(t))$ given by Eq. (22) plus their products are also shown. They all form the general output $w_{TSK}(a(t))$ of the adapted TSK model for eelgrass leaf biomass $w(t)$ in terms of related area $a(t)$ formalized by Eq. (21).

Meanwhile, Figs. 8 and 9 present a comparison of observed leaf biomass $w(t)$ values and corresponding surrogates projected by means of the weighted TSK output $w_{TSK}(a(t))$ respectively identified using the sets of raw and processed data.

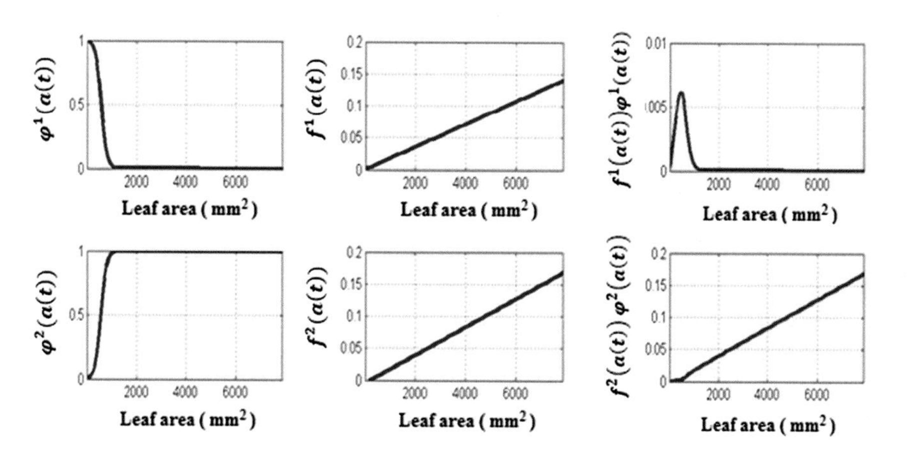

Fig. 7 Plots of linear consequents $f^i(a(t))$ and products $\varphi^i(a(t))f^i(a(t))$ for $i = 1, 2$ conforming the general TSK output $w_{TSK}(a(t))$ of Eq. (21)

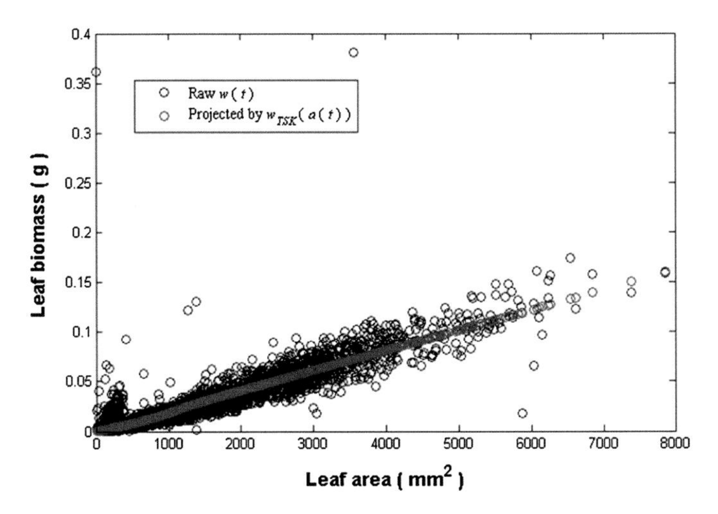

Fig. 8 Plot of individual leaf biomass values $w(t)$ values in the raw data and corresponding projections $w_{TSK}(a(t))$ produced by the general output of the TSK fuzzy adapted for eelgrass leaf biomass to area allometry and formalized by Eq. (21)

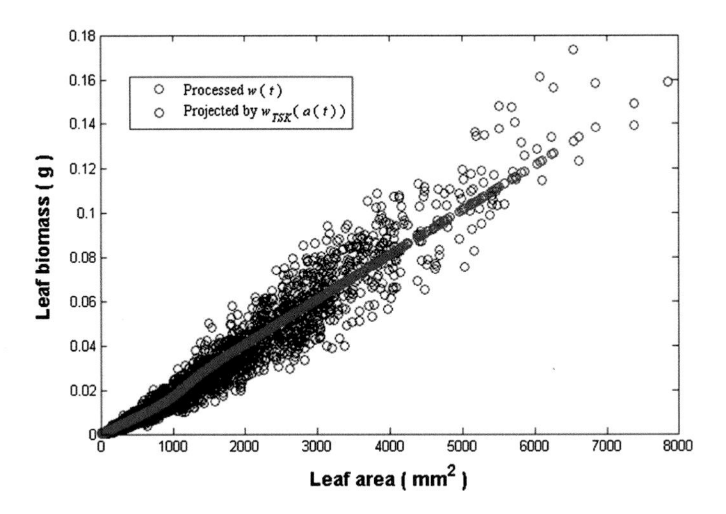

Fig. 9 Plot of individual leaf biomass $w(t)$ values in the processed data and corresponding projections $w_{TSK}(a(t))$ produced by the general output of the TSK fuzzy model adapted for eelgrass leaf biomass to area allometry and formalized by Eq. (21)

3.3 Assessment of the Predictive Strength of the Crisp $w_m(\alpha, \beta, t)$ and Fuzzy $w_{mTSK}(a(t))$ Surrogates

In Fig. 10 we provide the plot comparing observed $w_m(t)$ records with equivalent lines $w_m(\alpha, \beta, t)$ of Eq. (4) produced by using the raw data and the *Log LR*

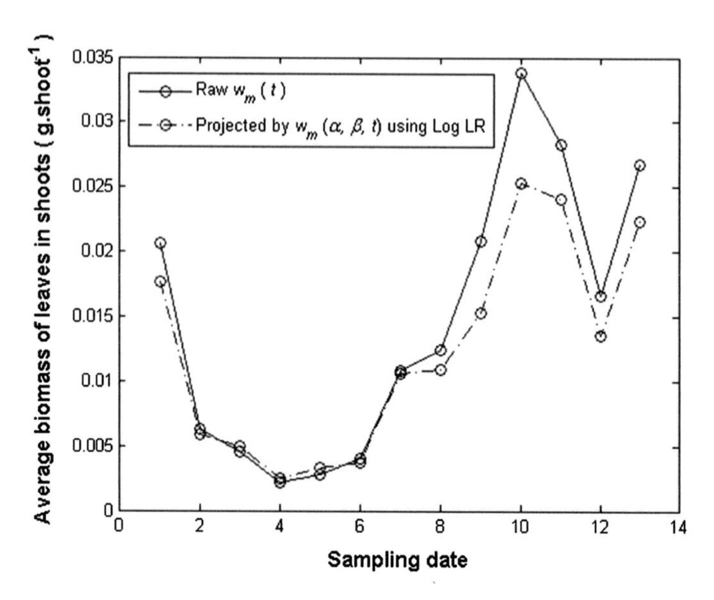

Fig. 10 Observed $w_m(t)$ averages compared with allometric projections $w_m(\alpha, \beta, t)$ obtained using the *LogLR* method and raw data. We can observe a marked bias specially for longer leaves

identification method. An assessment of observed and projected values produced $SEE_{Log\,LR} = 3.84 \times 10^{-3}$ and $\hat{\rho}_{Log\,LR} = 0.9285$ for the respective standard error of estimate and concordance correlation coefficient. We can assess in this plot that concordance between observed and projected values is poor [6].

Meanwhile, Fig. 11 shows a comparison of processed $w_m(t)$ averages and their allometric projections $w_m(\alpha, \beta, t)$ this time tied to processed data and the *LogLR* procedure. This yield values of $SEE_{LogLR} = 3.80 \times 10^{-3}$ and $\hat{\rho}_{LogLR} = 0.9489$. Although, we can record an improvement relative to the equivalent trial for the raw data set, the obtained gain is not significant enough as to grant suitability of projections. Particularly, we can assess that the marked bias observed for longer leaves in Fig. 10 still remains [6].

Figure 12 displays a comparison of observed $w_m(t)$ averages and their allomeric counterparts $w_m(\alpha, \beta, t)$ of Eq. (4) produced using the *NLR* identification method and the raw data set. This produced $SEE_{NLR} = 1.40 \times 10^{-3}$ and $\hat{\rho}_{NLR} = 0.9915$ for the respective standard error of estimate and concordance correlation coefficient. Moreover, for the raw data set we can observe that against the *LogLR* identification method the *NLR* procedure performed better. Indeed, *NLR* produced projections of a higher accuracy since the bias shown in Fig. 10 has greatly diminished [6].

Figure 13 shows a comparison of processed $w_m(t)$ averages versus allometric counterparts $w_m(\alpha, \beta, t)$ of Eq. (4) produced using the *NLR* identification method and processed data. This produced $SEE_{NLR} = 8.85 \times 10^{-4}$ and $\hat{\rho}_{NLR} = 0.9976$ for the respective standard error of estimate and concordance correlation coefficient. Moreover, the bias for longer leaves is no longer observed. And Table 2 allows to

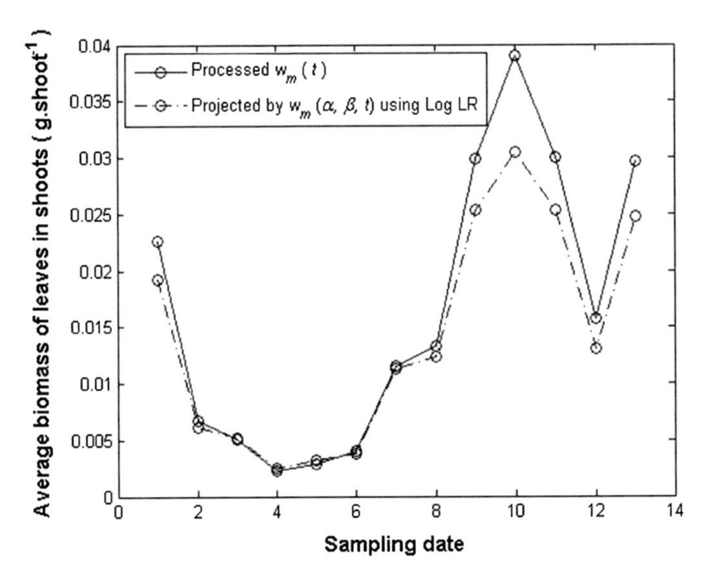

Fig. 11 Processed $w_m(t)$ averages compared with their allometric projections $w_m(\alpha, \beta, t)$ obtained using the *LogLR* method and processed data. We can observe that a marked bias specially for longer leaves still remains

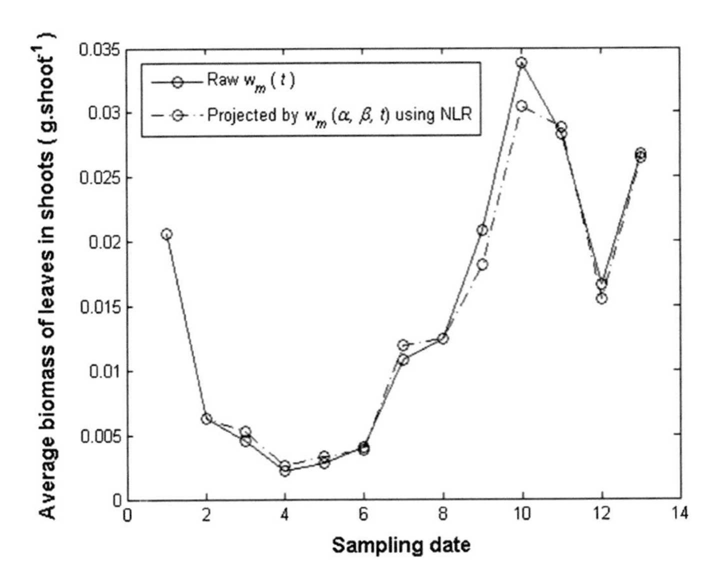

Fig. 12 Observed $w_m(t)$ averages compared with their allometric projections $w_m(\alpha, \beta, t)$ obtained using the *NLR* method and raw data. We can verify that the marked bias for longer leaves observed in Figs. 10 and 11 this time decreased noticeably

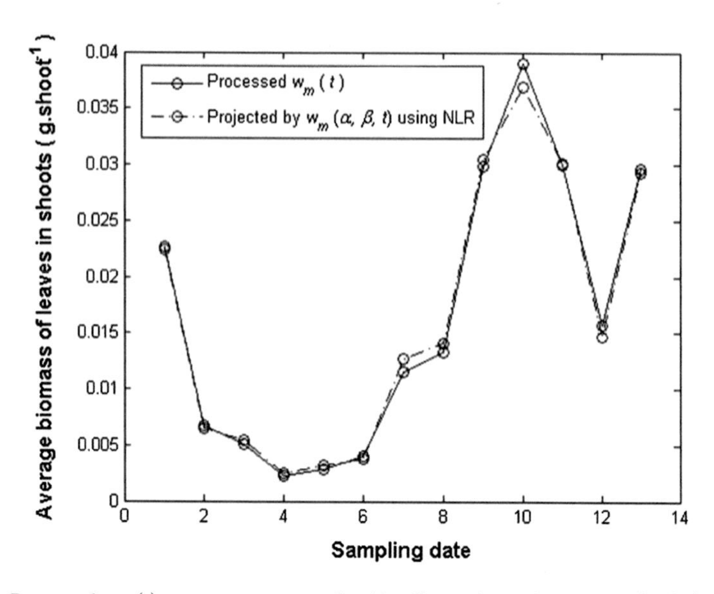

Fig. 13 Processed $w_m(t)$ averages compared with allometric projections $w_m(\alpha, \beta, t)$ obtained using the *NLR* analysis method and processed data. We can observe that in comparison with the *LogLR* method this combination yields a higher reproducibility level of the $w_m(\alpha, \beta, t)$ proxy

Table 2 Assessment of the reproducibly power of the $w_m(\alpha, \beta, t)$ and $w_{mTSK}(a(t))$ proxies

Data set	Analysis method	Standard error of estimate (SEE)	$\hat{\rho}$	95% confidence limits $(\hat{\rho}_{min}, \hat{\rho}_{max})$
Raw	Log LR	3.81×10^{-3}	0.9285	0.8871–0.9699
Processed	Log LR	3.80×10^{-3}	0.9489	0.9223–0.9755
Raw	NLR	1.40×10^{-3}	0.9915	0.9848–0.9982
Processed	NLR	8.85×10^{-4}	0.9976	0.9961–0.9991
Raw	TSK FM	1.33×10^{-3}	0.9924	0.9854–0.9995
Processed	TSK FM	9.09×10^{-4}	0.9976	0.9961–0.9991

We provide results for both the raw and processed data sets. Provided are standard error of estimate (SEE) values, as well as, calculated concordance correlation coefficients $(\hat{\rho})$ values along with their 95% confidence intervals

assess that relative to *LogLR* an improved goodness of fit for the combination of nonlinear regression and quality controlled data yields a higher reproducibility level of the $w_m(\alpha, \beta, t)$ proxy [6].

In turn, Fig. 14 shows the performances of the $w_{mTSK}(a(t))$ proxies for observed mean leaf biomass in shoots $w_m(t)$. In this comparison, we include $w_{mTSK}(a(t))$ proxies produced by the TSK output $w_{TSK}(a(t))$ of Eq. (21) identified using raw data. This produced $SEE_{TSK} = 1.33 \times 10^{-3}$ and $\hat{\rho}_{TSK} = 0.9924$ for the respective standard error of estimate and concordance correlation coefficient. We can observe that the outstanding approximation capabilities of the nonlinear dynamics exhibited

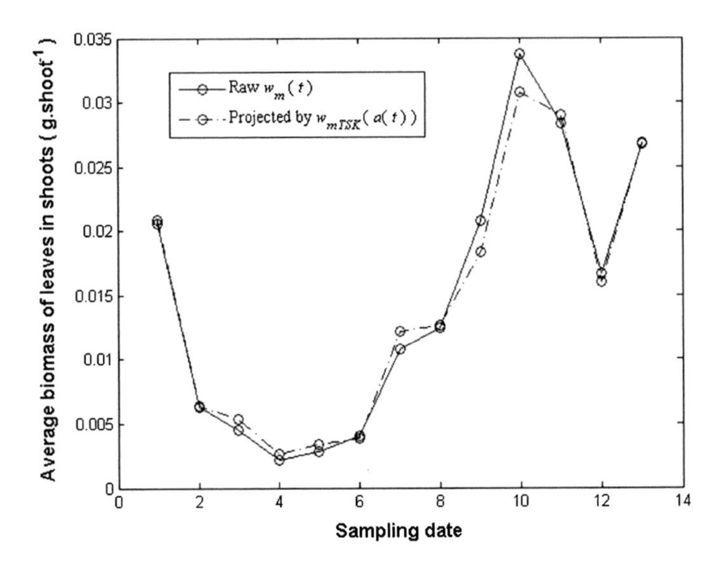

Fig. 14 Observed $w_m(t)$ averages compared with tied TSK projections $w_{mTSK}(t)$ identified on raw data

by the TSK paradigm can benefit eelgrass leaf biomass allometry. Indeed working on raw data the $w_{TSK}(a(t))$ proxy produced approximations for $w(t)$ that were of a higher accuracy than those tied to *NLR*, which explains the noted precision of the $w_{mTSK}(a(t))$ surrogates for $w_m(t)$ when using the raw data.

As expected, Fig. 15 shows that the efficacy of the $w_{TSK}(a(t))$ approximation improved when identification procedures were performed on the basis of processed data. This trial produced corresponding values of $SEE_{TSK} = 9.09 \times 10^{-4}$ and $\hat{\rho}_{TSK} = 0.9976$ for the standard error of estimate and the correspondence correlation coefficient respectively. Moreover, using the value of the correspondence correlation coefficient criteria for agreement, we can assess that the TSK method identified on the basis of processed data was able to deliver similar reproducibility capabilities than those sustained by the allometric method $w_m(\alpha, \beta, t)$ tied to nonlinear regression and quality controlled data. Table 2 summarizes the present results, which show that the fuzzy model approach supporting the $w_{TSK}(a(t))$ approximation to $w(t)$ warrants an efficient approximation alternative that bears efficient surrogates for $w_m(t)$ when raw data is used for model identification purposes. This entails the advantage of producing highly accurate non-destructive approximations to $w_m(t)$ without involving controversial data cleaning procedures.

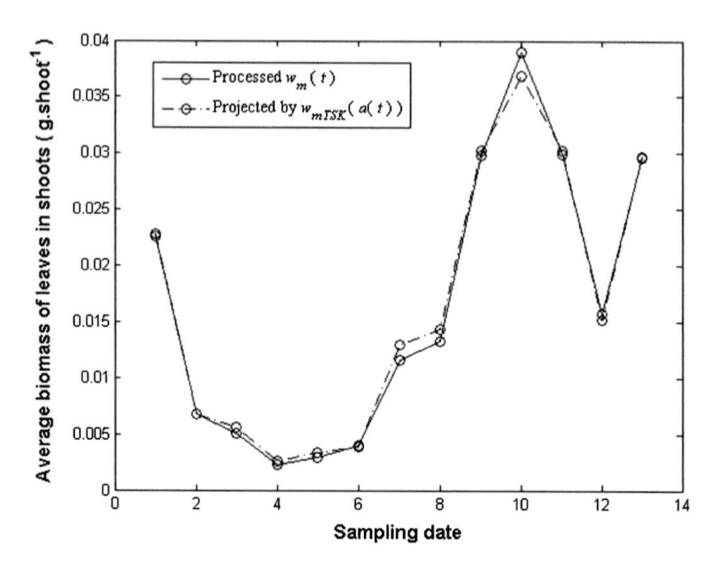

Fig. 15 Processed $w_m(t)$ averages compared with corresponding TKS projections $w_{mTSK}(t)$ identified on the basis of quality controlled data

4 Discussion

The acceptance of the scaling relationship of Eq. (3) as a suitable model for eelgrass leaf biomass $w(t)$ in terms of related area $a(t)$ is mainly justified by the simplicity of its mathematical form, and also by virtue of the copious documentation of straightforward statistical procedures for parameter identification available for practitioners of allometry. Moreover, the derivation of the $w_m(\alpha, \beta, t)$ substitute for $w_m(t)$ is a convenient by product that adds to the benefits that the use of allometric methods conveys for eelgrass conservation. The appropriateness of the $w_m(\alpha, \beta, t)$ constructs to yield truly nondestructive estimations of the $w_m(t)$ variable hangs on primarily on a time invariance of the parameters α and β. This because estimates of these allometric parameters that were previously fitted at a site along with currently measured leaf area values $a(t)$ could be used to project through the $w_m(\alpha, \beta, t)$ device indirect assessments for presently observed values of $w_m(t)$. Parameters α and β linked to an allometric scaling of eelgrass leaf biomass and length can indeed be considered as time invariant [7]. And since for eelgrass, length is an isometric descriptor of the concomitant area, a time invariance property may be also fore-seeing for the allometric parameters in Eq. (3). But, albeit α and β in Eq. (3) can be considered as statistically invariant, it can be anticipated that, local scale environ-mental forcing could induce a relative extent of variability on estimates through time. And since the accuracy of the mean response variable $E(w(t))$ in Eq. (3) is mainly resultant of uncertainty propagation of the parameters α and β, there are concerns on the suitability of the $w_m(\alpha, \beta, t)$ projections that relate to the influences that the uncertainties of the estimates of α and β may induce on precision.

Moreover, the results of Echavarría-Heras et al. [7] suggest that data quality, analysis method, as well as, sample size are factors that could sensibly affect the precision of estimates of the parameters α and β in Eq. (3), and also may concomitantly influence the reproducibility capabilities of the $w_m(\alpha, \beta, t)$ device. The present results endorse the views of Echavarría-Heras et al. [7] that factors like analysis method and data quality must be taken into consideration in order to enhance the precision of the $w_m(\alpha, \beta, t)$ device. Table 1 and the plots in the model identification section of results show that the traditional analysis method of allometry of using linear regression of log-transformed data yields parameter estimates that entail $w_m(\alpha, \beta, t)$ proxies for $w_m(t)$ values that exhibit poor reproducibility power, even for quality controlled data, while the alternate nonlinear regression fitting device produced consistent agreement levels between projected and observed values even for the raw data. As a matter of fact, that the most efficient conventional arrangement for the $w_m(\alpha, \beta, t)$ projection method comprises nonlinear regression as an analysis method and a quality controlled data set. This becomes a serious inconvenience for the allometrically supported $w_m(\alpha, \beta, t)$ device. Surely, using any data quality control procedure implicates subtleties mainly related to the outlier detection step. Moreover, accomplishing a compromise as to which subjective rejection threshold should be used is seemingly unattainable. And because of this, even how unjustified it might be, the use of any data quality control procedure will endure a doubt that a researcher picks the method that produces the most likely results [14]. This justifies the quest for proxies producing a highly consistent reproducibility of observed $w_m(t)$ averages while overcoming inconveniences of the allometric approach such as its dependence on data quality control procedures for assuring suitable accuracy levels. It is precisely at this stage that the present settings show, that on spite of an increase in complexity tied to a TSK fuzzy modeling approach the gain in suitability becomes evident. Indeed, the resulting reproducibility of the $w_{TSK}(a(t))$ device identified on the basis of raw data leveled off the most efficient settings of the $w_m(\alpha, \beta, t)$ approximation that involved data quality control procedures.

In summary, the proposed $w_{TSK}(t)$ surrogate for observed individual leaf biomass $w(t)$ entailed proxies $w_{mTSK}(t)$ that warrant highly precise approximations to observed $w_m(t)$ values, while making the controversial data quality control schemes required for the consistency of the $w_m(\alpha, \beta, t)$ device unnecessary. The dependence of the reproducibility features of the allometric approach on analysis method and data quality, along with an appraisal of the performance of the TSK paradigm here envisioned, favors the appeal of this approach as a necessary device for the approximation of true nonlinear functional mechanism that explains the observed variation pattern of eelgrass leaf biomass in terms of related area, which was formally hinted by Eq. (6).

Appendix 1: Data Quality Control Approach

Previous results demonstrate that it is reasonable to assume that for eelgrass the leaf area-to-leaf weight variation pattern should distribute about a power function-like trend [7]. Thus, by analyzing in a leaf area-to-weight plot, the spreading of leaf biomass values we can detect values that present severe deviations to a dominant power function trend. Since in the present settings leaf area is obtained by means of the length times width proxy [22] unduly deviations of replicates from an expected trend could be attributed to errors in leaf length or width estimations, imprecise gear for dry weight assessment or even due to inappropriate registering of measurements. In order to conceive a data cleaning criteria, we firstly observed that the set of n_{rw} leaves constituting the raw data can be arranged into several groups $G(a) = \{w_i(a)|1 \leq i \leq n(g)\}$ formed by the set of $n(g)$ leaf biomass replicates $w_i(a)$ that associate to an observed leaf area value a. Considering that the median of a group of data is totally immune to the sample size and a robust estimator of scale, for groups of ten or more replicates, we adapted a Median Absolute Deviation (MAD) data cleaning procedure [7, 14]. For each one of the groups $G(a)$ acquired median is denoted by means of the symbol $\mathrm{MED}\{w_1(a), \ldots, w_{n(g)}(a)\}$ or simply by means of $\mathrm{MED}(G(a))$ for short. Then, for each replicate $w_j(a)$ in $G(a)$ its absolute deviation $\delta_j(a)$ relative to the group median $\mathrm{MED}(G(a))$ is

$$\delta_i(a) = |w_i(a) - \mathrm{MED}(G(a))| \tag{27}$$

Similarly, we obtained the median of the set of absolute deviations denoted by the symbol $MED\{\delta_1(a), \ldots, \delta_{n(g)}(a)\}$. Following Huber [15] and by also eliciting that eelgrass leaf biomass values are log-normally distributed [7] then obtained Median Absolute Deviation of a group $G(a)$, denoted through $\mathrm{MAD}(G(a))$ is given by

$$\mathrm{MAD}(G(a)) = b\mathrm{MED}\{\delta_1(a), \ldots, \delta_{n(g)}(a)\}, \tag{28}$$

where $b = 1/Q(0.75)$, and $Q(0.75)$ stands for the 0.75 quantile associated to a lognormal distribution. Henceforth, for the deletion of uneven replicates in a group $G(a)$ we used the decision criterion

$$\begin{aligned} &\mathrm{MED}(G(a)) - T \cdot \mathrm{MAD}(G(a)) < w_j(a) < \\ &\mathrm{MED}(G(a)) + T \cdot \mathrm{MAD}(G(a)), \end{aligned} \tag{29}$$

where $w_j(a)$ stands for the *jth* leaf in the group $G(a)$ and T is the rejection threshold, which after Miller [23] we set at a value $T = 3$.

Appendix 2: Lin's Concordance Correlation Coefficient (ρ)

The value of ρ the concordance correlation coefficient [12, 24] is commonly used to assess how well a new set of observations y mimic an original set x. In other words, the value of ρ provides a measure of reproducibility as it evaluates the agreement between the variables x and y by assessing the extent to which they fall on the 45° line through the origin. Its value is characterized in terms of the ratio of the expected orthogonal squared distance from the diagonal $y = x$ to the expected orthogonal squared distance from the diagonal $y = x$ assuming independency. When ρ is computed on a m-length data set (i.e., two vectors $(x_1, x_2, \ldots, x_m)$ and $(y_1, y_2, \ldots, y_m)$ the resulting statistics is denoted by means of $\hat{\rho}$ and calculated through

$$\hat{\rho} = \frac{2s_{xy}}{s_x^2 + s_y^2 + (\bar{x} + \bar{y})^2}, \tag{30}$$

being

$$\bar{x} = \frac{1}{m}\sum_{j=1}^{m} x_j \tag{31}$$

$$s_x^2 = \frac{1}{m}\sum_{j=1}^{m} (x_j - \bar{x})^2 \tag{32}$$

$$s_y^2 = \frac{1}{m}\sum_{j=1}^{m} (y_j - \bar{y})^2 \tag{33}$$

and

$$s_{xy} = \frac{1}{m}\sum_{j=1}^{m} (x_j - \bar{x})(y_j - \bar{y}) \tag{34}$$

In the present work the value of $\hat{\rho}$ provides a criterion to assess to what extent the allometric proxies $w_m(\alpha, \beta, t)$ or the weighted output $w_{TSK}(a(t))$ of the considered. Takagi–Sugeno-Kang model reproduce observed $w_m(t)$ values. Agreement will be defined as poor whenever $\hat{\rho} < 0.90$, moderate for $0.90 \leq \hat{\rho} < 0.95$, good for $0.95 \leq \hat{\rho} \leq 0.99$, or excellent for $\hat{\rho} > 0.99$ [25].

Appendix 3: The Takagi-Sugeno-Kang Fuzzy Inference System

A Takagi-Sugeno-Kang inference system is a special characterization of what is known as a fuzzy inference system. Generally in setting a fuzzy inference system for $1 \leq j \leq n$ we initially consider input values x_j an input domain U. The collection of input values is denoted by the symbol

$$X = \{x_1, \ldots, x_n\} \tag{35}$$

and similarly, for $1 \leq s \leq p$ we consider output values $y_s(x_1, \ldots, x_n)$ in a range V, and the collection of output variables will be denoted by the symbol

$$Y = \{y_1(x_1, \ldots, x_n), \ldots, y_p(x_1, \ldots, x_n)\} \tag{36}$$

For each input variable x_j we associate a set L_j containing a number $a(j)$ of linguistic terms $A_k(x_j)$, namely

$$L_j = \{A_k(x_j) | k = 1, 2, \ldots, q(j)\} \tag{37}$$

The symbol A_X will stand for the collection of linguistic terms that associate to X, formally

$$A_X = \cup_j L_j, \tag{38}$$

The linguistic term $A_k(x_j)$ associates to a membership function $\mu A_k(x_j)$ that setting a mapping $\mu A_k(x_j) : U \to [0, 1]$ characterizes $A_k(x_j)$ as the fuzzy set:

$$A_k(x_j) = \{\mu A_{kj}(x_{j1})/x_{j1}, \mu A_{kj}(x_{j2})/x_{j2}, \ldots, \mu A_{kj}(x_{jb(j)})/x_{jb(j)}\} \tag{39}$$

where $x_{j1}, x_{j2}, \ldots, x_{jb(j)}$ stand for the values that x_j takes on.

The symbol μ_X will stand for the collection of membership functions describing the set of input variables X, that is,

$$\mu_X = \cup_{(k,j)} \{\mu A_k(x_j)\}, \tag{40}$$

Moreover, the pair (A_X, μ_X) will stand for what is known as a fuzzy partition of the input domain U.

Respectively, the output variable y_s with $s = 1, 2, .., p$ associates to a set L_s, of linguistic terms $B_k(y_s(x_1, \ldots, x_n))$ namely

$$L_s = \{B_k(y_s(x_1, \ldots, x_n)) | k = 1, 2, \ldots, c(s)\} \tag{41}$$

Similarly,

$$B_Y = \cup_s L_s \tag{42}$$

will stand for the collection of linguistic terms that characterize Y. Also, for $B_k(y_s)$ we associate a membership function $\mu B_k(y_s(x_1, \ldots, x_n))$ such that the mapping $\mu B_k(y_s) : V \to [0, 1]$ establishes the fuzzy set

$$B_k(y_s(x_1,\ldots,x_n)) = \{\mu B_k(y_{s1})/y_{s1}, \mu B_k(y_{s2})/y_{s2},\ldots, \mu B_k(y_{sc(s)})/y_{sc(s))}\} \quad (43)$$

where $y_{s1},\ldots,y_{sc(s)}$ denote the values that $y_s(x_1,\ldots,x_n)$ acquires. Concurrently, we have the collection of membership functions tied to Y,

$$\mu_Y = \cup_{(k,s)}\{\mu B_k(y_s(x_1,\ldots,x_n))\} \quad (44)$$

and concomitantly we may also say that the pair (B_Y, μ_Y) sets a fuzzy partition for the output domain V.

Now, for $n > 1$, we will consider the Cartesian product L_X of fuzzy sets that designate the set of input variables X, namely

$$L_X = \prod_1^n L_j, \quad (45)$$

and use the symbol m to denote its cardinality, that is,

$$m = \#(L_X). \quad (46)$$

This leads to

$$m = \prod_1^n q(j). \quad (47)$$

Moreover, the ith element of the Cartesian product L_X. with $i = 1,2,\ldots,m$ associates univocally with a $n - tuple$ of linguistic terms $A_k(x_j)$. That is, we have an ordering

$$i \rightarrow \left[A_{k(1,i)}(x_1),\ldots,A_{k(j,i)}(x_j),\ldots,A_{k(n,i)}(x_n)\right], \quad (48)$$

where in the above $n - tuple\left[A_{k(j,i)}(x_j)\right]_1^n$ the index $k(j,i)$ takes a value out of the set $\{1,2,\ldots,q(j)\}$. Moreover, the elements of the Cartesian product L_X have form of a conjunction $Q^i(x_1,\ldots,x_n)$, namely

$$Q^i(x_1,\ldots,x_n) : \left[x_1\ is\ A_{k(1,i)}(x_1), x_2\ is\ A_{k(2,i)}(x_2),\ldots,x_n\ is\ A_{k(n,i)}(x_n)\right]. \quad (49)$$

And it is by no way redundant to write down the equivalence

$$L_X = \cup_1^m\{Q^i(x_1,\ldots,x_n)\}. \quad (50)$$

Meanwhile, the set of membership functions associated to the conjunction $Q^i(x_1,\ldots,x_n)$ will be denoted by means of the symbol $\mu Q^i(x_1,\ldots,x_n)$ and formally expressed through:

$$\mu Q^i(x_1, \ldots, x_n) = \cup_j \{ \mu A_{k(j,i)}(x_j) \}. \tag{51}$$

Moreover, for each value of the index $i = 1, 2, \ldots, m$ we can consider a relationship

$$Q^i(x_1, \ldots, x_n) \to R^i \tag{52}$$

being R^i a rule that associates an antecedent $Q^i(x_1, \ldots, x_n)$ to a consequent on the output value y_i. In the general fuzzy inference system R^i takes the form

$$R^i : \left\{ \begin{array}{lll} if & : & [x_1 \ is\ A_{k(1,i)}(x_1), x_2 \ is\ A_{k(2,i)}(x_2), \ldots, x_n \ is\ A_{k(n,i)}(x_n)] \\ then & : & [y_i(x_1, \ldots, x_n) \ is\ B_{k(i)}(y_i(x_1, \ldots, x_n))] \end{array} \right\} \tag{53}$$

with the set of tied membership functions $\mu Q^i(x_1, \ldots, x_n)$ given by Eq. (51).

For the case $n = 1$, we have only one input variable x_1. Hence, according to Eq. (37) the input space can be characterized by a number $q(1) \geq 1$ of linguistic terms, i.e., in this case we could have

$$A_X = \{ A_k(x_1) | k = 1, 2, \ldots, q(1) \} \tag{54}$$

and correspondingly from Eq. (40) μ_X becomes

$$\mu_X = \{ \mu A_1(x_1), .., \mu A_{q(1)}(x_1) \} \tag{55}$$

and by virtue of Eq. (47) for this case we have

$$m = q(1). \tag{56}$$

And regarding the arrangement of the elements of the Cartesian product established by Eq. (48), in this case $i = 1, 2, \ldots, m$, it suffices to advance a correspondence $i \to A_{k(1,i)}(x_1)$, where $k(1, i) = i$. Therefore, we can consider antecedent conjunctions

$$Q^i(x_1) : [x_1 \ is\ A_i(x_1)], \tag{57}$$

sustaining inferential rules R^i:

$$R^i : \left\{ \begin{array}{lll} if & : & x_1 \ is\ A_i(x_1) \\ then & : & y_i \ is\ B_i(x_1) \end{array} \right\}. \tag{58}$$

We also have

$$\mu Q^i(x_1) = \{\mu A_i(x_i)\}. \tag{59}$$

In summary, for case $n \geq 1$ as stated by Eqs. (35)–(59) we may consider a general fuzzy inference system F as a construct including an application $F: X \to Y$ characterized by sets of fuzzy partitions (A_X, μ_X) and (B_Y, μ_Y), the set inference rules $R = \mathrm{U}_1^m\{R^i\}$ and a defuzzification operator D that associates to the fuzzy set $\left[y_i\, is\, B_{k(i)}(y_i)\right]$ in Eq. (53) or a crisp value y_i in V.

In the Takagi-Sugeno-Kang fuzzy inference system representation, we consider decision rules R^i having an antecedent $Q^i(x_1, \ldots, x_n)$ of the form given by Eq. (49) but with a consequent taking a crisp functional form $y_i = f^i(x_1, x_2, \ldots, x_n)$. That is, for $n > 1$, in a TSK system we may envision inference rules

$$R^i : \left\{ \begin{array}{lll} if & : & [x_1\ is\ A_{k(1,i)}(x_1), x_2\ is\ A_{k(2,i)}(x_2), \ldots, x_n\ is\ A_{k(n,i)}(x_n)] \\ then & : & y_i = f^i(x_1, x_2, \ldots, x_n) \end{array} \right\} \tag{60}$$

for $i = 1, 2, \ldots, m$, and with $\mu Q^i(x_1, \ldots, x_n)$ given by Eq. (51).

Particularly, in defining a TSK system the function $f^i(x_1, x_2, \ldots, x_n)$ could have a linear form, that is,

$$f^i(x_1, x_2, \ldots, x_n) = p_1^i x_1 + \cdots + p_n^i x_n + p_{n+1}^i, \tag{61}$$

where the numbers p_j^i stand for parameters that can be empirically identified from data and being at least one nonzero. Nevertheless, in the general Takagi-Sugeno-Kang settings $f^i(x_1, x_2, \ldots, x_n)$ can take a nonlinear form.

An important concept tied to a TSK fuzzy model is that of the firing strength $\vartheta^i(x_1, x_2, \ldots, x_n)$ of the antecedent of a rule R^i. For the case $n > 1$ where the rules involve in their antecedents conjunctions of the form $Q^i(x_1, \ldots, x_n)$ (cf. Eq. (62)) the firing strength $\vartheta^i(x_1, x_2, \ldots, x_n)$ is obtained through the algebraic product of involved membership functions $\mu Q^i(x_1, \ldots, x_n)$, formally

$$\vartheta^i(x_1, x_2, \ldots, x_n) = \prod_{j=1}^{n} \mu A_{k(j,i)}(x_j), \tag{62}$$

and the normalized firing strength $\varphi^i(x_1, x_2, \ldots, x_n)$ is defined through

$$\varphi^i(x_1, x_2, \ldots, x_n) = \frac{\vartheta^i(x_1, x_2, \ldots, x_n)}{\sum_i \vartheta^i(x_1, x_2, \ldots, x_n)}. \tag{63}$$

The final output y of the Takagi-Sugeno-Kang inference system, is the weighted average of all rule outputs, computed as

$$y = \sum_i \varphi^i(x_1, x_2, \ldots, x_n) f^i(x_1, x_2, \ldots, x_n), \tag{64}$$

and identified as the output variable of the modeled system. The steps of the derivation of the formulae for the case $n = 1$ of a TSK fuzzy inference system are alike those establishing Eqs. (54)–(59) of the general fuzzy inference system and are elucidated in the methods section.

Appendix 4: Parameter Estimation for the TSK Fuzzy Model

In doing the parameter identification tasks, we bear in mind that the specification of the structure and the estimation of parameters of the TSK fuzzy model interrelates in such a way that none of them can be independently achieved without considering the other. Moreover, in the first stage of the involved identification tasks we made use of Substractive Clustering [8, 9], an efficient procedure endowing interesting advantages because it does not depend on optimization methods. Subtractive Clustering mainly relies on a measure of density of each data in the input space X. The goal is finding regions in the input space with high data densities. The data with the highest number of neighbors is selected as the center for a group. The data of the selected group that fall within a pre-specified fuzzy radius are then withdrew, and the algorithm keeps on searching for new data with the largest number of neighbors. This process is recursively applied until all the data are examined. Once Subtractive Clustering is completed, the number of decision rules can be known as each group associates to one of them. Moreover, the parameters tied to the membership functions characterizing the fuzzy sets of the antecedents of the rules are also estimated at this SC stage, and these are plugged into a recursive minimum squared (RLS) routine to obtain estimates of the parameters of the linear functions of the consequents of the rules. Following Jang et al. [20], we now explain the steps involved in the formalization of the RLS algorithm.

In the general least-squares problem, the output of a model y is given by a linearly parameterized expression known as a regression function, formally

$$\boldsymbol{y} = \gamma_1 h_1(\boldsymbol{u}) + \gamma_2 h_2(\boldsymbol{u}) + \cdots + \gamma_n h_n(\boldsymbol{u}) \tag{65}$$

where $\boldsymbol{u} = [u_1, \ldots, u_p]^T$ is the model's input values vector, $h_1(\boldsymbol{u}), \ldots, h_n(\boldsymbol{u})$ are known functions of $\boldsymbol{u}$, and $\gamma_1, \ldots, \gamma_n$ called regression coefficients are parameters to be estimated.

Drooping the time variable for the sake of simplifying notation, we have that the general output of the TKS model of Eq. (7) becomes

$$w_{TSK}(t) = \varphi^1(a)f^1(a) + \varphi^2(a)f^2(a) \tag{66}$$

where $\varphi^i(a)$ are defined by Eq. (16),

$$f^1(a) = p_1^1 a + p_2^1 \tag{67}$$

$$f^2(a) = p_1^2 a + p_2^2. \tag{68}$$

Replacing Eqs. (67) and (68) into (66) we have

$$w_{TSK}(t) = \varphi^1(a)\left[p_1^1 a + p_2^1\right] + \varphi^2(a)\left[p_1^2 a + p_2^2\right]. \tag{69}$$

Then, rearranging we obtain that the regression function tied to model w_{TSK} expressed in form (65) becomes

$$w_{TSK} = \varphi^1(a)p_1^1 a + \varphi^1(a)p_2^1 + \varphi^2(a)p_1^2 a + \varphi^2(a)p_2^2 a, \tag{70}$$

where $\boldsymbol{a} = [a]^T$ stand for the model's input values vector, and p_1^1, p_2^1, p_1^2 and p_2^2 are the unknown parameters.

In order to get estimates for the involved parameters, we have to take into account that in the present settings the target system to be modeled involves an input-output relationship $a \to w(a)$ being a the descriptor variable leaf area and $w(a)$ standing for the leaf biomass response. Therefore, we have a training data set composed of data pairs $(a_i : w_i)$, for $i = 1, \ldots, m$ representing replicates of the addressed input-output relationship. Therefore, in order to identify the unknown parameters p_1^1, p_2^1, p_1^2 and p_2^2, in (70) we must fill in for each data pair $(a_i : w_i)$, into Eq. (70) in order to obtain the set of m linear equations:

$$\left\{\begin{array}{lcl} \varphi^1(a_1)p_1^1 a_1 + \varphi^1(a_1)p_2^1 + \varphi^2(a_1)p_1^2 a_1 + \varphi^2(a_1)p_2^2 & = & w_1 \\ \varphi^1(a_2)p_1^1 a_2 + \varphi^1(a_2)p_2^1 + \varphi^2(a_2)p_1^2 a_2 + \varphi^2(a_2)p_2^2 & = & w_2 \\ \vdots & \vdots & \vdots \\ \varphi^1(a_m)p_1^1 a_m + \varphi^1(a_m)p_2^1 + \varphi^2(a_m)p_1^2 a_m + \varphi^2(a_m)p_2^2 & = & w_m \end{array}\right\} \tag{71}$$

In seeking for a solution to the above system we notice that it can be equivalently written in the concise form

$$BP = w, \tag{72}$$

where B is the $m \times n$ matrix,

$$B = \begin{Bmatrix} \varphi^1(a_1)a_1\varphi^1(a_1)\varphi^2(a_1)a_1\varphi^2(a_1) \\ \varphi^1(a_2)a_2\varphi^1(a_2)\varphi^2(a_2)a_2\varphi^2(a_2) \\ \vdots \\ \varphi^1(a_m)a_m\varphi^1(a_m)\varphi^2(a_m)a_m\varphi^2(a_m) \end{Bmatrix} \tag{73}$$

and P the $n \times 1$ vector of unknown parameters,

$$P = \begin{Bmatrix} p_1^1 \\ p_2^1 \\ p_1^2 \\ p_2^2 \end{Bmatrix} \tag{74}$$

And being w the $m \times 1$ output values vector:

$$w = \begin{Bmatrix} w_1 \\ w_2 \\ \vdots \\ w_m \end{Bmatrix}. \tag{75}$$

The i-th row of the data matrix $\left[B \vdots \boldsymbol{w}\right]$ is denoted by $\left[b_i^T, w_i\right]$ and formally represented by,

$$b_i^T = \left[\varphi^1(a_i)a_i\varphi^1(a_i)\varphi^2(a_i)a_i\varphi^2(a_i)\right] \tag{76}$$

Then, Eq. (72) is modified to incorporate an error vector e in order to account for random noise or modeling error, that is,

$$BP + e = w. \tag{77}$$

Since $e = w - BP$ then $e^T e = (w - BP)^T (w - BP)$, and if we let $E(P) = e^T e$ we will have

$$E(P) = \sum_{i=1}^{m} \left(w_i - b_i^T P\right).^2 \tag{78}$$

We call $E(P)$ the sum of squared errors and we need to search for $\hat{P}$, a characterization of the vector P, which minimizes $E(P)$. Moreover, the vector $\hat{P}$ is called the least-squares estimator (LSE) of P and since $E(P)$ is in quadratic form, $\hat{P}$ is unique. It turns out that $\hat{P}$ satisfies the normal equation

$$B^T B\hat{P} = B^T w. \tag{79}$$

Furthermore, $\hat{P}$ is given by

$$\hat{P} = \left(B^T B\right)^{-1} B^T w \tag{80}$$

A k-order least squares estimator $\hat{P}_k$ of $\hat{P}$ defined by means of the expression

$$\hat{P}_k = \left(B^T B\right)^{-1} B^T w, \tag{81}$$

is a characterization of $\hat{P}$ that associates to k data pairs taken out of the training data set $(a_i : w_i)$. Once we have obtained $\hat{P}_k$ we can get the succeeding estimator $\hat{P}_{k+1}$ with a minimum of effort, by using use the recursive least-squares estimator (RLSE) technique, a procedure where the kth$(1 \leq k \leq m)$ row of $[B \vdots \boldsymbol{w}]$, denoted by $[b_k^T \vdots w_k]$, is recursively obtained. In what follows, we will explain the formulism sustaining the RLSE method.

A new data pair $\left(b_{k+1}^T ; w_{k+1}\right)$ becomes available as the $(k+1)$th entry in the data set, producing the $\hat{P}_{k+1}$ estimate,

$$\hat{P}_{k+1} = \left(\begin{bmatrix} B \\ b_{k+1}^T \end{bmatrix}^T \begin{bmatrix} B \\ b_{k+1}^T \end{bmatrix} \right)^{-1} \begin{bmatrix} B \\ b_{k+1}^T \end{bmatrix}^T \begin{bmatrix} \boldsymbol{w} \\ w_{k+1} \end{bmatrix}. \tag{82}$$

Further, in order to simplify the notation, the pair $\left(b_{k+1}^T ; w_{k+1}\right)$ will be symbolized by $(b^T ; w)$ and we also introduce the $n \times n$ matrices H_k and H_{k+1} defined by means of

$$H_k = \left(B^T B\right)^{-1}, \tag{83}$$

and

$$H_{k+1} = \left(\begin{bmatrix} B \\ b^T \end{bmatrix}^T \begin{bmatrix} B \\ b^T \end{bmatrix} \right)^{-1} \tag{84}$$

or equivalently

$$H_{k+1} = \left(B^T B + bb^T\right)^{-1}.$$

Then H_k and H_{k+1} are related through

$$H_{k+1} = \left(H_k^{-1} + bb^T\right)^{-1}, \tag{85}$$

Therefore, using H_k from Eq. (83) and H_{k+1} from Eq. (85), we explain why Eqs. (81) and (82) can be equivalently written in the form

$$\hat{P}_k = H_k B^T w \tag{86}$$

and

$$\hat{P}_{k+1} = H_{k+1}\left(B^T w + bw\right) \tag{87}$$

From (86) we have $B^T w = H_k^{-1}\hat{P}_k$, then replacing this result in Eq. (87) we get

$$\hat{P}_{k+1} = H_{k+1}\left(H_k^{-1}\hat{P}_k + bw\right). \tag{88}$$

Now, from Eq. (85) we have $H_k^{-1}\hat{P} = \left(H_{k+1}^{-1} - bb^T\right)\hat{P}_k$, so replacing this result in the above expression we get

$$\hat{P}_{k+1} = H_{k+1}\left[\left(H_{k+1}^{-1} - bb^T\right)\hat{P}_k + bw\right],$$

then simplifying yields

$$\hat{P}_{k+1} = \hat{P}_k + H_{k+1} b\left(w - w^T\hat{P}_k\right) \tag{89}$$

Thus $\hat{P}_{k+1}$ can indeed be recursively specified in terms of the previous estimate $\hat{P}_k$ and the new data pairs $(b^T; w)$. Moreover, the current estimate $\hat{P}_{k+1}$ is expressed as the previous one $\hat{P}_k$ plus a correcting term based on the new data $(b^T; w)$; this amending term can be interpreted as an adaptation gain vector H_{k+1} multiplied by a prediction error $\left(w - b^T\hat{P}_k\right)$ linked to the previous estimator $\hat{P}_k$.

Calculating H_{k+1} as given by Eq. (84) is computationally costly and requires the adaptation of a recursive formula. From Eq. (85), we have

$$H_{k+1} = \left(H_k^{-1} + bb^T\right)^{-1}$$

Applying the matrix inversion formula of Lemma 5.6 in Jang et al. [11] with $A = H_k^{-1}$, $B = b$, and $C = b^T$, we obtain the following recursive formula for H_{k+1} in terms of H_k:

$$H_{k+1} = H_k - H_k b\left(I + b^T H_k b\right)^{-1} b^T H_k,$$

equivalently,

$$H_{k+1} = H_k - \frac{H_k b b^T H_k}{I + b^T H_k b} \tag{90}$$

In summary, the **recursive least-squares estimator** for the problem of $AP + e = y$, where the kth$(1 \le k \le m)$ row of $\left[B \vdots w\right]$, denoted by $\left[b_k^T \vdots w_k\right]$, is sequentially obtained, can be calculated as follows:

$$\begin{cases} \hat{P}_{k+1} = \hat{P}_k + H_{k+1} b_{k+1} \left(w_{k+1} - b_{k+1}^T \hat{P}_k\right) \\ H_{k+1} = H_k - \frac{H_k b_{k+1} b_{k+1}^T H_k}{I + b_{k+1}^T H_k b_{k+1}} \end{cases}, \tag{91}$$

where in establishing Eq. (91) we have taken into account Eq. (90) and the fact that we had formerly established the convention that in order to shorten the presentation the pair $\left(b_{k+1}^T; w_{k+1}\right)$ would be symbolized by the simpler expression $(b^T; w)$.

References

1. R.M. McCloskey, R.K.F. Unworthy, Decreasing seagrass density negatively influences associated fauna, vol. 3 (PeerJ, 2015), p. e1053
2. M.L. Plummer, C.J. Harvey, L.E. Anderson, A.D. Guerry, M.H. Ruckelshaus, The role of eelgrass in marine community interactions and ecosystem services: results from ecosystem-scale food web models. Ecosystems **16**(2), 237–251 (2013)
3. E.I. Paling, M. Fonseca, M.M. van Katwijk, M. van Keulen, Seagrass restoration, ed. by M.E. Gerardo, E.W. Perillo, R.C. Donald, M.B. Mark. *Coastal Wetlands: An Integrated Ecosystem Approach*, 1st edn. (Elsevier Science, 2009) pp. 1–62
4. W.T. Li, Y.K. Kim, J.I. Park, X.M. Zhang, G.Y. Du, K.S. Lee, Comparison of seasonal growth responses of *Zostera marina* transplants to determine the optimal transplant season for habitat restoration. Ecol. Eng. **71**, 56–65 (2014)
5. M.S. Fonseca, Addy revisited: what has changed with seagrass restoration in 64 years? Ecol. Restor. **29**(1–2), 73–81 (2011)
6. H. Echavarría-Heras, C. Leal-Ramírez, E. Villa-Diharce, E. Montiel-Arzate, On the appropriateness of an allometric proxy for nondestructive estimation of average biomass of leaves in shoots of eelgrass (*Zostera marina*). Submitted, (2017)
7. H.A. Echavarría-Heras, C. Leal-Ramírez, E. Villa-Diharce, N.R. Cazarez-Castro, The effect of parameter variability in the allometric projection of leaf growth rates for eelgrass (*Zostera marina L.*) II: the importance of data quality control procedures in bias reduction. Theor. Biol. Med. Model. **12**(30), 2015
8. S.L. Chiu, Fuzzy model identification based on cluster estimation. J. Intell. Fuzzy Syst. **2**(3), 267–278 (1994)
9. J.R. Castro, O. Castillo, M.A. Sanchez, O. Mendoza, A. Rodríguez-Díaz, P. Melin, Method for higher order polynomial sugeno fuzzy inference systems. Inf. Sci. **351**, 76–89 (2016)
10. L.X. Wang, J.M. Mendel, Fuzzy basis functions, universal approximation, and orthogonal least-squares learning. IEEE Trans. Neural Netw. **3**(5), 807–814 (1992)

11. J.S.R. Jang, C.T. Sun, E.S. Mizutani, *Neuro-fuzzy and soft computing: a computational approach to learning and machine intelligence* (Prentice Hall, USA, 1997)
12. L.I.K. Lin, A concordance correlation coefficient to evaluate reproducibility. Biometrics **45**, 255–268 (1989)
13. C. Leal-Ramírez, H.A. Echavarría-Heras, O. Castillo, Exploring the suitability of a genetic algorithm as tool for boosting efficiency in monte carlo estimation of leaf area of eelgrass, ed. by P. Melin, O. Castillo, J. Kacprzyk. Design of Intelligent Systems Based on Fuzzy Logic, Neural Networks and Nature-Inspired Optimization. Stud. Comput. Intell. vol. 601, (Springer, 2015) pp. 291–303
14. C. Leys, O. Klein, P. Bernard, L. Licata, Detecting outliers: do not use standard deviation around the mean, use absolute deviation around the median. J. Exp. Soc. Psychol. **49**(4), 764–766 (2013)
15. P.J. Huber, *Robust statistics* (Wiley, New York, 1981)
16. M. Sugeno, G.T. Kang, Structure identification of fuzzy model. Fuzzy Sets Syst. **28**, 15–33 (1988)
17. J.R. Castro, O. Castillo, P. Melin, A. Rodríguez-Díaz, A hybrid learning algorithm for a class of interval type-2 fuzzy neural networks. Inf. Sci. **179**(13), 2175–2193 (2009)
18. D.W. Marquardt, An algorithm for least-squares estimation of nonlinear parameters. J. Soc. Ind. Appl. Math. **11**(2), 431–441 (1963)
19. M.K. Transtrum, J.P. Sethna, Improvements to the Levenberg-Marquardt algorithm for nonlinear least-squares minimization. Cornell University, USA, (2012). doi:arXiv:1201.5885
20. S. Jang, ANFIS: adaptive-network-based fuzzy inference system. IEEE Trans. Syst Man Cybern. **23**, 665–685 (1993)
21. D. Hui, R.B. Jackson, Uncertainty in allometric exponent estimation: a case study in scaling metabolic rate with body mass. J TheorBiol **249**, 168–177 (2007)
22. C. Leal-Ramírez, H.A. Echavarría-Heras, O. Castillo, E. Montiel-Arzate, On the use of parallel genetic algorithms for improving the efficiency of a monte carlo-digital image based approximation of eelgrass leaf area I: comparing the performances of simple and master-slaves structures, ed. by P. Melin, O. Castillo, J. Kacprzyk. Nature-Inspired Design of Hybrid Intelligent Systems, Volume 667 of the series Studies in Computational Intelligence, pp. 431–455, Springer (2016)
23. J. Miller, Reaction time analysis with outlier exclusion: Bias varies with sample size. Q. J. Exp. Psychol. **43**(4), 907–912 (1991)
24. L.I.K. Lin, Assay validation using the concordance correlation coefficient. Biometrics **48**, 599–604 (1992)
25. G.B. McBride, A proposal for strength-of-agreement criteria for lin's concordance correlation coefficient. NIWA Client Report: HAM2005-062; National Institute of Water & Atmospheric Research: Hamilton, New Zealand, May 2005. Available online: http://www.medcalc.org/download/pdf/McBride2005.pdf

Part VI
Optimization and Evolutionary Algorithms

Modeling and Project Portfolio Selection Problem Enriched with Dynamic Allocation of Resources

Daniel A. Martínez-Vega, Laura Cruz-Reyes, Claudia Gomez-Santillan, Nelson Rangel-Valdez, Gilberto Rivera and Alejandro Santiago

Abstract The problems of the real world, within which the variable *time* is present, have involved continuous changes. These problems usually change over time in their objectives, constraints or parameters. Therefore, it is necessary to carry out a readjustment when calculating their solution. This paper proposes an original way of approaching the project portfolio selection problem enriched with dynamic allocation of resources. A new mathematical model is proposed formulating this multi-objective optimization problem, as well as its exact and approximate solution, the latter based on four of the algorithms that in our opinion stand out in the state of the art: Archive-Based hybrid Scatter Search, MultiObjective Cellular, Nondominated Sorting Genetic Algorithm II and Strength Pareto Evolutionary

D. A. Martínez-Vega (✉) · A. Santiago
Tecnológico Nacional de México/Instituto Tecnológico de Tijuana, Tijuana, Mexico
e-mail: adalbertovega@gmail.com

A. Santiago
e-mail: alejandro.santiagopi@mail.uca.es

L. Cruz-Reyes · C. Gomez-Santillan
Tecnológico Nacional de México/Instituto Tecnológico de Ciudad Madero, Ciudad Madero, Mexico
e-mail: lauracruzreyes@itcm.edu.mx

C. Gomez-Santillan
e-mail: claudia.gomez@itcm.edu.mx

N. Rangel-Valdez
CONACYT, Tecnológico Nacional de México/Instituto Tecnológico de Ciudad Madero, Ciudad Madero, Mexico
e-mail: nrangelva@conacyt.mx

G. Rivera
Universidad Autónoma de Ciudad Juárez, Ciudad Juárez, Mexico
e-mail: gilberto.rivera@uacj.mx

A. Santiago
Universidad de Cádiz, Cádiz, Spain

O. Castillo et al. (eds.), *Fuzzy Logic Augmentation of Neural and Optimization Algorithms: Theoretical Aspects and Real Applications*, Studies in Computational Intelligence 749, https://doi.org/10.1007/978-3-319-71008-2_26

Algorithm 2. We experimentally demonstrate the benefits of our proposal and leave open the possibility that its study will apply to large-scale problems.

Keywords Dynamic allocation of resources · Dynamic portfolio Enriched problem · JMetal · ABYSS · MOCell · NSGA-II · SPEA 2

1 Introduction

In general, there is a growing need arising from a variety of factors—such as budget adjustments—that demand better output from the resources that are available, which in most cases are increasingly scarce, to generate a greater advantage competitively [1].

Dynamic allocation of resources is key to project portfolio management; this problem consists of monitoring and periodic adjustment of actions, these operations improve the quality of the portfolio due to the greater benefit they produce over time [2]. In Fig. 1 is represented the course of four years, in which the budget is reassigned for different activities, this is an example of what happens in the real world.

An important point within the problem of the dynamic allocation of resources is the need to achieve a correct selection of limited financial, human and technological resources that entail the financing of projects that confer a greater competitive advantage by the strategy adopted by the organization.

In practice, mathematical and heuristic models have limited utility because they do not consider, among others, the intrinsic dynamic nature of portfolio processes [3]. Among the few research efforts, there is a system for project portfolio generation based on a dynamic allocation of resources, which is presented in [2] as a patent. The system allows multi-objective optimization of a project portfolio with resource constraints as a function of time, such as labor and budget constraints. However, the available information is not enough to reproduce the proposed mathematical model and make comparison on it.

The problem in which we focus our study until our knowledge is unprecedented in a similar way. Due to the great thriving that the handling of dynamical problems has recently been having, especially due to the computational power that allows us

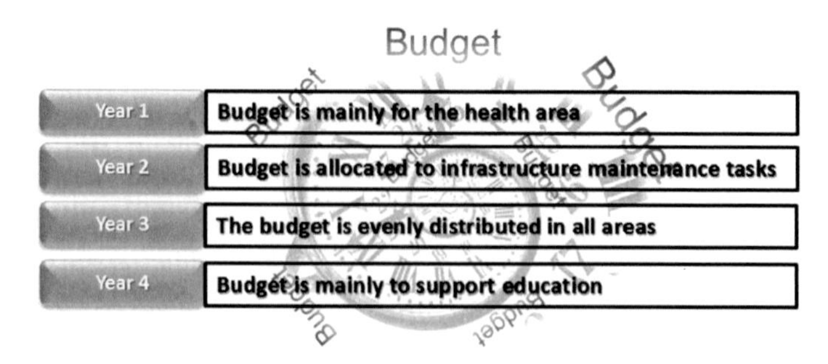

Fig. 1 Budget over time, dynamic allocation of resources

to simulate situations of the world more closely to the real thing, we decided to investigate in state of the art the case study problem. But, on the one hand, we note that to our knowledge there are few researchers who have tackled the subject, and on the other hand none of them see it from the point of view presented here, which we consider relevant due to that its range is very great.

2 Background

In this section, a basic definition of project portfolio selection problem is first given. The problem formulated below has been solve with different multi-objective algorithms: Archive-Based Hybrid Scatter Search (ABYSS) [4], MultiObjective Cellular (MOCell) [5], Nondominated Sorting Genetic Algorithm II (NSGA-II) [6], Strength Pareto Evolutionary Algorithm 2 (SPEA2) [7].

2.1 Project Portfolio Selection Problem

One of the main management tasks in public sector organizations, foundations, research centers and companies conducting research and development is to evaluate a set of projects that compete for financial support, to select those that contribute the maximum benefit to the organization. This subset constitutes a project portfolio [8].

The modeling of project portfolio selection problems is based on the following premises:

There is a set of N well-defined projects, each of them perfectly characterized from the point of view of the economic benefits it can provide and its budgetary requirements.

It is about deciding which subset of projects make up the ideal portfolio so that a certain measure of quality is optimized. If uncertainty and risk are ignored, and the benefit generated by each project is known, an attempt is made to maximize the net present value of each benefit associated with the portfolio [9]. It is assumed that if a project is accepted in the portfolio, it will receive all the support it requests.

2.2 Basic Formulation of the Project Portfolio Selection Problem

In any decision problem, the person making the final decision is known as the Decision Maker (DM). He is a person (or group), whose system of preferences is determinant in the solution of problems that consider several objectives, which possibly are in conflict with each other [10].

Let N projects of social interest that meet certain minimum requirements of acceptability to be supported. Each project has associated a region A, an area G and a cost C:

$$A = \langle a_1, a_2, \ldots, a_k \rangle,$$

$$G = \langle g_1, g_2, \ldots, g_r \rangle,$$

$$C = \langle c_{1,}c_{2,}, \ldots, c_N \rangle,$$

where c_j is an amount of money that fully satisfies the budget requirements of the j project [11].

Let $X = \langle x_1, x_2, \ldots, x_N \rangle$, the set of N projects where:

$$x_i = \begin{cases} 1, & \text{si el i} - \text{ésimo proyecto es soportado} \\ 0, & \text{en otro caso.} \end{cases} \quad (1)$$

One of the most complex tasks is the evaluation of the projects, which considers the contribution that each project has to each of the objectives set by the institution that provides the economic resources. The level of contribution (benefit) of each project x_i to the different objectives can be represented by the vector $f_{x_i} = \langle f_{x_i1}, f_{x_i2}, \ldots, f_{x_ip} \rangle$; which is called the benefit vector for project i, considering p objectives.

Let the matrix F of dimension $N \times p$ be the profit matrix (Table 1), where p is the total number of objectives and N is the total number of projects. Each row represents the benefit vector for the ith project.

Let P be the total amount of financial resources available for distribution to different projects. Since each project has a cost c_i, any project portfolio must comply with the following budget constraint:

$$\left(\sum_{i=1}^{N} x_i c_i \right) \leq P \quad (2)$$

Assume as possible that there are budgetary restrictions for each investment area. So if P_l is the budget dedicated to area l, and there is a minimum budget $P_{l_{min}}$ and a maximum budget $P_{l_{max}}$ established such that:

Table 1 Profit matrix F

Objectives	Projects			
	1	2	…	N
1	$f_{1,1}$	$f_{1,2}$	…	$f_{N,1}$
2	$f_{2,1}$	$f_{2,2}$	…	$f_{N,2}$
…	…	…	…	…
p	$f_{p,1}$	$f_{p,2}$	…	$f_{N,p}$

$$P_{l_{min}} \leq P_l \leq P_{l_{max}}. \tag{3}$$

The budget constraint by area that each portfolio must fulfill is given by:

$$P_l = \sum_{i=1}^{N} x_i c_i a_{i_l} \tag{4}$$

where

a_{i_l} is a binary variable that indicates whether project i belongs to the socio-economic area l

On the other hand, each project benefits a particular region, and as with the areas, there is a minimum budget $P_{r_{min}}$ and a maximum budget $P_{r_{max}}$ per established region such that:

$$P_{r_{min}} \leq P_r \leq P_{r_{max}} \tag{5}$$

where the budget by region for each portfolio is given by:

$$P_r = \sum_{i=1}^{N} x_i c_i g_{i_r} \tag{6}$$

where

g_{i_r} is a binary variable that indicates whether Project i belongs to region r or not.

The quality of a portfolio X depends on the benefits of its projects and is represented by the quality vector $Z(X)$, whose components are at the same time quality values in relation to each of p objectives of the projects:

$$Z(X) = \langle z_1, z_2, z_3, \ldots, z_p \rangle \tag{7}$$

where

$$z_j(X) = \sum_{i=1}^{N} x_i f_{j,i} \tag{8}$$

being f the profit matrix whose rows represent each of the p objectives and their columns each of the N projects.

Let R_F be the feasible portfolio space, the solution of the portfolio selection problem is to find one or more portfolios satisfying (9).

$$\max_{x \in R_F}\{z(x)\} \tag{9}$$

That is, the only accepted solutions will be those that meet the constraints established by (2) to (6).

3 Mathematical Model Proposed for Project Portfolio Selection Problem Enriched with Dynamic Allocation of Resources

In this section, we describe a new mathematical model which we propose to formulate the project portfolio selection problem enriched with dynamic allocation of financial resources. This model is an extension of the basic model presented in Sect. 2.

The project portfolio selection problem enriched with dynamic allocation of resources is a combination that frequently occurs in organizations because that person (or people) who decides in which the budget allocated for a certain period will be invested must be monitoring the results obtained from their decisions. Based on the results, he will have to make the decisions on future investments; this results in greater profits or greater losses, which can be accumulated over the years, due to the dependence that exists between them.

Decision variables:

$x_{i,t}$ = Binary matrix representing if project i is financed (1) at time t.
$P_{l,t}$ = Budget required by the area l in the year t.
$P_{r,t}$ = Budget required by the region r in the year t.

Constants:

N = Number of projects.
O = Number of objectives.
T = Number of years to calculate.
na = Number of areas.
nr = Number of regions.
i = Index for projects where $i \in \{1, 2, \ldots, N\}$.
a = Index of areas where $a \in \{1, 2, \ldots, na\}$.
r = Index of regions where $r \in \{1, 2, \ldots, nr\}$.
o = Index for objectives where $o \in \{1, 2, \ldots, O\}$.
$a_{i,l}$ = Binary matrix indicating whether project i belongs to area l.
$g_{i,r}$ = Binary matrix indicating whether project i belongs to region r.
P_t = Annual budget for year t.
$P_{l_{min},t}$ = Minimum budget for area l in year t.
$P_{l_{max},t}$ = Maximum budget for area l in year t.
$P_{r_{min},t}$ = Minimum budget for region r in year t.

$P_{r_{max},t}$ = Maximum budget for region r in year t.
$b^{i}_{o,t}$ = Benefit of project i, to objective o at time t.
$c_{i,t}$ = Matrix containing the costs of each project i in time t.

Objective function:

$$\max_{x \in R_F}\{Z(x)\} \tag{10}$$

where:

$$Z(\mathrm{x}) = \langle z_1(\mathrm{x}), z_2(\mathrm{x}), \ldots, z_N(\mathrm{x})\rangle, \tag{11}$$

$$z_i(x) = \sum_{o=1}^{O}\sum_{t=1}^{T} b^{i}_{o,t} x_{i,t}. \tag{12}$$

Constraints:

$$\left(\sum_{i=1}^{N} x_{i,t}c_{i,t}\right) \leq P_t \forall_t, \tag{13}$$

$$P_{l_{min},t} \leq P_{l,t} \leq P_{l_{max},t} \forall_{l,t}, \tag{14}$$

$$P_{r_{min},t} \leq P_{r,t} \leq P_{r_{max},t} \forall_{r,t}, \tag{15}$$

$$P_{l,t} = \sum_{i=1}^{N} x_{i,t}c_{i,t}a_{i,l} \forall_{l,t}, \tag{16}$$

$$P_{r,t} = \sum_{i=1}^{N} x_{i,t}c_{i,t}g_{i,r} \forall_{r,t}. \tag{17}$$

Equations (10) and (11) indicate that it is a problem that can be approached as a multi-objective problem in which the maximization of all the objectives is sought (where R_F is the feasible portfolio space). Eq. (12) breaks down the way of calculating the value of each of these objectives. Equations (13) to (17) show the constraints of this model:

Equation (13) deals with the total budget constraint per year.

Equation (14) indicates that the budget by area must be within a budget range for each year.

Equation (15) is similar to (11) but refers to the regions instead of the areas.

Equations (16) and (17) explain how the budget is calculated by area and region respectively. It is the accumulation of costs stored in the matrix c, costs will be added only when x(which represents those projects which are in the portfolio) and $\frac{a}{g}$

(which indicates which area or region respectively the project belongs) take the value of 1.

4 Experimentation and Results

In this section, we present the case of study and the results of the experimentation carried out.

4.1 Experimental Design

This work is the basis of a larger project that seeks to solve multi-objective realistic instances of the dynamic portfolio selection problem on the large scale. For feasibility purpose, the experimentation presented in this work used manually generated mono-objective instances, however, due to the intended final objective, the algorithms that were compared are those commonly used in the literature to solve multi-objective problems. They were adapted to solve the problem modeled mathematically in Sect. 3.

4.1.1 Hardware and Software

The hardware and software used in this work are shown in Table 2.

4.1.2 Description of the Metaheuristics

A pre-evaluation of the instances was carried out using the mathematical programming tool ILOG CPLEX, to obtain the optimal solutions. It allowed making a comparison based on the error between the results of the metaheuristics ABYSS, MOCell, NSGA-II, SPEA2. These algorithms were adapted to solve the dynamic portfolio problem using the framework JMetal 5.2 [12].

Table 2 Hardware and software

Hardware	Software
Processor Inside i3 2.1 GHz	S.O. Windows 10 × 64
6 GB RAM	Java language
HDD 7200 rpm HDD 7200 rpm	JDK 1.8.0
	IDE NetBeans 8.0.1
	Framework JMetal 5.2

It should be noted that the stop criterion was the number of evaluations of the objective function, it was set in 25,000, the rest of the parameters for each metaheuristic is left with the configuration that JMetal handles by default.

4.2 Case Study: The Project Portfolio Problem Enriched with Dynamic Allocation of Resources

In this section, the instances used in our experimentation are discussed, an example of one of the instances used, a summary of the results obtained, and the analysis of algorithms using statistical tests.

4.2.1 Description Instances

For the experimentation, manually created mono-objective instances were used, whose names, number of projects and years to calculate are shown in Table 3.

In Table 4 is shown a scalar example of what would be an instance for three projects, two years to calculate, two areas and three regions.

4.3 Results

For this experiment, 30 executions were performed on each metaheuristic for each of the instances described in Sect. 4.2.1. In Tables 5 and 6, the results and error averages of the executions are shown.

The worst results are shaded in light gray, while the best results are shaded in dark gray.

Next, a statistical analysis of the error obtained in each instance is presented to determine if there is a significant difference in the performance of the four algorithms compared.

Table 3 Instances

Name	Number of projects	No. of years to calculate
ADR_20p_5y	20	5
ADR_20p_10y	20	10
ADR_20p_20y	20	20
ADR_100p_5y	100	5
ADR_100p_10y	100	10
ADR_100p_20y	100	20

Table 4 Instance example

Line	Content
1	2 //Years to calculate
2	3 //Number of projects
3	1 //Number of objectives
4	2 //Number of areas
5	3 //Number of regions
6	
7	//For year 1
8	15000 //Total budget
9	
10	//Minimum and maximum budgets
11	2500 7000 //Area 1
12	3000 5000 //Area 2
13	3000 8000 //Region 1
14	2500 7000 //Region 2
15	0 5000 //Region 3
16	
17	//Cost Area Region Objective 1
18	2500 1 2 400
19	3000 2 3 200
20	4500 1 1 300
21	
22	//For year 2
23	10000 //Total budget
24	
25	//Minimum and maximum budgets
26	2500 5000 //Area 1
27	1000 3000 //Area 2
28	2000 4000 //Region 1
29	1500 6000 //Region 2
30	1000 5000 //Region 3
31	
32	//Cost Area Region Objective 1
33	2500 2 3 500
34	3000 1 1 100
35	4500 2 2 400

We used Statistical Tests for Algorithms Comparison (STAC) [13], a web platform for the analysis of algorithms using statistical tests, in this case, the Friedman non-parametric test was applied. In cases where significant differences were found in the Friedman test, we proceeded to apply the Post hoc Holm test, which is widely used in the scientific community.

Table 5 Average results of 30 executions by metaheuristic

Instance	Optimum value	Average results			
		ABYSS	MOCell	NSGA-II	SPEA2
20p-5a	307,425	271,684	287,236	288,733	290,552
20p-10a	1,051,550	865,615	989,490	992,341	992,142
20p-20a	2,103,100	1,483,233	1,868,003	1,896,677	1,896,007
100p-5a	2,347,300	1,599,723	2,048,696	2,090,016	2,074,784
100p-10a	4,694,600	2,673,296	3,827,630	3,935,976	3,896,230
100p-20a	9,389,200	5,333,722	6,788,788	7,117,285	7,070,974

Table 6 Error rate of 30 executions by metaheuristic

Instance	Optimum value	Error rate			
		ABYSS	MOCell	NSGA-II	SPEA2
20p-5a	307,425	0.116261	0.065673	0.060802	0.054885
20p-10a	1,051,550	0.176820	0.059017	0.056306	0.056496
20p-20a	2,103,100	0.294740	0.111786	0.098152	0.098471
100p-5a	2,347,300	0.318484	0.127212	0.109608	0.116098
100p-10a	4,694,600	0.430559	0.184674	0.161595	0.170061
100p-20a	9,389,200	0.431930	0.276958	0.241971	0.246903

To perform the Friedman test, we establish the null hypothesis (H_0): "the mean of the results of two or more algorithms is the same and a significance level of 0.05".

For the post hoc Holm test, we establish the null hypothesis (H_0): "the mean of the results of each pair of algorithms compared is equal and with a significance level of 0.05".

Instance 20p5a

As shown in Table 7, the result of the Friedman test rejects H_0, that is, it indicates that there is a significant difference between the performances of the algorithms for this instance.

To determine the cause of this significant difference, we proceeded to apply the Post hoc test; the results are presented below Table 8.

The pairs marked according to H_0 of the test indicate that the average of these pairs of algorithms compared is not the same, and the pairs MOCell-NSGA_II and SPEA2-NSGA_II have a similar performance.

Instance 20p5a, 20p10a, 20p20a, 100p5a, 100p10a and 100p20a

The two previous tests were carried out with the rest of the instances, and the results are in the Tables 9 and 10.

Table 7 Result of Friedman test for Instance 20p5a

Statistic	p-value	Result
72.79407	0	H_0 is rejected

Table 8 Result of Post hoc test for instance 20p5a

Comparison	Statistic	Adjusted p-value	Result
SPEA-2 vs. ABYSS	7.5	0	H_0 is rejected
ABYSS vs. NSGA-II	6.2	0	H_0 is rejected
ABYSS vs. MOCell	4.3	0.00007	H_0 is rejected
SPEA-2 vs. MOCell	3.2	0.00412	H_0 is rejected
MOCell vs. NSGA-II	1.9	0.11487	H_0 is accepted
SPEA-2 vs. NSGA-II	1.3	0.1936	H_0 is accepted

Table 9 Results of Friedman tests

Instance	Statistic	p-value	Result
20p10a	46.60834	0	H_0 is rejected
20p20a	81.40609	0	H_0 is rejected
100p5a	122.39211	0	H_0 is rejected
100p10a	189.22742	0	H_0 is rejected
100p20a	244.01255	0	H_0 is rejected

Table 9 shows that according to the results of the Friedman tests all the H_0 were rejected, then there is a significant difference between the performances of the algorithms for all instances.

In Table 10, shaded lines show that the average of these pairs of algorithms compared is not the same, and the unshaded lines indicate that the pair of algorithms compared each has a similar performance.

We conclude that there is a significant difference in performance between the algorithms, and in 5 of the 6 instances analyzed, the SPEA2 and NSGA-II algorithms presented similar performance, obtaining the best results.

5 Conclusions and Future Work

In this paper, a new mathematical model was proposed to formulate the project portfolio selection problem with dynamic allocation of resources, its operation was verified through the experimentation described in Sect. 4.

This work is a precedent and a basis for the conformation of a benchmark for the solution (optimization) of the dynamic project portfolio selection problem. Mono-objective instances were generated manually and given solutions with state-of-the-art algorithms for the small scale; a multi-objective instance generator is currently in the process of being developed to maximize the proposed mathematical model.

The final results show that there is no better algorithm for all test cases analyzed. For these cases, the algorithm NSGA-II obtained the best results in 5 of the 6 test instances, on the other hand, ABYSS presented the poorest performance.

Table 10 Results of Post hoc tests

Instance	Comparison	Statistic	Adjusted p-value	Result
20p10a	SPEA2 vs. ABYSS	6.4	0	H_0 is rejected
	ABYSS vs. NSGA-II	6.3	0	H_0 is rejected
	ABYSS vs. MOCell	5.3	0	H_0 is rejected
	SPEA2 vs. MOCell	1.1	0.814	H_0 is accepted
	MOCellvs. NSGA-II	1	0.814	H_0 is accepted
	SPEA2 vs. NSGA-II	0.1	0.92034	H_0 is accepted
20p20a	SPEA2 vs. ABYSS	7.3	0	H_0 is rejected
	ABYSS vs. NSGA-II	6.7	0	H_0 is rejected
	ABYSS vs. MOCell	4	0.00025	H_0 is rejected
	SPEA2 vs. MOCell	3.3	0.0029	H_0 is rejected
	MOCellvs. NSGA-II	2.7	0.01387	H_0 is rejected
	SPEA2 vs. NSGA-II	0.6	0.54851	H_0 is accepted
100p5a	ABYSS vs. NSGA-II	8	0	H_0 is rejected
	SPEA2 vs. ABYSS	6.3	0	H_0 is rejected
	MOCellvs. NSGA-II	4.3	0.00007	H_0 is rejected
	ABYSS vs. MOCell	3.7	0.00065	H_0 is rejected
	SPEA2 vs. MOCell	2.6	0.01864	H_0 is rejected
	SPEA2 vs. NSGA-II	1.7	0.08913	H_0 is accepted
100p10a	ABYSS vs. NSGA-II	8.4	0	H_0 is rejected
	SPEA2 vs. ABYSS	6.1	0	H_0 is rejected
	MOCellvs. NSGA-II	4.9	0	H_0 is rejected
	ABYSS vs. MOCell	3.5	0.0014	H_0 is rejected
	SPEA2 vs. MOCell	2.6	0.01864	H_0 is rejected
	SPEA2 vs. NSGA-II	2.3	0.02145	H_0 is rejected
100p20a	ABYSS vs. NSGA-II	8	0	H_0 is rejected
	SPEA2 vs. ABYSS	6.9	0	H_0 is rejected
	MOCell vs. NSGA-II	4.9	0	H_0 is rejected
	SPEA2 vs. MOCell	3.8	0.00043	H_0 is rejected
	ABYSS vs. MOCell	3.1	0.00387	H_0 is rejected
	SPEA2 vs. NSGA-II	1.1	0.27133	H_0 is accepted

The conclusion about the performance of the algorithms is not absolute, but it shows the feasibility of our proposal is not absolute; it should be noted that these results are for the set of test instances used in this work. To obtain more meaningful conclusions in future work, it must be done a more exhaustive experimentation with a greater number of larger instances regarding the number of objectives, the number of projects and periods.

References

1. K. Weicker, *Evolutionary algorithms and dynamic optimization problems* (Der Andere Verlag, Berlin, 2003)
2. C.P.A. Santos, I.A. Lopez-Sanchez. Portfolio Generation Based on a Dynamic Allocation of Resources. U.S. Patent Application No. 14/485,339 (2014)
3. J. Pajares, A. López, A. Araúzo, C. Hernández, Project Portfolio Management, selection and scheduling. Bridging the gap between strategy and operations, in XIII *Congreso de Ingeniería de Organización* (pp. 1421–1429) (2009, April)
4 A.J. Nebro, F. Luna, E. Alba, B. Dorronsoro, J.J. Durillo, A. Beham, AbYSS: adapting scatter search to multiobjective optimization. IEEE Trans. Evol. Comput. 12(4), 439–457 (2008)
5 A.J. Nebro, J.J. Durillo, F. Luna, B. Dorronsoro, E. Alba, Design issues in a multiobjective cellular genetic algorithm, *in International Conference on Evolutionary Multi-Criterion Optimization*. (Springer Berlin Heidelberg, 2007, March) (pp. 126–140)
6 K. Deb, A. Pratap, S. Agarwal, T.A.M.T. Meyarivan, A fast and elitist multiobjective genetic algorithm: NSGA-II. IEEE Trans. Evol. Comput. 6(2), 182–197 (2002)
7 E. Zitzler, M. Laumanns, L. Thiele, SPEA2: improving the strength Pareto evolutionary algorithm (2001)
8 A. Nebro, F. Luna, B. Dorronsoro, J. Durillo, Un algoritmomultiobjetivobasadoen búsqueda dispersa. Quinto Congreso Español de Metaheurísticas, Algoritmos Evolutivos y Bioinspirados (MAEB 2007), pp. 175–182 (2007)
9 K.R. Davis, R. Davis, P.G. Mckeown, Modelos cuantitativos para administración. Grupo Editorial Iberoamérica (1986)
10 P. Sánchez, Propuesta de anteproyecto de tesis: Nuevos métodos de incorporación de preferencias en metaheurísticas multiobjetivo para la solución de problemas de cartera de proyectos (2012)
11 H. Jain, K. Deb, An improved adaptive approach for elitist nondominated sorting genetic algorithm for many-objective optimization, *in International Conference on Evolutionary Multi-Criterion Optimization* (Springer Berlin Heidelberg, 2013, March) (pp. 307–321)
12 J.J. Durillo, A.J. Nebro, jMetal: a Java framework for multi-objective optimization. Adv. Eng. Softw, **42**(10), 760–771 (2011)
13 I. Rodríguez-Fdez, A. Canosa, M. Mucientes, A. Bugarín, STAC: a web platform for the comparison of algorithms using statistical tests. *in Fuzzy Systems (FUZZ-IEEE), 2015 IEEE International Conference* on (2015, August) (pp. 1–8). IEEE

A Grey Mathematics Approach for Evolutionary Multi-objective Metaheuristic of Project Portfolio Selection

Fausto Balderas, Eduardo Fernandez, Claudia Gomez-Santillan, Laura Cruz-Reyes, Nelson Rangel-Valdez and Maria Lucila Morales-Rodríguez

Abstract The aim of this chapter is to present the results of the comparison between the solutions obtained with the grey mathematics and the solutions obtained without the grey mathematics. The grey mathematics is used to represent the uncertainty associate with real-life decision-making. We define a multi-objective algorithm to perform the comparison between algorithms. The results obtained show that the approach using grey mathematics outperforms the results without grey mathematics.

Keywords Project portfolio selection · Decision problem · Uncertainty Grey systems

F. Balderas (✉) · C. Gomez-Santillan · L. Cruz-Reyes · N. Rangel-Valdez · M. L. Morales-Rodríguez
Tecnologico Nacional de Mexico, Instituto Tecnologico de Ciudad Madero, Ciudad Madero, Tamaulipas, Mexico
e-mail: fausto.balderas@itcm.edu.mx

C. Gomez-Santillan
e-mail: claudia.gomez@itcm.edu.mx

L. Cruz-Reyes
e-mail: lauracruzreyes@itcm.edu.mx

N. Rangel-Valdez
e-mail: nrangelva@conacyt.mx

M. L. Morales-Rodríguez
e-mail: lucila.morales@itcm.edu.mx

E. Fernandez
Universidad Autónoma de Sinaloa, Sinaloa, Mexico
e-mail: eddyf@uas.edu.mx

O. Castillo et al. (eds.), *Fuzzy Logic Augmentation of Neural and Optimization Algorithms: Theoretical Aspects and Real Applications*, Studies in Computational Intelligence 749, https://doi.org/10.1007/978-3-319-71008-2_27

1 Introduction

Portfolio Decision-Analysis (PDA) is the application of decision analysis to the problem of selecting a subset or projects to perform a portfolio [1]. All organizations whether public or privates have goals, that frequently present uncertainty that involve risk; for example, they seek the best distribution of resources according to the organizational objectives, subject to restrictions, which may be financial or human resources.

The difficulty of PDA-related problems comes from some of their following factors or their combination: large size of entry space, multidimensional consequences of projects and portfolio, qualitative, imprecise, or uncertain information.

Uncertainty and risk in project portfolio selection problem represents a fundamental component and needs to be model in order to obtain the best portfolio that takes into account one or more of the following types of uncertainty: (a) regarding future states of nature; (b) imprecision in the information, because it cannot be predicted or in the absence of information.

In this work, we carry out a comparative study between the Non-dominated Sorting Genetic Algorithm II (NSGA-II) [2] and using NSGA-II with grey mathematics denominated GM-NSGA-II. The purpose of the study is to model the uncertainty present project portfolio selection problem using a grey mathematics approach. The comparison was done in the study problem by using three objectives.

2 Background

In this section, the original NSGA-II [2] is described and some basic definitions about grey system theory [3] are given.

2.1 NSGA-II

The NSGA-II is considered one of the benchmarks of the multiobjective optimization to solve problems, preferably of two and three objectives. The NSGA-II (Algorithm 1) it is based on the creation of non-dominated fronts, establishing elitism on the first front; it also includes an indicator of diversity called crowding distance.

In the NSGA-II algorithm initially, an initial population is created (random or by an initialization technique) of parents P_0. The population is rank according to the levels of non-dominance ($F = F_0, F_1, \ldots$). Each solution is assigned with a fitness function according to its level of non-dominance (0 is the best level). Selection by tournament, crossover and mutation are used to create the offspring population Q_0 of size N.

Algorithm 1. NSGA − II [2]	
1: $R_T = P_T \cup Q_T$	combine parent and children population
2: $F = fast - non - dominated - sort\ (R_T)$	$F = (F_0, F_1, \ldots)$, all non-dominated fronts of R_T
3: $P_{T+1} = \emptyset\ or\ i = 1$	
4: while $\vert P_{T+1}\vert + \vert F_i\vert \le N$ do	till the parent population is filled
5: crowding − distance − $assignment(F_i)$	calculate crowding distance in F_i
6: $P_{T+1} = P_{T+1} \cup F_i$	include i-th non-dominated front in the parent pop
7: $i = i + 1$	
8: end while	
9: $SORT(F_{i,} \prec_i)$	sort in descending order using $\prec_i$
10: $P_{T+1} = P_{T+1} \cup F_i[1{:}(N - \vert P_{T+1}\vert)]$	choose the first N elements of P_{T+1}
11: $Q_{T+1} = make - new - pop(P_{T+1})$	use selection, crossover and mutation to create
12: $t = t + 1$	a new population Q_{T+1}

2.2 Grey System Theory

Grey system theory, developed by Deng in 1982 [3] provides a practical alternative to handle uncertainty [4]. For the problems that involve incomplete and poor information, grey systems provide a methodology using grey numbers. Deng in 1989 [5] adopts grey theory for decision making problems. A grey number is a number whose exact value is unknown, but a range within which the value lies is known [6].

A grey number is defined as $\otimes A = \left[\overline{A}, \underline{A}\right]$. The basic grey number operations (addition, subtraction, multiplication and division) are given as follows [7]:

$$\otimes A + \otimes B = \left[\underline{A} + \underline{B}, \overline{A} + \overline{B}\right] \tag{1}$$

$$\otimes A - \otimes B = \left[\underline{A} - \underline{B}, \overline{A} - \overline{B}\right] \tag{2}$$

$$\otimes A * \otimes B = \left[\min\left(\underline{AB}, \overline{AB}, \overline{A}\underline{B}, \underline{A}\overline{B}\right), \max\left(\underline{AB}, \overline{AB}, \overline{A}\underline{B}, \underline{A}\overline{B}\right)\right] \tag{3}$$

$$\otimes A / \otimes B = \otimes A * \left[\frac{1}{\overline{B}}, \frac{1}{\underline{B}}\right], 0 \notin \otimes B \tag{4}$$

In order to perform a comparison between two grey numbers Shi et al. in [8] define:

For two grey numbers $\otimes D = \left[\overline{D}, \underline{D}\right]$ and $\otimes E = \left[\overline{E}, \underline{E}\right]$

$$P(\otimes D \le \otimes E) = \frac{\max\left(0, L^* - \max\left(0, \overline{D} - \underline{E}\right)\right)}{L^*} \tag{5}$$

where $L(\otimes D) = (\overline{D} - \underline{D})$ is the length of interval number $\otimes D$ and $L^* = L(\otimes D) + L(\otimes E)$.

The sorting relation between $\otimes D$ and $\otimes E$ is determined as follows:

(i) If $\underline{D} = \underline{E}$ and $\overline{D} = \overline{E}$ that means $\otimes D$ is equal to $\otimes E(\otimes D = \otimes E)$ Then $P(\otimes D \leq \otimes E) = 0.5$.
(ii) If $\underline{E} > \overline{D}$ that means $\otimes E$ is greater than $\otimes D(\otimes E > \otimes D)$ Then $P(\otimes D \leq \otimes E) = 1$.
(iii) If $\overline{E} < \underline{D}$, that means is smaller than $\otimes D(\otimes E < \otimes D)$ Then $P(\otimes D \leq \otimes E) = 0$.
(iv) If there is an intercrossing part in them, when $P(\otimes D \leq \otimes E) > 0.5$ that means $\otimes E$ is greater than $\otimes D(\otimes E > \otimes D)$ When $P(\otimes D \leq \otimes E) < 0.5$ that means $\otimes E$ is smaller than $\otimes D(\otimes E < \otimes D)$.

2.3 Multi-objective Optimization Using Grey Mathematics

This section presents the basic concepts in multi-objective optimization using grey mathematics.

Definition 1 Grey multi-objective optimization problem (Grey-MOP)

Given a grey vector function $\otimes\overrightarrow{f}(\overrightarrow{x})[\otimes f_1(\overrightarrow{x}), \otimes f_2(\overrightarrow{x}), \ldots, \otimes f_k(\overrightarrow{x})]$ and its feasible solution space Ω, the Grey-MOP consists in find a vector $\overrightarrow{x} \in \Omega$ that optimizes the grey vector function $\otimes\overrightarrow{f}(\overrightarrow{x})$. Without loss of generality we will assume only maximization functions.

Definition 2 Grey Pareto dominance

A vector $\overrightarrow{x}$ dominates $\vec{x}'$(denoted by $\overrightarrow{x} \prec \vec{x}'$) if $\otimes f(\overrightarrow{x}) \leq \otimes f(\vec{x}')$ for all i functions in $\otimes\overrightarrow{f}$ and there is at least one i such that $\otimes f(\overrightarrow{x}) < \otimes f(\vec{x}')$.

Definition 3 Grey Pareto optimal

A vector $\overrightarrow{x}*$ is Pareto optimal if not exists a vector $\vec{x}' \in \Omega$ such that $\vec{x}' \prec \overrightarrow{x}*$.

Definition 4 Grey Pareto optimal set

Given a MOP, the Pareto optimal set is defined as $P* = \{\overrightarrow{x}* \in \Omega\}$.

Definition 5 Grey Pareto front

Given a MOP and its Pareto optimal set $P*$, the Pareto front is defined as $PF* = \{\otimes\overrightarrow{f}(\overrightarrow{x}) | \overrightarrow{x} \in P*\}$.

2.4 Grey Mathematics NSGA-II

We design a multiobjective evolutionary algorithm called GM-NSGA-II based on the algorithm present in the literature NSGA-II, to obtain the Grey Pareto front using the grey mathematics. The major changes were in the creation of the non-dominated fronts and in the crowding distance assignment, we will use the grey mathematics, and the definitions given in Sect. 2.2.

3 Experimentation and Results

In this section, we present the case of study and the results of the experimentation carried out.

3.1 Case of Study: The Project Portfolio Problem

In the project portfolio selection problem, let us consider that there is the role of the decision maker (DM), that is in charge of allocate the requirements and define the contributions of each project, he also is in charge of selecting a group of projects (a portfolio) that will be implemented by his/her organization.

Elements in Project portfolio selection problem:

N	total number of projects
$f(i) = \langle f_1(i), f_2(i), f_3(i), \ldots, f_p(i) \rangle$	p-dimensional vector representing the i th project
$f_j(i)$	contribution of project i to the jth objective
c_i	cost of the project
x_i	binary variable that identifies whether or not a project i is included in a portfolio
$x = \langle x_1, x_2, \ldots, x_N \rangle$	representation of a portfolio
B	total budget that the organization is willing to invest
a_i	area of the project i.

A feasible portfolio is when it satisfies the constraint of the total budget (Eq. 6) and for each area k, let us consider a lower and an upper limit, L_j and U_j respectively. Based on this, the constraint for each area k is (Eq. 7). In the same was as areas, each geographical region has a lower and upper limit as another constraint that must be fulfilled by a feasible portfolio (similar to Eq. 7).

$$\left(\sum_{i=1}^{N} x_i c_i \right) \leq B \tag{6}$$

$$L_j \leq \sum_{i=1}^{N} x_i g_i(j) c_i \leq U_j \tag{7}$$

where g may be defined as:

$$g_i(j) = \begin{cases} 1 & if \quad a_i = j, \\ 0 & otherwise \end{cases} \tag{8}$$

The union of the benefits of each of the projects that compose to a portfolio determine its quality. This can be expressed as:

$$z(x) = \langle z_1(x), z_2(x), z_3(x), \ldots, z_p(x) \rangle \tag{9}$$

where $z_j(x)$ in its simplest form, is determined by:

$$z_j(x) = \sum_{i=1}^{N} x_i f_j(i) \tag{10}$$

Therefore, considering R_F as the feasible region, the problem of project portfolio is to determine one or more portfolios that solve:

$$\max_{x \in R_F} \{z(x)\} \tag{11}$$

3.2 Results

In this section, we provide the experimental results. The conditions under which these experiments were carried out are described below:

(a) Testing environment for NSGA-II and GM-NSGA-II algorithm was implemented in Java programming language and executed in a computer with the following characteristics: Intel Core i7 3.5 GHz CPU, 16 GB of RAM, and Mac OS X Yosemite 10.10.4 operative system.
(b) The solutions were obtained from 30 independents runs of NSGA-II and GM-NSGA-II.
(c) The parameters of the evolutionary search were as follows, crossover probability = 0.9; mutation probability = 0.05 for both algorithm configurations.

We compared the results obtained from GM-NSGA-II and NSGA-II. We experimented with one instance addressing the project portfolio problem with three objectives.

To the results obtained from NSGA-II and for simplicity and without loss of generality, we will give a random number between 1 and 4% to the cost of the projects, and that it is symmetrically distributed around the value. For the imprecision in the benefit of projects is a random amount between 0.15 and 1.5% of its value. This is for the purpose to work with grey numbers.

The results from Table 1, show a cardinality of 38, that means it includes 38 projects in the portfolio, the quality of the objectives is less that the results obtained in Table 2, from NSGA-II with grey mathematics.

Table 2 shows the results with GM-NSGA-II. C is the non-dominated set in A and B.

In Table 3, show the results of the comparison of non-dominance between Table 1 (Set A) and Table 2 (Set B). The results show the advantages of using the grey numbers for the treatment of uncertainty.

Table 1 Results with NSGA-II

Id	Cost	Obj 1	Obj 2	Obj 3	Card (x)
1	249,015	1,286,010	325,675	1,395,355	38
2	249,445	1,356,335	315,520	1,433,560	38
3	249,570	1,292,315	323,325	1,417,510	38
4	249,250	1,315,880	321,550	1,407,435	38
5	249,925	1,338,335	318,560	1,421,450	38
6	249,990	1,321,930	319,755	1,456,465	38
7	249,435	1,339,375	316,735	1,351,335	38
8	248,450	1,332,030	318,920	1,423,320	38
9	249,125	1,354,485	315,930	1,437,335	38
10	249,365	1,343,320	316,010	1,381,235	38
11	247,115	1,308,620	321,895	1,422,150	38
12	249,315	1,299,475	322,745	1,442,450	38
13	249,690	1,308,465	322,685	1,409,370	38
14	249,185	1,326,315	319,310	1,413,865	38
15	248,190	1,302,010	322,710	1,403,625	38
16	249,755	1,341,225	316,325	1,347,560	38
17	249,525	1,325,420	319,735	1,404,795	38
18	249,230	1,376,180	289,345	1,319,295	36
19	248,405	1,369,875	298,640	1,397,755	37
20	248,925	1,367,305	301,340	1,421,570	37
21	248,360	1,359,795	303,275	1,349,245	37
22	247,160	1,364,150	302,790	1,418,125	37
23	247,730	1,367,300	302,300	1,388,090	37
24	247,025	1,360,675	303,040	1,379,915	37
25	248,875	1,370,935	296,730	1,347,530	36
26	249,975	1,111,715	327,190	1,356,145	38
27	249,995	1,181,380	326,770	1,485,250	38
28	249,985	1,221,830	326,620	1,449,765	38

Table 2 Results with GM-NSGA-II

Id	Costo total	Obj 1	Obj 2	Obj 3	Card (x)
1	[244.785, 255.185]	[1 205 310, 1 214 210]	[338 695, 347 895]	[1 611 280, 1 745 550]	40
2	[244.550, 255.350]	[1 261 865, 1 270 665]	[334 245, 343 845]	[1 633 255, 1 769 345]	40
3	[244.540, 255.440]	[1 138 115, 1 147 715]	[339 080, 348 980]	[1 516 285, 1 642 605]	40
4	[244.945, 254.845]	[1 424 610, 1 433 410]	[301 285, 311 085]	[1 432 010, 1 551 300]	38
5	[244.535, 255.435]	[1 336 055, 1 345 555]	[325 730, 335 830]	[1 594 790, 1 727 680]	40
6	[244.210, 254.910]	[1 292 090, 1 301 190]	[331 855, 341 655]	[1 632 055, 1 768 055]	40
7	[244.240, 255.140]	[1 288 330, 1 297 430]	[332 145, 342 445]	[1 585 690, 1 717 820]	41
8	[244.325, 255.425]	[1 312 100, 1 321 200]	[330 220, 340 420]	[1 627 030, 1 762 620]	41
9	[244.720, 255.220]	[1 333 325, 1 342 225]	[326 735, 336 735]	[1 564 010, 1 694 310]	40
10	[244.310, 254.710]	[1 320 265, 1 328 865]	[329 410, 339 210]	[1 584 355, 1 716 355]	40
11	[244.535, 254.835]	[1 403 180, 1 412 080]	[301 740, 311 340]	[1 478 410, 1 601 590]	38
12	[244.230, 254.930]	[1 330 590, 1 339 790]	[326 855, 336 955]	[1 560 955, 1 690 995]	40
13	[244.450, 254.650]	[1 282 205, 1 290 905]	[332 920, 342 520]	[1 673 595, 1 813 045]	40
14	[244.425, 254.525]	[1 259 890, 1 268 690]	[335 110, 344 510]	[1 627 190, 1 762 770]	40
15	[244.565, 254.865]	[1 246 085, 1 255 185]	[336 305, 345 905]	[1 612 050, 1 746 370]	40
16	[243.140, 253.840]	[1 232 775, 1 241 875]	[337 515, 347 515]	[1 560 395, 1 690 405]	40
17	[243.860, 254.960]	[1 294 275, 1 303 575]	[331 285, 341 285]	[1 515 115, 1 641 365]	40
18	[243.535, 254.535]	[1 252 590, 1 261 790]	[335 855, 345 555]	[1 571 555, 1 702 505]	40
19	[243.730, 254.530]	[1 258 015, 1 266 815]	[335 330, 345 330]	[1 604 585, 1 738 275]	40
20	[244.635, 254.935]	[1 300 085, 1 308 685]	[331 220, 341 220]	[1 652 895, 1 790 615]	40
21	[244.530, 255.130]	[1 278 095, 1 286 995]	[333 360, 343 160]	[1 640 110, 1 776 780]	40
22	[243.480, 254.280]	[1 341 620, 1 350 920]	[322 815, 332 615]	[1 594 890, 1 727 780]	40
23	[244.700, 255.300]	[1 339 995, 1 349 195]	[324 020, 333 620]	[1 647 120, 1 784 350]	40

(continued)

Table 2 (continued)

Id	Costo total	Obj 1	Obj 2	Obj 3	Card (x)
24	[244.275, 254.875]	[1 352 200, 1 361 700]	[321 590, 331 290]	[1 594 455, 1 727 315]	40
25	[243.450, 254.350]	[1 343 510, 1 352 510]	[322 580, 332 680]	[1 580 090, 1 711 760]	40
26	[244.580, 255.080]	[1 352 070, 1 360 970]	[322 105, 332 005]	[1 616 340, 1 751 000]	40
27	[244.380, 254.980]	[1 397 150, 1 406 250]	[309 335, 319 435]	[1 512 000, 1 637 980]	40
28	[243.425, 254.225]	[1 359 315, 1 368 215]	[320 635, 330 735]	[1 535 500, 1 663 420]	39
29	[242.955, 253.455]	[1 360 645, 1 369 345]	[318 215, 328 315]	[1 536 370, 1 664 380]	40
30	[244.035, 254.835]	[1 383 325, 1 392 425]	[310 990, 321 290]	[1 537 850, 1 665 970]	39
31	[244.625, 255.325]	[1 396 435, 1 405 335]	[310 005, 320 105]	[1 500 080, 1 625 050]	39
32	[244.145, 254.545]	[1 376 210, 1 384 910]	[315 115, 325 115]	[1 536 780, 1 664 820]	39
33	[244.060, 254.760]	[1 386 655, 1 395 955]	[310 815, 320 915]	[1 518 220, 1 644 710]	39
34	[244.640, 254.940]	[1 380 130, 1 388 930]	[312 120, 322 220]	[1 539 100, 1 667 330]	39
35	[244.360, 254.560]	[1 291 010, 1 299 910]	[332 290, 342 190]	[1 644 775, 1 781 825]	40
36	[244.095, 254.395]	[1 364 855, 1 373 455]	[317 240, 326 640]	[1 550 355, 1 679 535]	39
37	[244.605, 255.005]	[1 375 685, 1 384 785]	[316 695, 326 295]	[1 476 465, 1 599 475]	39
38	[244.130, 254.930]	[1 379 725, 1 389 325]	[314 475, 323 975]	[1 482 460, 1 605 990]	39
39	[243.775, 254.575]	[1 382 365, 1 391 665]	[312 030, 321 930]	[1 535 680, 1 663 640]	39
40	[244.720, 254.920]	[1 325 745, 1 334 145]	[327 005, 337 005]	[1 621 070, 1 756 160]	40

Table 3 The comparison of non-dominance between sets A and B

	$A \cap C$	$B \cap C$
Non-dominance solutions	A	0

4 Conclusions and Future Work

We have presented a comparative study between GM-NSGA-II and NSGA-II. The propose methodology validates that applying grey mathematics to treat uncertainty, finds results that remain non-dominated that the results obtained for treat the uncertainty without grey mathematics.

To validate our results, we designed an experiment that gives evidence of the quality of the solutions. The experiment is a comparison between the results obtained from NSGA-II and GM-NSGA-II. In conclusion, our proposal is able to obtain solutions that represent the best solutions to face the uncertainty.

As immediate work, we are going to work with nine objectives and applied a preferences system to find the best solution according to the preferences of the DM.

Acknowledgements This work has been partially supported by the following CONACyT projects: a) Fronteras de la Ciencias Project 1340; b) Consolidation National Lab Project 280712; c) Projects [236154, 269890]; d) Project 280081 Red Temática para el apoyo a la Decisión y Optimización Inteligente de Sistemas Complejos y de Gran Escala (OPTISAD) Universidad Autónoma de Nuevo León; and, e) Project 3058 from the program Cátedras CONACyT.

References

1. A. Salo, J. Keisler, A. Morton, An invitation to portfolio decision analysis, in *Portfolio Decision Analysis* (Springer, New York, 2011), pp. 3–27
2. K. Deb, A. Pratap, S. Agarwal, T.A.M.T. Meyarivan, A fast and elitist multiobjective genetic algorithm: NSGA-II. IEEE Trans. Evol. Comput. **6**(2), 182–197 (2002)
3. D. Ju-Long, Control problems of grey systems. Syst. Control Lett. **1**(5), 288–294 (1982)
4. B. Oztaysi, A decision model for information technology selection using AHP integrated TOPSIS-Grey: the case of content management systems. Knowl.-Based Syst. **70**, 44–54 (2014)
5. D. Julong, Introduction to grey system theory. J. Grey Syst. **1**(1), 1–24 (1989)
6. Y.H. Lin, P.C. Lee, H.I. Ting, Dynamic multi-attribute decision making model with grey number evaluations. Expert Syst. Appl. **35**(4), 1638–1644 (2008)
7. D. Yamaguchi, G.D. Li, K. Mizutani, T. Akabane, M. Nagai, M. Kitaoka, On the generalization of grey relational analysis. J. Grey Syst. **9**(1), 23–33 (2006)
8. J.R. Shi, S.Y. Liu, W.T. Xiong, A new solution for interval number linear programming. Syst. Eng.-Theory Pract. **2**, 016 (2005)

Statistical Comparative Between Selection Rules for Adaptive Operator Selection in Vehicle Routing and Multi-knapsack Problems

Jorge A. Soria-Alcaraz, Marco A. Sotelo-Figueroa and Andrés Espinal

Abstract Autonomous Search is an important field of artificial intelligence which focuses on the analysis and design of auto-adaptive systems capable to improving their own search performance on a given problem at runtime. An autonomous search system must can modify or select its internal components, improving its performance at execution time. A case of such algorithms is the Adaptive Operator Selection (AOS) method. This method uses a record of the latest iterations to propose which operator use in later iterations. AOS has two phases: Credit assignment and Selection rule. The first phase penalizes or rewards a specific operator based on their observed performance. The second phase makes the selection of the operator to use in subsequent iterations. This article shows the performance-based statistical comparison between two selection rules: Probability Matching and Adaptive Pursuit when using two different domains; namely, Vehicle Routing Problem and Multi-knap-Sack. The comparison is done with statistical rigor when integrating contrast algorithms such as no adaptation rule and random adaptation rule. This paper can be seen as an introduction on how to make proper statistical comparisons between two AOS rules.

Keywords Autonomous search · Adaptive operator selection · Selection rule
Vehicle routing problem · Multi-knapsack problem

J. A. Soria-Alcaraz (✉) · M. A. Sotelo-Figueroa · A. Espinal
Departamento de Estudios Organizacionales, División de Ciencias
Económico-Administrativas, Universidad de Guanajuato, Guanajuato, Mexico
e-mail: jorge.soria@ugto.mx

M. A. Sotelo-Figueroa
e-mail: masotelo@ugto.mx

A. Espinal
e-mail: aespinal@ugto.mx

O. Castillo et al. (eds.), *Fuzzy Logic Augmentation of Neural and Optimization Algorithms: Theoretical Aspects and Real Applications*, Studies in Computational Intelligence 749, https://doi.org/10.1007/978-3-319-71008-2_28

1 Introduction

Autonomous Search is an emerging field of artificial intelligence which focused on the analysis and design of systems capable to achieve one of the following two objectives: to select proper solving techniques or mechanisms form a pool of possible operators or to develop a strategy of self-tuning with respect to a given instance at runtime [1]. An autonomous search system must can modify or select its internal components, improving its performance at execution time; this is the case of Adaptive Operator Selection algorithms (AOS), where this kind of algorithms selects at execution time the most promising operator form an available pool to continue the search. Adaptive Operator Selection is a strategy that autonomously configures itself on the fly when solving a given instance of an optimization problem [2]. At the begin of the execution the AOS algorithm applies each available operator to the current solution storing their performance to then select the best reported operators with higher probability at later iterations. The proper selection is done by two mechanism; namely, Credit Assignment and Selection Rule. The first one penalizes or rewards a specific operator based on their observed performance. The second one makes the actual selection of the operator to use in subsequent iterations trough. This article shows the performance-based statistical comparison between two selection rules: Probability Matching and Adaptive Pursuit when using two different domains: Vehicle Routing Problem and Multi-Knapsack.

To make a fair comparative between two heuristic algorithms we use a methodology that first analyze the normality of the data to then apply a proper statistical test like Two Way ANOVA in the case our data belongs to a normal distribution.

Vehicle Routing Problem is a real-world problem which requires that a fleet of vehicles meets several requests minimizing costs. In this paper, we use a variant of this problem named Vehicle routing problem with time windows, whereby a customer request must be done in a specific interval of time for a solution to be valid.

Multi-Knapsack problem is a generalization of the binary Knapsack problem where a single knapsack and a set of objects are given. Each object has a weight and a profit. The objective of this problem is to fill the knapsack with those objects which maximize the profit.

The paper is organized as follows. Section 2 presents the problems definitions as well as important concepts related to AOS and Autonomous Search, Sect. 3 presents the solution approach and its justification. Section 4 contains the experimental set-up, results, their analysis and discussion. Finally, Sect. 5 contains our conclusions and describes some future work.

2 Background

In this Section, The Adaptive Operator selection process is detailed. We also describe each of the problem domains as well as the Statistical process used to compare the selection rules.

2.1 *Adaptive Operator Selection*

The Adaptive operator selection process consists of two well-identified mechanism: *credit assignment and selection rule.* The first one associates a reward value to the least applied operator. The second mechanist determines the next operator to be used at latter iterations based on its current reward value.

2.2 *Credit Assignment*

In this paper as credit assignment rule we use *the extreme credit assignment* approach, which is based on the idea that large improvements are more effective than small frequent improvements [2]. In the extreme credit assignment rule a reward is updated when an operator is applied to a current solution, then the new achieved fitness of the solution is computed and the difference of the fitness value before and after the operator execution is added to a FIFO list of size W. This list is unique and available to all operators. Finally, the operator reward is updated to the maximal fitness improvement in the list. Let be t the current step and $\delta(t)$ the fitness improvement at time t. Then the expected reward for our least applied operator is shown by Eq. 1.

$$r_t = argmax(\delta(t_i), t = 1...W) \tag{1}$$

As the reader can see in Eq. 1, parameter w must be selected carefully since it determines the scope of the reward assignation. A bigger value of W could cause that the selected reward is not relevant to the current state of the search, and a small value of W makes the reward information to small.

2.3 *Selection Rules*

Selection rules typically associate probabilities with operators via proportional selection. Let K denote the number of operators The selection mechanisms maintain a probability vector *(pi, t)i = 1,...,K*, and an estimate of the current operator credit denoted as $\hat{q}(i,t)$. At each iteration time *t*:

- Operator *i* is selected with probability *pi, t* according roulette-wheel selection scheme.
- The Selected operator is applied, and a credit r_t, is computed using the credit assignment mechanism (Extreme Value, Sect. 2.2)
- The quality estimate $\hat{q}(i,t)$ of the selected operator is updated according to the reward r_t.

The quality estimate $\hat{q}(i,t)$ is calculated depending which Selection rule is implemented. In this paper two selection rules, *Probability Matching and Adaptive Pursuit,* are implemented. Probability Matching [3] rule selects an operator using a *roulette wheel* selection, which is constructed with the current rewards values that each operator holds at current time. Equation 2 shows this process. To avoid operators being completely ignored, the selection rules normally assign a minimal selection probability *pmin > 0*.

$$P_{i,t+1} = P_{min} + (1-K * P_{min})\hat{q}_{(i,t+1)} / \sum_{j=1}^{k} \hat{q}_{(i,t+1)} \tag{2}$$

Adaptive pursuit follows a winner-take-all strategy selecting an operator with maximal reward and increasing its selection probability accordingly. Equations 3–5 describes this mechanism.

$$i^* = argmax\{\hat{q}\_(i,t), i = 1\ldots K\} \tag{3}$$

$$\mathrm{P}_{\mathrm{i}^*,\mathrm{t}+1} = \mathrm{P}_{\mathrm{i}^*,\mathrm{t}} + \beta\big(1-(\mathrm{K}-1)\mathrm{P}_{\min}-\mathrm{P}^*_{\mathrm{i}},\mathrm{t}\big), \quad (\beta > 0) \tag{4}$$

$$\mathrm{P}_{\mathrm{i},\mathrm{t}+1} = \mathrm{P}_{\mathrm{i},\mathrm{t}} + \beta\big(\mathrm{P}_{\min} - \mathrm{P}_{\mathrm{i},\mathrm{t}}\big), \mathrm{for\,i} \neq \mathrm{i}^* \tag{5}$$

2.4 Vehicle Routing Problem Domain

Vehicle routing problems (VRP) [4] require that a fleet of vehicles serves several requests in order to minimize costs. Many variants have been proposed, such as *capacitated VRP* where a fleet of identical vehicles located at a central depot must be optimally routed to supply a set of customers, *Vehicle Routing Problem with Time Windows* VRPTW whereby a customer must be served between two time points for a solution to be valid. The objective function balances the dual objectives of minimizing the number of vehicles needed and minimizing the total distance travelled. This latter variant (VRPTW) is which we use in our experiments. Equation 6 details VRPTW fitness function.

$$VrpFitness = A * num\_Vehicles + distance \quad (6)$$

where A is a constant empirically set to 1000 to give higher importance to the number of vehicles in a solution [4]. VRPTW instances used in this paper can be reach at the Transportation Optimization Portal maintained by SINTEF Applied Mathematics [5]. Operators used in these instances are taken from the work of Walker [4]. These 8 operators are briefly explained bellow:

- **Two-Opt** (Swaps two adjacent locations within a single route.)
- **Or-Opt** (Moves two adjacent locations to a different place, within a single route.)
- **Shift** (Moves a single location from one route to another.)
- **Interchange** (Swaps two locations from different routes.)
- **LocRR** (Removes several locations based on location proximity.)
- **TImeRR** (Removes a number of locations based on time window proximity.)
- **Two-opt** (Takes the end sections of two routes, and swaps them to create two routes.)
- **GENI** (A location is taken from one route and placed into another route. Re-optimization is then performed on the route.)

2.5 Multi-knapsack Problem Domain

Multi-Knapsack problem is a generalization of the binary Knapsack problem where a single knapsack and a set of objects are given. Each object has a weight and a profit. The objective of this problem is to fill the knapsack with those objects which maximize the profit.

The multiple version consists of m knapsacks of capacities c_1, c_2,…,c_m and n objects with profits pr_1, pr_2,…,pr_n. Each object has m possible weights: object i weighs w_{ij} when considered for inclusion in knapsack j ($i \leq j \leq m$). the objective is to find a vector $x = (x_1, x_2,\ldots,x_n)$ where $x_i \in \{0,1\}$ that guarantees that no knapsack is overfilled and yields maximum profit. Multi-Knapsack instances used in this paper are multi-modal constrained problems with sizes ranging from 50 to 105 objects and from 2 to 30 knapsacks. These problems are available online from the OR-library by Razali and Wah [6]. Operators used for these instances are: 1-flip, 3-flip, 5-flip and Worst-Flip. Description of this operators as well as more detailed information about this domain can be in the work of Soria-Alcaraz, 2014 [7, 8].

2.6 *Statistical Tests*

In this paper, several statistical tests are used to make a comparative between the two selection rules described in Sect. 2.3. Since selection rules are heuristic algorithms, we need robust tests to validate our conclusions. Two test are used to reach this objective; namely, Shapiro-Wilk Test and ANOVA omnibus test. Shapiro-Wilk Test [9, 10] is a test of normality in frequentist statistics. It was published in 1965 by Samuel Sanford Shapiro and Martin Wilk. Monte Carlo simulation has found that Shapiro–Wilk has the best power for a given significance, followed closely by Anderson–Darling when comparing the Shapiro–Wilk, Kolmogorov–Smirnov, Lilliefors, and Anderson–Darling tests [6]. We use Shapiro-Wilk test to find if our data has normal distribution; this is an important step since we need to select a proper omnibus test to gather evidence about of which of our selection rules have better performance.

In the case that our data presents normal distribution, it will be necessary to make the comparative using a proper test, in this paper we use ANOVA Omnibus Test to compare two heuristic algorithms both of which have presented normal distribution. In its simplest form, ANOVA [11] provides a statistical test of whether or not the means of several groups are equal, and therefore generalizes the t-test to more than two groups. ANOVAs are useful for comparing (testing) three or more means (groups or variables) for statistical significance.

3 Solution Approach

3.1 *High Level Algorithm*

As seen in Sects. 2.1, 2.2 and 2.3 the AOS approach gives to the user several tools to create a self-configuring algorithm capable to select the best operator to continue the search at runtime; However, we still need a high-level algorithm that calls AOS at each iteration. In this paper, we use a simple hill-climbing method in which we start with a uniform at random solution state, then at each iteration we apply an AOS call over the solution state to create a new one, we validate the fitness value of this new solution. Finally, we replace the old state only if the new solution presents a better fitness value. Algorithm 1 shows our high-level strategy.

Algorithm 1: High-Level Hill Climbing

Algorithm 1: High-Level Hill Climbing
Require: AOS Module *AOS()*, low-level operators *Opers*, Fitness Function *fit()*, Selection Rule *SR*, Credit assignment *CA*.
1: *S = Create_UniformAtRandomState();*
2: **while** stopping criterion not met **do**
3: *operator = AOS(Opers, CA, SR) ;*
4: *S* = operator.apply(S);*
5: **If** *fit(S*) isBetterThan fit(S)*
6: *S = S**
7: **end if**
8: **end while**
9: **return** S

Line 3 in Algorithm 1 executes the AOS module described on Sects. 2.2 and 2.3. this module selects the most promising operator from the operator pool to continue the search. Line 4 in Algorithm 1 shows the application of the selected operator to the current state producing a new temporal state. Finally, this offspring is compared in terms of fitness value against its parent fitness, if offspring fitness is better than its parent fitness it is replaced as shown in Line 6 Algorithm 1.

3.2 Comparative Methodology

Algorithm 1 in Sect. 3.1 can be considered as a simple yet complete example of an autonomous search algorithm that produces an optimized state at the end of its execution. Using this algorithm, we investigate which of the two selection rules presented in Sect. 2.3 presents better performance for VRPTW and multi-knapsack domains. To do this we run 35 independent executions of this algorithm with each selection rules and over each test instance (6 instances for each, Multi-knapsack and VRPTW). Afterwards, we analyze the 35 executions per instance and selection rule to find if the results belong to a normal distribution in each case. We do this using Shapiro-Wilk test presented in Sect. 2.6. Once we check if this assumption is true we apply ANOVA omnibus test to discern with statistical significance which configuration pair (Selection Rule-Credit assignment) achieves the bests results.

4 Experiments and Results

In this section we presents the results as well as the comparative between two selection rules; namely, Adaptive Pursuit and Probability Matching for Multi-knapsack and VRPTW domains following the methodology detailed in

Sect. 3.2. Also, we describe the instances for each domain and the parameters of our algorithm detailed in Sect. 3.1.

4.1 Vehicle Routing Problem with Time Windows Domain

For this domain, we use test instances obtained through SINTEF web page [5]. Specifically, instances used in this paper for VRPTW are shown in Table 1.

As stop criteria we follow the same configuration as the work of Walker et al. [4] using 300 s. Following this scheme, we obtain 35 results per instance and selection rule, then we apply Shapiro-Wilk normality test to the data obtaining the results show in Table 2.

Table 2 shows the results of Shapiro-Wilk test. It is important to consider that the null-hypothesis of this test is that the population is normally distributed. Thus, if the p-value is less than the chosen alpha level (usually 0.05), then the null hypothesis is rejected and there is evidence that the data tested are not from a normally distributed population, On the contrary, if the p-value is greater than the chosen alpha level, then the null hypothesis that the data came from a normally distributed population cannot be rejected. This latter case is the one we achieve with both selection rules and in each instance. Thus, we can apply ANOVA Omnibus Test to discern which of both rules have the best performance. Application of two-way ANOVA Omnibus test on VRPTW can be seen in Table 3.

Table 3 shows the application of two-way ANOVAs omnibus test to our VRPTW data. The ANOVAs null hypothesis is that there is no difference between

Table 1 VRPTW instances

Solomon/RC/RC207	Homberger/C/C1-10-1
Solomon/RC/RC103	Homberger/C/C1-10-5
Solomon/R/R106	Homberger/C/C1-10-8

Table 2 Shapiro-wilk results for VRPTW

Probability matching	Adaptive pursuit
Solomon/RC/RC207 p-value = 0.5406	Solomon/RC/RC207 p-value = 0.115
Solomon/RC/RC103 p-value = 0.0896	Solomon/RC/RC103 p-value = 0.3873
Solomon/R/R106 p-value = 0.2886	Solomon/R/R106 p-value = 0.7137
Homberger/C/C1-10-1 p-value = 0.7133	Homberger/C/C1-10-1 p-value = 0.2974
Homberger/C/C1-10-5 p-value = 0.261	Homberger/C/C1-10-5 p-value = 0.6845
Homberger/C/C1-10-8 p-value = 0.1967	Homberger/C/C1-10-8 p-value = 0.1919

Table 3 Two-way ANOVA test for VRPTW

ANOVA	DF	Sum Sq	Mean Sq	F value	PR (>F)
Algorithm	1	736576	736576	54.771	$\mathbf{3.15 \times 10^{-13}}$
Instance	9	1991740	7390945	549.584	2.2×10^{-16}
Residuals	455	6118956	13448		
T-tests	Adaptive pursuit				
Probability matching	**0.0003**				
P-value adjustment: Bonferroni					

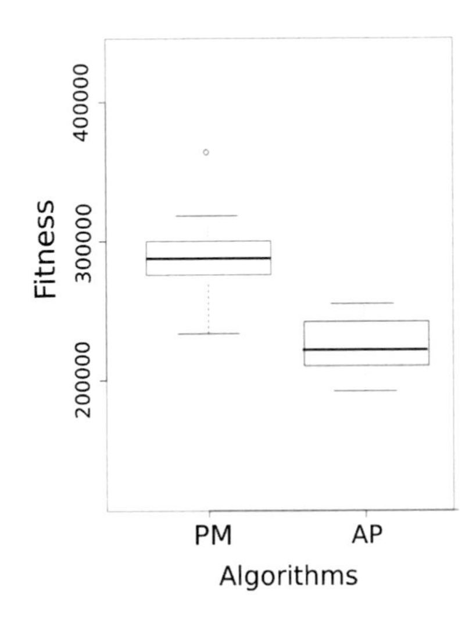

Fig. 1 Box-plot for VRPTW instance Solomon/RC/RC207 adaptive pursuit (AP) versus probability matching (PM)

means. Thus, If p-value < 0.05 we reject null hypothesis and If p-value $\geq$ 0.05 we don't have enough evidence to reject null hypothesis. From Table 3 we can identify a p-value lower than 0.05 in the case when both algorithms are compared this means the selection rule affects significantly our results. Since VRPTW sack is a Minimization problem (Sect. 2.4) we can produce a boxplot graphic to compare the performances that we know are significantly different between the two selection rules. Figure 1 shows this boxplot.

Figure 1 shows the performance of both selection rules over the same Solomon/RC/RC207 Instance. The same behavior was observed over all the test instances in VRPTW. With this evidence, we can conclude that Adaptive Pursuit is a better selection rule when used in an AOS module along to Extreme credit assignment for Solomon & Homberger VRPTW Instances. This conclusion is supported with the statistical evidence shown in Tables 2 and 3 since the ANOVAs test identifies differences in the performance of both rules with statistical significance.

4.2 Multi-knapsack Domain

For this domain, we use test instances obtained through OR library [12]. Specifically, instances used in this paper for Multi-knapsack are shown in Table 4.

As stop criteria we follow the same configuration as the work of Soria-Walker et al. [4] using 300 s. Following this scheme, we obtain 35 results per instance and selection rule, then we apply like in the previous domain the Shapiro-Wilk normality test to the data obtaining the results show in Table 5.

Again, Shapiro-wilk give us evidence about the normality of our data, using this evidence we apply two-way ANOVA test to multi-knapsack data. Results of ANOVA test can be seen in Table 6.

Table 4 Multi-knapsack instances

Sento1	Weing8
Weing1	Weish12
Weing7	Weish30

Table 5 Shapiro-Wilk results for VRPTW

Probability matching	Adaptive pursuit
Sento1 p-value = 0.2645	Sento1 p-value = 0.8949
Weing1 p-value = 0.7605	Weing1 p-value = 0.8538
Weing7 p-value = 0.1430	Weing7 p-value = 0.9032
Weing8 p-value = 0.5242	Weing8 p-value = 0.3661
Weish12 p-value = 0.2087	Weish12 p-value = 0.191
Weish30 p-value = 0.1080	Weish30 p-value = 0.9632

Table 6 Two-way ANOVA test for multi-knapsack

ANOVA	DF	Sum Sq	Mean Sq	F value	PR (>F)
Algorithm	1	736576	736576	54.771	$\mathbf{6.59 \times 10^{-13}}$
Instance	23	169991740	7390945	549.584	2.2×10^{-16}
Residuals	455	6118956	13448		
T-tests	Adaptive pursuit				
Probability matching	**0.0016**				
P-value adjustment: Bonferroni					

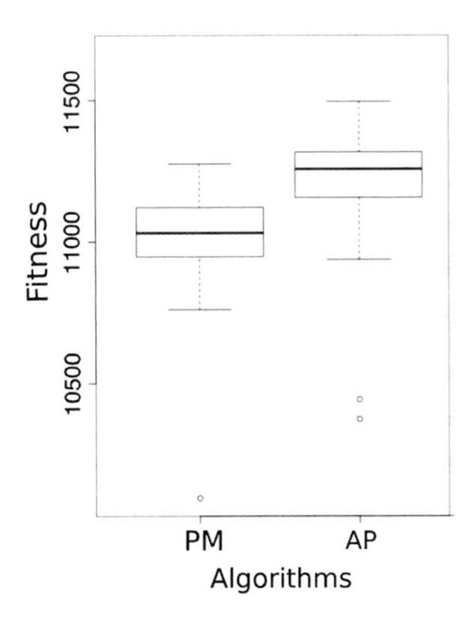

Fig. 2 Box-plot for multi-knapsack instance Sento 7 adaptive pursuit (AP) versus probability matching (PM)

Like VRPTW domain, we find statistical significance in the difference of performance between selection rules in Multi-Knapsack Domain. We present a boxplot in Fig. 2 with the data obtained by Sento7 instance, the same behavior was observed all across the domain. Since Multi-knapsack problem is a Maximization problem we can conclude that Adaptive Pursuit is a better selection rule when used in an AOS module along to Extreme credit assignment for OR-library Multi-knapsack Instances. This conclusion is supported with the statistical evidence shown in Tables 5 and 6.

5 Conclusions and Future Work

This paper presents a statistical comparative between two selection rules implemented in a AOS module for VRPTW and Multi- Knapsack domains. This comparative can be applied to other Autonomous search algorithms. In this comparative, the selection rule Adaptive Pursuit achieves best results, this was confirmed by a two-way ANOVA test and a Shapiro-Wilk test. This comparative is a robust statistical process that authors suggest be replicated in other similar cases. As future work, more selection rules will be analyzed as well as more statistical test will be explored when comparing heuristic rules/mechanism in Adaptive Operator Selection.

References

1. Y. Hamadi, E. Monfroy, F. Saubion, *AN Introduction to Autonomous Search. Autonomous Search* (Springer, Berlin Heidelberg, 2012), pp. 1–11
2. A. Fialho, L. Costa, M. Schoenauer, M. Sebag. Dynamic multi-armed bandits and extreme value-based rewards for adaptive operator selection in evolutionary algorithms, in *Learning and Intelligent Optimization*, volume 5851 of Lecture Notes in Computer Science (Springer, Berlin Heidelberg, 2009), pp. 176–190
3. D.E. Goldberg, Probability matching, the magnitude of reinforcement, and classifier system bidding. Mach. Learn. **5**(4), 407–425 (1990)
4. D. Walker, G. Ochoa, M. Gendreau, E.K. Burke, Vehicle routing and adaptive iterated local search within the HyFlex hyper-heuristic framework, in *International Conference on Learning and Intelligent Optimization (LION 6)*, Lecture Notes in Computer Science (Springer, 2012), pp. 265–276
5. M.M. Solomon, VRPTW benchmark problems. http://www.sintef.no/projectweb/top/vrptw/solomon-benchmark/100-customers/. (2003)
6. N. Razali, Y.B. Wah, Power comparisons of Shapiro–Wilk, Kolmogorov–Smirnov, Lilliefors and Anderson–Darling tests (2011)
7. J.A. Soria-Alcaraz, G. Ochoa, M. Carpio, H. Puga, Evolvability metrics in adaptive operator selection, in *Proceedings of the 2014 Annual Conference on Genetic and Evolutionary Computation* (GECCO'14) (ACM, New York, NY, USA, 2014), pp. 1327–1334
8. J.A. Soria-Alcaraz, G. Ochoa, J. Swan, M. Carpio, H. Puga, E.K. Burke, Effective learning hyper-heuristics for the course timetabling problem. Eur. J. Oper. Res. **238**(1), 77–86 (2014). ISSN 0377-2217
9. S.S. Shapiro, M.B. Wilk, An analysis of variance test for normality (complete samples). Biometrika **52**(3–4), 591–611 (1965)
10. M.B. Wilk, R. Gnanadesikan, Probability plotting methods for the analysis of data. Biometrika Trust **55**(1), 1–17 (1968)
11. R.A. Fisher, The correlation between relatives on the supposition of Mendelian inheritance. Philos. Trans. Roy. Soc. Edinb. **52**, 399–433 (1918)
12. http://people.brunel.ac.uk/~mastjjb/jeb/info.html

Performance Analysis of an a Priori Strategy to Elicitate and Incorporate Preferences in Multi-objective Optimization Evolutionary Algorithms

Laura Cruz-Reyes, Mercedes Perez-Villafuerte, Nelson Rangel, Eduardo Fernandez, Claudia Gomez and Patricia Sanchez-Solis

Abstract The project portfolio selection is one of the most important strategic problems, both in the private sector and in the public sector. This can become a complex activity due to several factors, as occurs in many real-world optimization problems in which many criteria must be considered simultaneously. The preferences of a *Decision Maker* (DM) are a relevant element for decision-making activities, in general, and in portfolio selection, in particular; they vary between decision-makers and evolve over time. A strategy is required that assists the DM in the identification of the best compromise solution that satisfies their preferences. In order to incorporate DM's preferences, given in examples, the methodology *Preferences Disaggregation Analysis* (PDA) is introduced to obtain the parameters of a preference model from examples. This model is the basis of a classifier that allows to a multi-objective optimization evolutionary algorithm lead the search towards the DM's region of interest. In this paper is analyzed the performance of two multi-objective optimization algorithms of the state of the art when preferences

L. Cruz-Reyes (✉) · N. Rangel · C. Gomez
National Mexican Institute of Technology/Madero Institute of Technology, Madero 89440, Tamaulipas, Mexico
e-mail: lauracruzreyes@itcm.edu.mx

N. Rangel
e-mail: nrangelva@conacyt.mx

C. Gomez
e-mail: claudia.gomez@itcm.edu.mx

M. Perez-Villafuerte · P. Sanchez-Solis
National Mexican Institute of Technology/Tijuana Institute of Technology, Tijuana 22500, Baja California, Mexico
e-mail: pvmercedes@gmail.com

P. Sanchez-Solis
e-mail: julia.sanchez@uacj.mx

E. Fernandez
Faculty of Civil Engineering, Autonomous University of Sinaloa, Culiacan 80040, Sinaloa, Mexico
e-mail: eddyf@uas.edu.mx

O. Castillo et al. (eds.), *Fuzzy Logic Augmentation of Neural and Optimization Algorithms: Theoretical Aspects and Real Applications*, Studies in Computational Intelligence 749, https://doi.org/10.1007/978-3-319-71008-2_29

are elicited indirectly through a PDA method. The experimental results showed the potential of the proposed method applied to small and medium scale instances.

Keywords Evolutionary algorithms · Multi-objective optimization
Preference disaggregation analysis · Preference incorporation

1 Introduction

In the multi-objective optimization is searched a good convergence and a well-distributed approximation of the Pareto frontier. Among the available methods for solving *Multi-objective Optimization Problems* (MOPs), the Multi-objective Evolutionary Algorithms (MOEAs) stand out, which provide a set of efficient solutions for the *Decision Maker* (DM).

According to the literature review, arises the need to incorporate the DM's preferences during the search for efficient solutions to direct it towards the *region of interest* (ROI) of the DM, avoiding guiding the search to the whole Pareto front. The main idea of *interactive optimization* is to help MOEAs to direct their search towards a subset of efficient and satisfactory DM solutions, as well as simplify decision making by integrating preferential information into the search process.

When DM preferences are requested, they can be related to objectives, constraints or solutions [1–3]. Introducing these preferences can become an excessive task for DM, this is why it is necessary to introduce these preferences indirectly so that this effort is lower and this information is obtained in an understandable and natural way for the DM.

One of the methods for obtaining preferences is the *Preferences Disaggregation Analysis* (PDA). These procedures are known as indirect because, instead of requiring an active participation of the DM to define the values of the parameters, they only require that he/she provide a set of actions that reflect its behavior in the decision making.

In this paper is analyzed the performance of two multi-objective algorithms with different schemes for the diversity preservation through the indirect elicitation of preferences using a PDA method. This work also concludes that it is possible to cover the entire optimization process, including the elicitation and incorporation of preferences, and the solution generation.

Section 2 introduces the concept of PDA, Sect. 3 describes an algorithm that addresses the incorporation of preferences through THESEUS, a multi-criteria classification method. Section 4 details an approach to integrate the elicitation and incorporation of preferences into the optimization process. Sections 5 and 6 describe the experimental process for this work and the results obtained in the comparison of the algorithms that are the object of study.

2 Preference Elicitation with an Evolutionary Disaggregation Method

Preference Disaggregation Analysis (PDA) are indirect methods for preference elicitation, use a reference set that is a set of solutions organized into categories that have a preferred order provided by a DM to infer values for the preferential parameters

The objective of PDA according to Doumpos and Zopounidis [4] is to analyze this reference set for specification of the model parameters in the most compatible way with DM policies and to provide greater flexibility with respect to the expression of these decisions.

Consequently, this approach minimizes the cognitive effort required by the DM, as well as the time required for the implementation of the process to aid the decision making; In other words, makes the task of defining parameter values as simple as possible for the DM [5, 6].

Rangel-Valdez et al. [7] proposed a new optimization model for PDA and its solution using an evolutionary algorithm. The definition of the model includes the use of the intensity effect (i.e., variations between the criteria used to evaluate decision alternatives) and new ways of combining the number of consistencies and inconsistencies with respect to the reference set.

In the research proposed in this article, we took the initial PDA version of Rangel-Valdez et al. [7] to create a basic prototype described in Sect. 4, this work was used to obtain and incorporate DM preferences into the optimization process.

3 Preference Incorporation with a Multicriteria Classification Method

3.1 *Preference Model of Fernandez et al. [8]*

Algorithms for solving MOPs can incorporate DM preferences using the preference model of Fernandez et al. [8]. The essence of the model is the degree of truth of the statement 'x is at least as good as y'. This is represented by $\sigma(x, y)$, and can be calculated using over classification methods such as ELECTRE [9] or PROMETHEE [10].

It is considered an acceptable credibility threshold λ, a parameter of asymmetry β, and a parameter of symmetry ε, where $0 \leq \varepsilon \leq \beta \leq \lambda$ and $\lambda > 0.5$. The model identifies one of the following preference relations for each pair of solutions (x, y):

Preference	Condition
Strict preference (*x*P*y*): The DM has clear and well-defined reasons to justify the choice of *x* over *y*	x dominates $y \vee$ $(\sigma(x, y) \geq \lambda \wedge \sigma(y, x) < 0.5) \vee$ $(\sigma(x, y) \geq \lambda \wedge [0.5 \leq \sigma(y, x) < \lambda]$ $\wedge [\sigma(x, y) - \sigma(y, x)] \geq \beta)$
Indifference (*x*I*y*): The DM has clear and positive reasons that justify an equivalence between the two options	$\sigma(x, y) \geq \lambda \wedge \sigma(y, x) \geq \lambda \wedge \lvert\sigma(x, y) - \sigma(y, x)\rvert \leq \varepsilon$
Weak preference (*x*Q*y*): The DM hesitates between *x*P*y* and *x*I*y*	$\neg x\mathrm{P}y \wedge \neg x\mathrm{I}y \wedge$ $\sigma(x, y) \geq \lambda \wedge \sigma(x, y) > \sigma(y, x)$
K-preference (*x*K*y*): The DM hesitates between *x*P*y* and *x*R*y*	$0.5 \leq \sigma(x, y) < \lambda \wedge \sigma(y, x) < 0.5 \wedge$ $\sigma(x, y) - \sigma(y, x) > \beta/2$
Incomparability (*x*R*y*): The DM perceives a high degree of heterogeneity between *x* and *y*, can not express a preference	$\sigma(x, y) < 0.5 \wedge \sigma(y, x) < 0.5$

The five situations of strict preference, indifference, weak preference, incomparability and *k*-preference are considered to establish a realistic model of decision-maker preferences with a more discriminatory capacity than other models in the literature.

3.2 THESEUS: A Multicriteria Classificator

The THESEUS method, developed by Fernández et al. [11], is based on outranked relations to solve multicriteria classification problems. The term classification refers to the assignment of a set of alternatives into ordered preference categories which are ordinally defined [12]. THESEUS assigns multicriteria objects to the categories of the set of ordered categories, comparing the object in question with the information of several preference and indifference relations. These relations are determined from a definite outranked relation on the universe of objects. The category assignment is not a result of the intrinsic characteristics of the object, it is rather the consequence of comparisons with other objects whose assignments (categories) are known. The THESEUS method is based on the following premises:

- There exists a limited number of categories $Ct = \{C_1, \ldots, C_M\}$, $(M \geq 2)$; where C_M indicates the best category.
- Let be U the universe of objects x which are characterized by a set of N criteria denoted by $G = \{g_1, g_2, \ldots, g_j, \ldots, g_N\}$, where $N \geq 3$.
- There exists a set of reference objects T (also called reference set or training set), which is formed by objects $b_{kh} \in U$ which are assigned to a category C_k, $(k = 1, \ldots, M)$.

- There is an outranked relation $\sigma(x, y)$ defined in $U \times U$ which models the degree of credibility of the statement "x is at least as good as y" from the decision maker's point of view.

3.3 Hybrid Multi-criteria Sorting Genetic Algorithm

Cruz-Reyes et al. [13] propose the algorithm Hybrid Multi-Criteria Sorting Genetic Algorithm (H-MCSGA). In this paper is presented the use of THESEUS, a multicriteria classification method, with a multiobjective algorithm that incorporates preferences; when it already has the Pareto front it is intended to guide a search for a better solution within a region of interest. H-MCSGA consists of two phases:

1. A multiobjective metaheuristic approach yields an approximation to the Pareto frontier.
2. The DM takes this approximation and classifies it into a set of ordered categories to construct a reference set that is used in the preference-based THESEUS classifier. In this second phase the information of the good solutions is used by a variant of the popular NSGA-II to guide the search towards the ROI, in this algorithm, the entire Pareto approximate front is divided into several fronts organized by dominance; The non-dominated solutions are in Front zero f_0. In each iteration, f_0 is redrawn using the THESEUS classifier. Thus, the new f_0 is conformed by the solutions of the class preferred by the DM, thus increasing the selective pressure towards the ROI.

In H-MCSGA two variants were created according to the way of preserving diversity in the population; H-MCSGA-CD using the crowding distance and H-MCSGA-RP that makes use of the reference points.

The *crowding distance* is a strategy used in the original algorithm NSGA-II. For a given solution, it provides an estimate of the density of solutions surrounding that solution. This distance is calculated by summing the distance between a solution and its two neighboring solutions (the anterior and posterior solutions). Considering these values, in H-MCSGA-CD, solutions having the largest crowding distance will be chosen to complement the new population.

Reference points were introduced in the A2-NSGA-III algorithm proposed by Jain and Deb [14]. This strategy guides the search in such a way that the resulting Pareto front has a graphic form very similar to the optimal Pareto front avoiding falling into local optima and also explores a larger space of solutions. The reference points can be provided by the DM or calculated in a structured manner. The process for creating reference points follows the following sequence of functions: *normalization*, *association*, *niching procedure* and *update*. For more information about this process, see the paper by Jain and Deb [14].

4 An Approach to Integrate Elicitation and Incorporation of Preferences in Optimization

4.1 The Proposal Approach

To perform the analysis proposed in this work, an architecture of a basic prototype of interactivity has been proposed. The proposed architecture (see Fig. 1) incorporates the integration of previous works. This architecture uses the PDA procedures, using a preference definer developed by Rangel-Valdez et al. [7], which uses a user-provided instance and a set of examples of alternatives compared in pairs that serve to incorporate the preferences of an emulated DM using the NO-ACO algorithm proposed by Cruz-Reyes et al. [15]. The definer transforms these preferences into equivalent configurations for the preferential model, takes these configurations, and creates instances with equivalent parameters of the preferential model. The H-MCSGA-CD and H-MCSGA-RP, which are variants from the optimizer proposed by Cruz-Reyes et al. [16] were selected to solve the instances of the portfolio problem. These optimizers use as input the instance created from PDA and a set of references (categorized examples) provided by the algorithm NO-ACO to emulate a DM, that are created from the instances with parameters of the preferential model obtained by PDA.

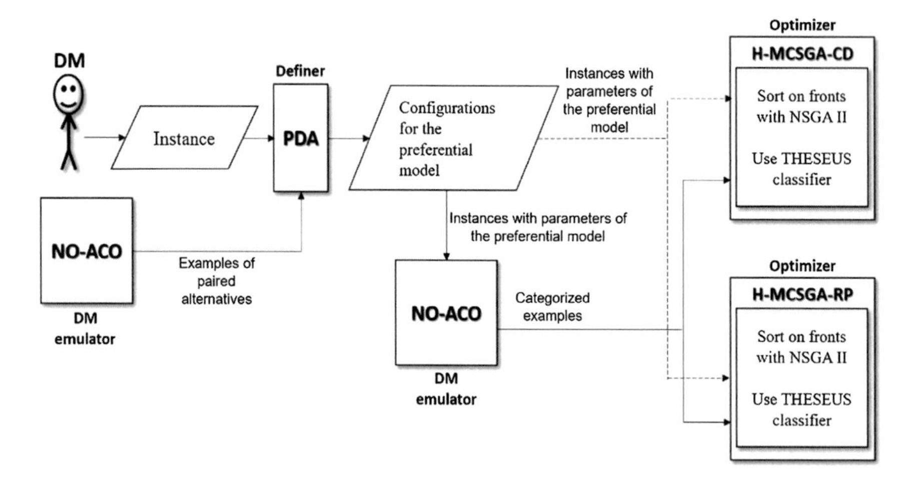

Fig. 1 The architecture proposed to elicit and incorporate preferences during the optimization process

4.2 Measures for Performance Evaluation

To evaluate the performance of MOEAs and to make comparisons between them, there are different metrics that have been proposed in the literature. Below, are described those that were used during the experimental process described in Sect. 5.

4.2.1 Dominance

To calculate this indicator, two sets of solutions are received (Approximated Pareto fronts of each algorithm) and are united in a single set to which a non-fast dominated sorting method is applied to sort and divide the solutions in fronts. Then it is calculated how many solutions of algorithm A are on the front zero f_0. The percentage of dominance is obtained by calculating the percentage representing the number of solutions that are in the f_0 and that belong to the algorithm, in relation to the total number of solutions of the algorithm. The higher the percentage of dominance, the better the performance of the algorithm.

$$\%\, dominance\, of\, A = \frac{\#\text{sol of } A \in f_0}{\#\text{sol of } A} * 100 \tag{1}$$

4.2.2 Non-strictly-Outranked

To calculate this indicator, are used two sets defined from the outranking model of Fernandez et al. [8]:

1. $S(O, x) = \{y \in O \mid yPx\}$ is the set composed by alternatives y that are strictly preferred over x.
2. $NS(O) = \{x \in O \mid S(O, x) = \varnothing\}$ is the set of alternatives x which has no relation yPx for any other y in O. This set is commonly referred as the *non-strictly-outranked frontier*.

In order to calculate this indicator, two sets of solutions are received (solutions non-strictly outranked for each algorithm) and they are united in a single set of which it is obtained a new the *non-strictly-outranked frontier* f_{NS} which will form the approximated ROI.

Then, it is calculated how many solutions of algorithm A belongs to f_{NS}. The percentage of *non-strictly-outranked* is obtained by calculating the percentage representing the number of solutions that are in the f_{NS} and belongs to the algorithm, relative to the total number of solutions of the algorithm that are non-strictly-outranked.

$$\% NS\, of\, A = \frac{\#sol\, NS\, of\, A \in f_{NS}}{\#sol\, NS\, of\, A} * 100 \tag{2}$$

4.2.3 Generational Distance

It measures the distance between the set of non-dominated solutions (FP_{aprox}) by the MOEA and the optimal Pareto front reference set (FP_{optimo}) [17]. A lower value of GD represents a better performance and is calculated with the following equation:

$$GD = \frac{\sqrt{\sum_{i=1}^{n} d_i^2}}{n} \tag{3}$$

where

n is the number of solutions in FP_{aprox} and d_i is the Euclidean distance between each individual x_i of FP_{aprox} and the nearest member y_j in FP_{optimo}. This indicator was used to measure the distance between the non-outranked solutions and the approximated ROI.

5 Case of Study: The Project Portfolio Problem

A particular multi-objective optimization problem known as the Project Portfolio Problem (PPP) consists in the selection of the public projects that provide the best benefits to the society, which will receive financial support to its realization. A feasible portfolio must satisfy the constraints of the total budget and the budget allocated to the area and region. Please refer to [18] for more details about the problem.

5.1 Experimental Process

This experiment was designed to evaluate the entire optimization process, from the definition of parameter values for the preferential model, to the solution of instances through the optimization algorithms. This experiment is summarized in the following steps described below. These steps were performed for 9 random instances of the project portfolio problem with 4 objectives and 25 projects, and for 7 random instances of the project portfolio problem, with 9 objectives and 100 projects.

The experiment was performed by the optimizer incorporating H-MCSGA preferences with its two variants, described in Sect. 3.3. The experimental process is described below:

1. Introduce to PDA the instance of the social project portfolio problem and a set of DM preferences in the form of paired comparisons;
2. Generate n configurations of the parameters for the preferential model using the PDA strategy, a configuration represents a concrete estimation of the parameters for the preferential model, in this case, in an outranking model based on ELECTRE III;
3. Associate each configuration with the portfolio problem instance, in this way an instance of the portfolio problem is created for each configuration, and incorporate the values of the previously obtained preferential model as part of its parameters. There will be as many instances as configurations were generated by the PDA;
4. Create the reference set using an emulated DM (NO-ACO), and categorize them, this reference set is created for the instances associated with the previously generated configurations;
5. Introduce the instance and the reference set into the two versions of H-MCSGA optimizer, and solve the instance 30 times.

In order to calculate the metrics with which the algorithms will be evaluated, the solutions obtained by the optimizers are collected as follows.

- To determine the dominance of the algorithms:
  1. Select a configuration generated by PDA.
  2. Collect the solutions from the 30 experiments on H-MCSGA-CD and H-MCSGA-RP for that configuration and eliminate repeated solutions.
  3. Submit the two sets of solutions to the dominance calculation to obtain the Pareto front.
  4. Calculate the percentage of dominance described in Sect. 4.2.1.
- To determine the non-strictly-outrank of the algorithms:
  1. Select a configuration generated by PDA.
  2. Collect the solutions from the 30 experiments on H-MCSGA-CD and H-MCSGA-RP for that configuration and eliminate repeated solutions.
  3. Obtain solutions non-strictly-outrank for each set.
  4. Calculate the non-strictly-outrank percentage described in Sect. 4.2.2.
- To determine the generational distance of the algorithms with respect to an approximated ROI:
  1. Select a configuration generated by PDA.
  2. Collect the solutions of the 30 experiments on H-MCSGA-CD and H-MCSGA-RP for that configuration and eliminate repeated solutions.
  3. Submit the two sets of solutions to the generational distance calculation described in Sect. 4.2.3.

The result of this experimental process for the random instances and the 30 experiments is presented in Sect. 6. The instances of 4 objectives and 25 projects that we considered small, from these were selected 4 random configurations for each instance. The instances of 9 objectives and 100 projects that we considered medium, from these were selected 5 random configurations for each instance.

6 Experimental Results

Table 1 shows the result of a configuration for a small instance. The solutions went through several filters and the table concentrates the results for each of them. The process and results are described below.

For each algorithm, we first determined the number of non-dominated solutions and the solutions non-strictly-outranked, these results are shown in the first 3 rows with the #Sol column.

After this, we took the solutions non-strictly-outranked of each algorithm to create a single set to obtain the front zero and with this same set obtain the solutions non-strictly-outranked, these conformed the ROI.

The GD-ROI row contains the generational distance from the initial solution set of the algorithm to the ROI.

The row Dominance contains the number of solutions that remains non dominated and that belongs to the front zero. The row NS contains the number of solutions that remains non-strictly-outranked and that below to the ROI.

The individual percentages column in Sol NDom and Sol NS rows where calculated considering the initial solution size. The individual percentages in Dominance and NS rows where calculated considering the Sol NS size.

The at the front percentages column where calculated considering the number of solutions on the f_0 in the case of Dominance row and with the number of solutions in the ROI in the case of NS row.

Tables 2 and 3 summarize the count of times that each algorithm won for each instance in dominance, non-strict-outrank, and generational distance. It should be considered that the maximum number of times an algorithm can win is 4 times in

Table 1 Results of a configuration for an instance of 4 objectives and 25 projects

	H-MCSGA-CD			H-MCSGA-RP		
	#Sol	%individual	%at front	#Sol	%individual	%at front
Sol size	10			11		
Sol NDom	10	100		11	100	
Sol NS	6	60		7	64	
Dominance	6	100	86	7	100	100
NS	6	100	86	7	100	100
GD-ROI	0.232			0.231		

Number of solutions on the $f_0 = 7$
Number of solutions in ROI = 7

Table 2 Count and average of wins by each instance of 4 objectives and 25 projects for algorithms H-MCSGA-CD and H-MCSGA-RP

Instance	Dominance		NS		GD-ROI	
	CD	RP	CD	RP	CD	RP
4.1	3	4	3	4	1	4
4.2	4	1	3	1	3	1
4.3	1	4	1	4	1	3
4.4	0	4	0	4	3	1
4.5	1	4	1	3	3	1
4.7	0	4	0	4	2	2
4.8	1	3	1	3	1	4
4.9	1	4	1	4	3	1
4.10	1	4	0	4	3	1
Total	12	32	10	31	20	18
Average	33.33	88.89	27.78	86.11	55.56	50.00

Table 3 Count and average of wins by each instance of 9 objectives and 100 projects for algorithms H-MCSGA-CD and H-MCSGA-RP

Instance	Dominance		NS		GD-ROI	
	CD	RP	CD	RP	CD	RP
9.1	5	4	5	4	4	5
9.2	5	5	5	5	5	5
9.3	5	3	5	3	5	3
9.4	5	4	5	4	4	5
9.5	4	2	4	3	3	3
9.8	5	0	4	1	4	1
9.10	5	4	5	4	5	4
Total	34	22	33	24	30	26
Average	97.14	62.86	94.29	68.57	85.71	74.29

the case of small instances and 5 in the case of the medium instances. At the end, a total count and an average of times gained by each indicator are made.

Tables 4 and 5 summarize the average of dominance, non-strict-outrank and generational distance to the ROI of all configurations selected for each instance. At the end, a general average is used to determine the performance of the algorithms according to each indicator.

Table 4 Averages of configurations for instances of 4 objectives and 25 projects

Instance	Dominance		NS		GD-ROI	
	CD	RP	CD	RP	CD	RP
4.1	96.4	100.0	96.4	100.0	0.1086	0.1054
4.2	83.5	71.8	92.9	68.8	0.1477	0.8341
4.3	81.3	100.0	81.3	100.0	5.6665	5.3623
4.4	72.3	98.0	71.5	100.0	0.7908	2.1380
4.5	63.2	90.0	73.2	83.3	1.5776	0.5162
4.7	76.2	96.7	75.2	100.0	0.5292	0.3503
4.8	64.9	85.3	61.3	92.5	0.0382	0.0228
4.9	74.9	98.3	74.4	100.0	0.0506	0.0363
4.10	91.2	98.1	91.2	99.0	0.0430	0.0402
Average	78.2	93.1	79.7	93.7	0.9947	1.0451

Table 5 Averages of configurations for instances of 9 objectives and 100 projects

Instance	Dominance		NS		GD-ROI	
	CD	RP	CD	RP	CD	RP
9.1	91.2	98.1	91.2	99.0	0.0430	0.0402
9.2	100.0	100.0	100.0	100.0	0.0000	0.0000
9.3	87.5	74.3	86.7	75.5	0.0247	0.0507
9.4	93.2	86.9	93.9	86.3	0.0074	0.0024
9.5	73.3	56.7	70.8	60.9	0.1960	0.6416
9.8	65.8	35.0	59.2	42.0	0.0082	0.0247
9.10	92.4	87.6	90.7	89.4	0.0019	0.0029
Average	86.2	77.0	84.6	79.0	0.0402	0.1089

7 Conclusions

The experimental analysis aims to explore the integration of PDA in the solution process of the SPP problem using two MOEAs with different diversity preservation schemes (H-MCSGA-CD and H-MCSGA-RP). The behavior derived from these algorithms was observed in instances with 4 and 9 objectives.

In the instances of 4 objectives, H-MCSGA-RP has a better performance, which is according to the literature since the Crowding distance works well even with three objectives.

In the case of 9 objectives, it was observed that H-MCSGA-CD had a better behavior, this behavior derived from the observation that the considered instances had a weight of zero in 7 of 9 objectives, which is equivalent to have instances of 2 objectives. From this analysis, it was possible to detect that it is necessary as future work, to make adjustments in PDA to treat large instances and to assign weights other than zero.

It was presented the result of the process to obtain and incorporate preferences in the optimization process using the technique of disaggregation analysis of preferences. The built prototype will be the basis for the development of interactive optimization methods since it has been statistically demonstrated that it can be used to cover the entire optimization process including the preference elicitation an incorporation, processing and solution generation.

References

1. C.A.C. Coello, Handling preferences in evolutionary multiobjective optimization: a survey, in *IEEE Congress on Evolutionary Computation*, vol. 1, pp. 30–37 (2000)
2. M. Kadzinski, S. Greco, R. Słowinski, Selection of a representative value function in robust multiple criteria ranking and choice. Eur. J. Oper. Res. **217**(3), pp. 541–553 (2012)
3. L. Rachmawati, D. Srinivasan, Preference incorporation in multi-objective evolutionary algorithms: a survey, in *IEEE Congress on Evolutionary Computation*, pp. 3385–3391 (2006)
4. M. Doumpos, C. Zopounidis, The robustness concern in preference disaggregation approaches for decision aiding: an overview, in *Optimization in Science and Engineering* (Springer, New York 2014), pp. 157–177
5. E. Jacquet-Lagrèze, Y. Siskos, Preference disaggregation: 20 years of MCDA experience. Eur. J. Oper. Res. **130**(2), pp. 233–245 (2001)
6. M. Doumpos, Y. Marinakis, M. Marinaki, C. Zopounidis, An evolutionary approach to construction of outranking models for multicriteria classification: the case of the ELECTRE TRI method. Eur. J. Oper. Res. **199**(2), pp. 496–505 (2009)
7. N. Rangel-Valdez, E. Fernandez, L. Cruz-Reyes, C.G. Santillan, R.I. Hernandez-Lopez, Multiobjective optimization approach for preference-disaggregation analysis under effects of intensity, in *Advances in Artificial Intelligence and Its Applications* (Springer International Publishing, 2015), pp. 451–462
8. E. Fernandez, E. Lopez, F. Lopez, C.A. Coello Coello, Increasing selective pressure towards the best compromise in evolutionary multiobjective optimization: the extended NOSGA method. Inf. Sci. **181**(1), pp. 44–56 (2011)
9. B. Roy, Nonconvex optimization and its applications, in *Multicriteria Methodology for Decision Aiding* (Springer, 1996)
10. J. Brans, B. Mareschal, Promethee methods, in *Multiple Criteria Decision Analysis: State of the Art Surveys*, volume 78 of International Series on Operations Research & Management Science (Springer, Berlin 2005), pp. 163–190
11. E. Fernandez, J. Navarro, A new approach to multi-criteria sorting based on fuzzy outranking relations: the THESEUS method. Eur. J. Oper. Res. **213**(2), pp. 405–413 (2011)
12. M. Doumpos, C. Zopounidis, *Multicriteria Decision Aid Classification Methods*, vol. 73 (Springer Science & Business Media, 2002)
13. L. Cruz-Reyes, E. Fernandez, P. Sanchez, C.A.C. Coello, C. Gomez, Incorporation of implicit decision-maker preferences in multi-objective evolutionary optimization using a multi-criteria classification method. Appl. Soft Comput. **50**, pp. 48–57 (2017)
14. H. Jain, K. Deb, An improved adaptive approach for elitist nondominated sorting genetic algorithm for many-objective optimization, in *International Conference on Evolutionary Multi-Criterion Optimization* (Springer Berlin Heidelberg, 2013), pp. 307–321
15. L. Cruz-Reyes, E. Fernandez, C. Gomez, G. Rivera, F. Perez, Many-objective portfolioop-timization of interdependent projects with 'a priori' incorporation of decision-maker preferences. Appl. Math. Inf. 8, pp. 1517–1531 (2014)

16. L. Cruz-Reyes, E. Fernandez, C. Gomez, P. Sanchez, Preference incorporation into evolutionary multiobjective optimization using a multi-criteria evaluation method, in *Recent Advances on Hybrid Approaches for Designing Intelligent Systems* (Springer International Publishing, 2014), pp. 533–542
17. G.G. Yen, Z. He, Performance metric ensemble for multiobjective evolutionary algorithms in *IEEE Transactions on Evolutionary Computation*, vol. 18, no. 1, pp. 131–144 (2014)
18. L. Cruz-Reyes, E. Fernandez, C. Gomez, P. Sanchez, G. Castilla, D. Martinez, Verifying the effectiveness of an evolutionary approach in solving many-objective optimization problems, in *Design of Intelligent Systems Based on Fuzzy Logic, Neural Networks and Nature-Inspired Optimization* (Springer International Publishing, 2015), pp. 455–464

A Methodology for Optimization of Visual Comfort of Multi-User Intelligent Systems Based on Genetic Algorithms

Wendoly J. Gpe. Romero-Rodríguez, R. Baltazar, Juan Martín Carpio Valadez, Héctor Puga, J. F. Mosiño and V. Zamudio

Abstract Lighting Standard in work spaces establish a range between a minimal and a maximum level of luminance depending of the task to provide visual comfort. In the state of the art, researches had been focusing to offer the adequate level of luminance just to a single user through the control of the artificial lighting systems using intelligent algorithms, without taking in account of daylight sources that can reduce energy costs. Nevertheless, an intelligent system has more than one user and a shared lighting system, there may be conflicts between users and their different activities, preferences, profiles and priorities, therefore a new approach is required. In this work, a novel methodology based on Genetic Algorithms is proposed, focusing on lights and blinds management of a multi-user scenario and it is presented. It is concentrated to find optimal configurations in energy savings, visual comfort, and conflict resolution between users based on Genetic Algorithms. Finally, the results of our proposal methodology are showed and discussed.

Keywords Visual comfort · Lighting systems · Energy management
Ambient intelligence · GA · Multi-user · Conflict resolution

1 Introduction

The need to guarantee comfort and generate savings in electrical energy mainly in offices, hotels, public and shopping centers, etc. has led to the development of various systems for energy control and comfort. This work will take into account visual comfort, which is achieved by providing adequate lighting (Lx) in the work areas to perform certain activities [1]. According to the Federal Occupational Safety and Health Regulations, article 35, one of the rules is to apply control measures

W. J. Gpe.Romero-Rodríguez (✉) · R. Baltazar · J. M. Carpio Valadez · H. Puga · J. F. Mosiño · V. Zamudio
Tecnológico Nacional de México, Instituto Tecnológico de León, León, Gto, Mexico
e-mail: wendolyjgrr@gmail.com

O. Castillo et al. (eds.), *Fuzzy Logic Augmentation of Neural and Optimization Algorithms: Theoretical Aspects and Real Applications*, Studies in Computational Intelligence 749, https://doi.org/10.1007/978-3-319-71008-2_30

when lighting levels are below or above the limit values determined in their areas and positions according to the corresponding Standard [2].

In the state-of-the-art have been reported different approaches of Intelligent Systems to provide visual comfort to a single user, using soft-computing techniques to provide the closest real-time answers. In [3, 4], Particle Swarm Optimization is used to tune the set points of a control system to provide comfort to users, but only is an average comfort of all users. In [5, 6], the energy saving and average comfort level are taken as two control objectives in the control system, and PSO is used but in Multi-Objective version to optimize a fuzzy controller. A multi-agent system is controlled and developed with stochastic optimization using genetic algorithm in [7].

The control systems available so far in the state of the art are focused on providing visual comfort by keeping the environment under certain lighting ranges that the user defines, but this can not be assured of energy savings. These works have been focused on single-user problems and only control of artificial lighting systems is done, without taking advantage of daylight sources that can help reduce energy costs.

In this research we are interested on model a Multi-User Intelligent System that will be able to find configurations of states of lamps and blinds, to get energy efficiency and provide visual comfort to more than a single user, using soft-computing techniques to set up the environment taking into account that each user have different activities, user's profiles and priority's activities.

2 Genetic Algorithm

A genetic algorithm is an optimization technique based on natural selection principles, belongs to the group of evolutionary algorithms (EA) [8]. It is commonly used in search problems and uses operations such as selection, mutation and crossover to obtain populations of individuals evolved in order to obtain better solutions.

The pseudocode of a simple genetic algorithm can be shown in Fig. 1.

Fig. 1 Pseudocode of a binary GA

```
Define fitness function, variables, Select GA parameters
Initialize a random population
Decode chromosome
        Evaluate fitness function for each chromosome
        Select mates
        Mating
        Mutation
        Crossover
        Convergence Check
Done
```

3 Multi-user Systems

In the multiuser category we are taking into account all systems that simultaneously meet the needs of two or more users, who share the same resources. They have the ability to perceive changes in the environment and adjust according to the specific activities of users, resolve conflicts through collaboration and negotiation, similar to the process of a team of humans [9].

In order for a system to be multiuser it must take into account the following:

- Number of users, because we can have more than one user.
- Hierarchy or User Profiles, in intelligent systems users are generally divided according to hierarchies or profiles, where the activity of one user can be given more importance or priority than another depending on their hierarchy.
- Priority of Activities for each user, tt is taken into account if one activity has more relevance than another. For example, relax is less important than that working on computer.
- Preferences on lighting levels for each activity, lighting levels must be within the limit values determined in their areas and positions according to the corresponding Standard.

4 System Definition

In this paper, we define a Multi-User Intelligent System whose objective is to provide the recommended luminance for multiple users with different profiles, activities and priorities. It should be mentioned that it is a static system, users can't be moving around the ambient while the Multi-User Intelligent System is searching a configuration of states that satisfy the visual comfort of multiple users.

This problem can be formulated using Eq. (1):

$$\min F(x) = \min\left(\sum_{i=1}^{m} abs(f_i(x))\right) \tag{1}$$

where

- X is a set of variables: User's variables, lighting variables and scenario variables, which are mentioned below.
- $f_i(x)$ is luminance granted (Lx) by lamps and blinds in Zone i.

We have defined 3 different kind of variables and their domain are shown below:

User's variables:

- Zone, Z = {1–21}

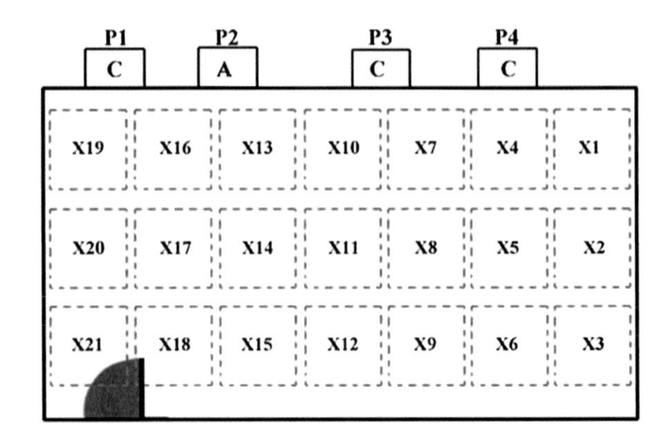

Fig. 2 Map of zones of the Multi-User Intelligent System

- Activity, A = {Absent, Reading, resting, working on computer, projecting, exposing}
- Activity Priority P = {0–3}.
- User Profile J = {Teacher, Student, Visitor}

Lighting variables:

- 21 Sensors Interior lighting.
- 4 Sensors Outdoor lighting.

Scenario variables:

- 21 lamps $(l_1, l_2, \cdots l_{21})$, L = {1-9}
- 4 blinds (P_1, P_2, P_3, P_4) B = {0-9}

This problem is subject to the following restriction:

- Cannot perform more than one activity in one zone.

This is problem can be posed as a constraint satisfaction problem (CSP), where we have a set of variables that must satisfy the visual comfort of different areas subject to restrictions.

In Fig. 2 we are showing a map of zones of the Multi-User Intelligent System, we divided in zones to be able to know how much visual comfort you have in each area occupied by a user. The level of illumination of each lamp grants different illumination to each zone in the ambient. If the level of a lamp changes, has an effect on the amount of illumination granted for each zone of the stage.

5 Proposed Model Multi-User Intelligent System

There are intelligent systems that provide visual comfort to a single user, they are known as Mono-user Intelligent Systems. The calculation of visual comfort for a single user and fitness function is shown below:

1. Lighting required for each activity is calculated by:

$$\Phi_T = \frac{(E \times S)}{(\eta \times f_m)} \quad (2)$$

where

Φ_T	Total Luminous Flux (Lm) required in a specific zone.
E	Desired average luminance (Lx), assigned value for each activity, taken from the standards [10].
S	Surface of the working plane (m^2).
η	Light Output (Lm/W).
f_m	Maintenance Factor.

2. In order to calculate how much illumination each lamp and blind provides depending on its state to each zone, the model proposed in [11] was used. In that model, measurements were made in an environment avoiding the influence of the external illumination, taking into account the distance and angle of the lamp or blind to the area of interest to describe the distribution of illumination.

$$\sum_{n=1}^{21} \Phi_{Ln}$$
$$\sum_{n=1}^{4} \Phi_{Pn} \quad (3)$$

3. Visual comfort of a zone will be the value from the total value of illumination granted by shutters and lamps shall be subtracted from the calculated optimum luminous flux value. The fitness is considered as optimum the closer it is to zero.

$$Cv_i = \Phi_T - \sum_{n=1}^{21} \Phi_{Ln} - \sum_{n=1}^{4} \Phi_{Pn} \quad (4)$$

where

- Cv is visual comfort and is in function of the required lighting Lx.
- Φ_T is the total luminous flux required (lx) for certain activity, area and lamp conditions.
- Φ_{Ln} is the illumination (lx) of the lamp n to the zone i.
- Φ_{Pn} is the illumination (lx) of the blind n to the zone i.

Do not forget that the last model is for a Mono-user Intelligent System, but can not be used for a Multi-User Intelligent System. For example, when the users are dispersed in the zones of the environment and the illumination of an activity does not generate conflict with another, does not interfere. In this case, we can apply the fitness function from Fig. 3 for each zone.

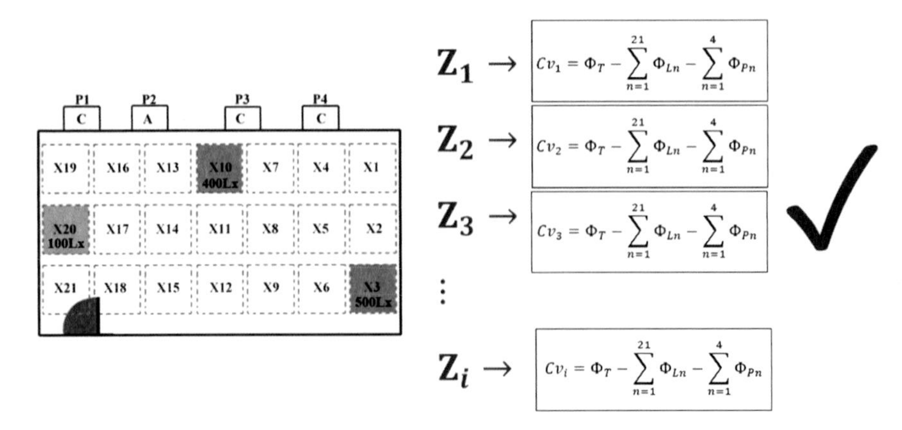

Fig. 3 Example, when the users are dispersed in the zones

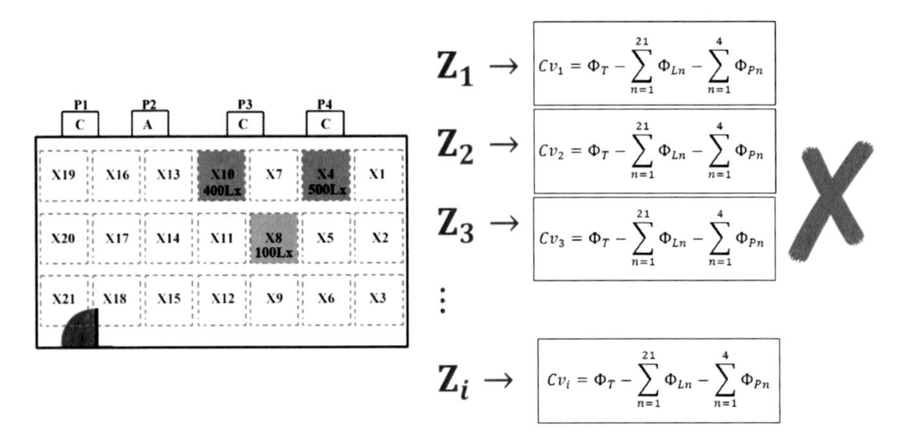

Fig. 4 Example, when the users are close to each other in the zones

Other example is when the users are close to each other in the zones and have different lighting needs, different priorities and user's profiles, we have conflicts because the Intelligent System needs to provide visual comfort to each user without interfering with the visual comfort of other users around them. This is shown in Fig. 4.

To model an Intelligent System for multiple users, we are proposing to use a weighting for the different zones mentioned in the equation, taking into account the priority of activities and user profile for each area occupied.

$$f_i(x) = Cv_i * \left(\frac{1}{P_i \cdot J_i \cdot 100} \right) \tag{5}$$

Table 1 Activity priority weighting

Activity	Priority (P)
Absent	0
Resting	1
Reading	2
Working on computer	2
Projecting	3
Exposing	3

Table 2 User profile weighting

User Profile	Hierarchy (J)
Teacher	3
Student	2
Visitor	1

where

- Cv_i is visual comfort of the User in the Zone i.
- P_i is activity priority in the Zone i.
- J_i is User Hierarchy in the Zone i.

We have assigned weights for each activity priority and user profile, taking into account how important one is in comparison to others. And it also helps the system to "negotiate" with the resources of the environment, granting maximum visual comfort to each user and thus address the problem of conflict resolution that can arise when users have different needs for visual comfort. This is shown in Tables 1 and 2. We consider that the assigned values could be these because we are giving importance to each activity, for example, the activity of resting could be considered less important than the activity of working in computer since perhaps the user must deliver some work at a certain time.

The assignment of values for user profiles in Table 2 was based on their hierarchy and taking into account that the activities of a teacher may be more important than those of a student or visitor in the system.

5.1 Chromosome Coding Scheme

According to the model proposed for the Intelligent Multiuser System, it was proposed to use a genetic algorithm to obtain the closest real-time responses. Due to this, it was proposed to use binary representation to facilitate in the future the application of binary soft-computing techniques.

The levels of the lamps are from 0 to 255 levels but were divided into 7 levels, while the blinds handle 7 lighting levels.

Figure 5 shows the representation of the genotype or possible solution, indicates the state of each lamp and blind taking into account the values in Table 3. This

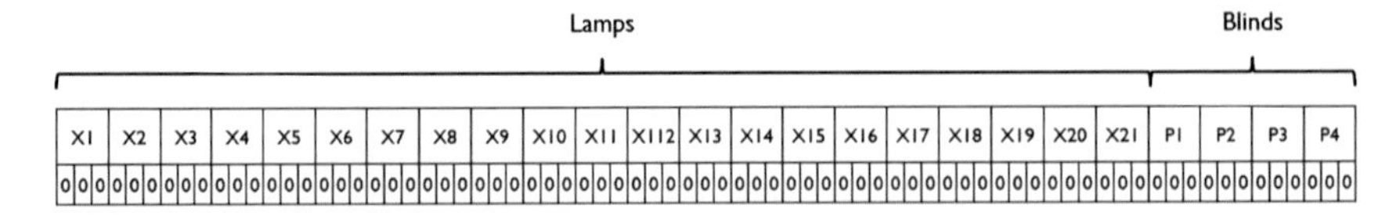

Fig. 5 Chromosome Coding Scheme

Table 3 Binary representation of level of the lamps

Level	Binary Representation		
0	0	0	0
1	0	0	1
2	0	1	0
3	0	1	1
4	1	0	0
5	1	0	1
6	1	1	0
7	1	1	1

binary representation is based on the work [12], where the states of each lamp and blind were represented in a binary way which will facilitate the application of binary soft-computing techniques.

5.2 *Model Multi-User Intelligent System*

Figure 6 shows how works the Multi-User Intelligent System, taking as inputs: the external illumination values delivered by 4 sensors located in each window of the environment, zones occupied, activities to be carried out with their respective priority and user profiles. Then starts a cycle where we get a set of the best solutions launched by the soft-computing technique and the weighting of priorities of activities and user profiles. In the end, the output is the best solution from the set of possible solutions generated. As we are minimizing, the best solution is the one that has the lowest evaluated value in its fitness function.

6 Experimental Results

For preliminary results, it shows how the model works with 5 test instances which in this case will be called scenarios. These test instances were constructed due to the fact that in the state of the art no test instance was found. These 5 instances have been built so that first scenarios are where the users are dispersed in different areas of the stage, having different lighting needs, activity priorities and user profiles. Then there are scenarios where the number of users is increasing and are closer to

Inputs: Outdoor lighting, Activities by zone, User's Profiles and Activities's Priority.
FOR each cycle to n
Soft-Computing Technique (Genetic Algorithm)
Evaluate Visual Comfort in each zone, $\{Cv_1, Cv_2, Cv_3, \ldots, Cv_i\}$
FOR each Evaluated Zone do:

$$f_i(x) = Cv_i * \left(\frac{1}{P_i \cdot J_i \cdot 100}\right)$$

END FOR

$$\min F(x) = \min\left(\sum_{i=1}^{m} abs(f_i(x))\right)$$

END Soft-Computing Technique (Genetic Algorithm)
Store $F(x) = \{f_1, f_2, f_3, \ldots, f_m\}$ a set of solutions $P = \{F_1(x), F_2(x), F_3(x), \ldots, F_n(x)\}$.
Search for new individuals;
cycle++;
END FOR
Output $F_n(x)$ with less value from P.

Fig. 6 Pseudocode Model Multi-User Intelligent System

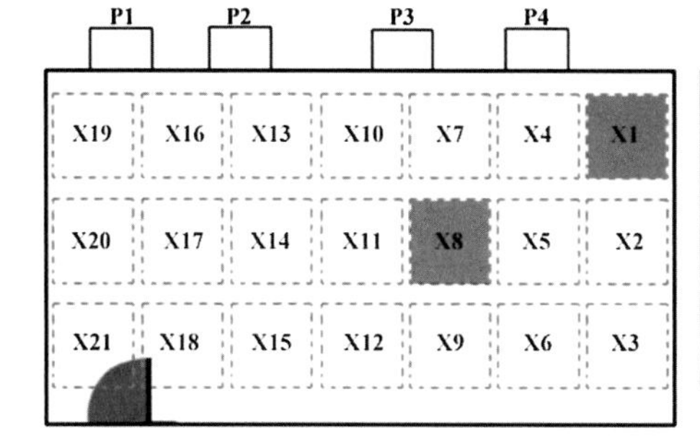

Zone	Activity (P)	User Profile (J)
I	Computer work (2)	Teacher (3)
8	Reading (2)	Visitor (1)

Fig. 7 Test Instance 1

each other, and may have conflicts in the moment to offer visual comfort to each user without affecting others.

The test instances are showed in Figs. 7, 8, 9, 10, and 11 where on the left side the diagram of areas occupied in the scenario is shown and on the right side the zone, activities and their activity priorities and user profile values are shown as a table.

Based on experiments in which the values of the input parameters were changed, the values shown in Table 4 were selected because they showed better results with values close to zero.

For each instance 100 runs were performed, of which the mean and standard deviation were obtained. As already mentioned throughout the article, a solution is considered good when its value is close to zero. Therefore, based on the results obtained in Table 5, it can be said that the results were good since they are quite close to zero. This means that for these first results, the intelligent multiuser system is giving visual comfort to each user, despite the conflicts that can be had by having different levels of visual comfort to offer for each user.

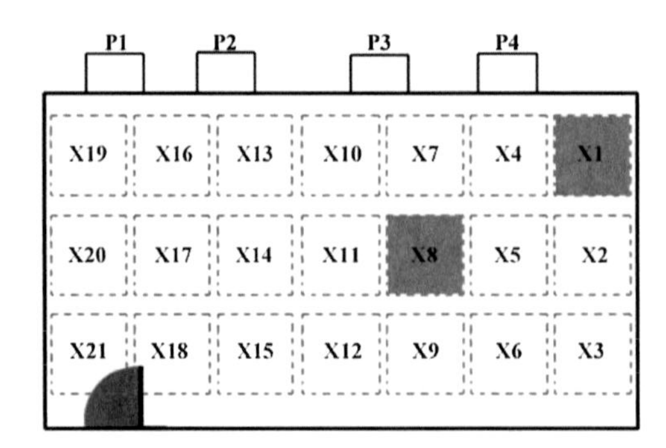

Zone	Activity (P)	User Profile (J)
1	Computer Work (2)	Teacher (3)
8	To Project (3)	Visitor (1)

Fig. 8 Test Instance 2

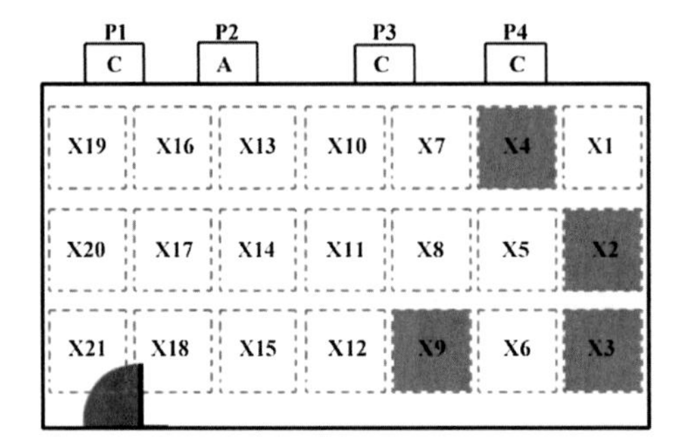

Zone	Activity (P)	User Profile (J)
2	Reading (2)	Teacher (3)
3	Computer Work (2)	Visitor (1)
4	Computer Work (2)	Teacher (3)
9	Exposition (3)	Visitor (1)

Fig. 9 Test Instance 3

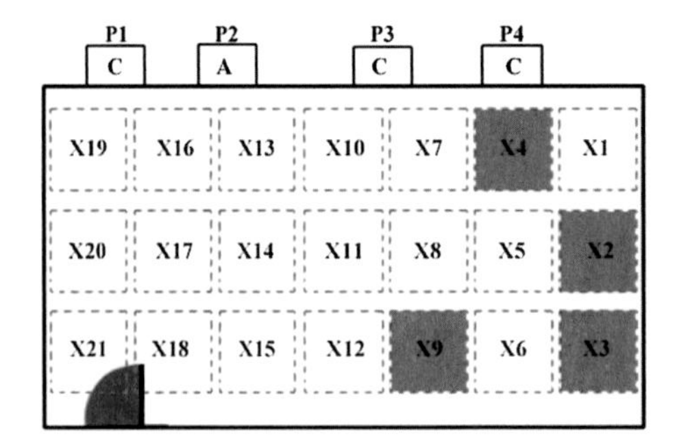

Zone	Activity (P)	User Profile (J)
2	Reading (2)	Teacher (3)
3	To Project (2)	Visitor (1)
4	Computer Work (2)	Teacher (3)
9	Exposition (3)	Visitor (1)

Fig. 10 Test Instance 4

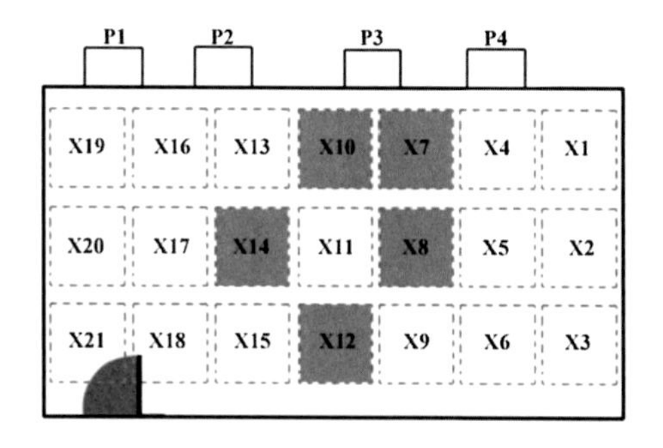

Zone	Activity (P)	User Profile (J)
7	Reading (2)	Teacher (3)
8	Absent (1)	Teacher (3)
10	Computer Work (2)	Student (2)
12	Exposition (3)	Teacher (3)
14	Computer Work (2)	Student (2)

Fig. 11 Test Instance 5

Table 4 Input parameters using the GA

Algorithm	Parameter	Value
Genetic Algorithm	No. generations	100
	Population size	80
	Mutation probability	0.8
	Elitism	0.1
	Iterations AG	100

Table 5 Results Test Instances

Test Instance	Mean	Standard deviation
I	0.0223085	0.10561445
II	0.0462556	0.09576603
III	0.1153521	0.23723311
IV	0.0254847	0.24837348
V	0.0304982	0.26931239

6.1 Representation of Solutions

According to the results table of the instances, results are very close to 0, which is considered as good results because for this problem is intended to minimize to be able to provide visual comfort to each user.

The Fig. 12 shows an example of a solution that threw the genetic algorithm and their respective representation of the states of lamps and blinds. The shading in each zone represents the level of illumination in which each lamp and blind. The X represents the area occupied by some user.

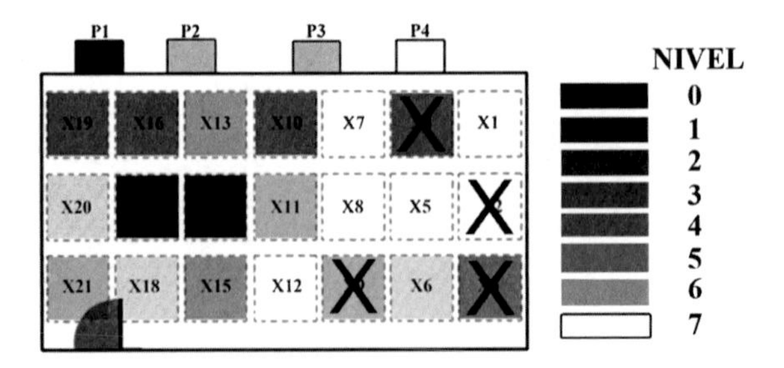

Fig. 12 Representation of the states of lamps and blinds for Instance 4

Table 6 Results Test Instance 4

Zone	Activity (P)	User profile (J)	$f_i(x)$	Lighting needed	Lighting awarded
2	Reading (2)	Teacher (3)	0.00051	299.59	299.89691
3	To project (2)	Visitor (1)	0.00005	50	50.01036
4	Computer work (2)	Teacher (3)	0.00017	199.28	199.38133
9	Exposition (3)	Visitor (1)	0.00018	159.78	159.83518
		$\sum f_i(\boldsymbol{x})$	**0.00092**		

Below are the solutions thrown by the results in Table 6, showing how the recommended lighting to offer visual comfort in each area is almost reached even in cases where users need different levels of illumination and are nearby. It is noteworthy that the different state configurations in blinds, helps that also the daylight can cooperate with providing adequate lighting to each area dependent on their needs, therefore could have a saving in energy consumption.

However, there may be different configurations that can give the same value of visual comfort in each zone, but with different energy consumptions. That is why it must be proposed as future work, to be able to measure the energy consumption in each configuration of states of shutters and windows so that it can offer energy savings according to the possible solutions that are obtained.

7 Conclusions and Future Work

Multiuser modeling in systems is important, since in real scenarios there is more than one user and each user has different visual comfort needs. Having different visual comfort needs, conflicts can arise because the system is not negotiating correctly with the system resources in order to offer maximum visual comfort to each user. In this work a proposed model was presented for a Multi-User Intelligent

System and the first experimental results obtained using a genetic algorithm. The results obtained with the 5 test instances show a desired behavior, since the visual comfort in all the zones of each instance tends to be reached equally taking into account the priorities of activities and hierarchies of users that is an important goal in a Multiuser System. Managing ranking hierarchies and priorities of activities has played an important role in being able to grant maximum visual comfort to each user because is a way to "negotiate" with the needs of multiple users regarding visual comfort. The average response of this proposal is 50 s with 55 ms, which plays an important role in the Intelligent Systems that needs to respond to the user in real time. Future work will include refinement to the proposed model by giving priority to the management of blinds to achieve taking advantage of daylight, with the intention of minimizing energy consumption. Improve the methodology to prioritize lighting to lamps that are close to user-occupied zones. Design and test with more test instances where there may be even more conflicts between users and their visual comfort. Research about what more soft-computing techniques can give us better results in even less time and compare results.

References

1. A.K. Yener, A method of obtaining visual comfort using fixed shading devices in rooms. Build. Environ. **34**(3), 285–291 (1998)
2. "Reglamento Federal de Seguridad y Salud en el Trabajo," SEGOB, D. Of. la Fed. Mèxico (2014)
3. L. Wang, Z. Wang, R. Yang, Intelligent multiagent control system for energy and comfort management in smart and sustainable buildings. IEEE Trans. Smart Grid **3**(2), 605–617 (2012)
4. R. Yang, Z. Wang, L. Wang, A GUI-based simulation platform for energy and comfort management in zero-energy buildings, in *NAPS 2011—43rd North American Power Symposium* (2011)
5. R. Yang, L. Wang, Multi-objective optimization for decision-making of energy and comfort management in building automation and control. Sustain. Cities Soc. **2**(1), 1–7 (2012)
6. R. Yang, L. Wang, Multi-zone building energy management using intelligent control and optimization. Sustain. Cities Soc. **6**(1), 16–21 (2013)
7. P.H. Shaikh, N. Bin, M. Nor, P. Nallagownden, I. Elamvazuthi, Optimized intelligent control system for indoor thermal comfort and energy management of buildings (2014)
8. T. Back, in *Evolutionary Algorithms in Theory and Practice: Evolution Strategies, Evolutionary Programming, Genetic Algorithms* (Oxford University Press, Oxford, 1996)
9. J. Lee, Conflict resolution in multi-agent based Intelligent Environments. Build. Environ. **45** (3), 574–585 (2010)
10. N.O. Mexicana, "NOM-025-STPS-2008," *Condiciones iluminación en los centros Trab.* (2008)
11. G. Mendez, M.A. Casillas, R. Baltazar, C. Lino, L. Mancilla, S. Lopez, Intelligent management system for the conservation of energy, in *2015 International Conference on Intelligent Environments*. (2015)
12. J.A.S. Romero-Rodríguez, W.J.G. Rodríguez, V.M.Z. Flores, R.B. Sotelo-Figueroa, M.A. Alcaraz, Comparative study of BSO and GA for the optimizing energy in ambient intelligence, in *Mexican International Conference on Artificial Intelligence*, (Springer, Berlin, 2011), pp. 177–188

Part VII
Hybrid Intelligent Systems

Translation of Natural Language Queries to SQL that Involve Aggregate Functions, Grouping and Subqueries for a Natural Language Interface to Databases

Rodolfo A. Pazos R., Andres A. Verastegui, José A. Martínez F., Martin Carpio and Juana Gaspar H.

Abstract Currently, huge amounts of information are stored in databases (DBs). In order to facilitate access to information to all users, natural language interfaces to databases (NLIDBs) have been developed. To this end, these interfaces translate natural language queries to a DB query language. For businesses, the main application of NLIDBs is for decision making by facilitating access to information in a flexible manner. For a NLIDB to be considered complete, it must deal with queries that involve aggregate functions: COUNT, MIN, MAX, SUM and AVG. The prototype developed at the Instituto Tecnológico de Cd. Madero (ITCM) can translate queries in natural language to SQL; however, it did not have a module for dealing with aggregate functions, grouping and subqueries. In this paper a new module of this NLIDB for dealing with aggregate functions, grouping and subqueries is described, and experimental results are presented, which show that this interface has a performance (recall) better than that of C-Phrase.

Keywords Natural language · Natural language interfaces to databases aggregate functions · Grouping · Subqueries

---

R. A. Pazos R. (✉) · A. A. Verastegui · J. A. Martínez F. · J. Gaspar H.
Instituto Tecnológico de Ciudad Madero, Ciudad Madero, Mexico
e-mail: r_pazos_r@yahoo.com.mx

A. A. Verastegui
e-mail: andres.verastegui@hotmail.com

J. A. Martínez F.
e-mail: jose.mtz@gmail.com

J. Gaspar H.
e-mail: jgasparhdz@gmail.com

M. Carpio
Instituto Tecnológico de León, León, Mexico
e-mail: jmcarpio61@hotmail.com

O. Castillo et al. (eds.), *Fuzzy Logic Augmentation of Neural and Optimization Algorithms: Theoretical Aspects and Real Applications*, Studies in Computational Intelligence 749, https://doi.org/10.1007/978-3-319-71008-2_31

1 Introduction

The development of natural language interfaces to databases (NLIDBs) is a complex task that requires the study of different aspects. The treatment of aggregate functions, grouping and subqueries is one of those aspects. The incorporation of this aspect improves the functionality of NLIDBs, allowing users to obtain answers for more complex queries.

In this paper a new module of the ITCM NLIDB [1] for dealing with aggregate functions, grouping and subqueries is described. Experimental results are presented, which show that this interface has a performance (recall) better than that of C-Phrase [2].

Next, we present some examples of queries for the Geobase database [3] with aggregate functions (AFs), grouping and subqueries; where the input is in natural language (NL) and the output is an SQL statement:

What is the area of the largest state?
SELECT MAX (*area*) FROM *state*

How many rivers per length?
SELECT COUNT (*river_name*), *length* FROM *river* GROUP BY *length*

What is the mountain with the smallest height?
SELECT *mountain_name*, *height* FROM *mountain* WHERE *height* =
(SELECT MIN (*height*) FROM *mountain*)

2 Description of the ITCM Interface

The module for the treatment of AFs, grouping and subqueries is embedded in the NLIDB developed at ITCM; therefore, it is necessary to explain the process carried out by the interface. This NLIDB was designed using functionality layers, which allows the process of translation to be straightforward and it also facilitates the scalability of the interface. The translation process is performed by three main modules: Lexical Analysis, Syntactic Analysis and Semantic Analysis [1]. Each module performs a specialized and systematic treatment to solve the inherent problems during the translation process. Figure 1 describes the interaction between the modules that constitute the functionality layers.

The ITCM NLIDB uses for the semantic analysis a structure that can be conceptually represented by a table, which stores all the information related to each lexical component (words or group of words) of the query. This structure allows analyzing individually each lexical component and additionally allows keeping all the semantic information organized. The use of the structure is critical for the treatment of AFs, grouping and subqueries.

The data structure for the semantic analysis is composed of the elements shown in Fig. 2. The attribute *Lexical component* stores the words present in the query.

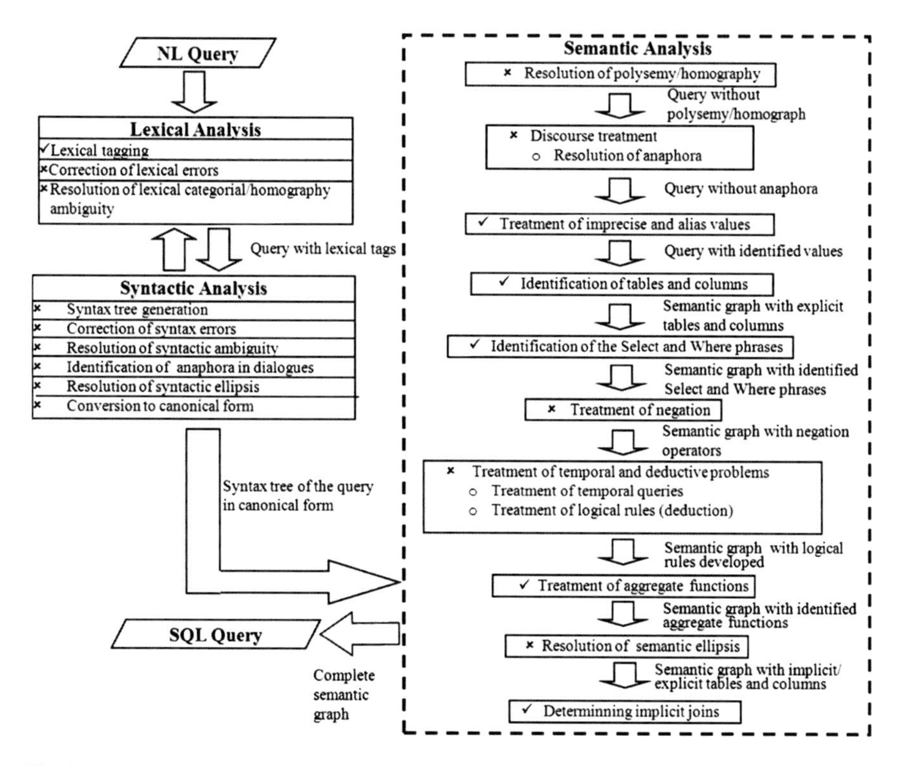

Fig. 1 Functionality layers of the ITCM NLIDB

Attribute	Query token										
Lexical component	*List*	*roundtrip*	*fares*	*from*	*Philly*	*to*	*DFW*	*arriving*	*in*	*the*	*evening*
Lemma	list	roundtrip	fare	from	?	to	?	arrive	in	the	evening
Syntactic category	verb	noun	noun	prep.		prep.		verb	prep.	art.	noun
Phrase											
Phrase ID											
Phrase type											
Column tag											
Table tag											
Final tag					Philly		DFW				
Table	false	false	false	false	false	false	false	false	false	false	false
Column	false	false	false	false	false	false	false	false	false	false	false
Value	false	false	false	false	true	false	true	false	false	false	false
View value	false	false	false	false	false	false	false	false	false	false	false
Imprecise value	false	false	false	false	false	false	false	false	false	false	false
Alias value	false	false	false	false	false	false	false	false	false	false	false
Marked	false	true	true	true	true	true	true	true	false	false	true

Fig. 2 Example of the information generated during the processing of a query

The attribute *Marked* acquires the Boolean value *true* or *false*, where *true* indicates that the lexical component is used in the construction of the SQL statement. The attribute called *Category* represents the syntactic category of the lexical component. The attribute *Phrase* is used when two or more lexical components constitute a phrase; for instance, *number of engines*.

The attribute *Phrase ID* stores a set of numbers, where each digit represents the position of a lexical component that belongs to a phrase. The attribute *Phrase type* refers to the phrase type of each lexical component; for instance, SELECT, WHERE, GROUP BY. The attribute *Column tag* is used only when the lexical component refers to a column, in this case the attribute stores the column tag for the SQL translation. The attribute *Table tag* (similar to column tag) stores the table for the SQL statement. Finally, the attribute *Final tag* is used to store the semantic equivalent in SQL of the lexical component, which implicitly indicates that the component will be used for the translation.

The kernel of the NLIDB developed at ITCM consists of three main modules. The first module called Lexical Analysis tags all the words that can be found in the lexicon with their syntactic category. If a word cannot be located in the lexicon, it is inferred that it is possibly a search value. The next module, called Syntactic Analysis, consists of a heuristics which performs an analysis to obtain only one syntactic category per lexical component, since there are lexical components that could have more than one syntactic category. Irrelevant lexical components are also ignored. The third module, called Semantic Analysis, deals with the identification of tables and columns according to the information stored in the semantic information dictionary (SID).

3 Problems in Queries with AFs, Grouping and Subqueries

In the Geobase corpus certain problems were detected which increase the complexity of query processing. In this section, we present the problems called *general*, which may appear in queries that involve AFs, grouping and subqueries. In the next subsections, the problems presented are particular to each of the three aspects presented. We present only the problems that were treated in this project. The general problems that exist in the Geobase corpus are described next.

Note: the examples are presented in Spanish because it is the language that the interface works with.

Ambiguity in column: *Combina la población de las ciudades del estado de Alabama.* (*Combine the population of the cities of the state of Alabama*). The word *población* (*population*) refers to a column that exists in the tables *city* and *state*.

Ambiguity COUNT/SUM: *¿Cuántos habitantes tienen las ciudades del estado de Texas?* (*How many inhabitants have the cities of the state of Texas?*). The word *cuántos* (*how many*) generates an ambiguity between the AFs COUNT and SUM, since *cuántos* may refer to counting or sum.

3.1 Problems in Aggregate Functions

Identification of greater than/smaller than: *¿Cuántos estados tienen área mayor que 1100?* (*How many states have area greater than 1100?*). The existence of the pair *mayor que* (*greater than*) indicates an inequality (>), in contrast, in Spanish the words *mayor*/*menor* are associated with AFs MAX/MIN.

Ellipsis in AF: *Suma [altura de] las montañas del estado de Washington.* (*Sum [height of] the mountains of the state of Washington*). The omission of the word *altura* (*height*) for referring to the DB column *height* needed for applying the aggregate function SUM.

3.2 Problems in Grouping

Phrase for specifying GROUP BY column: *¿Cuántos tipos de aeronave hay por número de motores*? (*How many aircraft types are per number of engines?*). The concatenation of three words *número de motores* (*number of engines*) that refer to a single DB column: *engines*.

Double grouping: *¿Cuántos estados hay por población por área?* (*How many states are per population per area?*). The grouping using two DB columns: *population* and *area*.

3.3 Joins and Subqueries

The prototype developed at ITCM, when translating a query that requires two or more tables, creates a graph using the tables identified in the query. Subsequently it creates the SQL statement using *joins*. The graph must be a connected graph in order to find the shortest path (or minimal tree) that involves all the tables of the query. Next, we show an example using Fig. 3 for the ATIS database [4].

Show me the cost of the business class for flight number 1.

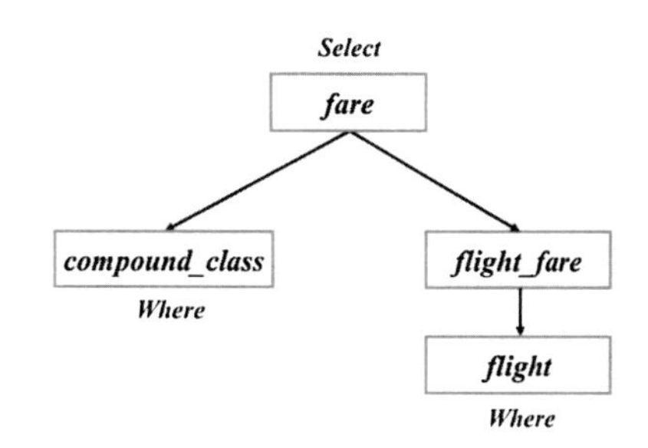

Fig. 3 Example of a semantic graph for a query that involves a join

The translation of the query using *joins* generated by the NLIDB is presented below:

SELECT *fare.one_way_cost*, *fare.rnd_trip_cost*
FROM *fare*, *compound_class*, *flight*, *flight_fare*
WHERE *compound_class.fare_class* = *fare.fare_class*
AND *fare.fare_code* = *flight_fare.fare_code*
AND *flight_fare.flight_code* = *flight.flight_code*
AND *flight.flight_number* = 1
AND *compound_class.class_type* LIKE 'Business'

As shown, the query using *joins* makes connections between tables using foreign keys. This method demands large amounts of memory and computing time, because it needs to find the path between all the tables involved. For this reason, a method was developed to generate an SQL expression using subqueries, when possible. Using this method makes it possible to detect queries that can be treated with subqueries or *joins*.

4 Solution for AFs, Grouping and Subquery Problems

The solutions of the problems presented in the previous section will be explained with examples to facilitate the comprehension of their treatment.

4.1 Treatment for General Problems

Ambiguity in column: *Combina la población de las ciudades del estado de Alabama*. (*Combine the population of the cities of the state of Alabama*). When starting the treatment of this problem, in the data structure (Table 1) the column *population* and the tables *city* and *state* have already been detected and have the lexical components marked as *true*. As can been seen highlighted in italic, there are two columns associated to the lexical component, because both tables were detected.

To solve the ambiguity the interface displays a dialog. The dialog shows the two options detected, so the user may choose the correct option according to the query; let us assume, *city* is chosen. Once the option has been selected, the final tag of the lexical component is modified and the ambiguity is solved (as shown in Table 2).

The detection of the AF is carried out analyzing all the lexical components marked as *false*, which indicates that they have not been detected so far as useful by the interface. Each *false* lexical component is looked for in the SID for determining its relevance. For this example, when looking for the lexical component *combina*,

Table 1 Example of a query with column ambiguity

Lexical component	Combina	la	población	de	las	ciudades	del	estado	de	Alabama
Phrase type			SELECT					WHERE		WHERE
Final tag			*city.population* *state.population*			city		state.state_name		Alabama
Marked	False	False	True	False	False	True	False	True	False	True

Table 2 Detection of the aggregate function SUM

Lexical component	Combina	la	población	de	las	ciudades	del	estado	de	Alabama
Phrase type	SELECT		SELECT					WHERE		WHERE
Final tag	*SUM*		*city.population*			city		state.state_name		Alabama
Marked	True	False	True	False	False	True	False	True	False	True

Table 3 Example of a query with COUNT/SUM ambiguity

Lexical component	Cuántas	ciudades	hay	en	el	estado	de	Texas
Phrase type	SELECT	SELECT				WHERE		WHERE
Final tag	*COUNT/ SUM*	city. city_name				state. state_name		Texas
Marked	True	True	False	False	False	True	False	True

an association between *combina* and the aggregate function SUM is detected. Thus, this component is tagged as SUM (Table 2).

From the information stored in the data structure (Table 2), the following SQL statement is constructed:

SELECT SUM (*city.population*)
FROM *city*, *state*
WHERE *city.state_abbreviation* = *state.abbreviation*
AND *state.state_name* LIKE 'Alabama'

COUNT/SUM ambiguity: *¿Cuántas ciudades hay en el estado de Texas?* (*How many cities are in the state of Texas?*). As shown in Table 3, the interface has already detected an AF; nevertheless, it cannot solve the ambiguity. Therefore, the final tag temporarily stays as COUNT/SUM.

To solve the COUNT/SUM ambiguity the interface displays a dialog so the user may solve the ambiguity; let us assume, counting is chosen. With the answer of the user the tag is modified and the ambiguity is solved.

From the information stored in the data structure, the following SQL expression is generated:

SELECT COUNT (*city.city_name*)
FROM *city*, *state*
WHERE *city.state_abbreviation* = *state.abbreviation* AND
state.state_name LIKE 'Texas'

4.2 Treatment for Aggregate Function Problems

Identification of greater than/smaller than: *¿Cuántos estados tienen área mayor que 1100?* (*How many states have area greater than 1100?*). As highlighted in italic in Table 4, the pair *mayor que* (*greater than*) refers to an inequality (>).

To solve the problem the interface analyzes the data structure (Table 4) searching for the lexical component *mayor*. Once located, it analyzes if the lexical component in the next position is *que*. If so, both lexical components are combined and form a phrase. Finally the final tag of the phrase is set to > (Table 5). If the

Table 4 Example of query with possible inequality

Lexical component	Cuántos	estados	tienen	área	*mayor*	*que*	1100
Phrase type	SELECT	SELECT		WHERE			WHERE
Final tag	count	state. state_name		state. area			1100
Marked	True	True	False	True	False	False	True

Table 5 Example of query with inequality identified

Lexical component	Cuántos	estados	tienen	área	*mayor que*	1100
Phrase type	SELECT	SELECT				WHERE
Final tag	count	state. state_name		state. area	>	1100
Marked	True	True	False	True	True	True

interface did not detect the word *que* after the lexical component *mayor*, then it would assume that it is not an inequality but the aggregate function MAX.

From the information stored in the data structure (Table 5), the following SQL query is constructed:

SELECT COUNT (*state.state_name*)
FROM *state*
WHERE *state.area* > 1100

Ellipsis in AF: *¿Cuál es la media [longitud] de ríos en el estado de Texas?* (*What is the mean [length] of the rivers in the state of Texas?*). As shown in italic in Table 6, an AF has been identified, but there is no column for applying the AF to. The AF has been detected already and has been assigned a final tag and phrase type.

The interface, when detecting that there is no numeric column for applying the AF to, uses the first detected table (*river*) to obtain all the numeric columns in it. Using the numeric columns a dialog is displayed so the user may select the correct column. In this case there are two options *length* and *river_id*. For this example, *length* is selected by the user and is added at the end of the lexical components (Table 7).

Using the semantic information obtained during the translation process (Table 7), the interface constructs the following SQL statement:

SELECT AVG (*river.length*)
FROM *river, state, riverState*
WHERE *river.river_id* = *riverState.river_id*
AND *riverState.state_abbreviation* = *state.abbreviation*
AND *state.state_name* LIKE 'Texas'

Table 6 Example of a query with ellipsis in AF

Lexical component	Cuál	es	la	*media*	de	los	ríos	en	el	estado	de	Texas
Phrase type				*SELECT*			SELECT			WHERE		WHERE
Final tag				*avg*			river.river_name			state. state_name		Texas
Marked	False	False	False	*True*	False	False	True	False	False	True	False	True

Table 7 Solution for query with ellipsis in AF

Lexical component	Cuál	es	la	media	de	los	ríos	en	el	estado	de	Texas	*length*
Phrase type				SELECT			SELECT			WHERE		WHERE	*SELECT*
Final tag				avg			river. river_name			state. state_name		Texas	*river. length*
Marked	False	False	False	True	False	False	True	False	False	True	False	True	*True*

4.3 Treatment for Grouping Problems

Phrase for specifying GROUP BY column: *¿Cuántas velocidades hay por tipo de aeronave*? (*How many speeds are per aircraft type?*). There is a grouping in this case *tipo de aeronave* (*aircraft type*), which is a phrase that refers to a single column: *aircraft_type* (Table 8).

From the semantic information gathered during the previously described process, the interface detects the lexical component *por* (*per*), which refers to a grouping. Therefore, its attributes *Phrase type*, *Final tag* and *Marked* are modified, as shown in Table 9. There are three lexical components that refer to a single column *aircraft_type*, which is used to perform the grouping; therefore, they are combined into a single lexical component (phrase).

Once the collection of semantic data is completed (Table 9), it is possible to generate the following SQL expression:

SELECT COUNT (*aircraft.cruising_speed*), *aircraft_type*
FROM *aircraft*
GROUP BY *aircraft_type*

Double grouping: *¿Cuántas montañas hay por estado por altura?* (*How many mountains are per state per height?*).

As seen in the query, there is a COUNT/SUM ambiguity, which was already explained; therefore, its treatment is omitted here. In this case the ambiguity is solved with the aggregate function COUNT. Additionally, there are two columns for grouping. To solve the double grouping, the same process used with one column is used. Once the interface has detected the columns and they are assigned their final tags, the lexical component that refers to the grouping (*por*) is detected. Then, it is inferred that the next lexical component must be a column associated to the grouping. In double grouping cases, it is simply detected that there are two lexical components that refer to two groupings; hence, the phrase type of both columns is modified to GROUP BY.

The information obtained (Table 10) is used to construct the following SQL query:

SELECT COUNT (*mountain.mountain_name*), *mountain.state_name*, *mountain.height*

Table 8 Example of query with phrase that specifies GROUP BY column

Lexical component	*Cuántas*	velocidades	hay	por	tipo	de	aeronave
Phrase type	*SELECT*	SELECT			SELECT	SELECT	SELECT
Final tag	*count*	aircraft. cruising_speed			aircraft. aircraft_type	aircraft. aircraft_type	aircraft. aircraft_type
Marked	*True*	True	False	False	True	True	True

Table 9 Solution for query with phrase that specifies GROUP BY column

Lexical component	Cuántas	velocidades	hay	*por*	*tipo de aeronave*
Phrase type	SELECT	SELECT		*GROUP BY*	*SELECT*
Final tag	count	Aircraft.cruising_speed		*group by*	*aircraft. aircraft_type*
Marked	True	True	False	*True*	*True*

FROM *mountain*
GROUP BY *mountain.state_name*, *mountain.height*

4.4 Treatment for Joins and Subqueries

When analyzing the queries that are treated with *joins* by the interface, we discovered that certain queries could be treated with subqueries, instead of using *joins*. When it is detected that a query has two or more tables in the Select clause, *joins* have to be used. Otherwise, when only one table is detected subqueries are used. Next, we present an example of a query treated with subqueries.

Show the cost of business class for flight number 1.

As mentioned in Sect. 3.3, the interface creates a semantic graph using the tables detected in the query: for this example, the graph of Fig. 4. The translation of the query using subqueries is presented below:

```
SELECT fare.one_way_cost, fare.rnd_trip_cost
FROM fare
WHERE fare.fare_class IN (
  SELECT compound_class.fare_class
  FROM compound_class
  WHERE compound_class.class_type LIKE 'Business')
  AND fare.fare_code IN (
    SELECT flight_fare.fare_code
    FROM flight_fare
    WHERE flight_fare.flight_code IN (
      SELECT flight.flight_code
      FROM flight
      WHERE flight.flight_number = 1))
```

As seen in the SQL statement, the root-node (*fare*) corresponds to the main query. The first subquery corresponds to the next node *compound_class*. It is important to highlight that because the node is the last one it must have a Where

Table 10 Example of query with double grouping

Lexical component	Cuántas	montañas	hay	*por*	*estado*	*por*	*altura*
Phrase type	SELECT	SELECT		*GROUP BY*	*GROUP BY*	*GROUP BY*	*GROUP BY*
Final tag	count	mountain.mountain_name		*group by*	*mountain.state_name*	*group by*	*mountain.height*
Marked	True	True		*True*	*True*	*True*	*True*

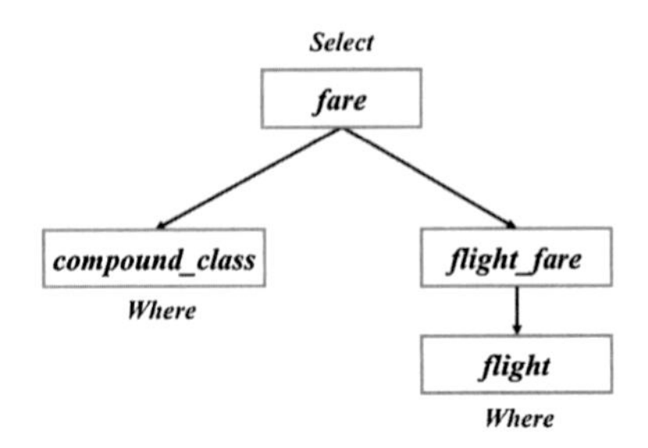

Fig. 4 Example of a semantic graph for a query that involves subqueries

clause associated, corresponding to the search value *business*. The process continues with the second subquery which is associated to the node *flight_fare*. Finally, the last subquery is built with the *flight* node, and since it is the last one, similarly to the *compound_class* node, the Where clause has to be associated to the second search value (*1*).

In the SQL expression with subqueries, it is possible to observe the path the interface must generate to cover all the *joins* between tables (see SQL statement using joins in Sect. 3.3). The path is divided in two parts, this is because the Select node has two branches that connect it to the two nodes with Where clause.

Next, we present an example of a query with subqueries and AF.

Show the cost of business class for the first flight of flight number 1.

The process is similar to the one described before. The only difference is that this query involves an AF (MIN), which is solved in a similar way as explained in Sect. 4.1. The SQL expression generated is shown below.

```
SELECT fare.one_way_cost, fare.rnd_trip_cost
FROM fare
WHERE fare.fare_class IN (
  SELECT compound_class.fare_class
  FROM compound_class
  WHERE compound_class.class_type LIKE 'Business')
  AND fare.fare_code IN (
    SELECT flight_fare.fare_code
    FROM flight_fare
    WHERE flight_fare.flight_code IN (
      SELECT MIN (flight.departure_time)
      FROM flight
      WHERE flight.flight_number = 1))
```

5 Experimental Results

For the comparative test with C-Phase, queries for the Geobase DB were used, since C-Phrase is customized (by its implementers) for this database. C-Phrase correctly answered 11 queries out of 58, which corresponds to 18% of the queries for Geobase. However, the results generated for three queries are considered partially correct. Even though it did not produce the correct answers, the results imply the expected results. Next, we present an example to clarify.

When inputting the query: *How many cities are in the state of Montana?* The C-Phrase output is: *Greatfalls* and *Billings*. The output of C-Phrase is partially correct, but it is not exactly the information requested, since the query is asking for a number, which is the count of the cities, not the names.

Additionally, it was found out that C-Phrase does not detect neither solves ellipsis in AFs, which significantly reduces its performance (*recall*). In the case of queries with grouping, C-Phrase does not detect the GROUP BY clause; therefore, it does not treat grouping. In regards to queries that involve subqueries, C-Phrase answered only three out of 10 queries of this type. Considering the correctly answered and the partially correct queries, the recall of C-Phrase is 24%. In contrast, the ITCM interface obtained a recall of 86%.

6 Final Comments

Even though NLIDBs have been in development for over 50 years, the complexity of man-machine interaction via natural language (NL) keeps this area in continuous investigation, in order to devise better procedures to increase the percentage of correctly answered queries (*recall*). With the prototype developed at ITCM, we have a solid foundation that can be improved with additional modules, that allows enhancing its performance and functionality.

As shown by the experimental results, the ITCM interface has better performance than that of C-Phrase, attaining 86% *recall* versus 24% obtained by C-Phrase. It is important to mention that C-Phrase cannot solve the ellipsis in aggregate functions and cannot deal with grouping in NL queries, which explains why C-Phrase has such a low *recall*.

The future work on this prototype will focus on improving the syntactic analyzer and deal with search values that are difficult to detect (such as language words that are used as search values). It is also necessary to deal with problems related to AFs, grouping and subqueries, which have been identified, but that have not been considered so far, in order to enhance the performance of the interface.

References

1. R. Pazos, M. Aguirre, J. Gonzalez, J. Martínez, J. Pérez, A. Verástegui, Comparative study on the customization of Natural Language Interfaces to Databases. SpringerPlus **5**, 553 (2016)
2. M. Minock, C-phrase: a system for building robust natural language interfaces to databases. Data Knowl. Eng. **69**, 290–302 (2010)
3. L. Tang, R. Mooney, Using multiple clause constructors in inductive logic programming for semantic parsing, in *Proceedings of the 12th European Conference on Machine Learning*, pp. 466–477 (2001)
4. Linguistic Data Consortium, "ATIS2". https://catalog.ldc.upenn.edu/ LDC93S5. Accessed 2017

Comparative Study of Computational Strategies for Protein Structure Prediction

Fanny G. Maldonado-Nava, Juan Frausto-Solís, Juan Paulo Sánchez-Hernández, Juan Javier González Barbosa and Ernesto Liñán-García

Abstract Protein Folding Problem (PFP) is one of the most challenging problems of combinatorial optimization with applications in bioinformatics and molecular biology. The aim of PFP is to find the three-dimensional structure of a protein, this structure is known as Native Structure (NS), which is characterized by the minimal energy of Gibbs and it is commonly the best functional structure. To find an NS knowing only the amino acids sequence (primary structure) of a protein is known as ab initio problem. A protein can take a huge number of different conformational structures from its primary structure to the NS. For solving PFP, several computational strategies are applied in order to search structures of protein on a huge space of possible solutions. In this work, the most popular methods and strategies are compared, and advantages and disadvantages of them are discussed.

Keywords Protein folding problem · Computational strategies
Ab initio · Threading · Homology

F. G. Maldonado-Nava (✉) · J. Frausto-Solís · J. J. González Barbosa
TecNM/Instituto Tecnológico de Ciudad Madero, Ciudad Madero, Mexico
e-mail: fanny_mn@hotmail.com

J. Frausto-Solís
e-mail: juan.frausto@itcm.edu.mx

J. J. González Barbosa
e-mail: jjgonzalezbarbosa@itcm.edu.mx

J. P. Sánchez-Hernández
Universidad Politécnica del Estado de Morelos, Jiutepec, Mexico
e-mail: juan.paulosh@upemor.edu.mx

E. Liñán-García
Universidad Autónoma de Coahuila, Saltillo, Mexico
e-mail: ernesto_linan_garcia@uadec.edu.mx

O. Castillo et al. (eds.), *Fuzzy Logic Augmentation of Neural and Optimization Algorithms: Theoretical Aspects and Real Applications*, Studies in Computational Intelligence 749, https://doi.org/10.1007/978-3-319-71008-2_32

1 Introduction

Proteins are molecules, which play a central role in our body. Proteins are needed to catalyze reactions, transport molecules, and other important functions. Proteins consist of smaller units named amino acids, attached to one another in long chains by peptide bonds. A functional protein has a specific three-dimensional structure, usually named Native Structure (NS), which takes when it is correctly folded. The NS is biologically active, in which the protein correctly performs its functions. The natural process of protein folding is not completely understood; this is because nature takes an unknown path to achieve the native structure in a very fast way [1].

The process of protein folding in living organisms (Natural process of protein folding, folding of proteins in vivo; or in short, folding) occurs within cells, which as is well known are prokaryotes in all bacteria and eukaryotes for animals, plants, and fungi. Understanding the process of protein folding is important because many human diseases are related to improper folding in vivo; some of these diseases are [2–4]: Alzheimer's, Parkinson's, Prion, Tauopathy, Huntington's disease, Creutzfeldt-Jakob disease, Cystic Fibrosis, Gaucher disease, and Sickle Cell Anemia. In fact, recent specialized publications have noticed that incorrect folding of proteins (or misfolded) is involved with most of the diseases not caused by infectious agents and is involved in the progression of hundreds of diseases [4, 5].

Protein Folding Problem has been studied for the last 50 years and is one of the biggest unsolved problems in science [3, 6]. PFP is an NP-Hard problem [7, 8], which consists in determining the native structure of a protein, this structure is the one in which the Gibbs free energy is the lowest [9]. Due to the amount of conformations that a protein can take, computational methods are becoming important. Some methods for the study of the tertiary structure of the proteins have been developed are X-ray Crystallography and Nuclear Magnetic Resonance (NMR). These methods are regularly very expensive and their processes can consume very long time [10, 11]. Thus, the NS prediction is necessary and it has become one of the most important challenges of modern computational biology [7]. Different computational approaches for finding the three-dimensional structure have been proposed over the last decades. These approaches can be classified into three categories: (a) ab initio, (b) homology, and (c) threading. The main challenge is to understand how the information included in the amino acids sequence can be translated into a three-dimensional structure (functional structure), in order to develop computational algorithms that can predict a protein structure correctly.

Over the last decades, many algorithms have been proposed and tested as a solution to PFP. Most common algorithms are Simulated Annealing (SA), Genetic Algorithms (GA), Ant Colony Optimization (ACO), Tabu Search (TS), and among other. The most successful algorithms for solving PFP are SAL algorithms (Simulated Annealing Like algorithms) [12]; these successful methods are usually hybridized with other heuristics. Despite the efforts made so far, just a little number of protein sequences have been solved, which has motivated the scientific community on working on more powerful algorithms [10]. Recently, new and more efficient SAL algorithms have been proposed; as Golden Ratio Simulated Annealing (GRSA), which is part of these successful SAL algorithms [13]. GRSA is important because has obtain very good results in the case of peptides, particularly the Met-enkephalin, which is commonly studied in PFP area.

This paper is organized as follows: in Sect. 2, three strategies for Protein Folding Problem are presented. Section 3, describes three important methods presented in CASP. In Sect. 4, PFP for ab initio approach is described and an energy function is presented. Finally, conclusions for this work are discussed.

2 Computational Strategies

Many computational methodologies and algorithms have been proposed as a solution to the PFP. Strategies used in these algorithms can be classified in three categories: ab initio, homology, and threading. The main difference between these strategies is the information they need to address the problem.

2.1 *Ab Initio Approach*

Ab initio strategy is perhaps the most difficult approach for protein structure prediction. As is shown in Fig. 1, ab initio looks for the three-dimensional structure using only the amino acids' sequence and it does not require other information of the target protein. Ab initio methods are based on basic physics and quantum mechanics, this is on the thermodynamic hypothesis which points out that the NS of a protein is the one for which the free energy achieves the global minimum [9].

Ab initio methods provide a natural approach to obtain structures from protein sequences without referring any information or any appropriate templates. This strategy is clearly the most difficult, but the most useful approach. As any other strategy, ab initio presents some advantages and disadvantages. Ab initio methods are useful when appropriate templates cannot be consulted, that is, when sufficiently homologous proteins have not been found or when the template does not provide an appropriate structure. New folds can be predicted by this strategy, since there are

Fig. 1 Ab initio approach

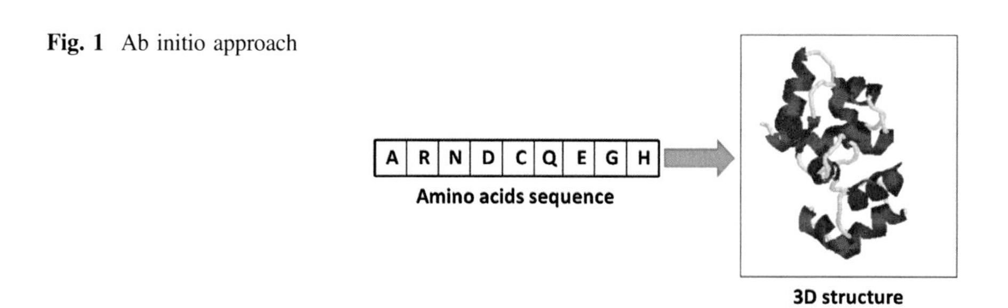

still proteins whose native structures have not been solved, an ab initio method does not need templates from any library. This strategy requires a lot of computational processing time because of the complexity of the problem. In addition, because PFP is an NP-hard problem [7], heuristic algorithms are currently considered as the best alternative; however, these algorithms do not guarantee to achieve exactly the optimal solution. As a consequence, the research of ab initio algorithms is focused on peptides and proteins with a limited number of amino acids (60–150). However, to study small proteins could lead to finding general algorithm solutions for solving the real challenge that is PFP.

For ab initio strategy, PFP is considered as an optimization problem, where the goal is identifying the values of the variables (angles) which describe the minimum energy of the protein. Ab initio methods simulate the protein conformational space using an energy function, which describes the internal energy of the protein and its interactions with the environment. An ab initio algorithm consists of three components: (1) a geometric representation, (2) an energy function, and (3) a searching technique.

2.2 Homology Approach

Known as comparative modeling or template-based modeling, this strategy is based on the understanding of protein evolution, mainly in two facts: (a) proteins that have a homologous sequence, will have similar three-dimensional structures, and (b) proteins structures are more conserved than their sequences. Many proteins can be solved by this approach. Figure 2 illustrates this strategy. Homology process starts with the identification phase, in which an identification of homologous proteins should be done from PDB (Protein Data Bank), phase two is an alignment, which is carried out between the target protein and its homologous (template), and next, a method for modifying the structure should be applied for optimizing the model and get to the final three-dimensional structure of the target protein.

Comparative modeling exploits the fact that evolutionarily related proteins with similar sequences, as measured by the percentage of identical residues at each

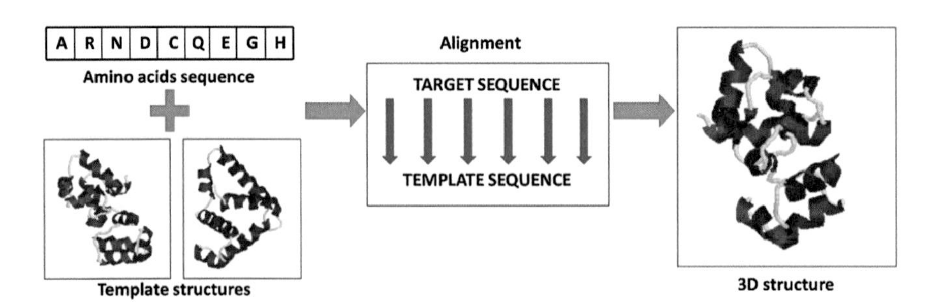

Fig. 2 Homology approach

position based on an optimal structural superposition, often have similar structures. The complexity of the problem becomes smaller than other strategies, since this approach takes advantage of the reduction of the conformational search space, because the process uses a template of a protein whose three-dimensional structure has already been found. When a homolog protein is found, this method is applicable to almost all proteins [14]. If the homology between proteins is high (bigger than 35%) the three-dimensional structure can be found in many cases [15]; however, the use of templates and heuristic algorithms may obtain the NS in almost of the cases.

One of the main disadvantages of this strategy is that only structures of proteins with known homologous sequences can be predicted. If the degree of homology is low, the method must use a more powerful algorithm to be able to find the three-dimensional structure, since with a lower homology the quality of the model will be smaller.

2.3 Threading Approach

Known as fold recognition, this strategy construct protein from known templates even if there is no homologous protein deposited in the Protein Data Bank. Threading models the protein with experimental structures as templates, is a different approach from the homology in terms of the methodology. In Fig. 3, this strategy is shown. The term threading is stand for the process of aligning a protein sequence into a backbone structure and evaluate the compatibility with a set of potential scores or energy functions. Threading is based on the observation that the number of unique protein folds in nature is much smaller than number of proteins.

During the process of threading, the target protein is placed, following the sequential order, into structural positions of a template three-dimensional structure in an optimal way. This process consists of two phases: (1) select a structural template from a library, and (2) find the correct replacement between the target protein against the structural models in the space of possible replacements. Threading has some advantages; it uses known protein structures as templates for

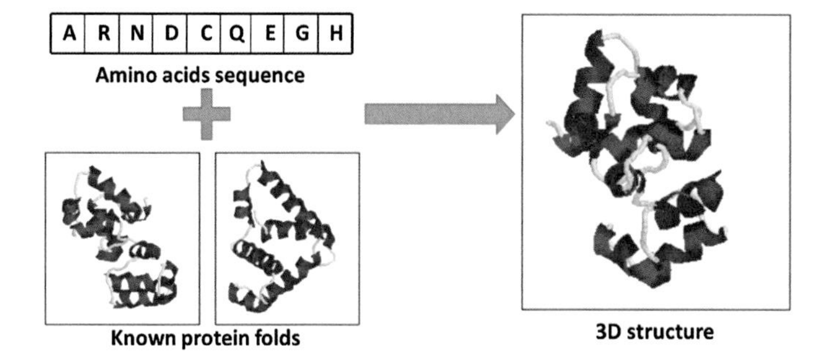

Fig. 3 Threading approach

sequences of unknown structures. Threading finds the most similar conformation to the NS that can be uses as an initial solution with other methods. Threading presents some limitations; these methods are computationally expensive. Identifying appropriate templates for a given protein is also a problem classified as NP-Hard [16]. In addition, the NS found with this approach could not be present in the space of possible conformations.

Many algorithms implement different metaheuristics to provide near optimal solutions for PFP, considering the limitations and the advantages of the approaches for protein structure prediction methods, researchers have developed hybrid methods in their algorithms, which combine principles of the three strategies presented in this paper.

3 Methods

There is a biannual competition named CASP (Critical Assessment of protein Structure Prediction), in this competition researchers test their structure prediction methods. Targets proteins for structure prediction are structures solved, but they are kept on hold by the Protein Data Bank. Here are presented three protein structure prediction methods, which use different approaches and different strategies for constructing three-dimensional protein models. These methods have been presented and tested in CASP, obtaining good results, so that they have obtained first places in lasts competitions.

I-TASSER

I-TASSER is a server for protein structure predictions, built by Zhang Lab. This server was ranked as the number one server for protein structure prediction in CASP7, CASP8, CASP9, CASP10, CASP11, and CASP12 experiments. The I-TASSER method is divided in three phases: threading, assembling, and refinement. In the first phase, I-TASSER identifies templates from the PDB (Protein Data Bank) by a threading approach using LOMETS (which combines algorithms to generate models by collecting their target-template alignments). In the second phase, fragments of the threading aligned regions are extracted from the template structures, and are used to assemble new structural conformations, while ab initio approach processes the unaligned regions. The assembly is performed by a replica-exchange Monte Carlo (REMC) Simulation. The low free-energy states are identified by SPICKER (algorithm to identify the near-native models) through clustering. In the third phase, a second assembly is performed, the purpose of the second iteration is to refine the global topology of the cluster centroids. The lowest energy structures are selected, and the final full-atomic models are obtained by REMO, and fragment-guided molecular dynamics [10, 17].

QUARK

QUARK is an ab initio structure prediction built by Zhang Lab, which construct 3D structures models. QUARK was ranked as the No 1 server in free-modeling in

CASP9 and CASP10 experiments. QUARK models are built from small fragments (1–20 residues long) by replica-exchange Monte Carlo simulation. This procedure can be divided into three steps. The first step is for multiple feature predictions and fragment generation starting from one query sequence. QUARK first predicts a variety of selected structural features by neural network (NN). In the second step, the global fold is generated by replica-exchange Monte Carlo (REMC) simulations by assembling the small fragments, these fragments in QUARK have multiple sizes from 1 to 20 residues. The third step is full-atomic refinement. QUARK simulations perform movements of free-chain constructions and fragment substitutions between decoy and fragment structures. These techniques have increased the efficiency of conformational search while taking the advantage of the reduction of the conformational search owing to fragment assembly [10, 18].

ROSETTA
ROSETTA is a fragment-based method for the three-dimensional protein structure prediction problem developed by Baker Lab. Is one of the best-established ab initio protein folding methods as demonstrated in the last CASP experiments. ROSETTA uses an assembly strategy to combine native-like structures of fragments of unrelated protein structures with similar local sequences using Bayesian scoring functions. The main goal of ROSETTA scoring function is to search for the most probable structure of a protein given the amino acid sequence. This algorithm predicts protein structures based on a library of residue fragments. The fragments are selected according to their sequence similarity with the target protein. The Rosetta method assumes that short sequence segments have strong local structural biases. In the first step, fragment libraries for each 3- and 9-residue segment of the target protein are extracted from the protein structure database. Then, tertiary structures are generated using a Monte Carlo search of the possible combinations, minimizing a scoring function [10, 19].

4 Protein Folding Problem

Protein folding problem is the process of finding the three-dimensional native structure of a protein, this structure is usually named Native Structure (NS). NS is the conformation in which the protein performs its biological role. As mentioned earlier, the PFP since the ab initio approach can be considered as an optimization problem, where the goal is identifying the set of values of the variables that satisfy an objective function, that in this case is the energy function. PFP is an enormous challenge because the space of possible conformations a protein can take is extremely large [7]. For an ab initio approach PFP can be defined as follows:

- A sequence of n amino acids; $a_1, a_2, a_3, \ldots, a_n$, that represents the primary structure of a protein, with a set of dihedral angles $\sigma^m = \sigma_1, \sigma_2, \sigma_3, \ldots, \sigma_m$,
- An energy function $f(\sigma_1 \sigma_2 \ldots \sigma_m)$ that represents the free energy.

The solution to this problem is to find the native structure such that $f^*(\sigma_1\sigma_2...\sigma_m)$ represents the minimum energy value, where the optimal solution $\sigma^* = \sigma_1\sigma_2...\sigma_m$ defines the best three-dimensional configuration.

The atoms of a protein are represented in three-dimensional Cartesian coordinates. There are four types of torsion angles or dihedral angles presented in Fig. 4, and defined below:

- Phi (ϕ) is the angle between the amino group and the alpha carbon. Represents the angle between the amino group (or NH_2) of the amino acid i, and the alpha Carbon C_i in the sequence; it represents the bond angle between the N_i atom of amino group and the alpha carbon (αC_i).
- Psi (ψ) is the dihedral angle between the alpha carbon and the carboxyl group. Psi represents the angle between the carboxyl $(COOH_i)$ group of the amino acid i, and the alpha carbon i (C_i) of the same amino acid. Psi measures the angle of the covalent bond between the C_i of the carboxyl group, and the alpha carbon (αC_i).
- Omega (ω) is defined for each two consecutive amino acids; it is the angle of the covalent bond between the atom N_i of amino acid i, and carbon $C_{(i-1)}$ of the carboxyl group of the amino acid $(i - 1)$.
- And, Chi (χ) is defined between the two planes conformed by two consecutive carbon atoms in the radical group.

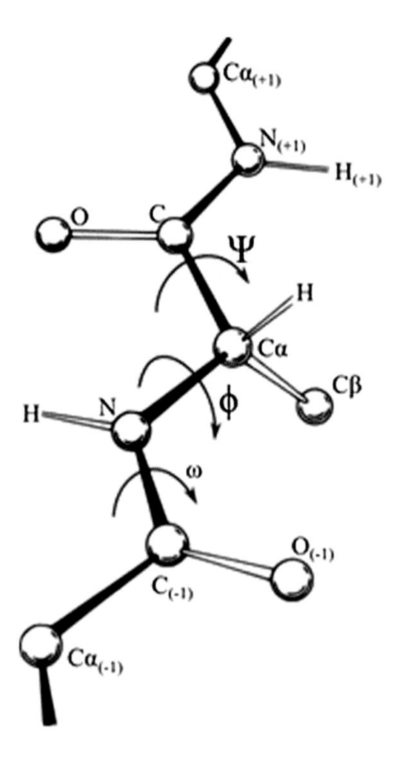

Fig. 4 Representation of the four dihedral angles

The PFP variables are the set of dihedral angles that satisfies the minimum energy value.

4.1 Energy Function

The protein's energy depends on the interaction among their atoms (angles and distance). Force fields are used to measure the energy of a protein; these include many interactions among atoms affecting different energies [20]. A force field includes terms associated with the bond interactions, and terms associated with no-bond interactions. Some of the most popular and successful force fields are CHARMM [21], AMBER [22], ECEPP/2 and ECEPP/3 [23].

One of the most used energy functions for PFP is ECEPP/2, that is a relatively simple force field based on rigid geometry (i.e., constant bond angles and lengths), with conformations thus defined solely by the backbone and side chain dihedral angles. In ECEPP/2 the potential energy is given by the sum of the electrostatic term E_{elect}, Lennard-Jones term E_{LJ}, and hydrogen-bond term E_{HB} for all pairs of atoms in the peptide together with the torsion term E_{tor} for all torsion angles [24]:

$$E_{bonded} = E_{elect} + E_{LJ} + E_{HB} + E_{tor} \tag{1}$$

These terms in Eq. (1) are expressed in Eq. (2) through which energy function ECEPP/2 minimize the energy [24].

$$\begin{aligned} E_{total} = & \sum_{j>i}\left(\frac{A_{ij}}{r_{ij}^{12}} - \frac{B_{ij}}{r_{ij}^{6}}\right) + 332\sum_{j>i}\frac{q_i q_j}{\varepsilon r_{ij}} + \sum_{j>i}\left(\frac{C_{ij}}{r_{ij}^{12}} - \frac{D_{ij}}{r_{ij}^{10}}\right) \\ & + \sum_{n} U_n(1 \pm \cos(k_n \varphi_n)) \end{aligned} \tag{2}$$

where:

- r_{ij} is the distance in Å between the atoms i and j.
- A_{ij}, B_{ij}, C_{ij} and D_{ij} are the parameters of the empirical potentials.
- q_i and q_j are the partial charges on the atoms i and j, respectively.
- ε is the dielectric constant which is usually set to $\varepsilon = 2$.
- 332 is a factor for using the energy units expressed in kcal/mol.
- U_n is the energetic torsion barrier of rotation about the bond n.
- k_n is the multiplicity of the torsion angle φ_n.

The energy function ECEPP/3 is a modify version of ECEPP/2. ECEPP/3 contains updated parameters for proline and oxyproline residues. This energy function is used until recently for PFP.

5 Conclusions

The study of protein folding problem for finding the three-dimensional structure is one of most important research problems in Bioinformatics. Over the last decades, computational methods, and algorithms have been developed for solving PFP. However, there is no method yet that can predict structures without the need of information about templates, this is because of the complexity and high conformational search space, so that the problem still challenges in bioinformatics and computer science. Three strategies were described in this paper, these strategies are now use in algorithms and methods for PFP. Some of these methods are ROSETTA, I- TASSER and QUARK, which have been three of the most successful predictors in the CASP competition. A common characteristic of these methods is that for some of their processes they use a Monte Carlo method.

Acknowledgements The authors would like to acknowledge with appreciation and gratitude to CONACYT. Also, acknowledge to Laboratorio Nacional de Tecnologías de la Información (LaNTI) of the Instituto Tecnológico de Ciudad Madero for the access to the cluster. Fanny Gabriela Maldonado-Nava would like to thank CONACYT for the support in the project 429028.

References

1. C. Levinthal, Are there pathways for protein folding. J. Chim. Phys. **65**(1), 44–45 (1968)
2. T.K. Chaudhuri, S. Paul, Protein-misfolding diseases and chaperone-based therapeutic approaches. FEBS J. **273**(7), 1331–1349 (2006)
3. K.A. Dill, J.L. MacCallum, The protein-folding problem, 50 years on. Science (80-.) **338** (6110), 1042–1046 (2012)
4. J.S. Valastyan, S. Lindquist, Mechanisms of protein-folding diseases at a glance. Dis. Model. Mech. **7**(1), 9–14 (2014)
5. C. Spiess, A.S. Meyer, S. Reissmann, J. Frydman, Mechanism of the eukaryotic chaperonin: protein folding in the chamber of secrets. Trends Cell Biol. **14**(11), 598–604 (2004)
6. K.A. Dill, S.B. Ozkan, T.R. Weikl, J.D. Chodera, V.A. Voelz, The protein folding problem: when will it be solved? Curr. Opin. Struct. Biol. **17**(3), 342–346 (2007)
7. W.E. Hart, S. Istrail, Robust proofs of NP-hardness for protein folding: general lattices and energy potentials. J. Comput. Biol. **4**(1), 1–22 (1997)
8. J.T. Ngo, J. Marks, M. Karplus, in *The Protein Folding Problem and Tertiary Structure Prediction.* Computational Complexity, Protein Structure Prediction, and the Levinthal Paradox (Birkhäuser Boston, Boston, 1994), pp. 433–506
9. C.B. Anfinsen, Principles that govern the folding of protein chains. Science (80-.) **181**(4096), 223–230 (1973)
10. M. Dorn, M.B. e Silva, L.S. Buriol, L.C. Lamb, Three-dimensional protein structure prediction: methods and computational strategies. Comput. Biol. Chem. **53**, 251–276 (2014)
11. A.A. Yee, A. Savchenko, A. Ignachenko, J. Lukin, X. Xu, T. Skarina, E. Evdokimova, C.S. Liu, A. Semesi, V. Guido, A.M. Edwards, C.H. Arrowsmith, NMR and X-ray crystallography, complementary tools in structural proteomics of small proteins. J. Am. Chem. Soc. **127** (47), 16512–16517 (2005)
12. L.B. Morales, R. Garduño-Juárez, D. Romero, Applications of simulated annealing to the multiple-minima problem in small peptides. J. Biomol. Struct. Dyn. **8**(4), 721–735 (1991)

13. J. Frausto-Solis, J.P. Sánchez-Hernández, M. Sánchez-Pérez, E.L. García, Golden ratio simulated annealing for protein folding problem. Int. J. Comput. Methods **12**(6), 1550037 (2015)
14. Y. Zhang, J. Skolnick, The protein structure prediction problem could be solved using the current PDB library. Proc. Natl. Acad. Sci. **102**(4), 1029–1034 (2005)
15. G. Helles, A comparative study of the reported performance of ab initio protein structure prediction algorithms. J. R. Soc. Interface **5**(21), 387–396 (2008)
16. R.H. Lathrop, The protein threading problem with sequence amino acid interaction preferences is NP-complete. Protein Eng. Des. Sel. **7**(9), 1059–1068 (1994)
17. J. Yang, R. Yan, A. Roy, D. Xu, J. Poisson, Y. Zhang, The I-TASSER Suite: protein structure and function prediction. Nat. Methods **12**(1), 7–8 (2015)
18. Y. Zhang, Interplay of I-TASSER and QUARK for template-based and ab initio protein structure prediction in CASP10. Proteins Struct. Funct. Bioinforma. **82**(SUPPL. 2), 175–187 (2013)
19. C.A. Rohl, C.E.M. Strauss, K.M.S. Misura, D. Baker, Protein structure prediction using rosetta. Methods Enzymol. **383**, 66–93 (2004)
20. K.A. Dill, Dominant forces in protein folding. Biochemistry **29**(31), 7133–7155 (1990)
21. B.R. Brooks, R.E. Bruccoleri, B.D. Olafson, D.J. States, S. Swaminathan, M. Karplus, CHARMM: a program for macromolecular energy, minimization, and dynamics calculations. J. Comput. Chem. **4**(2), 187–217 (1983)
22. J.W. Ponder, D.A. Case, Force fields for protein simulations. Adv. Protein Chem. **66**, 27–85 (2003)
23. F.A. Momany, R.F. McGuire, A.W. Burgess, H.A. Scheraga, Energy parameters in polypeptides. VII. Geometric parameters, partial atomic charges, nonbonded interactions, hydrogen bond interactions, and intrinsic torsional potentials for the naturally occurring amino acids. J. Phys. Chem. **79**(22), 2361–2381 (1975)
24. F. Eisenmenger, U.H.E. Hansmann, S. Hayryan, C.K. Hu, [SMMP] A modern package for simulation of proteins. Comput. Phys. Commun. **138**(2), 192–212 (2001)

Issues in Querying Databases with Design Anomalies Using Natural Language Interfaces

Rodolfo A. Pazos R., José A. Martínez F., Alan G. Aguirre L. and Marco A. Aguirre L.

Abstract Accessing information is a vital activity in businesses; therefore, databases (DBs) have become necessary tools for storing their information. However, for accessing the information stored in a database, it is necessary to use a DB query language, such as SQL. Natural language interfaces to databases (NLIDBs) allow inexperienced users to obtain information from a DB using natural language expressions without the need of using a DB query language. Despite the relative effectiveness of NLIDBs, most of the approaches proposed for designing NLIDBs ignore the possibility that the DB to be queried could be poorly designed; i.e., it could have design anomalies. Unfortunately, various experiments (described in this paper) show that DB anomalies degrade the performance (recall) of NLIDBs. The purpose of this paper is to analyze the most common DB design anomalies for proposing solutions to this problem and avoid performance degradation of NLIDBs when accessing such DBs.

Keywords Database · Database anomaly · Database schema · Natural language interfaces to databases · Database normalization

R. A. Pazos R. (✉) · J. A. Martínez F. · A. G. Aguirre L. · M. A. Aguirre L.
Tecnológico Nacional de México, Instituto Tecnológico de Ciudad Madero, Ciudad Madero, Mexico
e-mail: r_pazos_r@yahoo.com.mx

J. A. Martínez F.
e-mail: jose.mtz@gmail.com

A. G. Aguirre L.
e-mail: li.aguirre.lam@hotmail.com

M. A. Aguirre L.
e-mail: marco.aguirre@itcm.edu.mx

O. Castillo et al. (eds.), *Fuzzy Logic Augmentation of Neural and Optimization Algorithms: Theoretical Aspects and Real Applications*, Studies in Computational Intelligence 749, https://doi.org/10.1007/978-3-319-71008-2_33

1 Introduction

Nowadays, relational databases (DBs) are widely used in most businesses for storing important information. In many of the activities carried out at those businesses, it is necessary to access information swiftly and reliably. However, for a user to be able to obtain information stored in a DB, he/she must have knowledge of a DB query language, which makes it difficult obtaining the required information for most users.

There exist a large number of software tools in the market that facilitate obtaining information from DBs. Among the easiest tools to use by inexperienced users are natural language interfaces to databases (NLIDBs). Such interfaces allow users formulating queries in their own language, facilitating access to information without needing knowledge of a DB query language or programming skills.

There exist two types of NLIDBs: domain dependent and domain independent. Domain-dependent NLIDBs are designed for querying a specific DB; thus they are adapted to the structure of the DB for which they were designed. In contrast, domain-independent NLIDBs are designed for accessing any DB; therefore, their design must enable them to adapt to any DB.

Currently it is common to find DBs that are affected by design anomalies [1–3]. However, though domain-dependent NLIDBs perform adequately despite the design anomalies present in the DBs for which they were designed, most domain-independent NLIDBs have not considered their use with poorly designed DBs. Therefore, these interfaces would not perform adequately with a large number of DBs that are affected by this problem.

2 Design Anomalies in Databases

A problem that complicates query translation from NL to SQL by a domain-independent NLIDB is the existence of design anomalies in the schema of some DBs. For example, there exist DBs for real-life applications with the following anomalies: absence of primary and foreign keys for tables, lack of normalization of the DB schema, use of surrogate primary keys, and columns whose values can be calculated using aggregate functions (AFs): COUNT, SUM, AVG, MAX, MIN.

DB design anomalies may cause the following problems [4]:

- Applications require more complex SQL code, which is processed more slowly.
- The same information is stored in several tables. Therefore, when changing some value in one table it is necessary to make the same change in several tables.
- The information is inconsistent or ambiguous.
- The database becomes more complex.

- The database stores hidden information; for example, a sequence of rows in a table.
- The analysis of indexes in the DB becomes difficult, which results in redundant or missing indexes. This increases processing time and degrades performance.

Next the design anomalies considered in this paper are briefly described.

Absence of primary and foreign keys. The most important reason for defining primary and foreign keys is for identifying distinct rows in each table of a DB. An equally important function for primary and foreign keys is establishing the relationships between several tables of the DB. The absence of primary keys complicates the task of keeping data organized. Additionally, the absence of foreign keys hinders directly the referential integrity between tables and the translation by NLIDBs of queries that involve these tables.

Nonconformity with the first normal form. A table is not in the first normal form if at least one of its columns is not atomic. A column is atomic if its values cannot be divided. This column is said to show atomicity [4]. For example, in Fig. 1 the column *statestringlist* in table *river* is not atomic, because the values in the column *statestringlist* can be divided into state names. For complying with the first normal form, column *statestringlist* has to be divided into two tables (Fig. 1). This increases the flexibility for answering queries involving rivers and states.

Nonconformity with the second normal form. A table is not in the second normal form if at least one of its columns has a functional dependency on a column that is not the primary key [5].

Nonconformity with the third normal form. A table is not in the third normal form if at least one of its columns has a transitive dependency on a column that is not the primary key [6].

river

name	length	statestringlist
St. Francis	684	Missouri, Arkansas
Tombigbee	658	Mississipi, Alabama
Washita	805	Texas, Oklahoma
Wateree Catawba	636	North Carolina, South Carolina

river

name	Length
St. Francis	684
Tombigbee	658
Washita	805
Wateree Catawba	636

river_state

river_name	state_name
St. Francis	Missouri
St. Francis	Arkansas
Tombigbee	Mississipi
Tombigbee	Alabama
Washita	Texas
Washita	Oklahoma
Wateree Carawba	North Carolina
Wateree Carawba	South Carolina

state

name	abbreviation	...
Missouri	mo	...
Mississipi	ms	...
Oklahoma	ok	...
Texas	tx	...

Fig. 1 Example of a table that does not conform to the first normal form

Use of surrogate primary keys. A surrogate primary key is a simple column that has the following property: its values are useful only as substitutes for the entities (rows) they represent; i.e., they are useful for representing the fact that the corresponding entities exist and do not provide additional information about the entities [7]. An example of the use of surrogate primary keys is shown in Fig. 3.

Column duplicity in multiple tables. This problem is related to information redundancy, since these columns are neither foreign keys nor primary keys in the tables they belong to. An example of this type of columns occurs in the Geobase DB, where the column *state_name* is present in several tables (Fig. 2). In the realm of NLIDBs, this problem causes confusion, because for the interface might be difficult to decide from which table to extract state names.

Poor naming standards. The consistency in table and column descriptors is a very important aspect. The correct naming of descriptors for tables, columns and DB objects allows, not only users but programmers and developers, understanding and manipulating the DB more efficiently.

The anomalies mentioned above usually cause several problems that hinder the correct operation of the semantic analysis of NLIDBs. These problems are described next.

Difficulty for identifying DB tables and columns. When a NLIDB is trying to construct the Select and Where clauses of an SQL query, two situations may occur: (1) the NLIDB is not capable of finding associations of the words in the NL query with the objects of the DB schema, and (2) the NLIDB includes in the SQL translation columns that are not required by the NL query. For example, when a column is duplicated in multiple tables, it may happen that in the SQL statement the NLIDB includes one or several of the duplicate columns, which were not requested in the NL query.

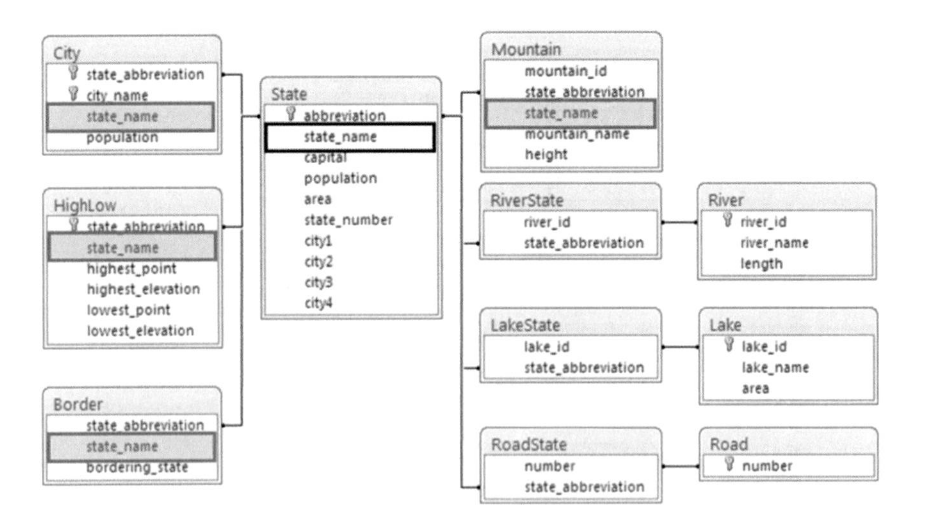

Fig. 2 Duplicate columns in Geobase

Table 1 DB design anomalies and their relation with different problems

DB design anomaly	Identification of tables and columns	Definition of joins between tables	Erroneous information
Absence of PKs and FKs		✓	
Nonconformity with 1NF	✓		✓
Nonconformity with 2NF	✓	✓	
Nonconformity with 3NF	✓	✓	
Use of surrogate PKs	✓	✓	
Column duplicity	✓		
Columns with AF values			✓
Poor naming standards	✓		

Problems when defining joins between tables: When in the semantic analysis of a NL query, two or more tables are detected as relevant for constructing the SQL statement, then it is necessary to include *joins* between the tables. The joins have to be defined in the Where clause of the query; however, if no foreign keys for these tables are defined in the DB schema, the NLIDB will not be able to construct the SQL statement.

Retrieval of erroneous information when querying a DB: Some DB design anomalies produce inconsistencies in the data stored in the DB; consequently, even though the NLIDB may correctly translate a NL query to SQL, the DB server retrieves inconsistent or erroneous information.

Table 1 shows the relations that exist between DB design anomalies and the problems related to semantic analysis. It is worth mentioning that the lack of normalization of DB tables may complicate the semantic analysis by a NLIDB, because the DB schema may not have the structure required by the NLIDB to carry out a successful analysis. Additionally, it can be observed that most of the design anomalies complicate finding the semantic information for a correct translation to SQL.

3 State of the Art

When carrying out an exhaustive survey on recently published works about NLIDBs, it was found out that none of these works has dealt with the problem of querying DBs with design anomalies. In fact, they do not even mention this

Table 2 Relevant works on NLIDBs

NLIDB	Year	Domain independence	Treatment of DB design anomalies
ELF [9]	2004	✓	Partial
Giordani, Moschitti [10]	2009	✓	?
C-Phrase [8]	2010	✓	Partial
SNL2SQL [11]	2013	✓	?
NaLIR [12]	2014	✓	?
Aneesah [13]	2015	✓	?
NL_2CM [14]	2015	✓	?
ILN Arab-Holy Quran [15]	2015	✗	?
ILN Hindi [16]	2015	✓	?

problem; therefore, it can be assumed that such NLIDBs are unable to correctly process NL queries on DBs with design anomalies, as revealed by the experimental tests described in Subsection 3.1.

Despite no NLIDB developed to date has dealt with the problems caused by DB design anomalies, for this section a pair of readily available interfaces (C-Phrase [8] and ELF [9]) have been considered, since they were tested for observing their behavior with poorly designed DBs. Additionally, other interfaces (such as Giordani and Moschitti NLI [10], SNL2SQL [11] NaLIR [12], Aneesah [13], NL2CM [14], NLI Arab-Holy Quran [15], NLI Hindi [16]) have been included because they are a sample of important works developed recently.

As can be observed in Table 2, most of NLIDBs considered are domain independent. Additionally, for all the interfaces (except ELF and C-Phrase) it is not possible to find out if they deal with DB design anomalies of some type, since this problem is not mentioned explicitly in their publications and there is no software available for testing.

Even though it is not mentioned in their publications or manuals, ELF and C-Phrase deal partially with problems caused by DB design anomalies, as shown in Subsection 3.1.

3.1 Experimental Tests

In the preceding paragraphs, several NLIDBs were mentioned. Unfortunately, the publications of most of them do not provide enough details of the DBs used for testing in order to determine if they have design anomalies. Therefore, it is unclear if the interfaces are able to deal with this problem.

However, for ELF and C-Phrase, their software is available for conducting tests. The NL queries used for testing involve some tables that have design anomalies of

Table 3 Test queries

NL query	Database	Design anomaly	ELF response	C-Phrase response
1. How many rivers does Colorado have?	Geobase	Absence of foreign keys	×	✓
2. Give me the population of the capital of the state of Alabama	Geobase	Absence of foreign keys	×	×
3. Give me the state with the highest point	Geobase	Duplicity of columns in multiple tables	×	×
4. Give me the population of Illinois	Geobase	Columns with FA values	partial	partial
5. Name the rivers in Alabama	Geobase	Use of surrogate keys	×	×

different types. Therefore, the process performed by the interface should deal with the problem for generating a correct translation to SQL; otherwise, it would not be able to make a correct translation.

Next the results of five test queries shown in Table 3 are discussed.

For query 1 (*How many rivers does Colorado have*?), ELF returns a correct answer when the original DB schema is used; however, if the same query is input with a modified schema (where the foreign keys are eliminated), ELF cannot generate the SQL statement, since it needs the foreign keys to construct the joins.

For query 2 (*Give me the population of the capital of the state of Alabama*), ELF gives an incorrect answer, since the result is the population of *Alabama* instead of the population of its capital (*Montgomery*). This is because the original schema of Geobase has no foreign key that relates *state.capital* to *city.city_name*; therefore, it is not possible to construct the *join* between the tables *state* and *city* for obtaining the population of *Montgomery*.

The answer returned by C-Phrase for query 2 is erroneous. Although it detects two possible translations for the query, the two are incorrect because they return the population of *Alabama*, instead of *Montgomery*, like ELF.

For query 3 (*Give me the state with the highest point*), ELF cannot answer correctly, because the values requested are those in column *state.state_name*, but there are duplicates of this column in several tables: *mountain.state_name*, *border.state_name*, *highlow.state_name*, *city.state_name*. In this case ELF constructs the SQL translation using *highlow.state_name*, instead of *state.state_name*, which yields the names and highest points of all the states.

The answer given by C-Phrase for query 3 is incorrect, since it ignores the fact that it is necessary to find first the highest point over all the states from table *highlow*.

For query 4 (*Give me the population of Illinois*), the answers obtained by ELF and C-Phrase are partially correct, because there are two alternatives for calculating the population: retrieving the population stored in table *state* for *Illinois*, or adding

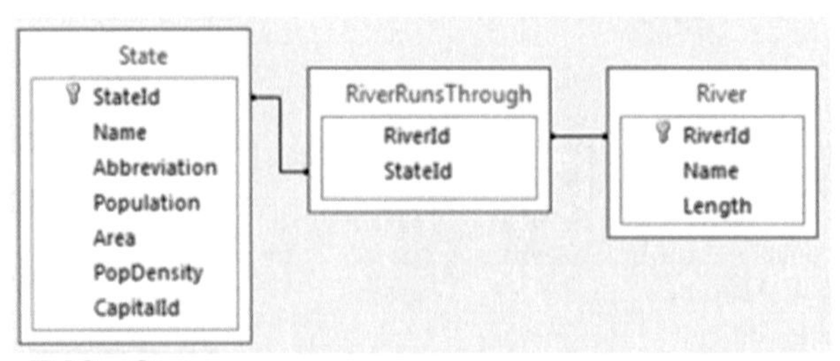

Table: State

StateId	Name	Abbreviation	Population
alabama	alabama	al	51700
alabama2	alabama	al	51700
alaska	alaska	ak	591000
arizona	arizona	az	114000

Table: riverRunsThrough

RiverId	StateId
tombigbee2	alabama
tennessee2	alabama
tombigbee2	alabama2
tennessee2	alabama2
colorado2	arizona
gila2	arizona

Fig. 3 Geobase with surrogate keys and data redundancy

the populations stored in table *city* for the cities in *Illinois*. It is unclear which alternative is the correct one; it depends on the particular DB and application.

For query 5 (*Name the rivers in Alabama*), a modified version of Geobase with surrogate keys was used (Fig. 3). The answer returned by ELF is incorrect, because in the Where clause it includes table *StateLoPoint* and the information retrieved includes redundant data.

The answer given by C-Phrase for query 5, though it is correct, has redundant data, because it includes duplicate rows from table *state* (shown in Fig. 3).

4 Treatment of Design Anomalies by a NLIDB

In this section several solutions are proposed for dealing with some of the DB design anomalies presented in Sect. 2. These solutions will be implemented in the NLIDB developed at ITCM [17].

The most important element of this interface is the semantic information dictionary (SID), which stores most of the semantic information relevant for the query

Student

ID	Name
105	Juan
106	Pedro
107	Maria
108	Luis

Student-Activity

ID	Activity	Cost
105	Football	80
105	Tennis	70
106	Basketball	60
107	Baseball	80
107	Basketball	60
108	Basketball	60
108	Tennis	70

Student

ID	Name
105	Juan
106	Pedro
107	Maria
108	Luis

Student-Activity

EID	Actividad
100	Football
100	Tennis
150	Basketball
175	Baseball
175	Basketball
200	Basketball
200	Tennis

Activity

Actividad	Cuota
Football	80
Tennis	70
Basketball	60
Baseball	80

Fig. 4 Example of table decomposition for complying with 2NF

translation process. Initially, such information is introduced in the SID by the interface through an automatic customization process. Later on, the DB administrator, by using the Customization Interface of the SID, can introduce/modify information on linguistic terms used in the specific domain of the DB, in order to increase the effectiveness of the translation process.

The SID is a database that consists of 9 tables and 151 columns, which contain semantic information on DB tables, columns, relationships between tables, imprecise values, alias values, aggregate functions, views, view tables, and view columns.

In this section, the solution for the most difficult DB design anomalies will be explained: nonconformity with the second normal form and use of surrogate primary keys. The rest of the anomalies are relatively easier to solve; therefore, their solutions will be explained briefly.

Next the solutions for the difficult anomalies are explained.

For correcting the **nonconformity with the 2NF**, usually a decomposition of the nonconforming table is performed in order to normalize it, as shown in Fig. 4.

The process for dealing with this design anomaly is explained next. To this end, let us consider a table T with the following characteristics: it has a primary key constituted by two columns (C_{PK1}, C_{PK2}), where C_{PK2} functionally determines a set of columns C_{DEP}; therefore, C_{DEP} is functionally dependent on the primary key.

First, the columns C_{PK2} and C_{DEP} must be identified. Then, a view V_1 that contains columns C_{PK2} and C_{DEP} is defined, where C_{PK2} becomes the primary key of V_1. The definition of V_1 in SQL is the following:

CREATE VIEW V_1 (C_{PK2}, C_{DEP}) AS
SELECT DISTINCT C_{PK2}, C_{DEP}
FROM T

Next, another view V_2 is defined, which contains the primary key of T (C_{PK1}, C_{PK2}), where C_{PK2} is used to define a foreign key that relates V_2 to V_1.

Afterwards, the DB administrator has to store in the SID the semantic information of these views (V_1, V_2) and has to *logically erase* from the SID the semantic information of the nonconforming table T. Then, the new DB schema will comply with the 2NF. When a NL query involves a table with this anomaly, the translation

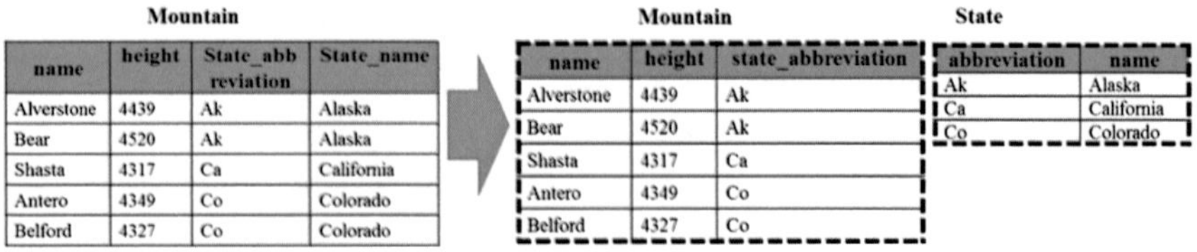

Mountain

name	height	State_abbreviation	State_name
Alverstone	4439	Ak	Alaska
Bear	4520	Ak	Alaska
Shasta	4317	Ca	California
Antero	4349	Co	Colorado
Belford	4327	Co	Colorado

Mountain

name	height	state_abbreviation
Alverstone	4439	Ak
Bear	4520	Ak
Shasta	4317	Ca
Antero	4349	Co
Belford	4327	Co

State

abbreviation	name
Ak	Alaska
Ca	California
Co	Colorado

Fig. 5 Example of table decomposition for complying with 3NF

process of the NLIDB (which assumes that there are no design anomalies) will consider the views instead of the original table and will generate an SQL statement involving the views. Finally, before submitting the statement to the DB server, it has to be transformed by the interface so the original table is considered instead of the views.

As mentioned in Sect. 2, a table does not comply with the third normal form if at least one of its columns has a transitive dependence on a column that is not the primary key. For correcting the **nonconformity with the 3NF**, usually a decomposition of the nonconforming table is performed. This decomposition allows eliminating the transitive dependency (Fig. 5).

The process for dealing with this design anomaly is explained next. Let us consider a table T and three columns: C_1, C_2, C_3, where C_1 is the primary key and $C_2 \rightarrow C_3$ (C_3 is functionally dependent on C_2).

First, a view V_1 is defined, which contains the columns that are involved in the transitive dependency $C_2 \rightarrow C_3$, where C_2 is the primary key of V_1. Then, a view V_2 is defined, which contains the primary key C_1 of table T and a foreign key C_2 that relates V_2 to V_1.

NL queries that involve a table with this anomaly will be processed similarly as queries involving tables that do not comply with the 2NF.

For solving the anomaly caused by the **use of surrogate primary keys**, let us consider a table T that has a surrogate primary key K_S. For solving this problem it is necessary that T_P has a candidate key. In this case, the problem can be solved using the Customization Interface of the SID in order to change the definition of K_S from primary key to surrogate key and to define the new primary key K_P for T_P.

This problem becomes more complex when there exists another table T_R that has a referential integrity relation with T_P. This problem has the three following cases:

1. In the DB schema, a foreign key from T_R to T_P is defined through the surrogate key K_S, and there is a column (or column combination) C_{IFK} that constitutes an implicit foreign key that relates to the natural primary key K_P of table T_P.
2. In the DB schema, a foreign key from T_R to T_P is defined through the surrogate key K_S, and there is no column (or column combination) C_{IFK} that constitutes an implicit foreign key that relates to the natural primary key K_P of T_P.
3. In the DB schema, no foreign key from T_R to T_P is defined through the surrogate key K_S, and there is a column (or column combination) C_{IFK} that constitutes an implicit foreign key that relates to the natural primary key K_P of table T_P.

The first case can be solved using the Customization Interface of the SID in order to define C_{IFK} as an implicit foreign key of T_R, which could be used for some semantic analysis that could be needed for other purposes. However, for the current semantic analysis (for generating joins and subqueries), the foreign key defined through K_S could be used.

The second case happens when the database does not have a mechanism for maintaining the referential integrity of table T_R with respect to table T_P. In order to solve this case, it is necessary first to repair the DB for satisfying the referential integrity, then this case can be solved like the first case.

Like the second case, the last case requires repairing the DB for satisfying the referential integrity. For solving this case it is necessary to use the Customization Interface of the SID for defining C_{IFK} as a foreign key of T_R. Unlike the first case, in this one C_{IFK} could be used in the semantic analysis for generating joins and subqueries.

Finally, the solution for the less complex design anomalies is explained.

For dealing with the **absence of primary and foreign keys**, first it is necessary that the DB administrator identifies the columns that need to have a relation through a foreign key, and then the DB administrator has to use the Customization Interface of the SID for defining the primary and foreign keys for the tables affected by this anomaly. In this way, it is not necessary to modify the DB schema in the DB server, and the NLIDB can use the information stored in the SID for generating the joins and subqueries between tables affected by this anomaly.

For solving the problem caused by **column duplicity in multiple tables**, it is necessary to modify the structure of the DIS. The modification consists of adding a new column to the table *columns* of the DIS, in order to specify if a duplicated column will be considered or ignored when performing the translation process of a query.

Poor naming standards for tables and columns do not cause any problem for the NLIDB, since the semantic analysis carried out by the interface does not use table and column names; instead it uses table and column descriptors, which are imported from the DB server or are defined/modified using the Customization Interface of the SID.

The problem caused by **columns with AF values** can be dealt with by modifying the structure of the DIS. The modification consists of adding news columns to the table *columns* of the DIS, in order to store the following information: an indication if the column will be considered or ignored, the AF related to the column and the column to which the AF is applied. When processing a query that involves an AF, the NLIDB will check the SID to find out if there is a column with AF values that can be used; otherwise, it will generate an SQL statement that involves the AF on the appropriate column.

5 Final Remarks

Many domain-independent NLIDBs have been developed to date. However, they have achieved unsatisfactory performance due to the complexity of the problems encountered in NL processing. Additionally, NLIDB developers (including ourselves) have assumed that the DBs to be queried are not affected by design anomalies, such as those described in Sect. 2.

The tests presented in Sect. 3 on the ELF and C-Phrase interfaces, show that these NLIDBs are not able to deal with most of the design anomalies. According to our survey of the specialized literature on NLIDBs, papers do not mention if the interfaces deal with DB design anomalies; therefore, considering this and the tests mentioned before, it can be conjectured that this problem has not been dealt with.

Since it is common to find DBs that are affected by design anomalies [1–3], and most NLIDBs are not designed for dealing with this problem; then these interfaces would not perform adequately with a large number of DBs that are affected by this problem.

The solutions to the DB design anomalies proposed in this paper can be implemented in the NLIDB developed at ITCM. This is possible because it has a semantic information dictionary (SID) and an architecture based on functionality layers [17], which allows integrating modules for dealing with previously untreated problems. The implementation of methods for solving DB design anomalies will enable our NLIDB to achieve approximately 90% recall with this type of DBs.

References

1. O. Pivert, H. Prade, Handling dirty databases: from user warning to data cleaning towards an interactive approach. in *Fourth International Conference on Scalable Uncertainty Management* (France, 2010)
2. M.L. Pedro de Jesus, P.M.A. Sousa, Selection of reverse engineering methods for relational databases. in *Proceedings of the European Conference on Software Maintenance and Reengineering*. (1999)
3. N. Mofourga, Extracting entity-relationship schemas from relational databases: a form-driven approach, in *Proceedings of the Fourth Working Conference on Reverse Engineering* (IEEE, 1997)
4. P. Rob, C. Coronel, in *Database Systems: Design, Implementation, and Management*. 8th edn. (Course Technology, 2009)
5. D. Kroenke, in *Database Processing: Fundamentals, Design, and Implementation*. Eighth edn. (Pearson Education, 2003)
6. C.J. Date, in *The Relational Database Dictionary*, Extended Edition (O'Reilly Media, 2008)
7. C.J. Date, in *An Introduction to Database Systems*, 8th edn. (Pearson Education, 2004)
8. M. Minock, C-phrase: a system for building robust natural language interfaces to databases. Data Knowl. Eng. **69**, 290–302 (2010)
9. S. Conlon, J. Conlon, T. James, The economics of natural language interfaces: natural language processing technology as a scarce resource. Decis. Support Syst. **38**(1), 141–159 (2004)

10. A. Giordani, A. Moschitti, Translating queries with generative parsers discriminatively reranked, in *Proceedings of Computational Linguistics* (Mumbai, 2012), pp. 401–410
11. I. Esquivel, R. Córdoba, D. González, E. Ogarita, SNL2SQL: Conversión de consultas en SQL al idioma Español, in Congreso internacional de investigación, ISSN:1946-5351, Vol. 5, No. 3, Mexico (2013)
12. F. Li, H.V. Jagadish, Constructing an interactive natural language interface for relational databases. Proc. VLBD Endow. **8**(1) (2014)
13. J.D. O'Shea, K. Shabaz, K.A. Crockett, Aneesah: a conversational natural language interface to databases, in *Proceedings of World Congress on Engineering 2015*, vol. 1 (London, 2015)
14. Y. Amsterdamer, A. Kukliansky, T. Milo, A natural language interface for querying general and individual knowledge. Proc. VLDB Endow. **8**(12) (2015)
15. K. ElSayed, An Arabic natural language interface system for a database of the Holy Quran. Int. J. Adv. Res. Artif. Intell. (IJARAI **4**(7) (2015)
16. A. Kataria, R. Nath, Natural language interface for databases in hindi based on karaka theory. Int. J. Comput. Appl. **122**(7), India (2015)
17. R.A. Pazos, M.A. Aguirre, J.J. González, J.A. Martínez, J. Pérez, A.A. Verástegui, Comparative study on the customization of natural language interfaces to databases, SpringerPlus, doi:10.1186/s40064-016-2164-y (2016)

Comparative Study of ARIMA Methods for Forecasting Time Series of the Mexican Stock Exchange

Javier A. Rangel-González, Juan Frausto-Solis, J. Javier González-Barbosa, Rodolfo A. Pazos-Rangel and Héctor J. Fraire-Huacuja

Abstract Predicting volatility in stock market price indices is a major economic problem. The idea of forecasting time series is that the patterns associated with past values in a data series can be used to project future values. The study of volatility can be applied to solving these economic problems, because volatility allows measuring the risk of asset portfolios, since it shows the behavior of the variation of asset prices. In order to be able to predict effectively the future behavior of a time series, it is necessary to know the attributes of the series with the correct prediction method and thus to be able to define training patterns. The accurate selection of the attributes evaluated in a time series defines the impact on prediction accuracy. In this work the study of kurtosis and the comparison between different ARIMA methods for the solution of time series of the Mexican Stock Exchange and the Makridakis contests are shown.

Keywords ARIMA · Computational strategies · Mexican stock exchange Time series · Forecast

1 Introduction

A Time series model uses the past values of the variables to forecast their future values. The time series model can make accurate forecasts, especially in case where there are multidimensional relationships among variables. Because of the complexity of international economic relations, large structural models are likely to

J. A. Rangel-González · J. Frausto-Solis · J. Javier González-Barbosa (✉) · R. A. Pazos-Rangel · H. J. Fraire-Huacuja
TecNM/Instituto Tecnológico de Ciudad Madero, Ciudad Madero, Mexico
e-mail: jjgonzalezbarbosa@itcm.edu.mx

J. A. Rangel-González
e-mail: javieralberto64@hotmail.com

J. Frausto-Solis
e-mail: juan.frausto@itcm.edu.mx

O. Castillo et al. (eds.), *Fuzzy Logic Augmentation of Neural and Optimization Algorithms: Theoretical Aspects and Real Applications*, Studies in Computational Intelligence 749, https://doi.org/10.1007/978-3-319-71008-2_34

suffer from omitted variable bias, misspecifications, simultaneous causality and other problems leading to substantial forecasting errors [1]. A time series model can still produce quite accurate forecasts if the regression explains much of the variation and is stable over time.

This paper presents classic ARIMA models, so that knowing only the time series, they efficiently forecast the values of the series several periods ahead. This means that a few classic models must be determined to find high quality solutions. The classic models used in this paper are Autoregressive model, Moving Average model, Autoregressive Moving Average model, Integrated Autoregressive model, Integrated Moving Average model and Autoregressive Integrated Moving Average model.

The classic time series models have a limited application in problems of predicting the volatility of investment instruments. Therefore, it is necessary to know them and evaluate their performance with the time series available in the research area to be able to propose new heuristic methods capable of improving the results obtained with the classical methods.

2 Random Walk Theory

The fundamental models of financial theory assume that returns on financial assets follow a random walk. The weak-form efficient market hypothesis states that it is not possible to predict future prices based on historical price information. On the other hand, the Black and Scholes model [2] for the valuation of options, assume that returns follow a geometric Brownian movement, which implies that its proportional variation of the same is independent.

In 1970, Fama [3] defined an efficient market as one in which prices always reflect all available information. Therefore, the following would be sufficient conditions for an efficient market:

- Non-existent transaction costs.
- All available information can be obtained free of charge by all market participants.
- All participants agree on the implications of existing information for the prices of each asset.

In a market that meets these conditions, the price of each asset fully reflects all the information. However, it is difficult for markets to meet all these conditions.

As Fama says [3], these conditions are sufficient but not necessary, because if the number of participants with access to available information is sufficiently broad, none of them can consistently analyze the existing information more closely than others, and although the transaction costs are high, they are also known and taken into account by the participants; thus, the above conditions are not strictly met, but the market will certainly be efficient. The set of available information can be classified into three subsets, which leads to consider three levels of market efficiency [4]:

- Weak efficiency: the set of information includes historical prices.
- Semi-strong efficiency: the information includes all publicly available information for all market participants such as profit announcements, dividend payments and share splits.
- Strong efficiency: all the information known by any market participant is known. In this level also, private information is known.

With respect to a set of information, the efficiency implies that it is not possible to obtain extraordinary gains when negotiating financial assets based on this information. Therefore, market efficiency tests are focused in that direction, seeking to determine if market participants can make extraordinary gains with available information.

Perfect efficiency is an idealization that is not economically feasible, but serves as a benchmark for measuring the relative efficiency of a market [5].

The martingale model is probably the oldest model of financial asset price behavior:

$$E[P_{t+1}|P_t, P_{t-1}, \ldots] = P_t \tag{1}$$

$$E[P_{t+1} - P_t|P_t, P_{t-1}, \ldots] = 0 \tag{2}$$

If P_t is taken as the price of a financial asset, the expected price of tomorrow is equal to today's price, given all the information about historical prices, i.e., the best forecast of the price of tomorrow is the price of today. The martingale model originated the development of the random walk model.

A first version of random walk, was called RW1, requires that the prices increase and follow an independent and identical distribution. The dynamics of P_t is given by:

$$alignedP_t = \mu + P_{t-1} + \varepsilon_t \varepsilon_t \sim IID(0, \sigma^2) aligned \tag{3}$$

where μ is the expected value of the change in price, and σ is the standard deviation.

Since the increments are independent, the random walk is more demanding than the martingale model, because independence implies that the increments are not only uncorrelated; however, nonlinear functions are not correlated either.

A second version of the random walk, RW2, only requires that the increments be independent, without requiring them to have the same distribution. Then, this version considers heteroscedasticity in increments, a common feature of financial time series. Finally, a third version, RW3, requires that the increments be uncorrelated, although it admits that there may be dependence between them.

3 Autoregressive Model

The autoregressive model (AR) defines values in a time series t through the observations corresponding to previous periods, by adding an error term. In other words, all Y_t can be expressed as the sum of its previous values plus an error term.

An AR model equation is expressed as follows:

$$Y_t = c + \varphi_1 Y_{t-1} + \varphi_2 Y_{t-2} + \cdots + \varphi_p Y_{t-p} + \varepsilon_t \tag{4}$$

where c is a constant, $\varphi_1, \ldots, \varphi_p$ are the model parameters, and ε_t is an error variable.

The order of the AR model expresses the number of observations from the time series that affect the equation. In an abbreviated way, an AR model can be expressed as:

$$\varphi_p(L) Y_t = c + \varepsilon_t \tag{5}$$

where $\varphi_p(L)$ is known as the polynomial lag operator:

$$\varphi_p(L) = 1 - \varphi_1 L - \varphi_2 L^2 - \cdots - \varphi_p L^p \tag{6}$$

where the term L is known as the lag operator.

When L is applied to an element Y_t of a time series it produces the previous element Y_{t-1}:

$$L Y_t = Y_{t-1} \tag{7}$$

Therefore, when it is applied p times, it lags the value of an element by p periods:

$$L^p Y_t = Y_{t-p} \tag{8}$$

The error variable ε_t in this model is also referred to as white noise when it meets 3 conditions:

- Null mean.
- Null constant variance.
- Zero covariance between errors corresponding to different observations.

4 Moving Average Model (MA)

A Moving Average model (MA) is one that determines a value of a variable in a period t as a function of an independent term and a succession of errors corresponding to previous periods. An MA(q) model is defined as follows:

$$X_t = \mu + \varepsilon_t + \theta_1 \varepsilon_{t-1} + \cdots + \theta_q \varepsilon_{t-q} \tag{9}$$

where μ is the mean of the time series values, ε_t is an error variable and θ_q are the parameters of the series. Also, this equation can be abbreviated for the AR model in the following way:

$$X_t = \theta_q(L)\varepsilon_t + \mu \tag{10}$$

A model of moving averages can be obtained from an autoregressive model without further successive substitutions [6]. For example, a first order AR model, without the independent term, would be formulated as follows:

$$Y_t = \varphi_1 Y_{t-1} + \varepsilon_t \tag{11}$$

When t is changed as $t - 1$, the model would be:

$$Y_{t-1} = \varphi_1 Y_{t-2} + \varepsilon_{t-1} \tag{12}$$

Replacing Eq. (12) in Eq. (11):

$$Y_t = \varepsilon_t + \varphi_1 \varepsilon_{t-1} + \varphi_2 Y_{t-2} \tag{13}$$

Replacing Y_{t-2} by the regressive expression, the AR model becomes an MA model without the independent term:

$$Y_t = \varepsilon_t + \varphi_1 \varepsilon_{t-1} + \varphi_2 \varepsilon_{t-2} + \cdots + \varphi_p \varepsilon_{t-p} \tag{14}$$

5 Autoregressive Moving Average Model

One of the most popular and important forecasting models is Autoregressive Moving Average model (ARMA), which is the combination of the Autoregressive model (AR) and the Moving Average model (MA) [7].

The ARMA model combines the equations above as follows:

$$Y_t = \varphi_1 Y_{t-1} + \varphi_2 Y_{t-2} + \cdots + \varphi_p Y_{t-p} + \varepsilon_t + \theta_1 \varepsilon_{t-1} + \theta_2 \varepsilon_{t-2} + \cdots + \theta_q \varepsilon_{t-q} \tag{15}$$

The ARMA (p, q) notation refers to a model with p autoregressive terms and q moving average terms. The preceding expression can be reduced as follows:

$$Y_t = \varepsilon_t + \sum_{i=1}^{p} \varphi_i Y_{t-i} + \sum_{j=1}^{q} \theta_j \varepsilon_{t-j} \tag{16}$$

6 Autoregressive Integrated Moving Average Model

The Autoregressive Integrated Moving Average model (ARIMA), developed by George Box and Gwilym Jenkins [8]. In 1970s ARIMA was a true revolution in time series area. Unlike the ARMA model, the ARIMA model includes the stationary component to the ARMA model [7]. When a time series is not stationary, it can be integrated by defining a new variable as follows:

$$z_t = y_t - y_{t-1} \tag{17}$$

where $t = 2, 3, \ldots, n$.

The second differences are determined for $t = 3, 4, \ldots, n$ by:

$$z_t = (y_t - y_{t-1}) - (y_{t-1} - y_{t-2}) \tag{18}$$

The resulting model, integrating seasonality, is an ARMA model with a non-seasonal component. This model is known as ARIMA (p, d, q), where:

- p: order of the AR model.
- d: differencing order of the regular and non-regular parts of a stationary series.
- q: order of the MA model.

The parameter d in ARI, IMA and ARIMA models is the differencing order, which gives stability to the data; this stability gives the time series a seasonality to facilitate its forecasting. Figure 1 shows the closing values of the Kimberly-Clark company; and Fig. 2 shows the time series once the stationary element is applied. Particularly, it is well known, that the seasonality factor facilitates the prediction of the time series by normalizing the data.

To make a good forecast of some time series it is necessary to know the characteristics of each of the ARIMA models presented. The information required for each model is:

- AR(p) models: The prediction tends to the average as the time horizon of the prediction increases.
- MA(q) models: Given the limited memory that characterizes these models, the prediction is equal to the average of the process when the time horizon of the prediction is greater than the order of the process (q).

Fig. 1 Closing values

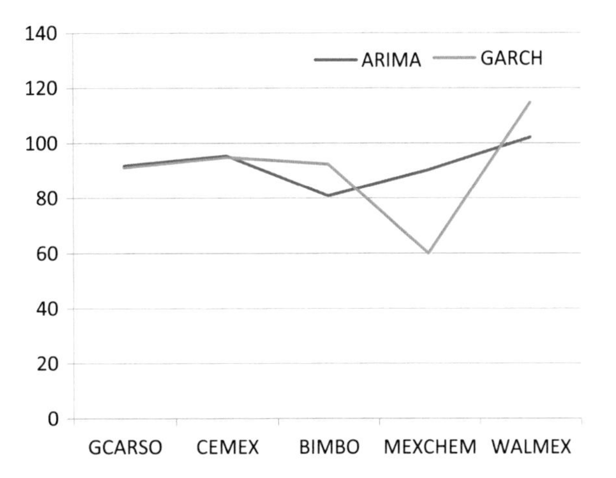

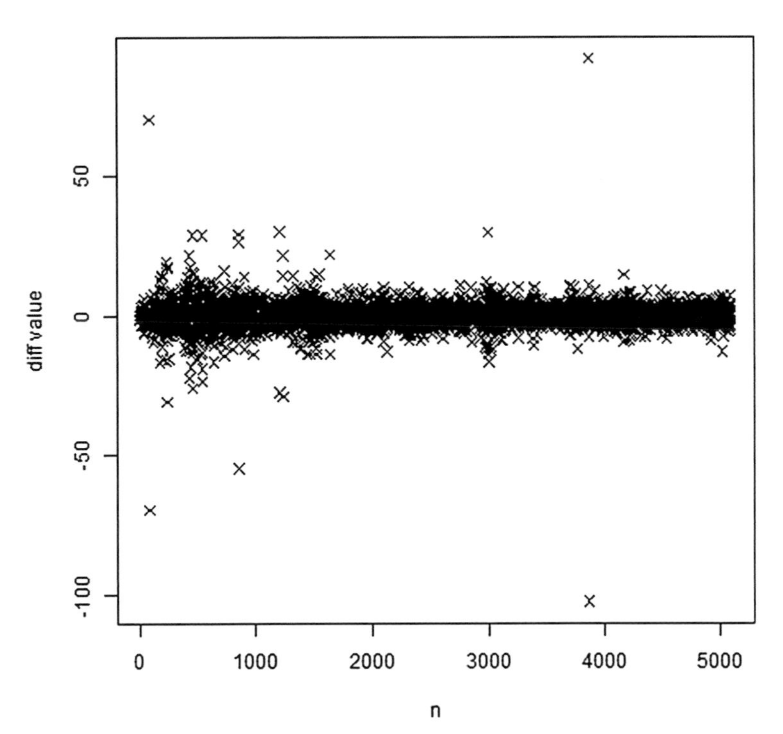

Fig. 2 Differenced values

- ARMA(p, q) models: From q future periods, the prediction tends to average as the time horizon of the prediction increases.
- ARI(p, d) and IMA(d, q) models: The prediction no longer tends to the average but will be a straight line that starts from the first data with slope equal to the average of the series resulting from the transformations necessary to make it stationary.
- ARIMA(p, d, q) models: The ARIMA predictions are adaptive and the results obtained for $t + 1$, with the information available up to period t, are the same as those obtained for the same period, taking as informative basis up to $t - 1$, and adding an error term.

7 ARIMA Models Experimentation

The implementation in this paper is based on the use of ARIMA models to predict the time series of the Mexican Stock Exchange. This implementation was developed in the R language using the *tseries* library.

The time series used in these experiments were taken from 6 time series companies of the Mexican Stock Exchange: CARSO, CEMEX, FEMSA, INVEX, IXE and Kimberly-Clark. In addition, 2 time series from the USA Stock Exchange were tested: NASDAQ and S&P 500. Finally, the Makridakis contests usually taken for testing were used: M1, M2, M3 and M5.

Table 1 shows the Mean Absolute Percentage Error (MAPE) generated by each model for the 13 time series.

The results in Table 1 show that for most cases the ARIMA model predicts with the smallest error. Even so, we can observe that for certain time series (CEMEX,

Table 1 MAPE results

Time series	AR	MA	ARMA	ARI	IMA	ARIMA
CARSO	**3.61**	44.01	4.81	4.68	3.77	**3.61**
CEMEX	74.73	201	84.08	42.36	**28.17**	28.3
FEMSA	24.41	39.7	24.33	34.08	**11.4**	12.04
INVEX	2.76	22.14	2.52	2.47	2.46	**2.36**
IXE	3.68	8.94	**3.66**	4.95	4.86	4.87
Kimberly-Clark	31.33	67.85	19.06	**9.31**	9.32	9.4
NASDAQ	35.17	74.94	32.1	8.63	22.35	**8.05**
S&P 500	33.66	76.18	32.59	32.35	29.11	**15.54**
M1	3.55	44.16	Inf.	5.74	**0.83**	**0.83**
M2_PANTER	31.62	**12.85**	32.70	38.74	33.08	25.92
M2_REALGNP	4.94	29.99	Inf.	**1.23**	2.56	1.31
M3	34.36	33.96	30.17	33.04	**29.03**	29.14
M5	99.29	100.48	**98.88**	103.37	104.11	**98.88**

FEMSA, IXE, Kimberly-Clark and M2_REALGNP) the error between the winning model for each case and ARIMA is very small. The only series where the ARIMA error is more than the double of the winning model is for the M2_PANTER series. This difference between the results depends on the characteristics of the time series used, their difficulty, and the advantages and disadvantages of each of the models used in this experimentation.

Table 2 shows a MAPE comparison and a percentage difference between the best and the second best models, and Table 3 shows the worst model. ARIMA is in most of the cases in the first position; when ARIMA is not in the first position it is in the second one, except for the IXE and Kimberly-Clark series. For the Kimberly-Clark series, the position of ARIMA is the third position with a minimal difference of 0.09 with respect to the first position and 0.08 with respect to the second position. The only time series where ARIMA has a larger MAPE is for the IXE series with a difference of 1.21 with respect to the first position and 1.23 with respect to the second position.

Table 4 shows the execution time spent by each ARIMA method for each time series. The execution times are quite short for this experimentation. It is noticeable that the difference in time between the models is not relevant because the times are very small.

Table 2 Comparative results

Time series	Model in the first best position		Model in the second best position		
	Model	MAPE	Model	MAPE	Δ best position (%)
CARSO	AR, ARIMA	3.61	IMA	3.77	4.43
CEMEX	IMA	28.17	ARIMA	28.3	0.46
FEMSA	IMA	11.4	ARIMA	12.04	5.6
INVEX	ARIMA	2.36	IMA	2.46	4.2
IXE	ARMA	3.66	IMA	4.86	32.78
Kimberly-Clark	ARI	9.31	IMA	9.32	0.10
NASDAQ	ARIMA	8.05	ARI	8.63	7.2
S&P 500	ARIMA	15.54	IMA	29.11	87.32
M1	ARIMA, IMA	0.83	ARI	5.74	591
M2_PANTER	MA	12.85	ARIMA	25.92	101.7
M2_REALGNP	ARI	1.23	ARIMA	1.31	6.5
M3	IMA	29.03	ARIMA	29.14	3.78
M5	ARIMA, ARMA	98.88	IMA	99.29	0.41

Table 3 Worst results

Model with worst results		
Model	MAPE	Δ best position (%)
MA	44.01	1119
MA	201	613
MA	39.7	248
MA	22.14	838
MA	8.94	144
MA	67.85	628
MA	74.94	830
MA	76.18	390
ARMA	Infeasible	Infeasible
ARI	38.74	201
ARMA	Infeasible	Infeasible
AR	34.36	18.36
IMA	104.11	5.28

Table 4 Execution time results

Time series	AR (s)	MA (s)	ARMA (s)	ARI (s)	IMA (s)	ARIMA (s)
CARSO	0.31	0.45	1.03	0.10	0.17	0.33
CEMEX	0.22	0.41	0.63	0.19	0.36	0.30
FEMSA	0.28	0.34	1.19	0.24	0.46	0.63
INVEX	0.21	0.12	0.28	0.12	0.11	0.16
IXE	0.16	0.1	0.14	0.13	0.12	0.14
Kimberly-Clark	0.26	0.32	0.5	0.23	0.53	0.55
NASDAQ	0.53	0.53	0.97	0.41	0.97	1.03
S&P 500	0.5	1.04	2.66	0.67	0.75	2
M1	0.20	0.15	0.31	0.08	0.14	0.22
M2_PANTER	0.11	0.13	0.14	0.16	0.11	0.14
M2_REALGNP	0.14	0.22	0.19	0.09	0.08	0.17
M3	0.10	0.11	0.36	0.11	0.16	0.20
M5	0.11	0.11	0.39	0.10	0.13	0.31

8 Conclusions

Six classic models used for the prediction of financial time series of the Mexican Stock Exchange, the USA Stock Exchange and the Makridakis contest were presented in this paper.

The results generated by the classic models show that the ARIMA model is the one that obtained the best results. However, to obtain the best performance of ARIMA models, is important to consider the characteristics of the model.

We emphasize that it is necessary to implement other heuristic methods to obtain better forecasting results. As future work, a support vector machine combined with fuzzy logic will be tested to compete with the classic methods.

Acknowledgements The authors would like to acknowledge the Consejo Nacional de Ciencia y Tecnología (CONACYT). Besides, they acknowledge the Laboratorio Nacional de Tecnologías de la Información (LaNTI) of the Instituto Tecnológico de Ciudad Madero for the access to the cluster. Also, Javier Alberto Rangel González thanks the scholarship 429340 received from CONACYT in his Ph.D.

References

1. A. Keck, A. Raubold, A. Truppia, Forecasting international trade: a time series approach. OECD J. J. Bus. Cycle Meas. Anal. **2009**(2), 157 (2009)
2. F. Black, M. Scholes, The pricing of options and corporate liabilities. J. Polit. Econ. **81**(3), 637–654 (1973)
3. E.F. Fama, Efficient capital markets: a review of theory and empirical work. J. Finance **25**(2), 383–417 (1970)
4. R.J. Shiller, From efficient markets theory to behavioral finance. J. Econ. Perspect. **17**(1), 83–104 (2003)
5. J.Y. Campbell, A.W.-C. Lo, A.C. MacKinlay, in *The Econometrics of Financial Markets* (Princeton University Press, New Jersey, 1997)
6. P.J. Brockwell, R.A. Davis, in *Introduction to Time Series and Forecasting* (Springer, Berlin, 2016)
7. J. Frausto-Solis, E. Pita, J. Lagunas, Short-term streamflow forecasting: ARIMA vs. Neural Networks, in *American Conference on Applied Mathematics (MATH'08), Harvard, Massachusetts, USA*, 2008, pp. 402–407
8. G.E.P. Box, G.M. Jenkins, in *Time Series Analysis: Forecasting and Control*, 1976. (1970) ISBN 0-8162-1104-3

A New Heuristic Algorithm for the Vertex Separation Problem

Norberto Castillo-García and Paula Hernández Hernández

Abstract The Vertex Separation Problem (VSP) belongs to a family of graph layout problems. VSP consists in finding a linear ordering of the vertices of a graph such that the maximum number of vertex separators at each position of the ordering is minimized. This problem has important practical applications in fields such as very large scale integration design, computer language compiler design or natural language processing. VSP has been proven to be NP-hard. In the literature reviewed, we found several heuristic and metaheuristic algorithms designed for solving large instances of VSP. As far as we are aware, these algorithms do not use fuzzy logic. In this chapter, we adapt two fuzzy logic classifiers (FLC) to a constructive algorithm from the literature. More precisely, the first FLC is used to select the vertex to be placed at the first position of the linear ordering according to the adjacency degree. The second FLC is used to select the following vertices according to the number of vertex separators. We have designed five variants of our fuzzy heuristic. The computational experiment indicates that the first four variants have a similar behavior in solution quality and execution time.

Keywords Vertex separation problem · Heuristics · Constructive algorithms
Fuzzy logic classifier

N. Castillo-García (✉)
Tecnológico Nacional de México, Instituto Tecnológico de Altamira,
Altamira, Mexico
e-mail: norberto_castillo15@hotmail.com

P. Hernández Hernández
CONACYT-Universidad Autónoma de Tamaulipas-Facultad de Ingeniería,
Tampico, Mexico

O. Castillo et al. (eds.), *Fuzzy Logic Augmentation of Neural and Optimization Algorithms: Theoretical Aspects and Real Applications*, Studies in Computational Intelligence 749, https://doi.org/10.1007/978-3-319-71008-2_35

1 Introduction

The Vertex Separation Problem (VSP) is a combinatorial optimization problem which belongs to a family of graph layout problems [1]. The goal of VSP is to find a permutation of the vertices of a generic graph that minimizes the maximum number of vertex separators at each position of the permutation. The vertex separators at the ith position of some permutation are those vertices placed to the left of i (including the vertex placed at position i) with at least one adjacent vertex to the right of i.

In order to elucidate the Vertex Separation Problem, let us consider the example shown in Fig. 1. Figure 1a depicts a simple graph with 5 vertices and 5 edges. Figures 1b, c show two different permutations of vertices (solutions). In Figs. 1b, c, the positions are shown at the top while the number of vertex separators at the bottom.

The first solution (Fig. 1b) is the permutation of vertices (`A, B, C, D, E`). This means that the vertex `A` is placed at position $i = 1$, the vertex `B` at $i = 2$, and so on. The number of vertex separators at $i = 1$ is 1 since there is only one vertex (`A`) with at least one adjacent vertex placed to the right of $i = 1$ (vertices `B` and `D`). The number of vertex separators at $i = 2$ is 2 since there are two vertices (`A` and `B`) placed before $i = 2$ that have at least one adjacent vertex after $i = 2$, that is, `D` is adjacent to `A` and `E` is adjacent to `B`. The same reasoning is used to compute the remaining numbers of vertex separators. The objective value of any solution for VSP is the largest number of vertex separators at each position. In the case of the solution shown in Fig. 1b, the objective value is $\max\{1, 2, 3, 2, 0\} = 3$.

The second solution (Fig. 1c) is the permutation (`A, B, D, C, E`). The difference between the first permutation and the second permutation resides in the positions of vertices `C` and `D`. This difference affects the number of vertex separators at position $i = 3$. Specifically, the number of vertex separators at this position is 2 since only the vertices `B` and `D` have at least one adjacent vertex to the right of i. The vertex `A` does not have any adjacent vertex to the right of $i = 3$ in this permutation, and hence, it does not contribute to the number of vertex separators at $i = 3$ of the second permutation. Finally, the objective value of the second permutation is

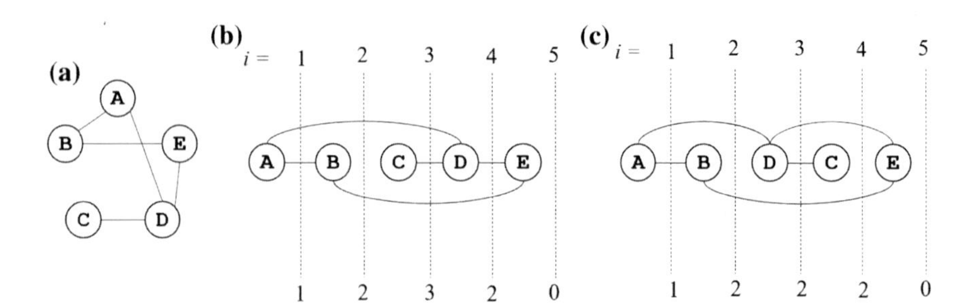

Fig. 1 Illustrative example of the Vertex Separation Problem

$\max\{1, 2, 2, 2, 0\} = 2$. Therefore, the best solution for this example is the one shown in Fig. 1c since VSP is a minimization problem.

The goal of the Vertex Separation Problem is to find the permutation that yields the minimum objective value. However, the number of possible solutions which must be evaluated to guarantee the optimality is $n!$, where n stands for the number of vertices of the input graph. Thus, the number of solutions for VSP (the cardinality of the solution space) grows exponentially with respect to the size of the instance.

The Vertex Separation Problem is NP-hard [2] with practical applications in very large scale integration (VLSI) design [3], computer language compiler design [4], natural language processing [5], processing of manufactured products [6], bioinformatics [7] and military logistics [8]. Due to its applications, the Vertex Separation Problem have been extensively studied and several exact and approximate methods have been proposed. Some solutions methods published in this decade can be found in [8–14]. Despite the relatively large number of methods proposed to solve VSP, we did not find any method that uses fuzzy logic.

In this chapter we adapt a Fuzzy Logic Classifier (FLC) to the constructive algorithm proposed by Castillo-García et al. in [12]. This constructive algorithm iteratively selects one vertex of the graph and places it at a determined position of the permutation. The selection of the most suitable vertex is performed taking advantage of the knowledge of the problem. The first vertex is selected according to its adjacency degree and the remaining vertices according to their numbers of vertex separators.

Our FLC assign a class to each vertex of the graph at each iteration of the algorithm. At the first iteration, we classify the vertices according to their adjacency degree. For the remaining iterations we use the numbers of vertex separators of each vertex. We have designed five classes for the vertices. The first class (C1) contains the vertices with a *very low* value of either the adjacency degree (for the first iteration) or the number of vertex separators (for the remaining iterations). Similarly, the second (C2), third (C3), fourth (C4) and fifth (C5) classes respectively contain the vertices with *low*, *medium*, *high* and *very high* values of either adjacency degree or number of vertex separators.

We have designed five variants of our fuzzy heuristic. The difference among the variants resides in the pool of candidate vertices to be selected. The pool of candidate vertices of the first variant is limited to the first class, that is, the first variant can only select one vertex v in the first class, i.e., $v \in C1$. The second variant can select a vertex in either the first class or the second class ($v \in C1 \cup C2$). The pool of candidate vertices of the third variant is $C1 \cup C2 \cup C3$, of the fourth variant is $C1 \cup C2 \cup C3 \cup C4$, and of the fifth variant is $C1 \cup C2 \cup C3 \cup C4 \cup C5$.

We have designed a computational experiment in order to assess the performance of each variant in practice. The experimental results indicated that the first, second, third and fourth variants had a similar behavior and the fifth variant had the worst performance.

The remainder of this chapter is organized as follows. Section 2 presents the formal definition of the Vertex Separation Problem. In Sect. 3 we detail our fuzzy heuristic FH along with an illustrative example. Section 4 reports the experimental evaluation of the five variants of FH. Finally, the conclusions of this study are presented in Sect. 5.

2 Formal Definition of VSP

We formally define the Vertex Separation Problem following the nomenclature used by Díaz et al. in [1]. Let $G = (V, E)$ be a connected simple graph without loops. The sets of vertices and edges of the graph are V and E, respectively. The number of vertices is denoted by $n = |V|$ and the number of edges by $m = |E|$.

A linear ordering is an arrangement of the vertices of G in a straight line, which is usually drawn horizontally. Mathematically, a linear ordering is a bijective function $\varphi : V \rightarrow [n] = \{1, \ldots, n\}$ that associates each vertex $u \in V$ with an integer number from 1 to n. The integer number associated with vertex u is known as the label of u and is represented by $\varphi(u)$. Notice that a linear ordering is actually a permutation of the vertices. Thus, the label of vertex u corresponds to the position where u is placed in the permutation.

Let $i = 1, \ldots, n$ be some position in permutation φ defined on graph G. The set of vertices placed to the left of position i is defined by $L(i, \varphi, G) = \{u \in V : \varphi(u) \leq i\}$. Notice that the vertex placed at position i is considered in the set of left vertices. The set of vertices placed to the right of position i is defined by $R(i, \varphi, G) = \{u \in V : \varphi(u) > i\}$. Notice that the definitions of the sets of left and right vertices imply that $L(i, \varphi, G) \cup R(i, \varphi, G) = V$ and $L(i, \varphi, G) \cap R(i, \varphi, G) = \phi$ for all $i = 1, \ldots, n$.

The vertex cut or separation at position i of permutation φ is defined as the number of vertices placed to the left of i with one or more adjacent vertices placed to the right of i. Formally:

$$\delta(i, \varphi, G) = |\{u \in L(i, \varphi, G) : \exists v \in R(i, \varphi, G) | (u, v) \in E\}|$$

This number is also known as cut value [9] or separation value [10]. Throughout this chapter, we will use the term cut value to allude to the value of $\delta(i, \varphi, G)$. Every permutation of vertices has exactly n cut values, one at each position. The first cut value is always 1 for connected graphs and the last cut value is always 0 since there are no vertices to the right of the last position $i = n$. The objective value of permutation φ (solution) is computed as the largest cut value, that is:

$$vs(\varphi, G) = \max_{i \in [|V|]} \delta(i, \varphi, G)$$

The Vertex Separation Problem consists in finding the permutation of vertices $\varphi^{\star}$ whose objective value is the minimum among the $n!$ possible permutations.

3 Solution Methods

3.1 Previous Approach HN_1

In 2015, Castillo-García et al. proposed two new constructive algorithms for the Vertex Separation Problem [12]. Each constructive can take four different configurations, totalizing eight different constructive algorithms. In this section we describe the constructive algorithm which yielded the best solutions for the instances evaluated, namely, HN_1.

Algorithm HN_1 constructs the permutation vertex by vertex. This constructive uses two subsets of vertices $U \subseteq V$ and $L \subseteq V$ to operate. The set U contains the unlabeled vertices, that is, those vertices which have not been selected to be in the permutation under construction. The set L contains the vertices which are already in the permutation, that is, the labeled vertices. Notice that, at each iteration of HN_1, both sets must satisfy $U \cup L = V$ and $U \cap L = \phi$.

At the beginning, the set U contains all the vertices of the graph while the set L is empty. Constructive HN_1 selects the vertex $u \in V$ with the lowest adjacency degree to place it at the first position of the permutation, i.e., $\varphi(u) = 1$. Then, the algorithm must update the sets $U = U \backslash \{u\}$ and $L = L \cup \{u\}$ since vertex u is now in the permutation.

The selection of the next vertex is performed as follows. Let $N = \{x \in U | \exists y \in L : (x, y) \in E\}$ be the set of unlabeled vertices with at least one adjacent vertex in the set of labeled vertices. Algorithm HN_1 iteratively places each vertex $v \in N$ at the next available position $k = |L| + 1$ of the permutation. Then, it computes the cut value at position k considering the set of left vertices $L(k, \varphi, G) = L \cup \{v\}$ and the set of right vertices $R(k, \varphi, G) = U \backslash \{v\}$. In simple words, the set of left vertices are the vertices already labeled (including v) and the set of right vertices are the unlabeled vertices (excluding v). In order to assign one label to each vertex, the constructive incrementally numbers each vertex in $L(k, \varphi, G)$ with an integer between 1 and k and each vertex in $R(k, \varphi, G)$ with an integer between $k + 1$ and n.

In order to determine the level of desirability of the vertices in N, constructive HN_1 computes the following greedy function:

$$f(v) = \delta(k, \varphi, G) \quad \forall v \in N \tag{1}$$

The algorithm then selects the vertex $v^{\star} \in N$ with the lowest f-value and places it at the next available position, i.e., $\varphi\left(v^{\star}\right) = k$. Once the vertex $v^{\star}$ has been selected, the sets U and L must be updated, i.e.,$U = U \backslash \left\{v^{\star}\right\}$ and $L = L \cup \left\{v^{\star}\right\}$. The algorithm continues this way with the remaining vertices until all the vertices of the graph are in the permutation. The pseudocode of constructive HN_1 is shown in Algorithm 1.

Algorithm 1 Pseudocode of HN_1 [12].

1. Let U and L be the sets of unlabeled and labeled vertices, respectively.
2. $U = V$ and $L = \phi$.
3. Select the vertex $u \in V$ with the lowest adjacency degree.
4. Assign label $l = 1$ to vertex u, i.e., $\varphi(u) \leftarrow l = 1$.
5. $U = U \backslash \{u\}$ and $L = L \cup \{u\}$.
6. **while** $U \neq \phi$ **do**
7. $\quad l \leftarrow l + 1$.
8. $\quad N = \{x \in U | \exists y \in L : (x, y) \in E\}$.
9. $\quad v^{\star} = \mathrm{argmin}_{w \in N}\{f(w)\}$.
10. $\quad$ Assign label l to selected vertex $v^{\star}$, i.e., $\varphi(v^{\star}) \leftarrow l$.
11. $\quad U = U \backslash \{v^{\star}\}$ and $L = L \cup \{v^{\star}\}$.
12. **end while**

3.2 *Proposed Approaches*

We propose to adapt a fuzzy logic classifier (FLC) to the constructive algorithm HN_1. We call this new fuzzy heuristic FH. The idea of our FLC is to assign a class to each vertex of the graph according to either the adjacency degree or the cut value. Thus, we have two FLCs, one for selecting the first vertex (based on the adjacency degree) and the other for selecting the remaining vertices (based on the cut value). We have designed five classes for our FLCs, namely, C1, C2, C3, C4 and C5. The description of each class is the following:

- C1 contains the set of vertices whose adjacency degrees (for selecting the first vertex) or cut values (for selecting the next vertices) are *very low*.
- C2 contains the set of the vertices whose adjacency degrees or cut values are *low*.
- C3 contains the set of the vertices whose adjacency degrees or cut values are *medium*.
- C4 contains the set of the vertices whose adjacency degrees or cut values are *high*.
- C5 contains the set of the vertices whose adjacency degrees or cut values are *very high*.

In order to classify all the vertices of the graph, two universes of discourse are required, one for the first FLC and the other for the second FLC. Both universes of discourse use the same membership functions: one L-function (for class C1), three triangular functions (for classes C2, C3 and C4) and one R-function (for class C5). Notice that the membership functions used in this study are all linear. Figures 2 and 3 depict both universes of discourse used for our fuzzy heuristic FH.

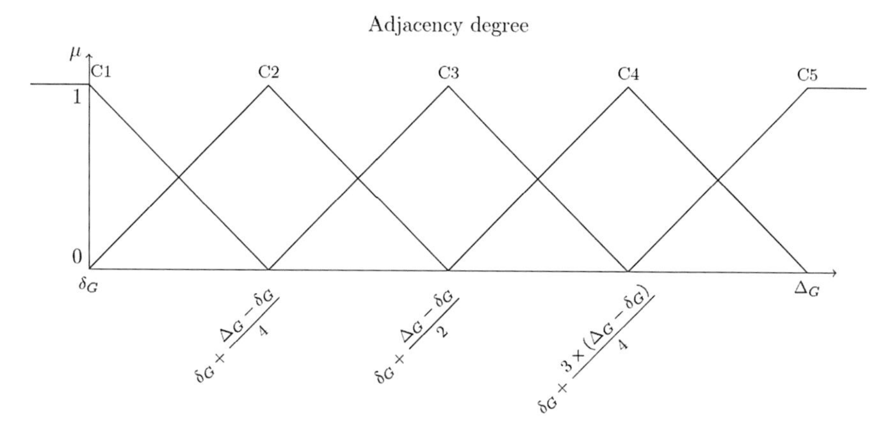

Fig. 2 Universe of discourse for the first Fuzzy Logic Classifier. δ_G and Δ_G represent the lowest and the largest adjacency degree of the graph, respectively

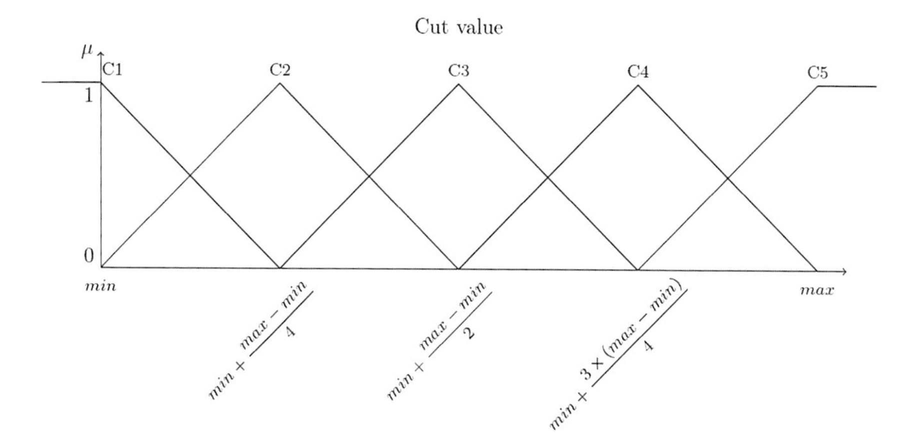

Fig. 3 Universe of discourse for the second Fuzzy Logic Classifier. *min* and *max* represent the lowest and the largest cut value from among the unlabeled vertices, respectively

The equations defining the membership functions used in this study were taken from [15].

The membership function for class C1 is completely defined by two parameters: b and c (with $b<c$). Parameter b represents the point where the membership function reaches its largest degree of membership. Conversely, parameter c is the point where the lowest degree of membership is reached. The membership function for class C1 is mathematically defined in Eq. 2.

$$\mu_{C1}(x) = \begin{cases} 1, & x \leq b \\ \frac{c-x}{c-b}, & x \geq 0 \\ 0, & x \geq c \end{cases}. \tag{2}$$

For the first FLC, x is the adjacency degree of some vertex, $b = \delta_G$, and $c = \delta_G + (\Delta_G - \delta_G)/4$. For the second FLC, x is the f-value obtained by Eq. 1, $b = min$, and $c = min + (max - min)/4$.

The membership functions for classes C2, C3 and C4 are triangular functions. These functions are fully defined by three parameters: $a < b < c$. Parameters a and c represent the lowest values of the triangular function while parameter b represents the peak value. Eq. 3 formally defines the triangular membership function.

$$\mu_{C2,C3,C4}(x) = \begin{cases} 0, & x \leq a \\ \frac{x-a}{b-a}, & a \leq x \leq b \\ \frac{c-x}{c-b}, & b \leq x \leq c \\ 0, & x \geq c \end{cases}. \tag{3}$$

We define the parameters for the class C2. For the first FLC, x is the adjacency degree, $a = \delta_G$, $b = \delta_G + (\Delta_G - \delta_G)/4$, and $c = \delta_G + (\Delta_G - \delta_G)/2$. For the second FLC, x is the f-value from Eq. 1, $a = min$, $b = min + (max - min)/4$, and $c = min + (max - min)/2$.

The parameters for the class C3 are the following. For the first FLC, x is the adjacency degree, $a = \delta_G + (\Delta_G - \delta_G)/4$, $b = \delta_G + (\Delta_G - \delta_G)/2$, and $c = \delta_G + 3 \times (\Delta_G - \delta_G)/4$. For the second FLC, x is the f-value from Eq. 1, $a = min + (max - min)/4$, $b = min + (max - min)/2$, and $c = min + 3 \times (max - min)/4$.

The parameters for the class C4 are defined as follows. For the first FLC, x is the adjacency degree, $a = \delta_G + (\Delta_G - \delta_G)/2$, $b = \delta_G + 3 \times (\Delta_G - \delta_G)/4$, and $c = \Delta_G$. For the second FLC, x is the f-value, $a = min + (max - min)/2$, $b = min + 3 \times (max - min)/4$, and $c = max$.

The membership function for class C5 is completely defined by two parameters: a and b (with $a < b$). Parameter a is the point where the membership function reaches the lowest degree of membership. Parameter b is the point where the membership function reaches the largest degree of membership. The membership function for class C5 is formally defined in Eq. 4.

$$\mu_{C5}(x) = \begin{cases} 1, & x \leq b \\ \frac{c-x}{c-b}, & x \geq 0 \\ 0, & x \geq c \end{cases}. \tag{4}$$

For the first FLC, x stands for the adjacency degree, $a = \delta_G + 3 \times (\Delta_G - \delta_G)/4$, and $b = \Delta_G$. For the second FLC, x is the f-valuefrom Eq. 1, $a = min + 3 \times (max - min)/4$, and $b = max$.

In this chapter, we have designed five variants of our fuzzy heuristic, namely, FH1, FH2, FH3, FH4 and FH5. The difference among the variants resides in the pool of candidate vertices. More precisely, FH1 randomly selects one vertex in C1; FH2 randomly selects one vertex in $C1 \cup C2$; FH3 randomly selects one vertex in $C1 \cup C2 \cup C3$; FH4 randomly selects one vertex in $C1 \cup C2 \cup C3 \cup C4$; and FH5 randomly selects one vertex in $C1 \cup C2 \cup C3 \cup C4 \cup C5$.

In order to illustrate how our fuzzy heuristic works, let us consider the following example. Given the graph depicted in Fig. 4a, generate a solution with FH1.

At the beginning, FH1 computes the adjacency degree of each vertex of the graph. In this example, vertex `A` has an adjacency degree of 2, vertex `B` of 4, vertex `C` of 2, vertex `D` of 3 and vertex `E` of 1. Thus, the lowest and the largest adjacency degree are $\delta_G = 1$ and $\Delta_G = 4$. As shown in Fig. 2, these values are the central values of the membership functions for classes C1 and C5, respectively. The central values of the membership functions for classes C2, C3 and C4 are respectively:

$$\delta_G + \frac{(\Delta_G - \delta_G)}{4} = 1 + \frac{(4-1)}{4} = 1.75,$$

$$\delta_G + \frac{(\Delta_G - \delta_G)}{2} = 1 + \frac{(4-1)}{2} = 2.5, \text{ and}$$

$$\delta_G + \frac{3(\Delta_G - \delta_G)}{4} = 1 + \frac{3(4-1)}{4} = 3.25.$$

The universe of discourse for the input adjacency degree is depicted in Fig. 5.

FH1 evaluates the degree of membership of each vertex of the graph in all the classes by applying Eqs. 2–4 with the corresponding parameters.

Figure 6 graphically shows the evaluation of the degrees of membership of vertex `A`.

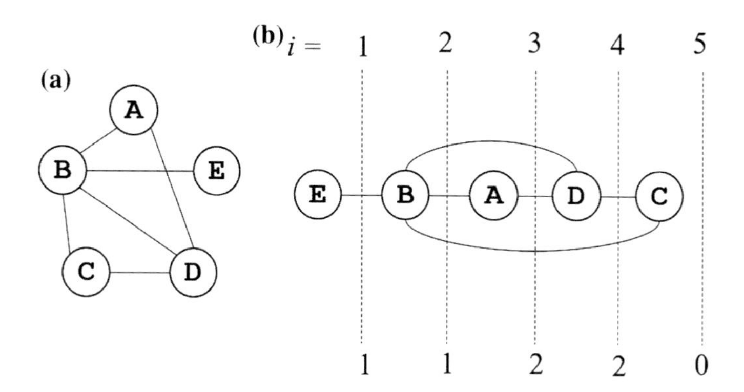

Fig. 4 **a** Connected simple graph with $n = 5$ vertices and $m = 6$ edges. **b** Solution obtained by the first variant of our fuzzy heuristic, FH1

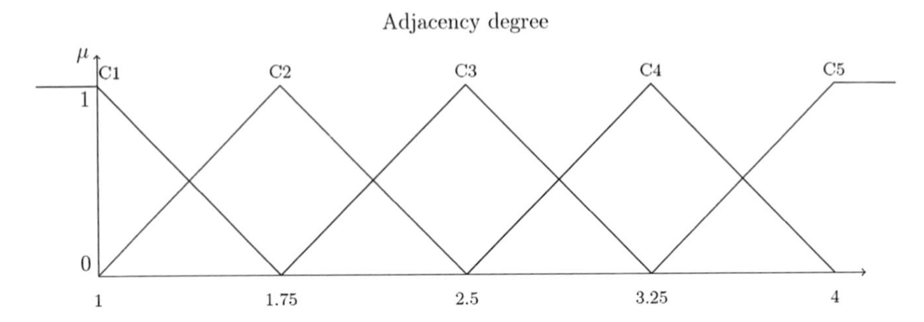

Fig. 5 Universe of discourse of the input adjacency degree for the graph shown in Fig. 4a

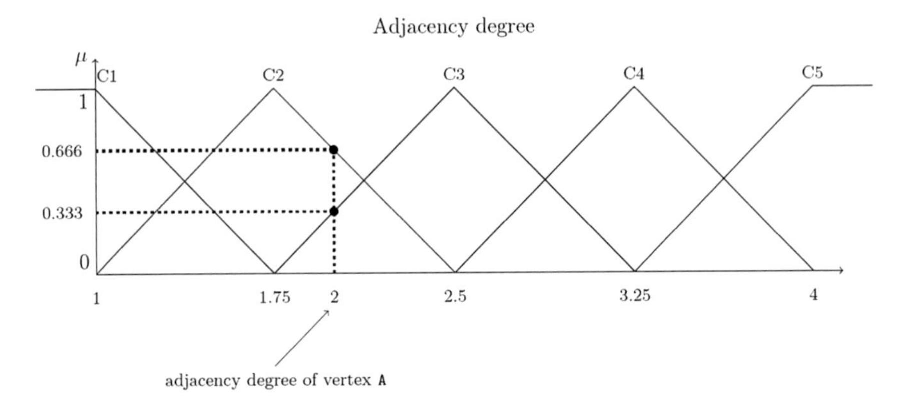

Fig. 6 Evaluation of the degree of membership of vertex A in all the classes

From Fig. 6 we can observe that vertex A only has two degrees of membership greater than zero (for classes C2 and C3) and three degrees of membership equal to zero (for classes C1, C4 and C5). This is so because the input $x = 2$ (adjacency degree of vertex A) only intersects the membership functions of classes C2 and C3. In order to obtain the degree of membership in class C2, we use Eq. 3 with $x = 2$, $a = 1$, $b = 1.75$, and $c = 2.5$ since $b \leq x \leq c$, that is:

$$\mu_{C2}(2) = \frac{c - x}{c - b} = \frac{2.5 - 2}{2.5 - 1.75} = 0.666.$$

Similarly, the degree of membership in class C3 can be computed by Eq. 3 with $x = 2$, $a = 1.75$, $b = 2.5$, and $c = 3.25$ since $a \leq x \leq b$, that is:

$$\mu_{C3}(2) = \frac{x - a}{b - a} = \frac{2 - 1.75}{2.5 - 1.75} = 0.333.$$

Table 1 Degrees of membership and classes assigned to the vertices

$v \in V$	μ_{C1}	μ_{C2}	μ_{C3}	μ_{C4}	μ_{C5}	Class
A	0	0.666	0.333	0	0	C2
B	0	0	0	0	1	C5
C	0	0.666	0.333	0	0	C2
D	0	0	0.333	0.666	0	C4
E	1	0	0	0	0	C1

Table 2 Degrees of membership and classes assigned to the vertices C and D at the fourth iteration of the algorithm FH1

$v \in V$	μ_{C1}	μ_{C2}	μ_{C3}	μ_{C4}	μ_{C5}	Class
C	0	0	0	0	1	C5
D	1	0	0	0	0	C1

Obviously, the degrees of membership in the remaining classes are all zero, that is, $\mu_{C1}(2) = \mu_{C4}(2) = \mu_{C5}(2) = 0$.

We compute the degrees of membership of the remaining vertices in all the classes. The class for some vertex $v \in V$ is the one with the largest degree of membership. Table 1 summarizes the degrees of memberships of each vertex and its corresponding class.

The algorithm FH1 then selects one vertex in the class C1 randomly. In this case C1 contains only one vertex (E), and hence, this vertex is selected and placed in the first position of the permutation. Thus, the sets of labeled and unlabeled vertices at this step are: $L = \{E\}$ and $U = \{A, B, C, D\}$ (see the line 5 of Algorithm 1).

The candidate vertices to be placed at the next available position of the permutation (position $k = |L| + 1 = 2$) are the unselected vertices adjacent to some vertex already labeled. In this case, the set of candidate vertices is $N = \{B\}$ (see the line 8 of Algorithm 1). Therefore, vertex B is placed at position $k = 2$. The algorithm updates the sets $L = \{B, E\}$ and $U = \{A, C, D\}$.

At the third iteration, the set of candidate vertices is $N = \{A, C, D\}$. In this case all the candidate vertex have an f-value equal to 2 (see Eq. 1).When this occurs, FH1 selects one vertex $v \in N$ randomly, for example the vertex A. Thus, vertex A is placed at position $k = 3$. The algorithm updates the sets $L = \{A, B, E\}$ and $U = \{C, D\}$.

At the fourth iteration, the set of candidate vertices is $N = \{C, D\}$. FH1 computes the f-value for each vertex in N. Thus, $f(C) = 3$ and $f(D) = 2$. Table 2 shows the degrees of membership and the classes for vertices C and D.

FH1 selects the vertex in C1 and places it at position $k = 4$ of the permutation. The last unselected vertex is placed at the last position. Finally, the solution constructed by FH1 is shown in Fig. 4b, i.e., $\varphi = (E, B, A, D, C)$.

4 Experimental Evaluation

In this section, the performance of our fuzzy heuristic proposed is assessed. We describe the test instances, the experimental environment and the experimental results in the following subsections.

4.1 *Test Cases*

In the next paragraphs, we describe four sets of test cases used in this experiment:

- **SMALL**. This set of test cases consists of 84 graphs whose optimal values are unknown. In this set the number of vertices ranges from 16 to 24 and the number of edges ranges from 18 to 49 [16].
- **TREE**. This set of test cases contains 50 different trees whose optimal values are known. The number of vertices varies from 22 to 202 and the number of edges varies from 21 to 201 [9].
- **HB**. This set of test cases consists of 62 instances whose optimal value is not known. The number of vertices and edges ranges from 24 to 960 and from 34 to 3721, respectively [9].
- **GRID**. This set of test instances consists of 52 instances of two dimensional meshes. The optimal objective value for these test cases ranges from 3 to 54. The number of vertices ranges from 9 to 2916 and the number of edges ranges from 12 to 5724 [9].

4.2 *Infrastructure*

The configuration shown below corresponds to the experimental conditions used in this study:

- **Software**: Microsoft Windows 10 Pro operating system; Java programming language (JRE 1.8.0_102) and Integrated Development Environment Eclipse 4.6.0.
- **Hardware**: Standard computer equipment with a processor Intel (R) Core (TM) i7-5500U CPU 2.4 GHz and RAM of 8 GB.

4.3 *Experimental Results*

Table 3 shows the comparison of the results obtained by the five variants of our fuzzy heuristic proposed. Each row of the table is a variant of FH. We measure the

Table 3 Average objective values and execution time obtained by our fuzzy heuristic

Heuristic		GRID (52)	SMALL (84)	TREE (50)	HB (62)
FH1	O.V.	28.50	3.68	6.17	58.28
	Time	0.04	0.00	0.00	0.01
FH2	O.V.	28.50	3.65	6.09	58.62
	Time	0.04	0.00	0.00	0.01
FH3	O.V.	28.51	3.65	6.11	59.22
	Time	0.04	0.00	0.00	0.01
FH4	O.V.	28.51	3.67	6.20	59.69
	Time	0.04	0.00	0.00	0.01
FH5	O.V.	121.56	6.28	10.73	141.83
	Time	0.11	0.00	0.00	0.02

solution quality, through the objective value (O.V.), and the execution time (Time) of each variant. The last four columns show each group of instances. We did not fix any kind of limitation for an early termination.

From the experimental results, we can see that our approach has a similar performance in the four first variants for all the datasets, according the two attributes measured. The variant FH5 obtained the worst solution quality.

5 Conclusions

In this chapter we have used fuzzy logic to solve the Vertex Separation Problem (VSP). As far as we know, this is the first time that fuzzy logic is used in the context of VSP. In particular, we propose a fuzzy heuristic (FH) to construct a solution for VSP. We have used fuzzy logic to classify the candidate vertices to be selected in the construction of the solution. The classifier uses five classes and their membership functions are all linear (L-function, R-function and triangular). We have proposed five variants of this approach, namely, FH1, FH2, FH3, FH4, FH5. The experimental results show that the first four variants are very close to each other in solution quality. The last variant (FH5) obtained the worst solution quality. The fuzzy heuristic proposed in this research can be easily adapted to other graph layout problems.

Acknowledgements The first author thanks Tecnológico Nacional de México and especially Instituto Tecnológico de Altamira for their support in this research. The second author would like to thank the CATEDRAS CONACYT program.

References

1. J. Díaz, J. Petit, M. Serna, A survey of graph layout problems. ACM Comput. Surv. (CSUR) **34**(3), 313–356 (2002)
2. T. Lengauer, Black-white pebbles and graph separation. Acta Informatica **16**(4), 465–475 (1981)
3. C.E. Leiserson, Area-efficient graph layouts, in *21st Annual Symposium on Foundations of Computer Science, 1980*, (IEEE, 1980, October), pp. 270–281
4. H. Bodlaender, J. Gustedt, J.A. Telle, Linear-time register allocation for a fixed number of registers. In *SODA*, vol. 98 (1998, January), pp. 574–583
5. A. Kornai, Z. Tuza, Narrowness, pathwidth, and their application in natural language processing. Discrete Appl. Math. **36**(1), 87–92 (1992)
6. I.C. Lopes, J.M. Carvalho, Minimization of open orders using interval graphs. Int. J. Appl. Math. **40**(4), 297–306 (2010)
7. G. Luque, E. Alba, Metaheuristics for the DNA fragment assembly problem. Int. J. Comput. Itell. Res. **1**(2), 98–108 (2005)
8. H.J.F. Huacuja, N. Castillo-García, Optimization of the Vertex Separation Problem with genetic algorithms. In *Handbook of Research on Military, Aeronautical, and Maritime Logistics and Operations* (IGI Global, 2016), pp. 13–31
9. A. Duarte, L.F. Escudero, R. Martí, N. Mladenovic, J.J. Pantrigo, J. Sánchez-Oro, Variable neighborhood search for the Vertex Separation Problem. Comput. Oper. Res. **39**(12), 3247–3255 (2012)
10. J. Sánchez-Oro, J.J. Pantrigo, A. Duarte, Combining intensification and diversification strategies in VNS. An application to the Vertex Separation Problem. Comput. Oper. Res. **52**, 209–219 (2014)
11. N. Castillo-García, H.J.F. Huacuja, R.A.P. Rangel, J.A.M. Flores, J.J.G. Barbosa, J.M.C. Valadez, On the exact solution of VSP for general and structured graphs: models and algorithms, in *Recent Advances on Hybrid Approaches for Designing Intelligent Systems* (Springer International Publishing, 2014), pp. 519–532
12. N. Castillo-García, H.J.F. Huacuja, R.A.P. Rangel, J.A.M. Flores, J.J.G. Barbosa, J.M.C. Valadez, Comparative study on constructive heuristics for the Vertex Separation Problem, in *Design of Intelligent Systems Based on Fuzzy Logic, Neural Networks and Nature-Inspired Optimization* (Springer International Publishing, 2015), pp. 465–474
13. H. Fraire Huacuja, N. Castillo-García, R.A. Pazos Rangel, J.A. Martínez Flores, J.J. González Barbosa, J.M. Carpio Valadez, Two new exact methods for the Vertex Separation Problem. Int. J. Comb. Optim. Prob. Inform. **6**(1), 31–41 (2015)
14. H.J. Fraire-Huacuja, N. Castillo-García, M.C. López-Locés, J.A.M. Flores, J.J.G. Barbosa, J. M.C. Valadez, Integer linear programming formulation and exact algorithm for computing pathwidth, in *Nature-Inspired Design of Hybrid Intelligent Systems* (Springer International Publishing, 2017), pp. 673–686
15. R. Sepúlveda, O. Montiel, O. Castillo, P. Melin, Fundamentos de Lógica Difusa. Ediciones ILCSA (2002)
16. J.J. Pantrigo, R. Martí, A. Duarte, E.G. Pardo, Scatter search for the cutwidth minimization problem. Ann. Oper. Res. **199**(1), 285–304 (2012)

The Deployment of Unmanned Aerial Vehicles at Optimized Locations to Provide Communication in Natural Disasters

Gabriela L. Rodríguez-Cortés, Anabel Martínez-Vargas and Oscar Montiel-Ross

Abstract Our economy and society depend on the continued operation of the internet and other networks. During a natural disaster, the communication infrastructure is affected and as a consequence interrupted. In such scenario, there is a vital need to maintain communication between first responders and victims. Recently, the use of Unmanned Aerial Vehicles (UAVs) has been proposed to deliver broadband connectivity since they are deployed quickly as aerial base stations to the affected area. However, figuring out the optimized locations of the UAVs is a difficult task due to a large number of combinations. To solve this, we apply a genetic algorithm with steady-state population model and binary representation with the aim of improving the network coverage.

Keywords Unmanned Aerial Vehicles · Genetic algorithms · Steady-state model Public safety communications

1 Introduction

Natural disasters have affected the population, infrastructure, environment, and economy. During these events the communication infrastructure is collapsed and as a consequence interrupted. For example, hurricane Katrina in 2005 destroyed up to

G. L. Rodríguez-Cortés · A. Martínez-Vargas
Universidad Politécnica de Pachuca, Carretera Pachuca - Cd. Sahagún km 20
Ex-Hacienda de Santa Bárbara, 43830 Zempoala, Hidalgo, Mexico
e-mail: lizeth.rodriguez.cortes@gmail.com

A. Martínez-Vargas
e-mail: anabel.martinez@upp.edu.mx

O. Montiel-Ross (✉)
Instituto Politécnico Nacional, Centro de Investigación y Desarrollo
de Tecnología Digital (IPN-CITEDI), Av. Instituto Politécnico Nacional
No. 1310 Col. Nueva Tijuana, 22435 Tijuana, Baja California, Mexico
e-mail: oross@ipn.mx

O. Castillo et al. (eds.), *Fuzzy Logic Augmentation of Neural and Optimization Algorithms: Theoretical Aspects and Real Applications*, Studies in Computational Intelligence 749, https://doi.org/10.1007/978-3-319-71008-2_36

2000 base stations (BSs) and the emergency 911 service was severely damaged, the devastation of communications infrastructure leave first responders without a reliable network for coordinating operations [1]. The Great East Japan earthquake and tsunami in 2011 damaged 1.9 million fixed telephone lines and 29,000 BSs. The emergency restoration took one month, while full restoration took 11 months [2]. In México, Odile in 2014 was the first major hurricane to strike the Baja California peninsula in 25 years and the most destructive tropical cyclone on record to affect this region. It devastated the region's electrical infrastructure and communications, therefore, the people were not able to request emergency services during the first critical hours or communicate with friends and family members for several weeks [3]. In such scenarios, there is a vital need to maintain communication between first responders and victims to save lives and reduce material damages. One of the solutions proposed is the installation of temporal networks that can be rapidly moved to the disaster zone and deployed within a reasonably short time to provide connectivity. In this context, the Unmanned Aerial Vehicles (UAVs) are well suited to achieve those commitments due to their mobility and self-organization capabilities [4, 5]. They can be raised in the sky in the aftermath of an emergency offering an agile and low-cost communication infrastructure [6]. However, one of the challenges is to determine the best possible UAVs placement to provide the largest number of services [7]. This problem belongs to the NP-hard class due to a large number of possible combinations of locations that a UAV can have. Metaheuristics allow tackling large-size problem instances by delivering satisfactory solutions in a reasonable time [8]. Among them, the population-based metaheuristics are iterative processes to improve a population of solutions that stop when a given condition is satisfied [8]. Genetic algorithms (GAs) [9] belong to this class; they are inspired by the evolution of populations. They evolve a population of solutions by applying selection, crossover and mutation operations.

Some works have addressed the deployment of UAVs at optimized locations. For example, work in [4] describes a scenario in which the infrastructure communication was damaged by a natural disaster. Hence, the UAVs are used to deliver communication between the first responders and the victims, and a brute force search technique is applied to find the optimized locations of the UAVs with the aim of maximizing the capacity of the network. However, it is well known that brute force is computationally costly due to it performs all possible combinations of locations to find a solution. In [10], the problem of finding the optimal position of an UAV for maximizing the data rate is posed. The UAV is used to extend the wireless coverage of the public safety network, optimizing the data rate between the BS and terminal device by applying an exhaustive search technique. Such exhaustive examination of all possibilities is prohibitively large. The aforementioned is a disadvantage particularly in scenarios that the time to provide a solution is critical such as in the case of a natural disaster. On the other hand, some other works as [11–13] show approaches for optimizing BS locations. Although the context is different aside from a natural disaster, the problem to solve is the same *i.e.*, to find out the optimal positions. GAs with a generational model are applied in works [11–13] to find the best location that maximizes coverage and minimize the number of BSs. The above objectives are treated separately *i.e.*, as a mono-objective problem. Additionally, in [13], the optimization problem of minimizing the BS power is also solved. Those works do not consider Quality of Service (QoS), an aspect that is needed to ensure reliable communication.

In this work, we solve the next problem: Given a number of limited resources (the UAVs), how many mobile users can be covered to provide wireless communication? To solve the problem we used a GA with a steady-state population model and binary representation with the aim of maximizing the network coverage. The GA with a generational model is the approach which after each generation, the whole population is replaced by offspring [14]. Each individual exists for just one cycle, and the parents are simply discarded, to be replaced by the entire set of offspring [14]. Unlike the GA with a generational model, we used the GA with a steady-state model in which the entire population is not changed at once, but rather a part of it. It overlaps populations where a single offspring is created and inserted in the population in each cycle [14]. By applying the steady-state model, we save time to provide a solution. That is advantageous in situations in which time to offer a solution is critical. We also consider Quality of Service (QoS) constraints to ensure communication.

This paper is organized as follows: Section 2 describes the problem formulation. Section 3 is devoted to the genetic algorithm used for solving the problem. Section 4 shows the achieved experiments and results. Finally, Sect. 5 concludes the work.

2 Problem Formulation

The typical scenario aftermath of an earthquake, tsunami, or hurricane is shown in Fig. 1. The figure illustrates a number of BSs collapsed, and the emergency services have some UAVs that can be used to maintain connectivity during a natural

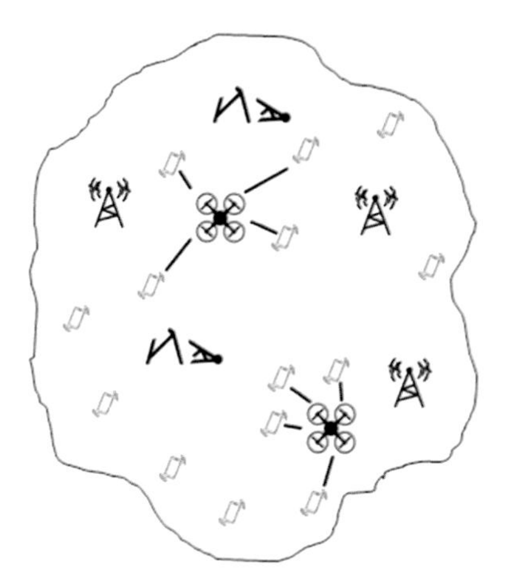

Fig. 1 A typical scenario aftermath of a natural disaster

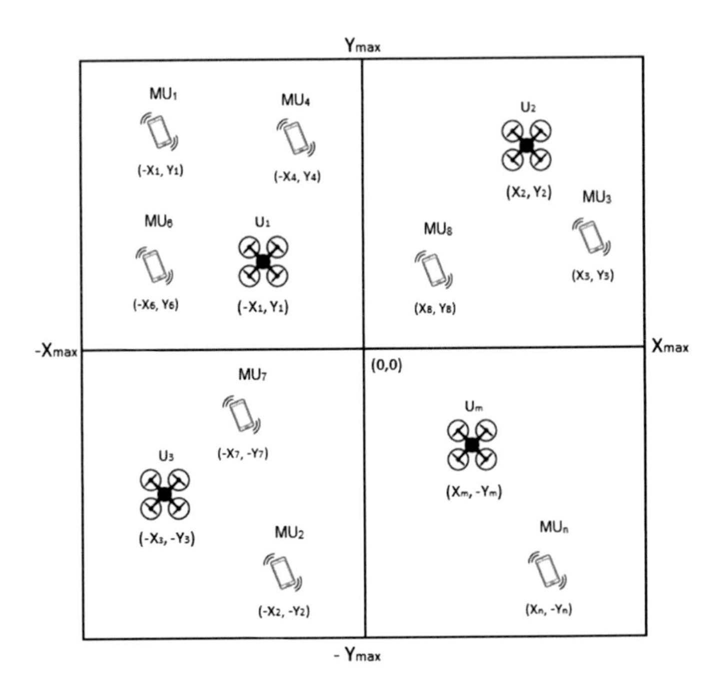

Fig. 2 Simulation scenario

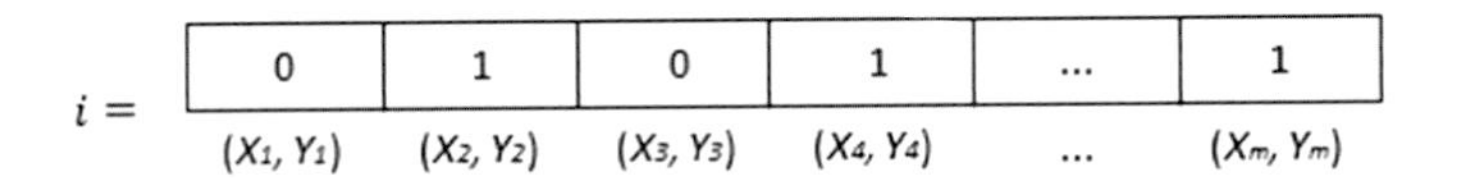

Fig. 3 Definition of individuals

disaster. To exploit efficiently the UAVs, they should be deployed at optimized locations. Once, the optimized locations are found, they are deployed quickly as LTE aerial BSs to the affected area. It is expected that soon the LTE technology becomes in the most widely deployed broadband communication system, that enables send and receive critical voice, video, and data to save lives [5].

Figure 2 shows the simulation scenario. We consider a two-dimensional area limited by $[-X_{max}, X_{max}]$ and $[-Y_{max}, Y_{max}]$ with the origin at (0, 0). It represents the affected area by a natural disaster. A mobile user, MU_n, is randomly placed over the area; therefore $MU_n = (X_n, Y_n)$ where n is the index of mobile users. In contrast, an UAV, U_m, is also deployed randomly over the area; therefore $U_m = (X_m, Y_m)$ where m is the index of UAVs.

The individual i is represented by a binary string where each bit denotes if the mth UAV is selected ($U_m = 1$) *or* if it remains out ($U_m = 0$). The Fig. 3 illustrates the individual i; each one has a different configuration (placement) of UAVs. The coordinates of every U_m in the individual i are kept in an additional vector.

To find the locations that maximize UAVs coverage c, we define the following objective function:

$$\text{Maximize} \quad c = \frac{V}{N} \tag{1}$$

where V is the number of mobile users covered by UAVs, and N is the total number of mobile users inside the affected area.

To calculate the number of mobile users covered by a UAV, it is necessary to analyze the phenomena affecting a signal. In this work, we perform the downlink analysis and consider the path loss Lp (dB) and Signal-to-Interference ratio SIR (dB). The path loss between MU_n and U_m is computed by using the Hata's model for an urban area:

$$Lp_{nm} = A + B\log_{10}(d_{nm}) \tag{2}$$

where d_{nm} is the Euclidian distance between MU_n and U_m. A and B are constant values defined by the carrier frequency and antenna heights [15]. The SIR is a measure that expresses how much interference is perceived by a receiver on a current link from the set of active transmitters, which use the same channel. In any wireless network, SIR level indicates the signal quality. Wireless applications must meet minimum signal quality levels in order to work properly. The above is known as QoS, *e.g.*, the voice for multimedia requires SIR values between 3 and 10 dB to guarantee service. Then, the SIR at receiver MU_n can be expressed as:

$$SIR_n = \frac{PR_n}{I_{Total}} \tag{3}$$

where PR_n is the power reception of MU_n and I_{Total} is the total interference, *i.e.*, the interference generated by others MU_n that use the same channel. We assume that a U_m transmits at power PT_m, then, the PR_n is calculated by:

$$PR_n = PT_m - Lp_{nm} \tag{4}$$

A MU_n is associated with an U_m as long as it meets the following restrictions:

$$R_{nm} \leq 2 \text{ km} \tag{5}$$

$$Lp_{nm} < 120 \text{ dB} \tag{6}$$

$$SIR_n \geq 3 \text{ dB} \tag{7}$$

$$MU_n = 0 \tag{8}$$

The coverage radius R_{nm} in (5) is the Euclidian distance between MU_n and U_m. The values thresholds R_{nm} in (5) and Lp_{nm} in (6) were derived by a prior characterization of several scenarios to reduce the interference. Restriction in (7) ensures a successful voice communication. In contrast, restriction in (8) ensures that a MU_n is not associated with more than one U_m. If a MU_n has zero value, indicates that it is not associated with any U_m. On the other hand, if it has a value different to zero (the index of UAVs, m), it indicates that the MU_n is already associated with an U_m.

3 Algorithm to Optimize Locations Based on GA

A GA represents a particular class of evolutionary algorithm that uses techniques inspired by evolution [14]. It is a population-based model that applies selection and recombination operators to generate new sample points in a search space [16]. Therefore, in order to find solutions to optimization problems, a GA evolves a population of solutions by applying selection, crossover and mutation operations.

Mainly, two different models of population management are applied to GAs: the generational model and the steady-state model [14]. In the generational model, the population, mating pool and offspring are all the same size, so that each generation is replaced by the offspring. In contrast, in the steady-state model, the entire population is not changed at once, but rather a part of it. The proportion of the population that is replaced is known as the generational gap and is equal to offspring/population. The steady-state model was first introduced by Whitley's GENITOR algorithm [17]. The idea is to iteratively breed a new child or two and reintroduce then directly into the population; therefore, there are no generations.

We use the steady-state GA proposed in [18], there the tournament selection is applied to select two parents. Next, two offspring are created, and then the two best individuals out of the two parents and two offspring are reintroduced into the population.

The complete procedure of deploying UAVs at optimized locations is proposed as follows:

Input: Total number of mobile users N, Total number of UAVs M, population size P, crossover probability P_c, mutation probability P_m, the total number of cycles T_c.
Output: Maximum coverage c.

1: Initialize population with random candidate solutions.
2: **do**
3: Select two parents by using tournament selection.
4: Recombine pairs of parents.
5: Mutate the two-resulting offspring.
6: Evaluate parents and offspring in Eq. (1).

7: Select the two best individuals out of the two parents and two offspring. Call those best individuals, *best1* and *best2*.
8: Replace parents with *best1* and *best2* respectively.
9: Keep the fittest individual from the population.
10: **while** (*number_of_cycles* < T_c)
11: Select the fittest individual from the total number of cycles.

Firstly, a scenario is created by locating randomly N mobile users and M UAVs over a grid. After that, in step 1, the individuals are initialized randomly with 0's and 1's. Then in step 3, two individuals are randomly picked and the winner of these two individuals is selected as a parent. The above is the tournament selection [14] and it is performed to select two parents. Next, in step 4, two new individuals (offspring) are created using the information of the two parent solutions as long as a random number in [0, 1] < P_c. In this context, we performed two-point crossover [14]. Some kind of randomized change is introduced at the offspring in step 5 by allowing each bit to flip with a small probability P_m. As indicated in step 6, the two parents and the two offspring are evaluated according to the objective function posed in (1). Then, in step 7, the two best individuals out of the two parents and two offspring are keeping. Those best individuals are called *best1* and *best2*. In step 8, parents are replaced by *best1* and *best2* respectively. In step 9, the fittest individual from the population is kept. Finally, in step 10, if the total number of cycles T_c is not accomplished, another cycle is performed. When the total number of cycles T_c is met, the solution of the problem is the fittest individual from the total number of cycles T_c. That solution contains the optimized locations of the UAVs.

4 Simulation Results

This section shows numerical results to demonstrate the performance of the proposed algorithm to deploy UAVs at optimized locations based on GA. The simulation parameters used for the GA are shown in Table 1, we applied the GA parameters configuration proposed by DeJong [19].

Table 2 shows the experimental specific problem parameters setup used in the experiments.

We performed six experiments as shown in Table 3. Experiments A, B and C represent the worst case or condition, *i.e.*, the situation in which each U_m has one channel to serve the mobile users. Those experiments differ from each other just for the number of UAVs provided to cover the affected area. In contrast, experiments D, E and F are those which have more resources or channels per U_m, *i.e.*, more than one channel to serve the mobile users. There is a relation between the number of channels and the number of mobile users served: the larger the number of channels in a UAV is, the larger the number of mobile users served is. Each experiment was

Table 1 Parameters used for GA

Parameters	Values
Representation	Bit-strings
Crossover	Two-point crossover
Mutation	Bit flip
Parent selection	Tournament
Crossover probability P_c	0.6
Mutation probability P_m	0.001
Population size P	50
Total number of cycles T_c	1000

Table 2 Parameters used for experiments

Parameters	Values
Area (km^2)	10 km × 10 km
Total number of mobile users N	500
Total number of UAVs M	5, 7, 10
Total number of channels per U_m	1, 5
U_m antenna height	150 m
MU_n antenna height	1.5 m
Carrier frequency	700 MHz

Table 3 List of experiments performed

Experiment	M	Total number of channels per U_m
A	5	1
B	7	1
C	10	1
D	5	5
E	7	5
F	10	5

executed 100 times and the locations of the mobile users were kept fixed over the affected area. As we mentioned in Sect. 2, each individual had a different placement of UAVs.

Table 4 shows the results for the worst cases. From each experiment was selected the best solution found and the worst solution found, additionally, the average was calculated from the 100 runs. The best solution found represents the run with the best fitness from the 100 runs; therefore, it is the best deployment. The worst solution found is the run which has the worst fitness; therefore, it is the worst placement. In Table 4 can be observed that as the number of UAVs increases, the number of mobile users served increases as well. That is an expected behavior since more UAVs were deployed at each experiment to serve more mobile users.

On the other hand, Table 5 shows the results when the UAVs are provided with better conditions *i.e.*, more than one channel. In such experiments, the results

Table 4 Results under the worst cases

Experiment	Best solution found	Worst solution found	Average
A	0.38	0.28	0.35
B	0.49	0.39	0.45
C	0.68	0.49	0.58

Table 5 Results with more than one channel

Experiment	Best solution found	Worst solution found	Average
D	0.48	0.32	0.40
E	0.54	0.39	0.49
F	0.89	0.51	0.70

calculated by the algorithm based on GA, also behave accordingly. As the number of UAVs is increased in the affected area, more mobile users are covered. Experiment *F* represents the best deployment since it has a coverage rate = 0.89. In average, it provides a coverage rate = 0.70.

Figure 4 shows the optimized locations of the UAVs for the experiments *A*, *B*, and *C* for the best solution found, reported in Table 4. In Fig. 5 the optimized locations of the UAVs for the experiments *D*, *E*, and *F* for the best solution found (reported in Table 5) are shown. As the number of UAVs and channels increase, the coverage also does.

In Fig. 6, a scenario for verification is proposed to testify the quality of the solutions provided by the algorithm based on GA. We analyzed the ideal scenario *i.e.*, a number of UAVs were placed to cover the complete affected area (10 km × 10 km). Then the next parameters values are used: the total number of mobile users $N = 500$, the total number of UAVs $M = 25$, the U_m antenna height = 150 m, the MU_n antenna height = 1.5 m, the carrier frequency = 700 MHz, the coverage radius $R_{nm} = 1$ km, and one channel per U_m. After 100 runs, the 25 UAVs covered the 95% (0.95) of the affected area. Therefore, one UAV covers approximately the 3.8% of the whole scenario under the ideal conditions. Now, considering the experiment *A* with the worst condition (just one channel per U_m and $M = 5$) in average from the 100 runs, one UAV offers a coverage equals to 7.13% of the whole scenario. Clearly, the above is superior to the average offered per UAV in the ideal scenario.

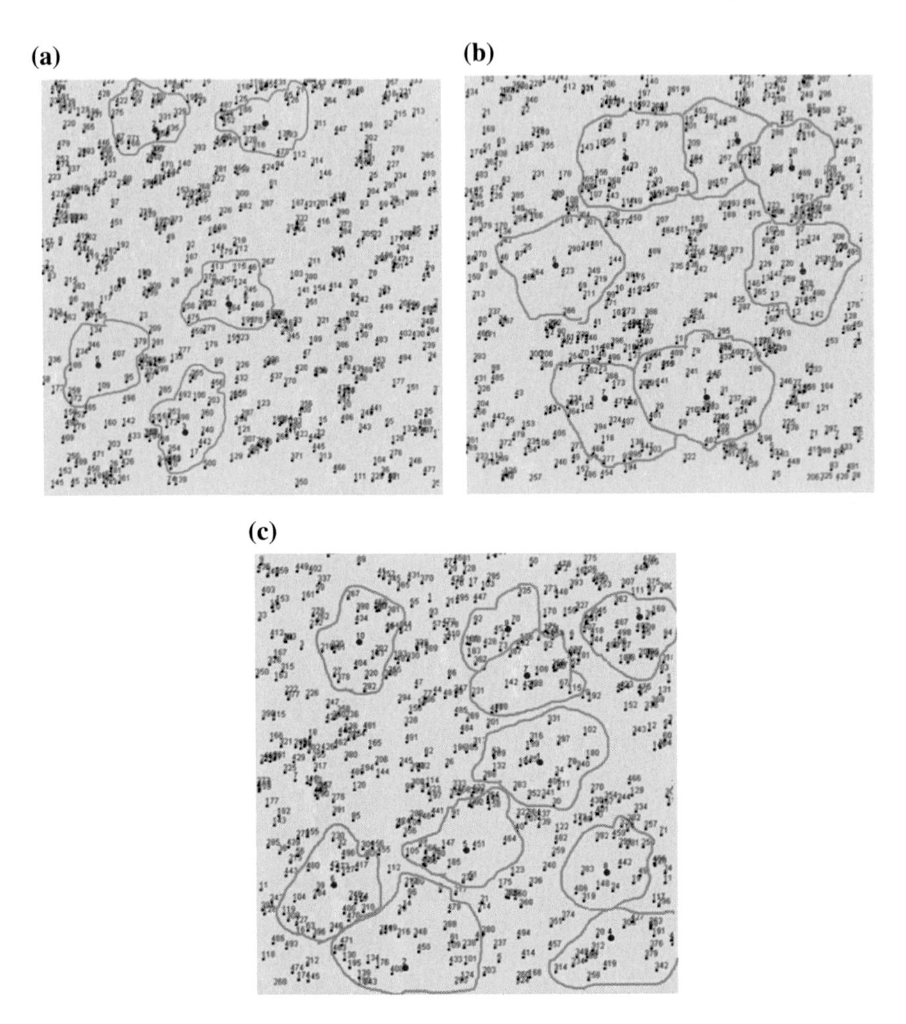

Fig. 4 Simulation results of the experiments *A*, *B*, *C*. **a** Best solution found at experiment: *A*, coverage: 0.38, **b** Best solution found at experiment: *B*, coverage: 0.49, **c** Best solution found at experiment: *C*, coverage: 0.68

5 Conclusions

In this work, we presented a solution to the problem of providing communication to the maximal number of mobile users with a limited number of UAVs. This could be of vital importance in the context of a natural disaster in which the UAVs can be

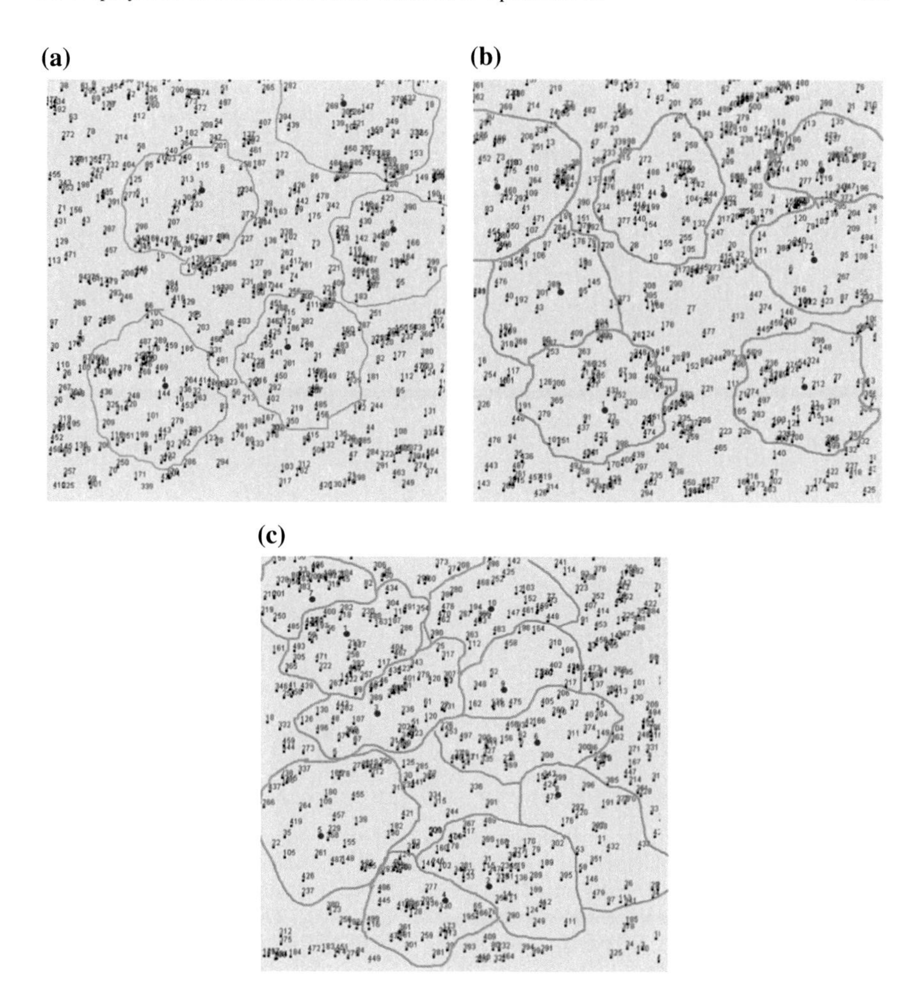

Fig. 5 Simulation results of the experiments *D*, *E*, *F*. **a** Best solution found at experiment: *D*, coverage: 0.48, **b** Best solution found at experiment: *E*, coverage: 0.54, **c** Best solution found at experiment: *F*, coverage: 0.89

raised in the sky to offer an agile and low-cost communication infrastructure to the affected area. We developed a solution procedure based on GA with a steady-state population model and binary representation that was evaluated. It showed to produce better results considering the ideal scenario. As a future work, we will implement floating-point representation in order to compare the quality of the solutions' performance.

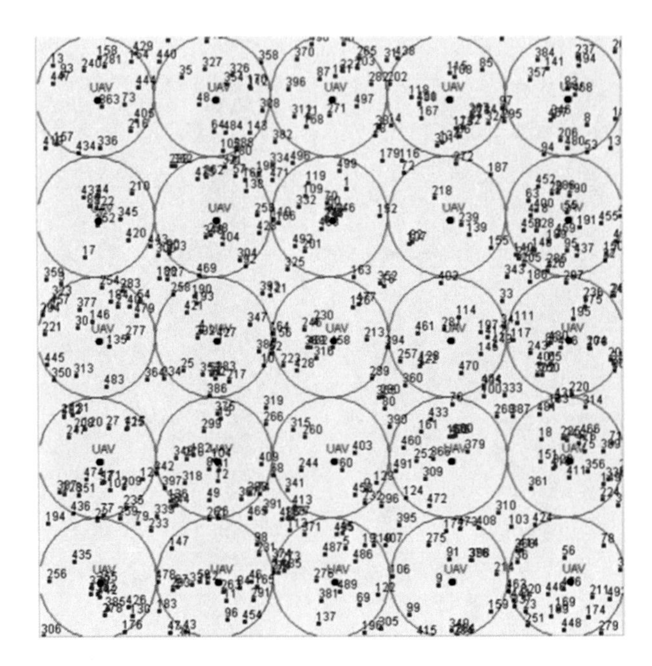

Fig. 6 Scenario for verification

References

1. R. Miller, Hurricane Katrina: communications & infrastructure impacts. (2006)
2. T. Sakano, Z.M. Fadlullah, T. Ngo, H. Nishiyama, M. Nakazawa, F. Adachi, N. Kato, A. Takahara, T. Kumagai, H. Kasahara, S. Kurihara, Disaster-resilient networking: a new vision based on movable and deployable resource units. IEEE Netw. **27**, 40–46 (2013)
3. J.P. Cangialosi, T.B. Kimberlain, Hurricane Odile (EP152014). National Hurricane Center (2015)
4. A. Merwaday, I. Guvenc, UAV assisted heterogeneous networks for public safety communications, in 2015 IEEE Wireless Communications and Networking Conference Workshops (WCNCW). pp. 329–334 (2015)
5. A. Kumbhar, F. Koohifar, İ. Güvenç, B. Mueller, A survey on legacy and emerging technologies for public safety communications. IEEE Commun. Surv. Tutor. **19**, 97–124 (2017)
6. S. Chandrasekharan, K. Gomez, A. Al-Hourani, S. Kandeepan, T. Rasheed, L. Goratti, L. Reynaud, D. Grace, I. Bucaille, T. Wirth, S. Allsopp, Designing and implementing future aerial communication networks. IEEE Commun. Mag. **54**, 26–34 (2016)
7. Y. Zeng, R. Zhang, T.J. Lim, Wireless communications with unmanned aerial vehicles: opportunities and challenges. IEEE Commun. Mag. **54**, 36–42 (2016)
8. E.-G. Talbi, Metaheuristics: from design to implementation. (Wiley, 2009)
9. J.H. Holland, *Adaptation in Natural and Artificial Systems: An Introductory Analysis with Applications to Biology, Control and Artificial Intelligence* (MIT Press, Cambridge, MA, USA, 1992)
10. X. Li, D. Guo, H. Yin, G. Wei, Drone-assisted public safety wireless broadband network, in 2015 IEEE Wireless Communications and Networking Conference Workshops (WCNCW). pp. 323–328 (2015)

11. B.-S. Park, J.-G. Yook, H.-K. Park, The determination of base station placement and transmit power in an inhomogeneous traffic distribution for radio network planning, in *Proceedings IEEE 56th Vehicular Technology Conference*, vol. 4, pp. 2051–2055 (2002)
12. C.A. Brizuela, E. Gutiérrez, An experimental comparison of two different encoding schemes for the location of base stations in cellular networks, in *Applications of Evolutionary Computing*, ed. by S. Cagnoni, C.G. Johnson, J.J.R. Cardalda, E. Marchiori, D.W. Corne, J.-A. Meyer, J. Gottlieb, M. Middendorf, A. Guillot, G.R. Raidl, E. Hart (Springer, Berlin Heidelberg, 2003), pp. 176–186
13. Y.S. Choi, K.S. Kim, N. Kim, The displacement of base station in mobile communication with genetic approach. EURASIP J. Wirel. Commun. Netw. **2008**, 580761 (2008)
14. A.E. Eiben, J.E. Smith, *Introduction to Evolutionary Computing* (Springer Berlin Heidelberg, Berlin Heidelberg, 2015)
15. S.S. Haykin, M. Moher, in *Modern Wireless Communications* (Pearson Prentice Hall, 2005)
16. D. Whitley, A genetic algorithm tutorial. Stat. Comput. **4**, 65–85 (1994)
17. D. Whitley, J. Kauth, Genitor: a different genetic algorithm, in *Rocky Mountain Conference on Artificial Intelligence* (Denver, CO, 1988)
18. N. Holtschulte, M. Moses, Should every man be an island? Presented at the GECCO 2013, Amsterdam, The Netherlands (2013)
19. S. Sumathi, S. Paneerselvam, *Computational Intelligence Paradigms: Theory & Applications Using MATLAB* (CRC Press Inc, Boca Raton, FL, USA, 2010)

A Comparison of Image Texture Descriptors for Pattern Classification

Valentín Calzada-Ledesma, Héctor José Puga-Soberanes, Alfonso Rojas-Domínguez, Manuel Ornelas-Rodriguez, Martín Carpio and Claudia Guadalupe Gómez

Abstract Texture classification is a problem widely studied in computer vision, there exist two fundamental issues: how to describe texture images and how to define a similarity measure. The texture descriptors are mainly used to extract and represent the features of texture images and their performance is usually measured using a classification algorithm. In this paper, some of the most referenced texture descriptors, such as Gabor filter banks, Wavelets, and Local Binary Patterns, are compared using non-parametric statistical tests to know if there is a difference in performance. The descriptors are applied to five well-known texture image datasets, in order to be classified. Three classification algorithms, with a cross-validation scheme, are used to classify the described texture datasets. Finally, a Friedman test with multiple comparisons is used to compare the whole performance of the texture descriptors on a statistical basis. The statistical results suggest that for these tests there is a difference in performance, so it was possible to determine statistically, for the considered experimental settings, the best texture descriptor.

Keywords Texture classification · Local binary patterns · Discrete wavelet transform · Gabor filter banks

1 Introduction

A texture can be seen as an arrangement of intensities in an image whose spatial distribution creates visual patterns with distinctive characteristics, such as roughness, repetitiveness, directionality, and granularity [1]. There exist different

V. Calzada-Ledesma (✉) · H. J. Puga-Soberanes · A. Rojas-Domínguez · M. Ornelas-Rodriguez · M. Carpio
Tecnológico Nacional de México- Instituto Tecnológico de León, León Guanajuato, Mexico
e-mail: valecalzada@hotmail.com

C. G. Gómez
Tecnológico Nacional de México- Instituto Tecnológico de Ciudad Madero, Cd. Madero Tamaulipas, Mexico

O. Castillo et al. (eds.), *Fuzzy Logic Augmentation of Neural and Optimization Algorithms: Theoretical Aspects and Real Applications*, Studies in Computational Intelligence 749, https://doi.org/10.1007/978-3-319-71008-2_37

approaches to extract and represent textures; they can be classified into texture signatures, frequency-based models and space-based models [2]. These approaches are known as texture descriptors and depending on textures problem to classify, a proper selection of the descriptors can significantly improve the classification performance [3–5]. Thus, a comparative evaluation between different texture descriptors is necessary in order to choose the most appropriate descriptor for a particular application.

This paper presents a comparative evaluation and statistical study of four texture descriptors: Gabor Filter Banks (GFB), Haar Wavelet Transform (HWT), Daubechies-4 wavelet transform (Daub-4) and Local Binary Patterns (LBP). The comparison is made by computing the classification accuracy of numerical datasets generated by the texture descriptor, and by means of three classification algorithms: Support Vector Machine (SVM), *K*-Nearest Neighbor (*K*-NN) and Naïve Bayes (NB). Then, a statistical study is performed by means of the Friedman test to compare the performance of the texture descriptors.

Texture descriptors are applied in texture datasets containing, in some cases, more than 4000 images with interesting features and challenges, such as viewpoint changes, differences in scale, illumination, and rotation.

The rest of this paper is organized as follows. In Sect. 2 the texture descriptors definitions and the theoretical basis of the used descriptors are presented. In Sect. 3 the methodology of texture description and the classification process are described. Section 4 shows the experimental results and the statistical significance. Section 5 presents a discussion of the results. Finally, in Sect. 6, the conclusions are presented.

2 Texture Descriptors

A texture descriptor is a way of quantitatively representing the texture characteristics of an image. In general, a descriptor D is made up by the tuple (γ_D, δ_D) [6], where:

$\gamma_D : \{I\} \rightarrow \mathbb{R}^n$ is a function which extracts a feature vector $\vec{V} \in \mathbb{R}^n$ from an imagen I.

$\delta_D : \mathbb{R}^n \times \mathbb{R}^n \rightarrow \mathbb{R}$ is a similarity function that computes the similarity between two images through the similarity between their corresponding feature vectors.

The next sections describe the texture descriptors used in this research.

2.1 Gabor Filter Banks

A 2-D Gabor filter is a linear filter which impulse response is a sinusoidal function modulated by a Gaussian envelope [7]. The Gabor functions are defined within the spatial domain as well as the frequency domain. The general form of a Gabor function in the spatial domain is given by:

$$g(x,y) = \frac{F^2}{\pi\gamma\eta} e^{-F^2\left[\left(\frac{x'}{\gamma}\right)^2 + \left(\frac{y'}{\eta}\right)^2\right]} e^{i2\pi F x'}. \tag{1}$$

with:

$$\begin{aligned} x' &= x \cos\theta + y \sin\theta, \\ y' &= -x \sin\theta + y \cos\theta. \end{aligned} \tag{2}$$

where:

F: is the central frequency of the filter.
θ: is the angle between the direction of propagation of the sinusoidal wave and the x axis.
γ and η: are the standard deviations of the Gaussian envelope in dimensions x and y, respectively.

A Gabor filter bank (GFB) consists of applying several Gabor filters with different orientations and frequencies to extract features of an image, by the proper selection of the filter parameters mentioned above. Depending on the parameters, the filters will analyze in different ways the spatial and frequency domain [7, 8].

Different approaches have been used in the state-of-the-art to tackle the problem of Gabor parametrization and its influence in different tasks, such as classification and segmentation [8, 9].

2.2 Wavelet Transform

Wavelets are functions which are formed by two resulting coefficients, given by the wavelet function $\psi(t)$ and the scaling function $\varphi(t)$ in the time domain. The wavelet function applies a low pass filter to the signal $u(t)$ resulting in the *approximation* coefficients and the scaling function applies a high pass filter which results in the *detail* coefficients [10].

In general, the wavelet transform tries to express a continuous signal $u(t)$ in time, by means of an expansion of coefficients, proportional to the inner product between the signal and different scaled and translated versions of the “prototype” function, better known as mother wavelet [11], given by:

$$\psi_{a,b} = \frac{1}{\sqrt{a}}\psi\left(\frac{t-b}{a}\right). \tag{3}$$

where a and b are the scaling and shifting parameters, respectively.

The coefficients of the transform are the result of the inner product (Eq. 4) between the function to transform and the considered base function [11].

$$W_u(a,b) = \langle u(t), \psi_{a,b}(t) \rangle \tag{4}$$

In computer science, the wavelet transform has been employed through discretely sampled wavelets (Discrete Wavelet Transform), such as the Haar and Daubechies wavelet transforms.

2.2.1 Haar Wavelet Transform

The Haar Wavelet Transform (HWT) is the simplest of the discrete wavelet transformations. In HWT a function is multiplied against the Haar wavelet with different shifts and scales [12]. The Haar wavelet function is given by:

$$\psi(t) = \begin{cases} 1 & 0 \le t < \frac{1}{2}, \\ -1 & \frac{1}{2} \le t < 1, \\ 0 & \text{otherwise.} \end{cases} \tag{5}$$

Its scaling function is given by:

$$\phi(t) = \begin{cases} 1 & 0 \le t < 1, \\ 0 & \text{otherwise.} \end{cases} \tag{6}$$

2.2.2 Daubechies Wavelet Transform

The Daubechies wavelets are a family of orthogonal wavelets with compact support and a varying number of vanishing moments. These define discrete wavelet transformations [12]. The Daubechies scaling coefficients are defined by:

$$\sum_{k=0}^{h_k-1} (-1)^k h_k k^m = 0 \tag{7}$$

where:

k : Is the parameter of vanishing moments.
h_k : Are the scaling coefficients.

For integers $m = 0,\ 1,\ 2, \ldots \frac{N_k}{2} - 1$.

The scaling equation for a four-coefficients wavelet is [12]:

$$\phi(t) = h_0\phi(2t) + h_1\phi(2t-1) + h_2\phi(2t-2) + h_3\phi(2t-3) \tag{8}$$

And, the corresponding wavelet function is given by:

$$\psi(t) = h_3\phi(2t) - h_2\phi(2t-1) + h_1\phi(2t-2) - h_0\phi(2t-3) \tag{9}$$

For the Daubechies-4 wavelet transform (Daub-4), to find h_k, the following nonlinear system of equations must be solved:

$$\begin{aligned} h_0 + h_1 + h_2 + h_3 &= 2 \\ h_0^2 + h_1^2 + h_2^2 + h_3^2 &= 2 \\ h_0 - h_1 + h_2 - h_3 &= 0 \\ -1h_1 + 2h_2 - 3h_3 &= 0 \end{aligned} \tag{10}$$

2.3 *Local Binary Patterns*

The Local Binary Patterns (LBP) descriptor extracts micropatterns that are invariant to local greyscale variations in the image. Given a pixel c with gray value g_c, the LBP is computed by thresholding the difference between the gray values of its p neighbors $\{g_n\}_{n=0}^{p-1}$ (distributed on a circle of radius r pixels) and g_c by using the step function $s(x)$. A binary vector of p bits is extracted by concatenating the binary gradient directions [13, 14]. The LBP operator is given by:

$$LBP_{r,p}(c) = \sum_{i=0}^{p-1} s(g_i - g_c)2^i \tag{11}$$

where

$$s(x) = \begin{cases} 1 & x \geq 0 \\ 0 & x < 0 \end{cases} \tag{12}$$

3 Methodology

GFB, HWT, Daub-4 and LBP descriptors were compared against each other. These descriptors were applied in the next well-known benchmark texture datasets [15]:

UIUC database: The database includes textured surfaces such as wood, marble, gravel, and fur, as well as a mixture of them.

UMD tex: Is a dataset of high-resolution images, including floor textures, fruits, various plants, buckets and shelves of bottles.
Kylberg texture: This dataset contains textured surfaces including fabrics, stone, grains, sesame seeds and lentils.
KTH-TIPS: Images of materials are presented in this dataset, such as crumpled aluminum foil, sandpaper, sponge, styrofoam, linen, corduroy, cotton, brown bread, orange peel, and cracker.
KTH-TIPS 2b: This dataset is an extension of KTH-TIPS dataset and provides multiple different samples of the different materials in the latter.

Table 1 shows the description of the datasets.

The experimental methodology used in this paper consists of the following steps:

1. All images were resized to 256 × 256 pixels by means of linear interpolation, to reduce computational cost.
2. All datasets were processed by all of the texture descriptors considered. The parameter values of the texture descriptors are shown in Table 2.
3. Depending on the texture descriptor, the feature vectors are computed and from these, a numerical dataset is created and labeled. The feature representation used for each texture descriptor is shown below.

GFB: Given the filter parameters mentioned above, the number of Gabor Filters in the GFB (T_f) is given by $n_f \times n_o = 40$. The Gabor transform of an input image (I) is computed for each filter f_i of the bank, where $i \in \{1, \ldots, T_f\}$. The mean u_{f_i} and the standard deviation σ_{f_i} of each transformed image are used as elements of the feature vector $(\vec{V})$ [7].

$$\vec{V} = \left(u_{f_1}, \sigma_{f_1}, u_{f_2}, \sigma_{f_2}, u_{f_3}, \sigma_{f_3}, \ldots, u_{f_{T_f}}, \sigma_{f_{T_f}}\right). \tag{13}$$

Table 1 Description of the texture datasets

Dataset	Number of images	Number of classes	Resolution	Characteristics
UIUC	1000	25	640 × 480	Viewpoint change, rotations, differences in scale and illumination
UMD tex	1000	25	1280 × 960	High resolution, viewpoint changes, differences in illumination
Kylberg texture	4480	28	576 × 576	Rotations
KTH-TIPS 1	810	10	200 × 200	Rotations, differences in scale and illumination
KTH-TIPS 2b	4752	44	200 × 200	Rotations, differences in scale and illumination

Table 2 Parameter values of the texture descriptors

GFB	Num. of frequencies (n_f)	5
	Maximum frequency	0.25
	Frequency ratio	$\sqrt{2}$
	Orientations (n_o)	0°, 23°, 45°, 68°, 90°,113°, 135°, 158°
HWT	Scale	3
Daub-4	Scale	3
LBP	Radius (r)	1

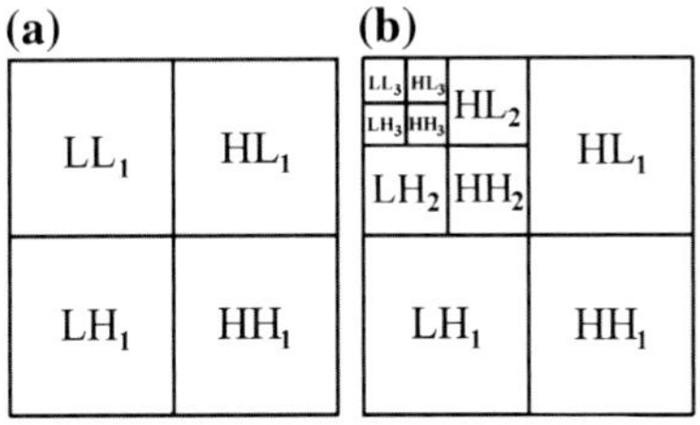

Fig. 1 **a** The resulting wavelet coefficients contained in the four bands, **b** Pyramidal decomposition of the LL band at scale $k = 3$

HWF and Daub-4: The application of the HWF or Daub-4 in a texture image results in a transformed image, as shown in Fig. 1a. The resulting 2-D array of wavelet coefficients contains 4 bands of data *LL*, *HL*, *LH* and *HH*, where *L* is a Low-pass filter and *H* is a High-pass filter [16].

The *LL* band can be processed once again by the wavelet transform, producing even more subbands. At scale $k = 3$, the resulting pyramidal decomposition is shown in Fig. 1b.

The mean $u_{subband}$ and the standard deviation $\sigma_{subband}$ of each subband are used as elements of the feature vector $(\vec{V})$ [16, 17], given in Eq. 14.

$$\vec{V} = (\mu_{LH_1}, \sigma_{LH_1}, \mu_{HH_1}, \sigma_{HH_1}, \mu_{HL_1}, \sigma_{HL_1}, \ldots \\ \ldots, \mu_{LH_k}, \sigma_{LH_k}, \mu_{HH_k}, \sigma_{HH_k}, \mu_{HL_k}, \sigma_{HL_k}). \tag{14}$$

LBP: Given a texture image, an LBP pattern can be computed at each central pixel *c* in a neighborhood. The texture image can be characterized by the distribution of LBP patterns. The histogram of these binary numbers is then used as feature vector [13, 14].

4. The labeled numerical datasets are classified by three classification algorithms: SVM with a linear kernel, *K*-NN (with $K = 1$) and NB. To ensure that the models are robust, a cross-validation scheme is used with 10-folds.

4 Experimental Results

Tables 3, 4, and 5 show the average classification accuracy of the 10-folds for each dataset, computed by SVM, 1-NN, and NB respectively. The values in bold typeface are the best results for each dataset.

The above results show that for these experiments, the LBP descriptor could be the best. However, to provide a statistical support of the results, a Friedman Test is performed.

4.1 Statistical Significance of the Results

The Friedman test is used to determine if the classification accuracy values reported for the different methods are drawn from the same distribution [18]. The null hypothesis in this test is that these values follow the same distribution (there is no difference in performance, Eq. 15) when different texture descriptors are used to characterize texture datasets.

Table 3 Average classification accuracy obtained by SVM

SVM				
	HWT	Daub-4	LBP	GFB
UIUC	47.6	50.5	**78.1**	50.2
UMD Tex	76.6	77.4	**97.5**	72.37
Kylberg Tex	93.37	91.89	**99.91**	80.66
KTH-TIPS	61.72	53.08	**93.95**	73.08
KTH-TIPS2	39.01	36.09	**86.01**	45.95

Table 4 Average classification accuracy obtained by 1-NN

1-NN				
	HWT	Daub-4	LBP	GFB
UIUC	69.1	44.4	**70.4**	53.1
UMD Tex	92.2	69	**96.7**	75.9
Kylberg Tex	98.81	78.39	**99.48**	85.44
KTH-TIPS	87.16	65.06	**95.92**	86.17
KTH-TIPS2	71.33	39.96	**85.50**	67.23

Table 5 Average classification accuracy obtained by Naïve Bayes

NB				
	HWT	Daub-4	LBP	GFB
UIUC	51.5	44.4	**54.8**	38.9
UMD Tex	71.8	70.1	**85.6**	52.1
Kylberg Tex	91.56	84.55	**92.34**	68.61
KTH-TIPS	48.14	35.30	**74.93**	52.71
KTH-TIPS2	31.25	22.51	**62.77**	27.86

Table 6 Friedman test parameters

Columns (***k***)	Rows (n)	Significance level (α)	Critical value
4	5	0.05	7.8

Table 7 Average ranking of the descriptors using different classification algorithms

Texture descriptor	Ranking (SVM)	Ranking (1-NN)	Ranking (NB)
HWT	3	2	2.2
Daub-4	3	4	3.3999
LBP	**1**	**1**	**1**
GFB	3	3	3.4

Table 8 Friedman test results

	SVM	1-NN	NB
F_{value}	**9**	**15**	**11.88**
p_{value}	0.0293	0.0018	0.0078

$$\begin{array}{l} H_0 : Difference\ between\ Medians = 0 \\ H_1 : Difference\ between\ Medians \neq 0 \end{array} \tag{15}$$

Table 6 shows the Friedman test parameters.

In the Friedman test, the texture descriptors are ranked according to the classification accuracy obtained for each of them. The best descriptor gets rank 1, the second best gets rank 2, and so on. In this test, LBP descriptor ranked best (Table 7) among the four descriptors.

The Friedman value (F_{value}) is computed from each table of results shown above (Tables 3, 4 and 5). If $F_{value} > Critical_{value}$ will be taken to indicate that the samples are drawn from different distributions [18]. The results of the Friedman test are shown in Table 8.

The values in bold typeface show that the corresponding F_{values} are higher than the $Critical_{value} = 7.8$, this indicates that the performance of the texture descriptors is different. However, these results do not give much information in order to choose the best descriptor. To do this, we consider a post hoc procedure based on Adjusted p-values (APV) [18] and multiple comparisons (Friedman test ($1 \times N$)). LBP descriptor ranked best (Table 7) is used as a control method and compared against the remaining ones. The APV is computed by Holm's procedure [18], the results are shown in Table 9.

The values in bold typeface show that the APV (Holm) are lower than $\alpha = 0.05$, it means that the post hoc procedure is significant.

Table 9 Unadjusted *p*-values and APV Holm

	SVM		1-NN		NB	
Texture descriptor	*p*-value	Holm	*p*-value	Holm	*p*-value	Holm
HWT	0.01430	**0.0429**	0.2206	0.2206	0.1416	0.1416
Daub-4	0.01430	**0.0429**	2.38E–4	**7.15E–4**	0.0032	**0.0098**
GFB	0.01430	**0.0429**	0.0143	**0.0286**	0.0032	**0.0098**

5 Discussion

The results of the Friedman test (Table 8) show that F_{value} is higher than the $Critical_{value}$ and the p_{values} are lower than $\alpha = 0.05$. This means that there is statistical evidence to say that classification accuracies are different to each other and the results are significant when different texture descriptors are used over each classification algorithm.

In the post hoc procedure, the Holm's values (Table 9) in bold typeface rejects all null hypotheses when the LBP descriptor is combined with an SVM classifier. When the classification algorithm is 1-NN or NB, there is not enough statistical evidence to reject the hypothesis that the performance of LBP and HWT is equivalent.

6 Conclusions and Future Work

In this work we presented a statistical comparison of the performance of four image texture descriptors for pattern classification: HWT, Daub-4, LBP, and GFB, tested on well-known texture datasets, with different classification algorithms. The Friedman test ($1 \times N$) was conducted on the target data. Therefore, we have found sufficient evidence to say that, for the considered experimental settings and metrics, using SVM as pattern classification algorithm, the best descriptor was LBP. Using 1-NN or NB, the performance of LBP was better than Daub-4 and GFB, but in the case of HWT, the performance of both descriptors was similar.

The results allow to observe that for a given texture dataset to be processed, an adequate selection of descriptors is necessary, leading to good classification results, where the knowledge of the expert is fundamental. This allows us to think that in order to solve the problem of the selection of descriptors, it is necessary to work on generic strategies that allow the selection and automatic integration of suitable descriptors for different image datasets, depending as little as possible on the knowledge of an expert.

Acknowledgements This work was partially supported by the National Council of Science and Technology (CONACYT) of Mexico, Grant numbers: 263,129 (V. Calzada) and CATEDRAS-2598 (A. Rojas).

References

1. L.G. Shapiro, C. George, Stockman, computer vision, upper saddle river (Prentice–Hall, 2001)
2. M. Petrou, P.G. Sevilla, in *Image Processing Dealing with Texture* (Wiley, New York, 2006)
3. F. Xu, Y.-J. Zhang, Evaluation and comparison of texture descriptors proposed in MPEG-7. J. Visual Commun. Image Represent. **17**, 701–716, Elsevier (2006)
4. O.A.B. Penatti et al., Comparative study of global color and texture descriptors for web image retrieval. J. Visual Commun. Image Represent. **23**, 359–380, Academic Press, Inc. (2012)
5. B.S. Manjunath et al., Color and texture descriptors. IEEE Trans. Circ. Syst. Video Technol. **11**(6) (2001)
6. Ricardo Da Silva Torres and Alexandre Xavier Falcão, Content-based image retrieval: theory and applications. Revista de Informática Teórica e Aplicada **13**, 161–185 (2006)
7. F. Bianconi, A. Fernández, Evaluation of the effects of Gabor filter parameters on texture classification. Pattern Recogn. **40**, 3325–3335, Elsevier (2007)
8. Á. Serrano et al., Analysis of variance of Gabor filter banks parameters for optimal face recognition. Pattern Recogn. Lett. **32**, 1998–2008, Elsevier (2011)
9. Weitao Li et al., Designing compact Gabor Filter banks for efficient texture feature extraction, in *11th International Conference on Control Automation Robotics & Vision* (IEEE, 2010), pp. 1193–1197
10. R.K. Young, in *Wavelet Theory and Its Applications*, 1st edn. (Springer Science + Business Media, 1993)
11. P.J. Van Fleet, in *Discrete Wavelet Transformations An Elementary Approach with Applications* (Wiley, 2008)
12. P.S. Addison, in *The Illustrated Wavelet Transform Handbook, Introductory Theory and Applications in Science, Engineering, Medicine and Finance* (Institute of Physics (IoP), 2002)
13. L. Liu et al., Local binary features for texture classification: taxonomy and experimental study. Pattern Recogn. **62**, 135–160, Elsevier (2017)
14. L. Nanni et al., Survey on LBP based texture descriptors for image classification. Expert Syst. Appl. **39**, 3634–3641, Elsevier (2012)
15. S. Hossain, S. Serikawa, Texture databases—a comprehensive survey. Pattern Recogn. Lett. **34**, 2007–2022, Elsevier (2013)
16. S Arivazhagana, L. Ganesanb, Texture classification using wavelet transform. Pattern Recogn. Lett. **24**, 1513–1521, Elsevier (2003)
17. S. Sidhu, K. Raahemifar, Texture classification using wavelet transform and support vector machines, in *Canadian Conference on Electrical and Computer Engineering*, (2005)
18. J. Derrac et al., A practical tutorial on the use of nonparametric statistical tests as a methodology for comparing evolutionary and swarm intelligence algorithms. Swarm Evol. Comput. **1**, 3–18, Elsevier (2011)

Comparative of Effectiveness When Classifying Colors Using RGB Image Representation with PSO with Time Decreasing Inertial Coefficient and GA Algorithms as Classifiers

Martín Montes, Alejandro Padilla, Juana Canul, Julio Ponce and Alberto Ochoa

Abstract Several transformations from basic RGB representation in digital color images have been developed, CIELab and HSV are commonly applied for color classification, because in this colors spaces there is only a single value adjusted for a specific color detection, nevertheless this transformation require high computational power for transforming every single pixel in a picture. Artificial intelligence (AI) algorithms have been applied before for color classification, but using indistinctly RGB, CIELab and HSV representations among other color transformations even when this transformation can be omitted since they were developed for color classification without AI algorithms. In this paper, is proposed an algorithm for optimizing line equations obtained from three spaces directly generated as a dimensional reduction of the RGB space and we show the comparison of the achieved results optimizing these equations with a GA and PSO algorithms.

Keywords Color classification · PSO · GA · Color spaces

M. Montes (✉)
Universidad Politécnica de Aguascalientes, Aguascalientes, Mexico
e-mail: martin.montes@upa.edu.mx

A. Padilla · J. Ponce
Universidad Autónoma de Aguascalientes, Aguascalientes, Mexico

J. Canul
Universidad Juárez Autónoma de Tabasco, Villahermosa, Mexico

A. Ochoa
Universidad Autónoma de Ciudad Juárez, Ciudad Juárez, Mexico

O. Castillo et al. (eds.), *Fuzzy Logic Augmentation of Neural and Optimization Algorithms: Theoretical Aspects and Real Applications*, Studies in Computational Intelligence 749, https://doi.org/10.1007/978-3-319-71008-2_38

1 Introduction

Human color vision together with brain processing allow humans to classify colors, this skill is exploded in several areas like education, security, transport, medicine, mining, computer sciences, among others [1, 2].

Computer vision and image processing are computer sciences related to take advantage of human vision capabilities using computers [3].

Color representation and classification are research fields of interest in computer vision, where several transformations and processes are required for detecting and representing color in digital pictures [2, 3].

Digital cameras like human eyes obtain pictures of their surroundings by sensing light in the visible spectrum (a limited bandwidth where electromagnetic radiation its measurable by cameras or human eyes) [2, 4].

Three basic colors Red (R), Green (G) and Blue (B) are perceived in digital cameras, with cells adapted for transforming to voltage the quantity of light in their wavelengths and later convert them into digital values [2, 4].

There are several representations for coding digital values of color images like RGB, HSV, YCbCr, YIQ, YUV, CIElab, among others. All of them are used in certain applications, for example RGB, YCbCr and YUV are directly obtained from digital cameras, but YCbCr and YUV are used when transmitting to analogue systems, like the cathode ray tube [1, 3, 5].

RGB and YUV representations are transformed to different color spaces in order to improve the classification of colors, which is commonly used in color segmentation for object recognition [3, 5].

Hue, Saturation and Intensity (HSI) and Hue, Saturation and Value (HSV) are no linear transformations of RGB and YUV spaces that simplify color identification, nevertheless this transformation increase computational power while transforming every pixel (basic units that constitute an image and take their values from measurement cells in digital cameras) in a picture [2–4].

Object recognition in real time, requires to reduce as much as possible computational power, while maintaining good accuracy, sorting with light changes, obstacles and object orientation, this has encourage to find other alternatives for color representations and color classification methods. This alternatives include new color representations and color detection algorithms using bio-inspired techniques, fuzzy inference systems and Artificial Neural Networks (ANN), among others [1, 4, 6].

Transformations for color representation or algorithms for color classification generally involves at least 3 parameters to work with, resulting in an $\mathbb{R}^3$ space, like is shown in [1, 3, 4, 7–10].

Since is required to classify colors in $\mathbb{R}^3$, we propose in [4] to work with equations of planes for separating different color regions without need any transformation from RGB space.

This alternative has not been explored previously with new optimizing techniques, like Artificial Intelligence (AI) methodologies, since color representations like HSV and CIElab has become very popular for classifying colors by changing a single value, even when this transformations imply several no linear and discontinuous operations [3, 4].

Soft Computing algorithms are AI heuristic algorithms and has been used before with color segmentation, but commonly HSI or HSV representations are used with AI algorithms even when this transformations were designed for color identification without use AI [4].

Evolutionary Algorithms (EA) are Monte Carlo Algorithms used in SC that take advantage of meta-heuristic schema for obtaining good quality solutions in different applications. This algorithms create a virtual environment with candidate solutions evaluated with objective functions and different rules depending of the algorithm [11, 12].

Recently two EA that have become popular for their capabilities in optimization problems are Genetic Algorithms (GA) and Particle Swarm Optimization (PSO). Both algorithms work with randomly initialized populations, but GA uses the natural selection principles proposed by Darwin while PSO imitates the birds and fish behavior to allow synchronized movement [12].

In this paper we propose a dimension reduction for allowing to change planes into lines in the separation of colors and is show a comparison of the results implementing GA and PSO for tuning this line equations required to classify colors in RGB space.

2 Backgrounds

Color classification with classical methodologies in computer vision, require transformations from RGB space to different representations and thresholds adjustment according to the desired color, while color classification with SC uses this transformations in some works and in others directly obtain color classification from RGB space but adding other extra information, the majority of this researches use ANN with at least 1 hidden layer, increasing the computational power required.

2.1 Color Classification Using Color Transformations

The works in [3, 13] describe Munsell, RGB, CMYK, YIQ, YUV, YCbCr, HSI, HSV, HSL, CIE XYZ, CIE L*U*V* and CIE L*a*b* color representations, their transformations from RGB and to RGB space, their applications, mention color families designed for color classification in computer vision and discuss their advantages and disadvantages in color detection.

Al-Tairi et al. in [7] propose an algorithm for skin detection where RGB space images are transformed to YUV space, the classification is performed by ignoring Y channel for avoid luminosity effects and selecting thresholds with a parameter CH related with the channels U and V, and the color histogram (diagram that express the quantity of different colors in a picture).

Theo Gevers and Arnold W.M. Smeulders test object recognition in [6] using classical RGB and HSV representations and their new proposed $c_1c_2c_3, l_1l_2l_3, m_1m_2m_3$ color representations which report robustness and better results in object recognition.

Severino and Gonzaga propose in [8] Hue Saturation and Mixture (HSM) a new color space for color classification by introducing a new parameter called color mixture and they test their color model against HSV, YCbCr and TSL representations in color skin detection.

The work of Bogdan et al. in [14], shows an application using classical RGB, HSV, HSL and CIE color representation where color parameters are obtained using a digital scanner for spectrophotometric measurements in electrophoresis interpretation.

A color edge detector is proposed in [15], where RGB representation is used together with canny detector to perform edge detection in the 3 channels of a picture resulting in accuracy improvement of a gray scale common edge detector even without use any transformation to other color spaces.

An other application with color detection by applying color transformation from RGB space to HSV representation is shown in [10], where texture or the quantity of like absorbed by a surf is detecting based in color identification.

2.2 Color Classification Using AI Techniques

An application in mining for mineral identification color detection based is shown in [1], in this paper the results obtained for mineral classification with 2 input images of crossed-polarized and plane-polarized lights are represented in RGB, HSV and CIElab spaces and then applied to an ANN, the compared results with all color spaces show a minor training error when RGB space is used.

The work in [16] propose a method for detecting drying levels in apples by transforming images into CIELab space for applying them to a multilayer perceptron ANN for monitoring and controlling drying state of apple slices dried with heat.

Color detection of plastic for pigments adjustment to produce specific colors in plastics is applied using CIELab representation and a multilayer perceptron in [17].

Human skin detection is made with ANN in [18], by taking 1.2 millions of pixels of known human skin and 1.2 millions of pixels of known no skin, then all pixels are transformed to YCbCr, YIQ, YDbDr and CIELab color spaces for use them as training inputs in the ANN together with texture parameters skewness and kurtosis. Better results are obtained with YIQ representation.

The work in [19] obtain a relation between color preference and color eye classification by applying CIELab representation pictures together with the eyes color of the tested individuals as input data for training an ANN.

Alessia and Clara develop an algorithm in [20] where GA is applied together with graph-base segmentation for develop its own algorithm for color segmentation based in graphs.

Human skin detection is performed using RGB images transformed to CIELab space, then are processed with Hiliclimb segmentation with k-means for obtaining local clusters and later optimized with PSO algorithm [21]

Nasiri and Yazdi propose in [22] an algorithm for lips identification based in color pictures transformed to CIELab representation and threshold adjustment optimized with PSO algorithm.

3 Plane Equations for Color Classification in RGB Space

In the work in [4], we propose to perform color classification directly applied in RGB representation, since avoid transformations to color models like HSV or CIELab models result in saving processing time which is very important in real time color segmentation and using IA techniques like GA, ANN and PSO for optimizing plane equations would maintain accuracy in color classification.

Several transformations from RGB color representation have been developed for simplifying color classification in pictures, but identify a single color in RGB space omitting other colors like it is done in transformations like HSV model, is more complex, that can be seen when both spaces are plotted in $\mathbb{R}^3$ for RGB (Fig. 1) and HSV (Fig. 2) using specific colors, nevertheless several planes with different positions and orientations could separate regions for an specific color detection [4].

The results with normalized parameters in RGB and HSV spaces with red, orange, yellow green, blue, purple and brown colors obtained in [4], show that several plane equations with the form in Eq. 1 could be used for detecting specific

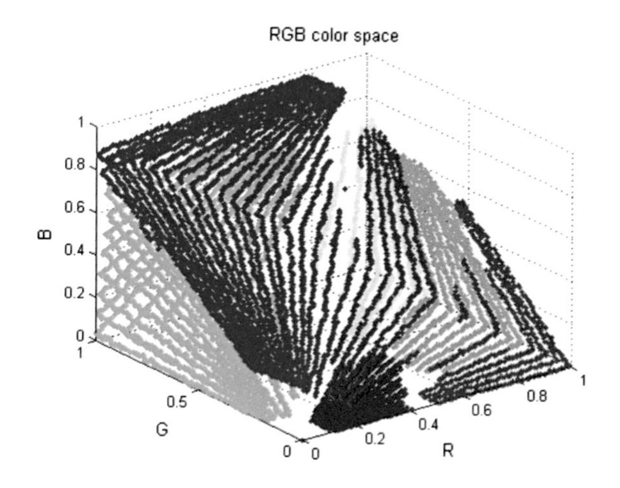

Fig. 1 RGB representation of red, orange, yellow, green, blue, purple and brown colors in RGB space obtained from [4]

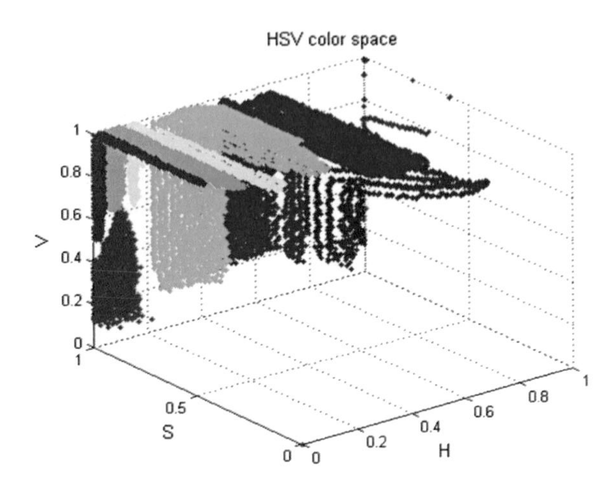

Fig. 2 HSV representation of red, orange, yellow, green, blue, purple and brown colors in RGB space obtained from [4]

colors and all parameters ($\alpha, \beta, \gamma, R_0, G_0$ and B_0) for each required plane could be obtained with GA or PSO algorithms.

$$\alpha(R - R_0) + \beta(G - G_0) + \gamma(B - B_0) = 0 \quad (1)$$

4 Genetic Algorithms

GA are EA algorithms that follow natural selection principles in [23], i.e. its main operators for search space exploration are population initializing, fitness value calculation, fitness-based selection, crossover and mutation, this operators are n_g generations evaluated, which is the number of iterations of a GA (Fig. 3) [4, 11, 12].

4.1 GA Population Initializing

GA chromosomes or candidate solutions are commonly represented like "1010100111" containing all the required alleles or bits (n_b), separated in sets or

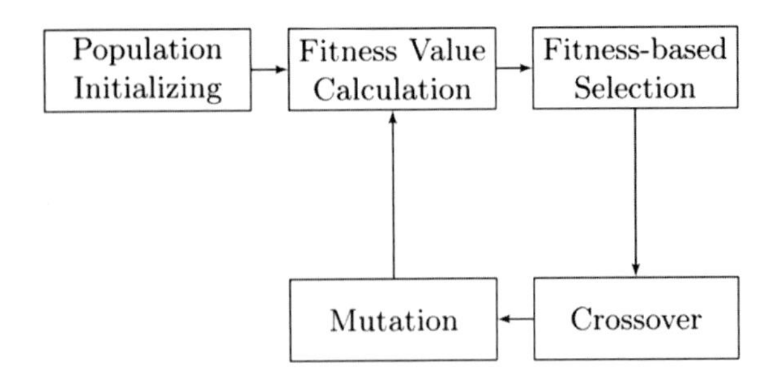

Fig. 3 GA main operators

genes for optimizing the numerical elements ($\alpha, \beta, \gamma, R_0, G_0$ and B_0) in the equations of planes described in Sect. 3 [11, 12].

Population initializing is made by generating p_s digital random numbers and converting them into binary representation with n_b length, where p_s is the population size and each binary element contain the numerical data required in a possible plane equation like those described in Sect. 3.

4.2 GA Fitness Function and Fitness Value

The fitness value $F_v \subseteq \mathbb{R}$ in GA is an scalar value that measures the quality of chromosomes in the optimizing problem, it is obtained with and objective function f which is subject to optimization and in GA is called fitness function [11, 24].

4.3 GA Fitness Based Selection

There are several selection methods used in GA, like proportional selection, rank selection, truncation selection, orderer selection, among others, each one with its own advantages and disadvantages, in this work we apply tournament selection which is a very common selection schema applied because selection pressure is easily controlled by changing T_s the tournament size, allowing to reduce or increase convergence speed for obtaining the desire solution (fast convergence could result in a bad quality solution) [11, 25].

Tournament selection is perform by obtaining a matting pool of size T_s containing T_s elements of the population P randomly selected, then the best individuals in the matting pool are used in crossover operation for offspring generation [11, 25].

4.4 GA Crossover Operation

There several algorithms to perform crossover operation two common kinds of crossover operations are haploid (single gamete crossover) and diploid (two gametes crossover) like in real organisms, but must common schema in GA is haploid reproduction which for binary chromosomes, include single crossover point, multiple crossover point, and variable length distance between crossover points [11, 24].

In this paper haploid crossover operation is performed with a single crossover point (C_p) randomly selected in ranges $U[0, n_b]$. Then two new candidate solutions are obtained from two winner parents P_a and P_b as the best two elements in the tournament selection. The first offspring element is obtaining with P_a alleles from 0 to position C_p and P_b alleles from C_p to n_b, second element is obtained similarly but changing positions of P_a and P_b, like is shown in Eqs. 2 and 3 [11, 24].

$$offspring_1 = [P_a(0, C_p)P_b(C_p, n_b)] \tag{2}$$

$$offspring_2 = [P_b(0, C_p)P_a(C_p, n_b)] \tag{3}$$

4.5 GA Mutation Operation

The population operation its used to add new extra genetic material that was not in the original population and preserve the diversity while algorithm is iterated n_g generations [11, 24].

Mutation is performed by making minor changes in the mutated chromosomes. There are different kinds of mutation but common algorithms change aleatory selected alleles (single gene mutation) or several alleles (multiple gene mutation) [11, 24]. In this paper single gene mutation is applied.

When chromosomes are binary arrays, mutation only requires toggle mutated bits, which are selected to achieve a desired mutation probability p_m and the total mutations in a run are obtained with Eq. 4 for binary single gene mutation, i.e. p_m is multiplied by the total number of alleles in a run [11, 24].

$$t_m = (p_s + 2 * n_g) * n_b * p_m \tag{4}$$

5 Particle Swarm Optimization Algorithm

The PSO is an EA developed by Eberhart and Kennedy in 1995, this optimization algorithm explores the search space with n_d dimensions simulating birds, bees and fish social behavior, where swarm movement depends from personal and social components [26–28].

5.1 PSO Population Initializing

Population initializing in PSO is made by creating i particles containing information of its position and velocity in a n_d space, which means that every particle has a position $X_i = [x_{i1}, x_{i2}, \ldots, x_{in_d}]$ and velocity $V_i = [v_{i1}, v_{i2}, \ldots, v_{in_d}]$. X_i must be randomly initialized in ranges $U[x_{min}, x_{max}]$ and V_i is in this first stage initialized as a n_d zeros array but later its values are calculated in ranges $U[v_{min}, v_{max}]$, this ranges are the max and min bounds for position and velocity in the algorithm. Best particle known position Xb_i is initialized equal to X_i in the initialization stage [26].

Social component is added with the global best known position of the swarm X_{gb} which is a n_d array initialized when fitness value described in Sect. 5.2 is calculated for all the particles after initialize their positions [26].

In this paper position and velocity dimensions are the number of numerical values required in the equations of planes for color classification, like its described in Sect. 3.

5.2 PSO Fitness Function

Fitness functions in PSO algorithms are objective functions that map a fitness value $F_v \subseteq \mathbb{R}$ according to the quality of the achieved position by a particle in the simulation [26, 28].

5.3 Velocity and Position Update Equations

After initialization stage, velocity V_i, position X_i, fitness value F_v, personal best position Xb_i and global best position X_{gb} in every particle are updated per iteration [27].

The velocity and position at iteration t are updated with Eqs. 5 and 6 respectively.

$$V_i(t) = wV_i(t-1) + c_1 r_1 [Xb_i - X_i(t-1)] + c_2 r_2 [X_{gb} - X_i(t-1)] \quad (5)$$

$$X_i(t) = X_i(t-1) + V_i(t) \quad (6)$$

where $t-1$ is the previous iteration, c_1 and c_2 are the individual and global learning rates, r_1 and r_2 are uniformly random numbers in ranges $U[0, 1]$ and w is the inertia weight.

The values of c_1 and c_2 are selected commonly equal to 2, so when multiply per r_1 and r_2 the particles approach the target about half of the difference between individual and social components [27].

Intertia Weigth w is commonly used in PSO implementation with dynamical environment algorithm. w is a mechanism for controlling exploration and exploitation, by adding the particle momentum, i.e. the contribution of the previous velocity, w is commonly initialized using large values near to 1 for extensively exploration and is time reduced for more exploitative exploration in the later iterations, in this paper w is time reduced calculating again per generation according to Eq. 7 [27].

$$w = 0.99\,w \tag{7}$$

6 Development

In this section is shown how is possible to reduce RGB space dimension for color classification, the general fitness function used in both GA and PSO algorithms, as well as the GA and PSO implementation applied for color classification performed in the RGB space.

6.1 Decreasing Dimension of Candidate Solutions

Equations of planes could be used for implementing region boundaries that separate different colors, like is described in 3, but following this process if two planes are required in a specific desired color detection, then 6 different numerical values must be calculated. Nevertheless, computational power required to test if the color belongs to a region inside several planes will be greater than if lines where used substituting a $\mathbb{R}^3$ space by 3 $\mathbb{R}^2$ spaces.

Following this principle the RGB $\in \mathbb{R}^3$ space was transformed into 3 $\mathbb{R}^2$ spaces, RG, GB and RB and plotting the same colors described in [4] (one plot per $\mathbb{R}^2$ space), was found that there was line equations that were capable to separate three different regions in the $\mathbb{R}^2$ spaces.

Since its only required to know the relation from r to g and g to b, for obtaining the relation from r to b, there are only 2 lines required and one space $\mathbb{R}^2$ could be omitted but if a third line is used to establish r, g or b admissible levels then equations required for color classification will consider luminosity. Equations 8 and 9 allow to obtain the lines in RG and GB spaces and Eq. 10 allows to control the luminosity with the red value.

$$r \leq \alpha_1 * g + \beta_1 \tag{8}$$

$$g \leq \alpha_2 * b + \beta_2 \tag{9}$$

$$\gamma_1 \leq r \leq \gamma_2 \tag{10}$$

The parameters $\alpha_1, \alpha_2, \beta_1, \beta_2, \gamma_1$ and γ_2 are the optimization parameters to explore in a $\mathbb{R}^6$ search space.

6.2 General Fitness Function

GA and PSO algorithms use the same fitness functions which perform evaluation of two images represented with raw RGB values.

First image considered as an input data (*inputs*) contain several pictures with different colors, and second picture or desired output data (out_d) contain the same number of pixels in *inputs*, but all colors different from the desired detected color are changed to black and all pixels that correspond to the desired detected color (in this paper red, orange, yellow, green, blue and purple colors) are changed to white.

Using if then rules, the input data of the r, g and b values of the *ij* pixel are passed as inputs, if they fulfill Eqs. 8, 9 and 10, then $out(i,j) = 1$ else $out(i,j) = 0$.

Fitness value is the total relative error between $out(i,j)$ and $out_d(i,j)$, calculated by summing every absolute difference between the outputs ($out(i,j)$) value and the desired outputs value ($out_d(i,j)$), like is shown in Eq. 11.

$$F_v = \frac{1}{rows \times cols} \sum_{i=1}^{rows} \sum_{j=1}^{cols} abs[out_d(i,j) - out(i,j)] \tag{11}$$

6.3 Implementing GA

The GA implementation for all the desired detected colors (in this paper red, orange, yellow, green, blue, and purple) is made using algorithm described in Sect. 4 and the same initial parameters, this values are shown in Table 1.

6.4 Implementing PSO

PSO implementation for all the desired detected colors (in this paper red, orange, yellow, green, blue, and purple) is made using algorithm described in Sect. 5 and the same initial parameters are used for all the training colors, this values are shown in Table 2.

Table 1 GA initial parameters

GA parameter	Value
p_s	1000
n_b	36
t_s	15
p_m	0.06
Genes	6

Table 2 PSO initial parameters

PSO parameter	Value
p_s	50
w	1
c_1	2
c_2	2
x_{max}	1
x_{min}	−1
v_{max}	0.4
v_{min}	−0.4
Iterations	200

7 Results

In this section are shown the results obtained for color dimensional reduction from $\mathbb{R}^3$ in the RGB space to RG, RB and GB spaces in $\mathbb{R}^2$, the training results with GA and PSO algorithms and the results when applying a not trained picture for color recognition.

7.1 Color Dimensional Reduction

The results obtained for red color dimension reduction from RGB space to RG, RB and GB spaces using algorithm in Sect. 3, are show in Figs. 4, 5, 6 and 7, where is clear that by adding correct lines as boundaries it is possible to separate or detect a desired color.

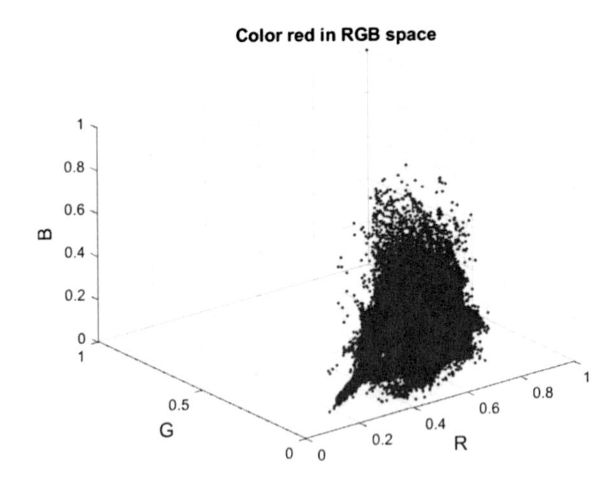

Fig. 4 Red color represented in RGB space obtained with algorithm in [4]

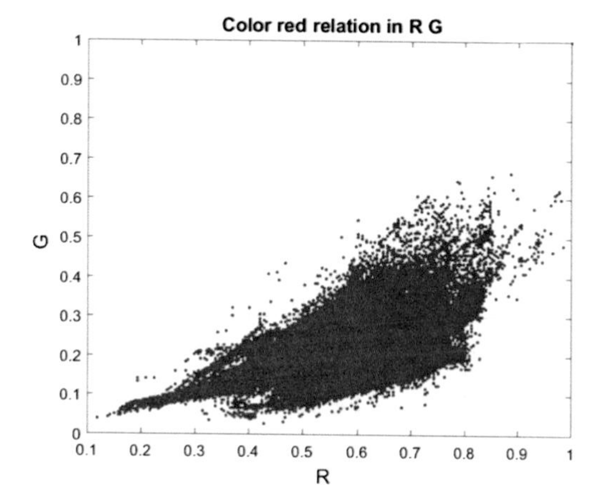

Fig. 5 Red color represented in RG space obtained with algorithm in Sect. 3

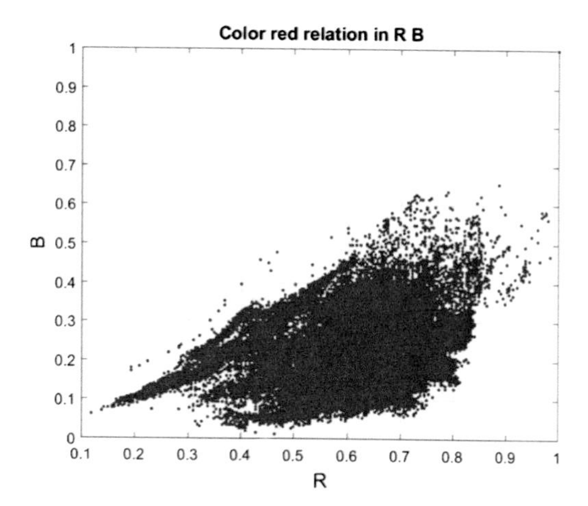

Fig. 6 Red color represented in RB space obtained with algorithm in Sect. 3

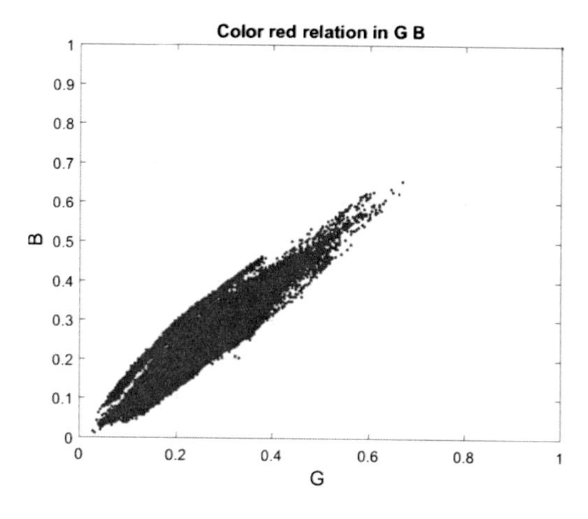

Fig. 7 Red color represented in GB space obtained with algorithm in Sect. 3

Following the same process for the other 5 colors, similar plots are obtained, where lines are clear boundaries for color classification, but in this work only red color is shown.

7.2 GA Training Results

The training results applying algorithm in Sect. 4 for all the detected colors using fitness function in Sect. 6.2 is shown in Figs. 8, 9, 10, 11, 12 and 13.

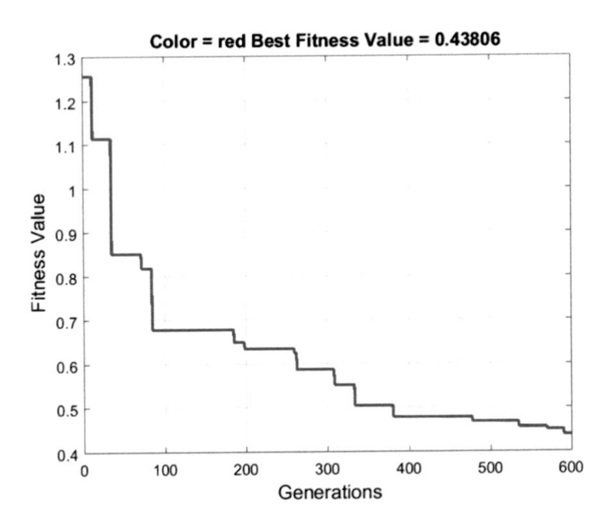

Fig. 8 Red color training with GA

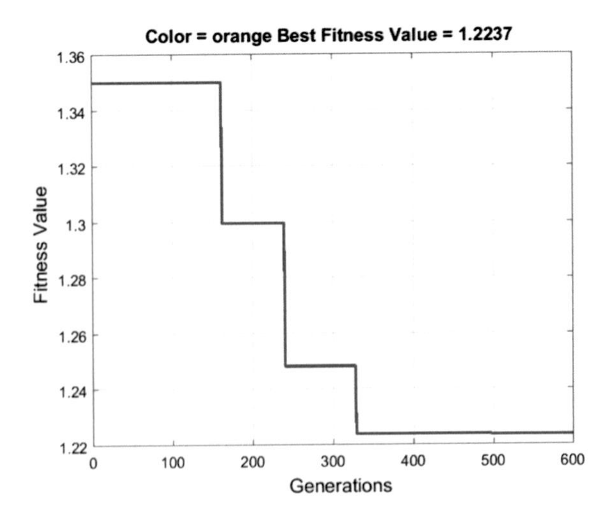

Fig. 9 Orange color training with GA

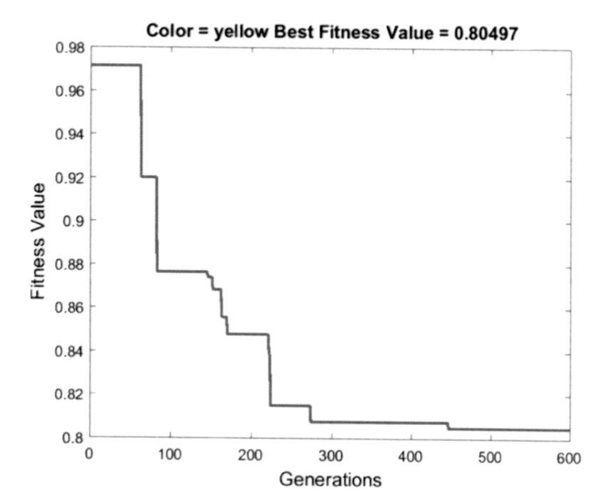

Fig. 10 Yellow color training with GA

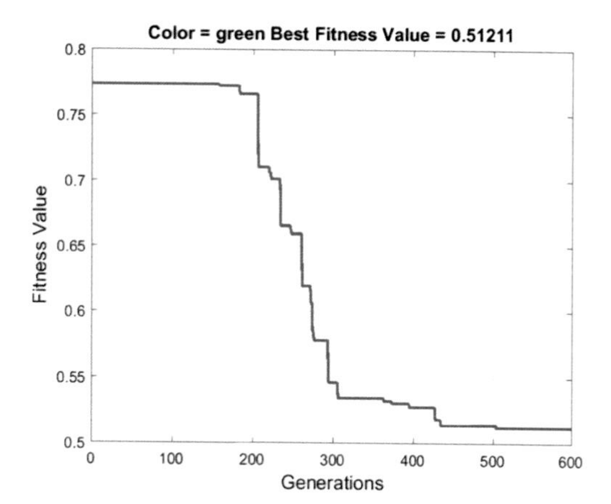

Fig. 11 Green color training with GA

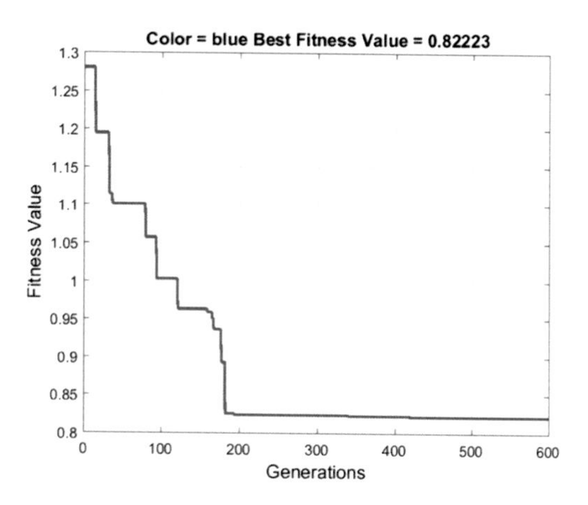

Fig. 12 Blue color training with GA

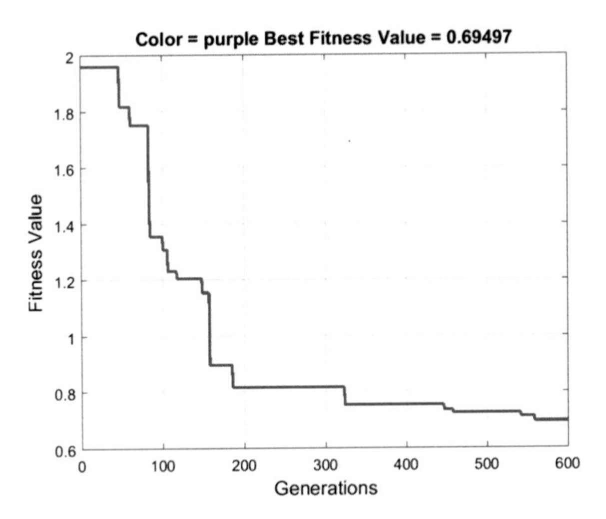

Fig. 13 Purple color training with GA

7.3 *PSO Training Results*

PSO training results applying algorithm in Sect. 5 for all the detected colors using fitness function in Sect. 6.2 is shown in Figs. 14, 15, 16, 17, 18 and 19.

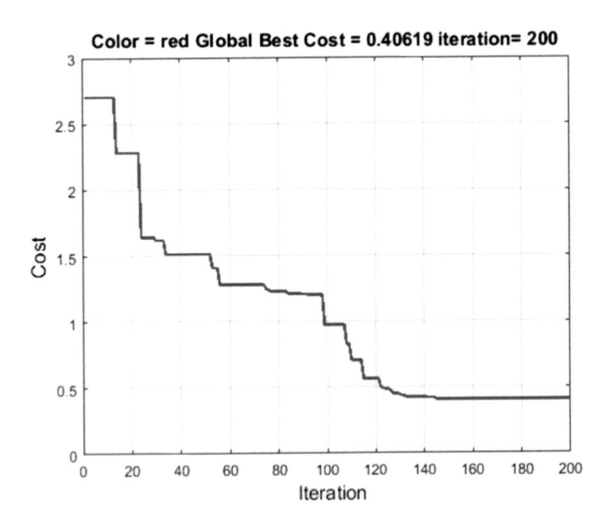

Fig. 14 Red color training with PSO

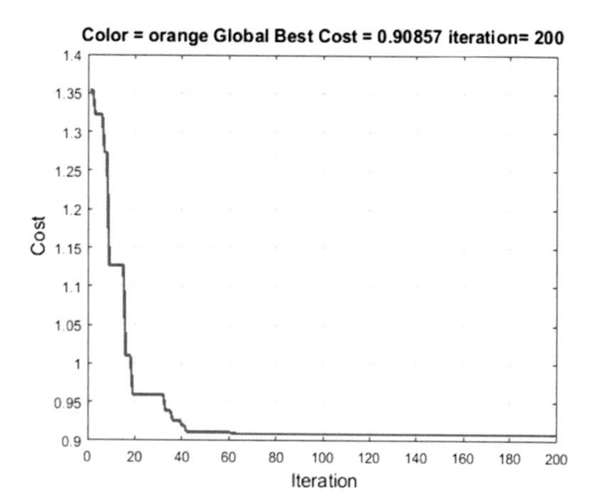

Fig. 15 Orange color training with PSO

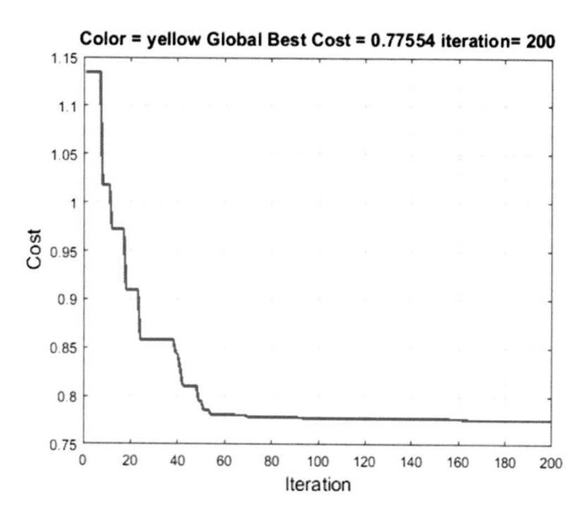

Fig. 16 Yellow color training with PSO

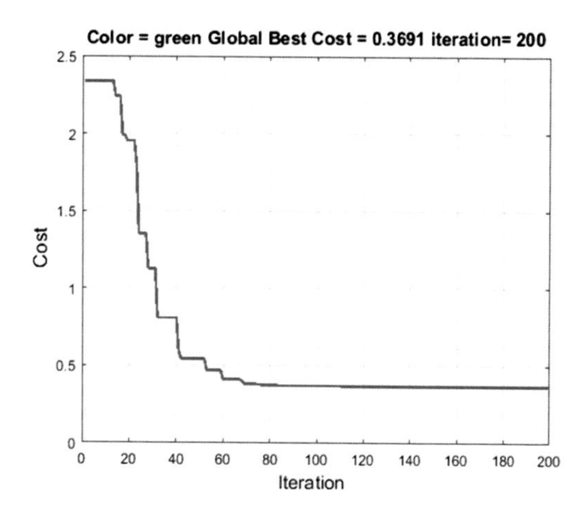

Fig. 17 Green color training with PSO

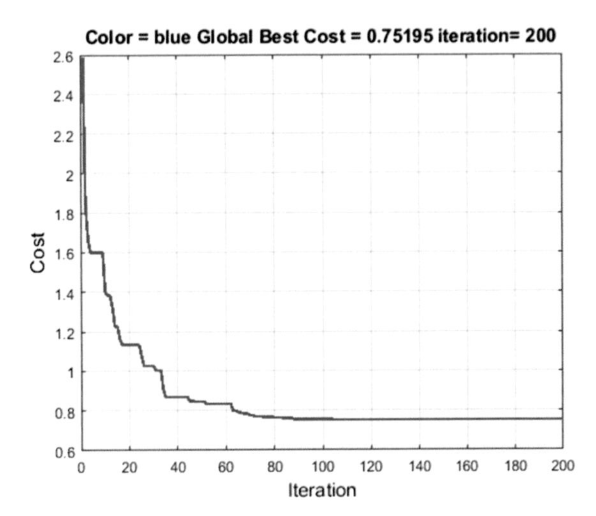

Fig. 18 Blue color training with PSO

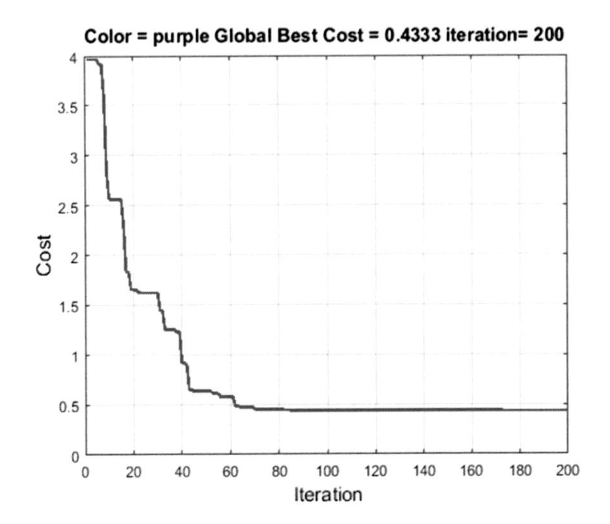

Fig. 19 Purple color training with PSO

8 Conclusions

The proposed algorithm achieves good results in the training stage with both GA and PSO algorithms when optimizing line equations for color classification in RGB space, like is shown in Sect. 7.

Training with PSO algorithm using time reduced inertial coefficient shows better results in every color training color, but specifically in orange color PSO shows significantly better results than GA.

Although both algorithms show total relative error minor to 1.2% in the worst case using GA, they require several generations and individuals to achieve this accuracy, which imply more training time. In general was found that with our proposed method for optimizing line equations PSO obtains better results in the training stage.

References

1. N.A. Baykan, N. Yılmaz, G. Kansun, Case study in effects of color spaces for mineral identification. Sci. Res. Essays **5**(11), 1243–1253 (2010)
2. A. Bovik, *Handbook of Image and Video Processing* (Academic Press, Department of Electrical and Computer Engineering. Austin, Texas 2000)
3. N.A. Ibraheem, M.M. Hasan, R.Z. Khan, P.K. Mishra, Understanding color models: a review. ARPN J. Sci. Technol. **2**(3), 265–275
4. M. Montes Rivera, A. Padilla Díaz, J. Canul Reich, J.C. Ponce Gallegos, A. Ochoa Zezzatti, Comparative between RGB and HSV color representations for color segmentation when it is applied with artificial neural networks and evolutionary algorithms. Avances en las tecnologías de la información ANIEI 2016
5. M.C. Murillo, F. García Lamont, A.D. Cuevas Rasgado, Segmentación de imágenes de color imitando la percepción humana del color. Res. Comput. Sci. **114**, 71–81 (2016)
6. T. Gevers, A.W.M. Smeulders, Color-based object recognition. Pattern Recognit. **32**, 453–464 (1999)
7. Z.H. Al-Tairi, R.W. Rahmat, M.I. Saripan, P.S. Sulaiman, Skin segmentation using YUV and RGB color spaces. J. Inf. Process Syst. **10**(2), 283–299
8. O. Severino Jr., A. Gonzaga, HSM: a new color space used in the processing of color images. RITA **XVI** Número 2 (2009)
9. R. Nayyer, B. Sharma Use and analysis of color models in image processing. Int. J. Adv. Sci. Res. **1**(8), 329–330 (2015)
10. M. Deswal, N. Sharma, A fast HSV image color and texture detection and image conversion algorithm. Int. J. Sci. Res (IJSR). **3**(6) (2014)
11. T. Weise, Global optimization algorithms: theory and application. E-Book obtained from: http://www.it-weise.de/projects/book.pdf. 27 Jan 2016, published 2009
12. K.O. Jones, Comparison of genetic algorithm and particle swarm optimization, in *International Conference on Computer Systems and Technologies CompSysTech* (2005)
13. A. Hanbury, J. Serra, A 3D-polar coordinate color representation suitable for image analysis. Pattern Recognition and Image Processing Group Institute of Computer Aided Automation Vienna University of Technology Technical Report December 16 2002
14. A. Bogdan, Haifa, V. Bacârea, O. Iacob, T. Călinici, A. Schiopu, Comparison between digital image processing and spectrophotometric measurements methods. Application in electrophoresis interpretation. Appl. Med. Inf. **28**(1), 29–36 (2011)
15. S. Dutta, B.B. Chaudhuri, A color edge detection algorithm in RGB color space, in *2009 International Conference on Advances in Recent Technologies in Communication and Computing* (2009)
16. M.H. Nadian, S. Rajiee, M. Aghbashlo, S. Hosseinpour, S.S. Mohtasebi, Continuous real-time monitoring and neural network modeling of apple slices color changes during hot air drying. Food Bioprod. Process. **94**, 263–274 (2015)
17. U. Saeed, S. Ahmad, J. Alsadi, D. Ross, G. Rizvi, Implementation of neural network for color properties of polycarbonates, in *Proceedings of PPS-29 AIP Conference Proceedings*, vol. 1593, pp. 56–59 (2014)
18. H.K. Al-Mohair, J. Mohamad-Saleh, S.A. Suandi, Color space selection for human skin detection using color-texture features and neural networks, in *International Conference on Computer and Information Sciences (ICCOINS)* (2014)

19. C. Cengiz, E. Köse, Modelling of color perception of different eye colors using artificial neural networks. Neural Comput. Appl. **23**, 2323–2332 (2013)
20. A. Amelio, C. Pizzuti, A genetic algorithm for color image segmentation, in *LNCS EvoApplications*, ed. by A.I. Esparcia-Alcázar et al., pp. 314–323 (2013)
21. R. Vijayanandh, G. Balakrishnan, Performance measure of human skin region detection based on hybrid particle swarm optimization. Int. J. Comput. Theory Eng. **4**(5) (2012)
22. J.A. Nasiri, H.S. Yazdi, A PSO tuning approach for lip detection on color images, in *Second UKSIM European Symposium on Computer Modeling and Simulation* (2008)
23. C. Darwin, *The Origin of Species* (John Murray, Penguin Classics, 1985 edition, 1859)
24. M. Melanie, *An Introduction to Genetic Algorithms* (A Bradford Book The MIT Press Cambridge, Massachusetts, London, England Fifth printing, 1999)
25. B.L. Miller, D.E. Goldberg, Genetic algorithms, tournament selection, and the effects of noise. Complex Syst. **9**, 193–212 (1995)
26. A. Lazinica: Particle swarm optimization. In-Tech Kirchengasse 43/3, A-1070 Vienna, Austria Hosti 80b, 51000 Rijeka, Croatia 2009
27. D.P. Rini, S.M. Shamsuddin, S.S. Yuhaniz: Particle swarm optimization: technique, system and challenges. Int. J. Comput. Appl. **14**(1) (2011)
28. R. Poli, J. Kennedy, T. Blackwell, *Particle Swarm Optimization an overview* (Springer Science Business Media, 2007)

Printed in the United States
By Bookmasters